T0189241

Lecture Notes in Computer Science   13081

More information about this subseries at http://www.springer.com/series/7409

Wenjie Zhang · Lei Zou ·
Zakaria Maamar · Lu Chen (Eds.)

Web Information Systems Engineering – WISE 2021

22nd International Conference
on Web Information Systems Engineering, WISE 2021
Melbourne, VIC, Australia, October 26–29, 2021
Proceedings, Part II

 Springer

Editors
Wenjie Zhang
School of Computer Science
and Engineering
The University of New South Wales
Sydney, NSW, Australia

Zakaria Maamar
Zayed University
Dubai, United Arab Emirates

Lei Zou
Peking University
Beijing, China

Lu Chen 🆔
Swinburne University of Technology
Melbourne, VIC, Australia

ISSN 0302-9743   ISSN 1611-3349 (electronic)
Lecture Notes in Computer Science
ISBN 978-3-030-91559-9   ISBN 978-3-030-91560-5   (eBook)
https://doi.org/10.1007/978-3-030-91560-5

LNCS Sublibrary: SL3 – Information Systems and Applications, incl. Internet/Web, and HCI

This Springer imprint is published by the registered company Springer Nature Switzerland AG
The registered company address is: Gewerbestrasse 11, 6330 Cham, Switzerland

Preface

Welcome to the proceedings of the 22nd International Conference on Web Information Systems Engineering (WISE 2021), held in Melbourne, Australia, during October 26–29, 2021. The series of WISE conferences aims to provide an international forum for researchers, professionals, and industrial practitioners to share their knowledge in the rapidly growing area of web technologies, methodologies, and applications. The first WISE event took place in Hong Kong, China (2000). Then the trip continued to Kyoto, Japan (2001); Singapore (2002); Rome, Italy (2003); Brisbane, Australia (2004); New York, USA (2005); Wuhan, China (2006); Nancy, France (2007); Auckland, New Zealand (2008); Poznan, Poland (2009); Hong Kong, China (2010); Sydney, Australia (2011); Paphos, Cyprus (2012); Nanjing, China (2013); Thessaloniki, Greece (2014); Miami, USA (2015); Shanghai, China (2016); Puschino, Russia (2017); Dubai, UAE (2018); Hong Kong, China (2019); Amsterdam and Leiden, The Netherlands (2020); and this year, WISE 2021 was held in Melbourne, Australia.

A total of 229 research papers were submitted to the conference for consideration, and each paper was reviewed by at least three reviewers. Finally, 55 submissions were selected as regular papers (an acceptance rate of 24% approximately), plus 29 as short papers. The research papers cover the areas of blockchain, social networks, graph neural networks, graph query, crowdsourcing, knowledge graph and entity linking, spatial temporal data analysis, service computing, cloud computing, text mining, recommender systems, database systems, workflow, deep learning, data mining, and applications. In addition to regular and short papers, the WISE 2021 program also featured tutorial and demo sessions.

We would like to sincerely thank our keynote speakers:

- Munindar P. Singh, North Carolina State University, USA
- Jie Lu, University of Technology Sydney, Australia
- James B. D. Joshi, University of Pittsburgh, USA
- Xiaokui Xiao, National University of Singapore, Singapore

In addition, special thanks are due to the members of the international Program Committee and the external reviewers for a rigorous and robust reviewing process. We are also grateful to Springer and the International WISE Society for supporting this conference. The WISE Organizing Committee is also grateful to the demo organizers for their great efforts in helping promote web information system research to broader domains.

We expect that the ideas that have emerged in WISE 2021 will result in the development of further innovations for the benefit of scientific, industrial, and social communities.

October 2021

Wenjie Zhang
Lei Zou
Zakaria Maamar
Lu Chen

Organization

General Co-chairs

Xiaofang Zhou   Hong Kong University of Science and Technology, Hong Kong
Yannis Manolopoulos   Aristotle University of Thessaloniki, Greece

Program Co-chairs

Wenjie Zhang   University of New South Wales, Australia
Lei Zou   Peking University, China
Zakaria Maamar   Zayed University, Dubai, United Arab Emirates

Publication Chair

Lu Chen   Swinburne University of Technology, Australia

Publicity Co-chairs

Xiaohui Tao   University of Southern Queensland, Australia
Georgios Kambourakis   University of the Aegean, Greece
Manik Sharma   DAV University, India
Xin Wang   Tianjin University, China

Diversity and Inclusion Chair

Wenny Rahayu   La Trobe University, Australia

PhD School Chair

Shazia Sadiq   University of Queensland, Australia

Demo Co-chairs

Weiguo Zheng   Fudan University, China
Dong Wen   University of Technology Sydney, Australia

Tutorial and Workshop Chair

Guandong Xu   University of Technology Sydney, Australia

Industry Relationship Chair

Jian Yang   Macquarie University, Australia

Finance Chair

Sudha Subramani   Victoria University, Australia

Website Chair

Yong-Feng Ge   La Trobe University, Australia

Senior Program Committee

Yanchun Zhang   Victoria University, Australia
Qing Li   Hong Kong Polytechnic University, Hong Kong
Xiaohua Jia   City University of Hong Kong, Hong Kong
Elisa Bertino   Purdue University, USA
Athman Bouguettaya   University of Sydney, Australia

WISE Steering Committee Representatives

Yanchun Zhang   Victoria University, Australia
Qing Li   Hong Kong Polytechnic University, Hong Kong

Program Committee

Marco Aiello   University of Stuttgart, Germany
Mohammed Eunus Ali   Bangladesh University of Engineering and Technology,
                             Germany
Toshiyuki Amagasa   University of Tsukuba, Japan
Bernd Amann   Sorbonne Université - LIP6, France
Chutiporn Anutariya   Asian Institute of Technology, Thailand
Boualem Benatallah   University of New South Wales, Australia
Djamal Benslimane   Université Claude Bernard Lyon 1, France
Devis Bianchini   University of Brescia, Italy
Mohamed Reda Bouadjenek   Deakin University, Australia
Athman Bouguettaya   University of Sydney, Australia
Bin Cao   Zhejiang University of Technology, China
Jinli Cao   La Trobe University, Australia
Xin Cao   University of New South Wales, Australia
Barbara Catania   University of Genoa, Italy
Richard Chbeir   Université de Pau et des Pays de l'Adour - LIUPPA,
                             France
Cindy Chen   University of Massachusetts Lowell, USA
Lu Chen   Zhejiang University, China

Lu Chen	Swinburne University of Technology, Australia
Xiaojun Chen	College
Xiaoshuang Chen	University of New South Wales, Australia
Dickson K. W. Chiu	University of Hong Kong, Hong Kong
Theodoros Chondrogiannis	University of Konstanz, Germany
Dario Colazzo	Université Paris-Dauphine - LAMSADE, France
Alexandra Cristea	Durham University, UK
Hai Dong	RMIT University, Australia
Schahram Dustdar	Vienna University of Technology, Austria
Abdelaziz Elfazziki	University of Marrakech, Morocco
Nora Faci	Université Claude Bernard Lyon 1, France
Zhang Fan	Peking University, China
Yixiang Fang	Chinese University of Hong Kong, Shenzhen, China
Xiaoming Fu	University of Goettingen, Germany
Yunjun Gao	Zhejiang University, China
Dimitrios Georgakopoulos	Swinburne University of Technology, Australia
Azadeh Ghari Neiat	Deakin University, Australia
Xiangyang Gou	Peking University, China
Daniela Grigori	Université Paris-Dauphine - LAMSADE, France
Tobias Grubenmann	University of Bonn, Germany
Viswanath Gunturi	IIT Ropar, India
Armin Haller	Australian National University, Australia
Kongzhang Hao	University of New South Wales, Australia
Yu Hao	University of New South Wales, Australia
Tanzima Hashem	Bangladesh University of Engineering and Technology, Bangladesh
Md Rafiul Hassan	King Fahd University of Petroleum and Minerals, Saudi Arabia
Yizhang He	University of New South Wales, Australia
Lin Hu	Peking University, China
Chenji Huang	University of New South Wales, Australia
Hao Huang	Wuhan University, China
Xin Huang	Hong Kong Baptist University, Hong Kong
Yilun Huang	University of Technology Sydney, Australia
Zhisheng Huang	Vrije Universiteit Amsterdam, The Netherlands
Zi Huang	University of Queensland, Australia
Dawei Jiang	Zhejiang University, China
Jiawei Jiang	ETH Zurich, Switzerland
Jyun-Yu Jiang	University of California, Los Angeles, USA
Lili Jiang	Umeå University, Sweden
Peiquan Jin	University of Science and Technology of China, China
Eleanna Kafeza	Athens University of Economics and Business, Greece
Georgios Kambourakis	University of the Aegean, Greece
Verena Kantere	University of Ottawa, Canada
Georgia Kapitsaki	University of Cyprus, Cyprus
Panagiotis Karras	Aarhus University, Denmark

Kyoung-Sook Kim	National Institute of Advanced Industrial Science and Technology, Japan
Hong Va Leong	Hong Kong Polytechnic University, Hong Kong
Binghao Li	University of New South Wales, Australia
Hui Li	Xiamen University, China
Jianxin Li	Deakin University, Australia
Youhuan Li	Hunan University, China
Xiang Lian	Kent State University, USA
Kewen Liao	Australian Catholic University, Australia
Dan Lin	University of Missouri, USA
Qingyuan Linghu	University of New South Wales, Australia
Sebastian Link	University of Auckland, New Zealand
An Liu	Soochow University, China
Boge Liu	University of New South Wales, Australia
Guanfeng Liu	Macquarie University, Australia
Cheng Long	Nanyang Technological University, China
Hua Lu	Roskilde University, Denmark
Siqiang Luo	Nanyang Technological University, China
Jianming Lv	South China University of Technology, China
Fenglong Ma	Pennsylvania State University, USA
Jiangang Ma	Federation University Australia, Australia
Zakaria Maamar	Zayed University, UAE
Murali Mani	University of Michigan–Flint, USA
Yannis Manolopoulos	Open University of Cyprus, Cyprus
Yuren Mao	University of New South Wales, Australia
Xiaoye Miao	Zhejiang University, China
Sajib Mistry	Curtin University, Australia
Natwar Modani	Adobe Research, India
Amira Mouakher	Corvinus University of Budapest, Hungary
Tsz Nam-Chan	Hong Kong Baptist University, Hong Kong
Mitsunori Ogihara	University of Miami, USA
Mourad Oussalah	University of Oulu, Finland
M. Tamer Ozsu	University of Waterloo, Canada
George Pallis	University of Cyprus, Cyprus
Yue Pang	Peking University, China
George Papastefanatos	Athena Research Center, Greece
Peng Peng	Hunan University, China
Zhiyong Peng	Wuhan University, China
Francesco Piccialli	University of Naples Federico II, Italy
Dimitris Plexousakis	Institute of Computer Science - FORTH, Greece
Nicoleta Preda	Université de Versailles, France
Tieyun Qian	Wuhan University, China
Lu Qin	University of Technology Sydney, Australia
Yu-Xuan Qiu	University of Technology Sydney, Australia
Jarogniew Rykowski	Poznan University of Economics and Business, Poland
Dimitris Sacharidis	Université Libre de Bruxelles, Belgium

Detian Zhang	Jiangnan University, China
Ji Zhang	University of Southern Queensland, Australia
Jiujing Zhang	Guangzhou University, China
Wenjie Zhang	University of New South Wales, Australia
Yanchun Zhang	Victoria University, Australia
Ying Zhang	University of Technology Sydney, Australia
Lei Zhao	Soochow University, China
Kai Zheng	University of Electronic Science and Technology of China
Xiangmin Zhou	RMIT University, Australia
Yuqi Zhou	Peking University, China
Yi Zhuang	Zhejiang Gongshang University, China
Lei Zou	Peking University, China

Additional Reviewers

Abeysekara, Prabath
Abusafia, Amani
Akram, Junaid
Alharbi, Ahmed
Allani, Sabri
Alnazer, Ebaa
Aryal, Sunil
Aytimur, Mehmet
Bahutair, Mohammed
Bornholdt, Johann
Chaki, Dipankar
Chan, Harry Kai-Ho
Chen, Dong
Chen, Xiaocong
Chen, Xiaoshuang
Du, Jing
Efthymiou, Vasilis
Fang, Uno
Georgievski, Ilche
Haghshenas, Kawsar
Han, Keqi
Hotz, Manuel
Huang, Guanjie
Islam, Fariha Tabassum
Islam, Khandker Aftarul
Joan, Yin
Kassawat, Firas
Kelarev, Andrei
Kunchala, Jyothi

Lan, Michael
Li, Chunbo
Li, Huan
Li, Meng
Li, Xinghao
Liao, Ningyi
Liao, Zhibin
Liesaputra, Veronica
Liu, Boge
Lumbantoruan, Rosni
Maaradji, Abderrahmane
Maheshwari, Ayush
Mahmood, Md Tareq
Maroulis, Stavros
Mountantonakis, Michalis
Nicewarner, Tyler
Papadakos, Panagiotis
Papoutsoglou, Maria
Paschalides, Demetris
Qiu, Yu-Xuan
Qu, Liang
Rashid, Syed Mukit
Salman, Muhammad
Sarwar, Kinza
Sassi, Salma
Setz, Brian
Sha, Alyssa
Shahzaad, Babar
Shao, Yachao

Sikdar, Sagor
Song, Xiangyu
Stamatopoulos, Vasileios
Taoufik, Yeferny
Tripto, Nafis Irtiza
Wang, Haixin
Wang, Hanchen
Wang, Jiaqi
Wu, Xiaoying
Wu, Yingpei
Yan, Qian

Yang, Peilun
Ye, Zesheng
Yin, Hui
Yu, Yuanhang
Zeginis, Chrysostomos
Zhang, Boyu
Zhang, Han
Zhang, Junhua
Zhang, Yunxiao
Ziaur, Rahman

Contents – Part II

Text Mining (1)

Text Mining (2)

Service Computing and Cloud Computing (1)

Service Computing and Cloud Computing (2)

Tutorial and Demo

Contents – Part I

Graph Neural Network

Graph Query

Social Network

Spatial and Temporal Data Analysis

Deep Learning (1)

Efficient Feature Interactions Learning with Gated Attention Transformer

Chao Long, Yanmin Zhu$^{(\boxtimes)}$, Haobing Liu, and Jiadi Yu

Shanghai Jiao Tong University, Shanghai 200240, China
{longchao,yzhu,liuhaobing,jdyu}@sjtu.edu.cn

Abstract. Click-through rate (CTR) prediction plays a key role in many domains, such as online advising and recommender system. In practice, it is necessary to learn feature interactions (i.e., cross features) for building an accurate prediction model. Recently, several self-attention based transformer methods are proposed to learn feature interactions automatically. However, those approaches are hindered by two drawbacks. First, Learning high-order feature interactions by using self-attention will generate many repetitive cross features because k-order cross features are generated by crossing $(k–1)$-order cross features and $(k–1)$-order cross features. Second, introducing useless cross features (e.g., repetitive cross features) will degrade model performance. To tackle these issues but retain the strong ability of the Transformer, we combine the vanilla attention mechanism with the gated mechanism and propose a novel model named Gated Attention Transformer. In our method, k-order cross features are generated by crossing $(k–1)$-order cross features and 1-order features, which uses the vanilla attention mechanism instead of the self-attention mechanism and is more explainable and efficient. In addition, as a supplement of the attention mechanism that distinguishes the importance of feature interactions at the vector-wise level, we further use the gated mechanism to distill the significant feature interactions at the bit-wise level. Experiments on two real-world datasets demonstrate the superiority and efficacy of our proposed method.

Keywords: Feature interactions · CTR prediction · Transformer

1 Introduction

Click-through rate (CTR) prediction is a critical task in many domains, such as online advertising and recommender systems, since a small improvement to the task will bring a lot of revenue [7, 21]. Taking online advertising as an example, when publishers have displayed the advertisements provided by the advertisers, their revenue depends on whether the users will click on these advertisements. That is to say, advertisers only pay for ads that users have clicked on. Considering the large amounts of existing users, a minor improvement to the prediction model means millions of additional users will click on advertisements, which will bring

© Springer Nature Switzerland AG 2021
W. Zhang et al. (Eds.): WISE 2021, LNCS 13081, pp. 3–17, 2021.
https://doi.org/10.1007/978-3-030-91560-5_1

a large amount of revenue. Thus, this field attracts more and more interest from both academia and industry.

Cross features, especially high-order cross features (i.e., feature interactions), are the critical factors of a successful CTR prediction model. For example, suppose we have a record including three features {Age = 20, Gender = Male, ProductCategory=Auto advertisement}, the two-order cross features ⟨Age = 20, Gender = Male⟩ is obviously more predictive than the raw features ⟨Gender = Male⟩ for the auto advertisement. Thus, using only raw features often leads to sub-optimal results [18].

To model high-order cross features, recently several self-attention based methods are proposed, such as AutoInt [24] and InterHat [16]. Despite the superior performance of these methods, two drawbacks will hinder model performance. First, these methods generate k-order feature interactions by crossing $(k\text{-}1)$-order feature interactions and $(k\text{-}1)$-order feature interactions, which is counter-intuitive and will introduce large amount repetitive cross features. Second, introducing useless feature interactions will degrade model performance.

To cope with the problems above, we propose a novel method named Gated Attention Transformer. The key idea is to generate k-order feature interactions by crossing $(k\text{-}1)$-order feature interactions and 1-order features, which is more intuitive and efficient. In addition, since each representation vector represents multiple cross features when modeling feature interactions automatically, the attention mechanism that distinguishes the importance of feature interactions at the vector-wise level loses its efficiency. Thus, the gated mechanism that distills the significant feature interactions at the bit-wise level is used to complement the attention mechanism. Specifically, the core of the proposed method is a transformer-like encoder, which consists of a gated multi-head attention layer and a gated distillation network. The gated multi-head attention layer first crosses the $(k\text{-}1)$-order cross features and the 1-order features by using the vanilla attention mechanism. Afterwards, it learns an update gate for each head, which merges different head adaptively. Though cross features can be learned by the gated multi-head attention layer, each vector represents multiple cross features. Thus, the vanilla attention mechanism loses its efficiency for distinguishing the importance of feature interactions at the vector-wise level. To tackle these problems, a gated distillation network that alleviates the effects of useless cross features at the bit-wise level is used. By stacking multiple such encoders, the method can learn feature interactions of different orders.

To summarize, we make the following contributions:

- We propose a novel method to learn feature interactions efficiently by combining the gated mechanism and the Transformer architecture.
- We design a Transformer-like encoder, which not only retains the Transformer's architecture but tackles the drawbacks of the self-attention mechanism. Besides, we combine the attention mechanism with the gated mechanism for overcoming the limitation of the attention mechanism in the process of learning feature interactions automatically.

- We conduct extensive experiments on two real-world datasets to evaluate our method, and the results demonstrate that our model outperforms several representative state-of-art methods.

The rest of this paper is organized as follows: Sect. 2 gives a brief review of the related work in the field of CTR prediction. Section 3 provides some important concepts and a problem formulation. Section 4 introduces our proposed method. Section 5 presents the results of extensive experiments on two public datasets. Section 6 concludes and points out some future research directions.

2 Related Work

In this section, we briefly review the existing approaches in the field of CTR prediction.

2.1 Traditional Models for CTR Prediction

Linear Regression (LR) is widely used in CTR prediction for its simplicity and efficiency, which learns a unique weight for each feature. However, human efforts are needed to generate high-order cross features, which is unacceptable for its high cost. To model feature interactions automatically, Factorization machine (FM) [23] and many variants are proposed [11–13]. FM models two-order feature interactions based on LR, which learns a latent vector for each raw feature and uses inner product operation to model two-order feature interactions. Field-aware FM (FFM) [13] considers the field information and learns a field-aware latent vector for each raw feature. Attentional FM (AFM) [28] considers that different two-order cross features have different significance. Therefore, it uses an attention network to acquire a significance score for each cross feature [1]. Though HOFM [3] can extend FM to model high-order feature interactions by efficient training algorithms, it is hard to apply it to real predictive systems [3].

2.2 Deep Neural Network Based Models for CTR Prediction

With the success of deep neural network (DNN) in various fields [8,19,30], a lot of methods using DNN to model high-order feature interactions are proposed. For example, Product-based Neural Network (PNN) [22] concatenates the raw features and the two-order cross features and feeds them into a DNN to model high-order cross features. Neural FM (NFM) [9] uses a Bi-interaction pooling layer to model two-order feature interactions and feeds them into a DNN to model high-order feature interactions. Considering both low-order and high-order cross features are important for prediction, Wide&Deep [4] uses a linear part to model low-order feature interactions and a deep part, which is a DNN, to model high-order feature interactions. DeepFM [7] combines FM with DNN to model both low- and high-order cross features jointly.

Neural Architecture Search (NAS) is proposed to search the most proper architecture of a neural network. Inspired by NAS, some methods utilizing

AutoML to seek useful feature interactions are proposed. For example, Auto-Group [18] first find multiple subsets of features, where the feature interactions among each feature subset are effective. Then, it models high-order feature interactions based on these effective feature subsets.

2.3 Self-attention Based Models for CTR Prediction

The Attention mechanism learns a function that measures the importance of different things. It is originally proposed for neural machine translation (NMT) [6] and is widely used in different domains, such as recommender system [16,20], due to its capability to learn the importance of things.

Self-attention is a form of the attention mechanism. Researchers from Google have designed the Transformer [25], which is entirely based on multi-head self-attention and achieves state-of-the-art performance on multiple NLP tasks. Inspired by the success of Transformer in different domains, several self-attention based methods are proposed to model high-order feature interactions. AutoInt [24] uses the multi-head self-attention mechanism to learn high-order feature interactions. By stacking the different number of self-attention layers, it can learn different orders of feature interactions. InterHat [16] first uses a self-attention based transformer encoder to model the polysemy of features and then learns different orders of feature interactions by hierarchical attention. Though self-attention based methods achieve state-of-the-art performance, some drawbacks remain to be tackled.

3   Preliminaries

In this section, we briefly introduce some notations and the problem statement.

Definition 1: Field and Feature. In this paper, the term *field* refers to the name of attributes, such as *gender, city*. The term *feature* refers to values of the corresponding field, such as the feature *male* for the field *gender* and the feature *Washington* for the field *city*.

Definition 2: P-order Cross Features. Given an input vector $x \in \mathbb{R}^m$ that x_i is the feature of field i, a p-order cross feature is defined as $g(x_{i1}, \ldots, x_{ip})$, where x_{ik} is an arbitrary feature of x, p is the number of distinct features involved, and $g(\cdot)$ is a non-additive interaction function, such as inner product [16,21,23] and outer product [17]. A p-order cross feature models the p-order feature interaction among the corresponding features.

Problem Statement. We assume that the training dataset composes of m categorical fields (i.e., user id, item id, etc.) and N records. Therefore, each record of the dataset is consisted of m categorical features and an associated label $y \in \{0, 1\}$. We suppose that all fields involved here are categorical because most fields in this problem are either categorical or can be made categorical through discretization. Let $x \in \mathbb{R}^m$ denotes the concatenation of categorical features of a record, where x_i is the feature of field i. Our goal is to predict a $\hat{y} \in (0,1)$ for the record based on the input vector x.

4   Method

In this section, we first give an overview of our proposed model depicted in Fig. 1. Afterward, we present a comprehensive description of each component.

4.1   Overview

We design a novel model to learn feature interactions efficiently with the following considerations: (1) Though Transformer is successfully applied to various domains, the self-attention mechanism fails to learn feature interactions efficiently in CTR prediction. (2) Useless feature interactions will degrade model performance. The attention mechanism loses its efficiency when learning feature interactions automatically.

Figure 1 shows an overview of our method. The sparse input features are first mapped into a low-dimensional vector space by an embedding layer. Thus, each feature is represented with a dense vector. Afterward, we feed these embedding vectors into several encoders to model feature interactions of different orders. The details of this encoder are introduced in Sect. 4.4. Finally, we employ an aggregation layer to aggregate all cross features of different orders and predict the result based on it. Next, we introduce the details of each component.

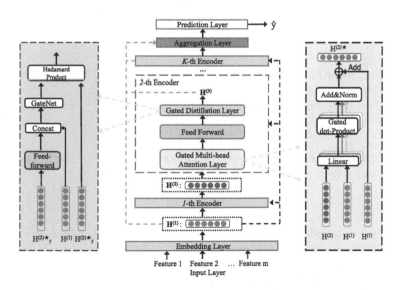

Fig. 1. The overview of our proposed method. By stacking multiple encoders, multiorder feature interactions can be modeled.

4.2   Input Layer

Suppose each record contains m fields, then the input is the concatenation of all fields. Specifically,

$$x = [x_1, x_2, \ldots, x_m]^T, \tag{1}$$

where x_i is the feature of the i-th field. x_i is a one-hot vector if the i-th field is categorical. Otherwise, x_i is a scalar value if the i-th field is numerical.

4.3   Embedding Layer

Since the one-hot representation of features is sparse and high-dimensional, it is necessary to represent each feature as a dense and low-dimensional vector. Taking the record {UserId=95, Gender=male, ItemId=27} as an example, the record becomes sparse and high-dimensional after one-hot encoding:

$$[\underbrace{0, 0, \cdots, 1, \cdots, 0, 0}_{UserId}, \underbrace{1, 0}_{Gender}, \underbrace{0, 0, \cdots, 1, \cdots, 0, 0}_{ItemId}], \tag{2}$$

which is hard to train. Thus, we apply an embedding layer to represent each feature as a dense and low-dimensional vector. Specifically,

$$e_i = V_i \times x_i, \tag{3}$$

where V_i is the embedding lookup table of the i-th field, and x_i the one-hot representation of the feature of the i-th field. If the field is multivalent, we take the average of the corresponding feature representations as the result. Take the field *Genre* as an example, a movie having multiple genres (e.g., Comedy and Romance) corresponds to a multi-hot representation. We consider the Comedy and Romance as a new feature of the filed *Genre*, and represent it as

$$e_{CR} = \frac{1}{2} V_{genre} \times x_{CR}, \tag{4}$$

where V_{genre} is the embedding lookup table of the field *Genre*, and x_{CR} is the multi-hot representation of the feature Comedy and Romance. In this way, the result of the embedding layer is an matrix:

$$H^{(1)} = [e_1, e_2, \ldots, e_m]^T, \tag{5}$$

where $H^{(1)} \in \mathbb{R}^{m \times d}$, $e_i \in \mathbb{R}^d$ denotes the embedding vector of the feature of the i-th field, and d denotes the dimension of feature embedding.

4.4   Encoder

The k-th encoder models $(k+1)$-order feature interactions. It takes $H^{(1)} \in \mathbb{R}^{m \times d}$ and $H^{(k)} \in \mathbb{R}^{m \times d}$ as inputs, and output $H^{(k+1)} \in \mathbb{R}^{m \times d}$, where $H^{(1)}$ denotes the embedding vectors of raw features, $H^{(k)}$ denotes the representations of k-order cross features, and $H^{(k+1)}$ denotes the representations of $(k + 1)$-order cross features. As shown in Fig. 1, each encoder is composed of two different layers. Next, we introduce each layer in detail.

Gated Multi-head Attention Layer. The multi-head self-attention mechanism is first proposed by Google [25] and proved effective in different domains, such as machine translation [6] and recommender systems [20,27]. Considering the drawbacks of the self-attention mechanism, the vanilla attention mechanism is more efficient here. In addition, inspired by GaAN [29] that learns a importance score for each head since heads are not equally important, we combine the attention mechanism with the gated mechanism and propose a novel gated multi-head attention layer. Different from the GaAN, the gated multi-head attention layer learns a gated vector for each head and weighted each head at the bit-wise level, since each head models multiple feature interactions in each subspace.

Specifically, for the gated multi-head attention layer of the k-th encoder, the inputs are the representations of k-order cross features and the embedding vectors of raw features, which are denoted as $H^{(k)} \in \mathbb{R}^{m \times d}$ and $H^{(1)} \in \mathbb{R}^{m \times d}$ respectively. In contrast to the vanilla multi-head attention network, we integrate a neural gating structure:

$$g_i = sigmoid(GateNet_1([H^{(k)}||H^{(1)}])), \tag{6}$$

where $||$ denotes the concatenation operation, $GateNet_1$ is a feed-forward neural network, and g_i is the gate learned for head i. Then, we distinguish the importance of feature interactions by using both the attention mechanism and the gating mechanism:

$$\alpha_i = softmax(\frac{Q_i K_i^T}{\sqrt{d_K}}), \tag{7}$$

$$Q_i = W_i^{(Q)} H^{(k)}, K_i = W_i^{(K)} H^{(1)}, \tag{8}$$

where $W_i^{(Q)} \in \mathbb{R}^{d_K \times d}, W_i^{(K)} \in \mathbb{R}^{d_K \times d}$ are parameters to learn for head i, d_K denotes the dimension of each head, and α_i is the final importance score of cross features. Finally, we obtain the representations by concatenating the outputs of all heads:

$$H_i^{(k)*} = g_i \odot \alpha_i V_i, \tag{9}$$

$$H^{(k)*} = Norm([H_1^{(k)*}||H_2^{(k)*}||\cdots||H_h^{(k)*}] + H^{(1)}), \tag{10}$$

$$V_i = W_i^{(V)} H^{(k)}, \tag{11}$$

where $W_i^{(V)} \in \mathbb{R}^{d_K \times d}$ is the parameter of head i, h is the number of heads, and *Norm* represents the normalization.

Gated Distillation Layer. Insignificant features may lead to suboptimal results [18]. Thus, how to minimize the influence of those useless features becomes valuable. The attention mechanism can distinguish the importance of features and give an importance score at the vector-wise level. However, in the process of the high-order cross features modeling, each representation vector represents multiple cross features. Take the representation matrix $H^{(k)} \in \mathbb{R}^{m \times d}$ as an example, $H_i^{(k)}$ represents all the k-order cross features related to the feature of the

i-th field. Thus, the importance score at the vector level fails to distinguish those mixed cross features.

The gating mechanism is widely used in LSTM [10] and GRU [5] to control the information transmission. The core idea of it is to learn an update gate for the input at the bit-wise level, which controls how much the information can be transmitted. In this paper, we adopt the gating mechanism to alleviate the effect of useless cross features. Specifically, we learn a gate for each cross feature by a feed-forward neural network and employ the gate to control the transmission of information.

For the gated distillation layer of the k-th encoder, we take the output of the corresponding feed-forward neural network and the embedding vectors of raw features as inputs. Then, the output of the gated distillation layer is obtained by:

$$gate = sigmoid(GateNet_2(W_{gh}H_f^{(k)\star} + W_{go}H^{(1)}), \tag{12}$$

$$H^{(k+1)} = gate \odot H_f^{(k)\star}, \tag{13}$$

where $W_{gh} \in \mathbb{R}^{d \times d}, Wgo \in \mathbb{R}^{d \times d}$ are trainable parameters of the gated distillation layer, $GateNet_2$ is a feed-forward neural network, and $\odot$ is the Hadamard Product. The matrix $H^{(k+1)} \in \mathbb{R}^{m \times d}$ is also the output of the k-th encoder.

4.5   Aggregation Layer and Prediction Layer

Aggregation Layer. Each encoder generates a representation matrix $H^{(t)} \in \mathbb{R}^{m \times d}$, where t ranges from 2 to M, M is the number of encoders, and the matrix $H^{(t)}$ denotes the representations of t-order cross features. We regard the raw features as the one-order cross features. Thus, we aggregate these matrices by an aggregation layer:

$$H = \sum_{t=1}^{M} W_t H^{(t)}, \tag{14}$$

where $W_t \in \mathbb{R}$ are trainable parameters, $H^{(1)} \in \mathbb{R}^{m \times d}$ is the embedding matrix of raw features, and $H \in \mathbb{R}^{m \times d}$.

Prediction Layer. The output of the aggregation layer is the matrix $H \in \mathbb{R}^{m \times d}$, in which $H_i \in \mathbb{R}^d$ represents all cross features related to the raw feature of field i. For each H_i, we obtain the attention score by an attention network:

$$\alpha_i = \frac{exp(AttentionNet(H_i))}{\sum_s^m exp(AttentionNet(H_s))}, \tag{15}$$

where $AttentionNet$ is a single hidden layer feed-forward neural network. Then, we obtain the final prediction by:

$$\hat{y} = sigmoid(\sum_i^m \alpha_i PredictionNet(H_i)), \tag{16}$$

where $PredictionNet$ is also a single hidden layer feed-forward neural network.

Table 1. Statistics of evaluation datasets

Datasets	#Samples	#Fields	#Features
Frappe	288,609	10	5,382
Avazu	40,428,967	22	1,544,488

4.6 Training

We adopt Logloss to optimize the model parameters. Formally, the objective function of our method is defined as follows:

$$Logloss = -\frac{1}{N}\sum_{j=1}^{N}(y_j log(\hat{y}_j) + (1-y_j)log(1-\hat{y}_j)) + \frac{\lambda}{2}(\|\Theta\|_F^2), \qquad (17)$$

where y_j and $\hat{y}_j$ are ground truth of the j-th sample and the estimated CTR respectively, j is the index of samples, N is the number of samples, and Θ denotes all the parameters of our method.

5 Experiments

In this section, we aim to answer the following questions:

- **RQ1:** Can our proposed method perform better than the baselines?
- **RQ2:** Are the critical components (e.g., gated multi-head attention layer) really effective for improving the model performance?
- **RQ3:** Is our proposed method more efficient than the self-attention based method?
- **RQ4:** How do the critical hyper-parameters (e.g., the number of encoders) affect the model performance?

5.1 Experiment Setup

Datasets. To answer the three questions, we conduct extensive experiments on two public real-world datasets: Frappe[1] and Avazu[2]. The Frappe [2] dataset contains records about users' app usage behaviors that whether a app is used by a user. The Avazu dataset contains records about users' ads click behaviors that whether a displayed mobile ad is clicked by a user. All the fields involved in both datasets are categorical. For the two datasets, we randomly select 80% of all samples for training, 10% for validating and 10% for testing. The statistics of Frappe and Avazu are shown in Table 1.

---

[1] http://baltrunas.info/research-menu/frappe.
[2] https://www.kaggle.com/c/avazu-ctr-prediction.

Table 2. Hyper-parameters of each model.

Model	Frappe			Avazu		
	emb	lr	L2	emb	lr	L2
FM	32	1e–3	0	32	1e–3	1e–4
PNN	32	1e–3	0	32	1e–3	1e–4
DeepFM	32	1e–3	0	32	1e–3	1e–4
xDeepFM	32	1e–3	1e–4	32	1e–3	1e–4
AutoInt	32	1e–3	1e–4	16	1e–3	1e–5
InterHat	32	5e–3	1e–4	32	1e–3	5e–5
Our model	32	5e–3	1e–4	16	1e–3	1e–4

Note: emb = dimension of the embedding vectors, lr=learning rate, $L_2 = l_2$ regularization.

Evaluation Metrics. We use AUC (Area Under the ROC Curve) and Logloss (cross entropy) to evaluate the performance of all methods. AUC measures the probability that a randomly selected positive record will be ranked higher than a randomly selected negative record. Higher AUC indicates better performance. Logloss measures how much the difference between the predicted score and the ground truth. Lower Logloss indicates better performance. It is noticeable that a improvement of AUC at **0.001-level** is regarded as significant for CTR prediction [4, 7, 15, 24, 26].

Baselines. As described in Sect. 2, we categorize the existing approaches into three types: (A) Tradional models; (B) Deep learning-based models; (C) Self-attention based models. We select the following representative models of the three types to compare with ours.

FM [23] learns a latent vector of each raw feature and models two-order cross features by inner product operation.

PNN [22] uses DNN to model high-order cross features.

DeepFM [7] combines FM with DNN to model low-order cross features and high-order cross features jointly.

xDeepFM [17] utilizes multiple Compressed Interaction Networks to model high-order cross features in an explicit fashion.

AutoInt [24] utilizes multi-head attention mechanism to explicitly model high-order cross features.

InterHat [16] utilizes the hierarchical attention mechanism to explain cross features, which aggregates cross features first and then takes inner product operation to model higher-order cross features at each step.

Implementation Details. All methods are implemented by Pytorch. We adopt early-stopping to avoid overfitting and implement FM, PNN, DeepFM, and xDeepFM following [24]. We use Adam [14] to optimize all methods, and the batch size is 1024 for both Frappe and Avazu. For our method, the $GateNet_1$

Table 3. Results of model comparision, the best results are highlighted.

Model type	Model	Frappe		Avazu	
		AUC	Logloss	AUC	Logloss
Traditional	FM	0.9719	0.1889	0.7356	0.4041
DNN-based	PNN	0.9827	0.1447	0.7440	0.3969
	DeepFM	0.9772	0.1787	0.7414	0.3983
	xDeepFM	0.9808	0.1631	0.7443	0.3980
Self-attention based	AutoInt	0.9810	0.1862	0.7441	0.3974
	InterHat	0.9805	0.1638	0.7417	0.3990
	Our model	**0.9848**	**0.1506**	**0.7468**	**0.3957**

and $GateNet_2$ are two single hidden layer feed-forward neural networks. Besides, the number of heads is 8 for the gated multi-head attention layer. The other hyper-parameters are summarized in Table 2, which are tuned on the validation dataset to obtain the best result of each model.

5.2   Performance Comparison(RQ1)

In this section, we compare our method with six representative baselines. The results on the Frappe and the Avazu datasets are summarized in Table 3, from which we have the following observations:

- All methods of modeling high-order feature interaction are superior to FM that only models second-order cross features, which indicates that high-order cross features are essential for a successful CTR prediction model.
- An interesting observation is that some self-attention based models are inferior to some DNN-based models on the Frappe and Avazu datasets. It indicates that there is no significant performance difference between the DNN-based models and the self-attention based models.
- Our method achieves the best performance on both Frappe and Avazu datasets, which indicates that combining the attention mechanism with the gated mechanism can boost the model performance.

5.3   Ablation Study(RQ2)

In this section, we aim to explore whether the critical components of our method are effective for improving the model performance. Thus, we conduct an ablation study.

- ours(-g/a): on the one hand, the encoder uses a vanilla multi-head key-value attention layer that removes the gate neural network from the gated multi-head attention layer. On the other hand, the encoder removes the gated distillation layer.

- ours(-g): the encoder removes the gated distillation layer but retains the gated multi-head attention layer.
- ours(-a): the encoder uses a vanilla multi-head attention layer and retains the gated distillation layer.

The results of each variant of our method and the baseline are shown in Fig. 2, from which we can obtain the following observations: (1) ours(-g/a) performs better than AutoInt, which indicates that the vanilla attention mechanism is proper than the self-attention mechanism in CTR prediction. (2) ours(-g) outperform ours(-g/a), which indicates that the gated multi-head attention mechanism is more effective than vanilla multi-head attention mechanism. (3) ours(-a) outperform ours(-g/a), which indicates that the gated distillation layer is effective for boosting model performance.

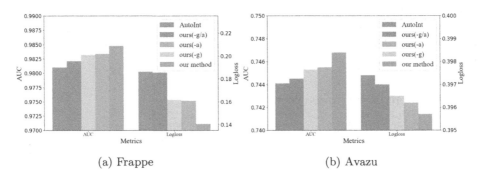

(a) Frappe   (b) Avazu

Fig. 2. Results of the ablation study.

5.4   Efficiency Comparison(RQ3)

In real-world scenarios, efficiency is an important factor that affects whether a model could be used. Thus, we record the training cost (running time per epoch) as the criterion of measuring the efficiency of methods.

For a fair comparison, all the models are trained in the same machine with a TITAN Xp GPU. Table 4 shows the training cost of attention-based methods. As can be seen, our method is significantly more efficient than self-attention based methods, which proves that self-attention mechanism may be the bottleneck of the method.

5.5   Hyper-Parameter Study(RQ4)

In this section, we study the impact of hyper-parameters on our proposed method, including the number of encoders and the dimension of embedding vectors.

Table 4. Efficiency comparison on two datasets.

Method	Dataset	Training cost(s)
AutoInt	Frappe	70.69
	Avazu	5985.26
InterHat	Frappe	60.89
	Avazu	5564.63
Ours	Frappe	44.51
	Avazu	4918.48

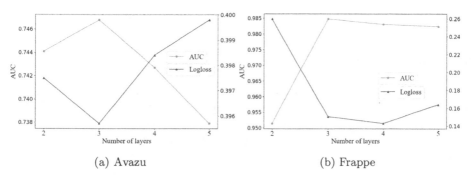

(a) Avazu   (b) Frappe

Fig. 3. The impact of number of encoders on our method with respect to AUC and Logloss.

Number of Modules. Figure 3 shows the impact of the number of encoders. We observe that the model performance shows an increasing trend, followed by a decreasing trend when the number of encoders is larger than 3. Because the number of encoders determines the maximal order of feature interactions that our method can learn, it indicates that 4-order feature interactions are good enough to predict the final result. Besides, an interesting observation is that for both the Frappe dataset and the Avazu dataset, our method thinks that 4-order feature interactions are enough to obtain a good prediction.

Dimension of Embedding Vectors. As shown in Fig. 4, the performance of our method first grows with the dimension of embedding vectors, which is attributed to the better learning ability of our method. The performance starts to degrade when the dimension is larger than 32 for the Frappe dataset and 16 for the Avazu dataset, which is caused by the overfitting.

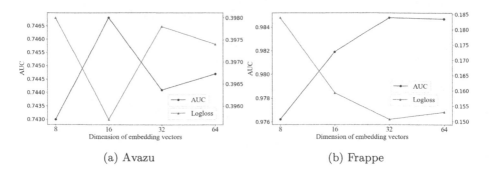

(a) Avazu   (b) Frappe

Fig. 4. Performance of our method affected by the dimension of embedding vectors.

6 Conclusions

In this paper, we point out that the self-attention mechanism is less efficient to learn feature interactions and propose a novel Transformer-like method. The method combines the key-value attention mechanism with the gated mechanism, which makes it more efficient and effective to learn feature interactions.

For future works, we are interested in proposing new technologies to alleviate the effect of irrelevant cross features. For example, a hot direction is automatic feature selection.

Acknowledgements. This research is supported in part by the 2030 National Key AI Program of China (2018AAA0100503), Shanghai Municipal Science and Technology Commission (No. 19510760500, and No. 19511101500), National Science Foundation of China (No. 62072304, No. 61772341), and Zhejiang Aoxin Co. Ltd.

References

1. Bahdanau, D., Cho, K., Bengio, Y.: Neural machine translation by jointly learning to align and translate. In: ICLR (2015)
2. Baltrunas, L., Church, K., Karatzoglou, A., Oliver, N.: Frappe: understanding the usage and perception of mobile app recommendations in-the-wild (2015). CoRR abs/1505.03014
3. Blondel, M., Fujino, A., Ueda, N., Ishihata, M.: Higher-order factorization machines. In: NIPS, pp. 3351–3359 (2016)
4. Cheng, H., et al.: Wide & deep learning for recommender systems. In: RecSys, pp. 7–10 (2016)
5. Chung, J., Gülçehre, Ç., Cho, K., Bengio, Y.: Empirical evaluation of gated recurrent neural networks on sequence modeling (2014). CoRR abs/1412.3555
6. Cui, H., Iida, S., Hung, P., Utsuro, T., Nagata, M.: Mixed multi-head self-attention for neural machine translation. In: EMNLP-IJCNLP, pp. 206–214 (2019)
7. Guo, H., Tang, R., Ye, Y., Li, Z., He, X.: Deepfm: a factorization-machine based neural network for CTR prediction. In: IJCAI, pp. 1725–1731 (2017)
8. He, K., Zhang, X., Ren, S., Sun, J.: Deep residual learning for image recognition. In: CVPR, pp. 770–778 (2016)

9. He, X., Chua, T.: Neural factorization machines for sparse predictive analytics. In: SIGIR, pp. 355–364 (2017)
10. Hochreiter, S., Schmidhuber, J.: Long short-term memory. Neural Comput. **9**(8), 1735–1780 (1997)
11. Hong, F., Huang, D., Chen, G.: Interaction-aware factorization machines for recommender systems. In: AAAI, pp. 3804–3811 (2019)
12. Jiao, L., Yu, Y., Zhou, N., Zhang, L., Yin, H.: Neural pairwise ranking factorization machine for item recommendation. In: Nah, Y., Cui, B., Lee, S.-W., Yu, J.X., Moon, Y.-S., Whang, S.E. (eds.) DASFAA 2020. LNCS, vol. 12112, pp. 680–688. Springer, Cham (2020). https://doi.org/10.1007/978-3-030-59410-7_46
13. Juan, Y., Zhuang, Y., Chin, W., Lin, C.: Field-aware factorization machines for CTR prediction. In: RecSys, pp. 43–50 (2016)
14. Kingma, D.P., Ba, J.: Adam: a method for stochastic optimization. In: ICLR (2015)
15. Li, Z., Cui, Z., Wu, S., Zhang, X., Wang, L.: Fi-gnn: modeling feature interactions via graph neural networks for CTR prediction. In: CIKM, pp. 539–548 (2019)
16. Li, Z., Cheng, W., Chen, Y., Chen, H., Wang, W.: Interpretable click-through rate prediction through hierarchical attention. In: WSDM, pp. 313–321 (2020)
17. Lian, J., Zhou, X., Zhang, F., Chen, Z., Xie, X., Sun, G.: xdeepfm: combining explicit and implicit feature interactions for recommender systems. In: KDD, pp. 1754–1763 (2018)
18. Liu, B., et al.: Autogroup: automatic feature grouping for modelling explicit high-order feature interactions in CTR prediction. In: SIGIR, pp. 199–208 (2020)
19. Liu, H., Zhu, Y., Xu, Y.: Learning from heterogeneous student behaviors for multiple prediction tasks. In: Nah, Y., Cui, B., Lee, S.-W., Yu, J.X., Moon, Y.-S., Whang, S.E. (eds.) DASFAA 2020. LNCS, vol. 12113, pp. 297–313. Springer, Cham (2020). https://doi.org/10.1007/978-3-030-59416-9_18
20. Luo, X., Sha, C., Tan, Z., Niu, J.: Multi-head attentive social recommendation. In: Cheng, R., Mamoulis, N., Sun, Y., Huang, X. (eds.) WISE 2020. LNCS, vol. 11881, pp. 243–258. Springer, Cham (2019). https://doi.org/10.1007/978-3-030-34223-4_16
21. Pan, J., et al.: Field-weighted factorization machines for click-through rate prediction in display advertising. In: WWW, pp. 1349–1357 (2018)
22. Qu, Y., et al.: Product-based neural networks for user response prediction. In: ICDM, pp. 1149–1154 (2016)
23. Rendle, S.: Factorization machines. In: ICDM, pp. 995–1000 (2010)
24. Song, W., et al.: Autoint: automatic feature interaction learning via self-attentive neural networks. In: CIKM, pp. 1161–1170 (2019)
25. Vaswani, A., et al.: Attention is all you need. In: NIPS, pp. 5998–6008 (2017)
26. Wang, R., Fu, B., Fu, G., Wang, M.: Deep & cross network for ad click predictions. In: KDD, pp. 12:1–12:7 (2017)
27. Wu, C., Wu, F., Ge, S., Qi, T., Huang, Y., Xie, X.: Neural news recommendation with multi-head self-attention. In: EMNLP-IJCNLP, pp. 6388–6393 (2019)
28. Xiao, J., Ye, H., He, X., Zhang, H., Wu, F., Chua, T.: Attentional factorization machines: learning the weight of feature interactions via attention networks. In: IJCAI, pp. 3119–3125 (2017)
29. Zhang, J., Shi, X., Xie, J., Ma, H., King, I., Yeung, D.: Gaan: gated attention networks for learning on large and spatiotemporal graphs. In: UAI 2018, Monterey, California, USA, 6–10 August 2018, pp. 339–349. AUAI Press (2018)
30. Zhang, J., Zheng, Y., Qi, D.: Deep spatio-temporal residual networks for citywide crowd flows prediction. In: Thirty-first AAAI conference on artificial intelligence (2017)

Interactive Pose Attention Network for Human Pose Transfer

Di Luo[1], Guipeng Zhang[1], Zhenguo Yang[1(✉)], Minzheng Yuan[2], Tao Tao[2], Liangliang Xu[2], Qing Li[3], and Wenyin Liu[1,4(✉)]

[1] Guangdong University of Technology, Guangzhou, China
{yzg,liuwy}@gdut.edu.cn
[2] Guangzhou Metro Group Company Ltd., Guangzhou, China
{yuanminzheng,taotao,xuliangliang}@gzmtr.com
[3] The Hong Kong Polytechnic University, Hong Kong, China
qing-prof.li@polyu.edu.hk
[4] Cyberspace Security Research Center, Peng Cheng Laboratory, Shenzhen, China

Abstract. In this paper, we propose an end-to-end interactive pose attention network (IPAN) to generate the person image in a target pose, where the generator of the network comprises a sequence of interactive pose attention (IPA) blocks to transfer the attended regions regarding to intermedia poses progressively, and retain the texture details of the unattended regions for subsequent pose transfer. More specifically, we design an attention mechanism by interacting with image and pose pathways to transfer the regions of interest based on the human pose, and capture the uninterested regions in the current IPA block against the uncertainty of the intermedia poses. In particular, we devise long-distance residual to inject the low-level features of the person image into the IPA blocks to keep its appearance characteristics. In terms of adversarial training, the generator exploits reconstruction loss, perceptual loss and contextual loss, and the discriminator exploits the adversarial loss. Quantitative and qualitative experiments conducted on the DeepFashion and Market-1501 datasets demonstrate the superior performance of the proposed method (e.g., FID value is reduced from 36.708 to 22.568 and 15.757 to 12.835 on Market-1501 and DeepFashion datasets, respectively).

Keywords: Human pose transfer · Interactive pose attention · Long-distance residual

1 Introduction

Human pose transfer aiming to synthesize a person image in a target pose has been applied in enormous scenarios, e.g., movie making [8], human-computer interactions [10], motion prediction [2], video generation [27], etc.

In the context of human pose transfer, the diversity of substantial appearances and various spatial layout of clothes and body parts that may be occluded are difficult to be captured, making the existing works usually generate

© Springer Nature Switzerland AG 2021
W. Zhang et al. (Eds.): WISE 2021, LNCS 13081, pp. 18–33, 2021.
https://doi.org/10.1007/978-3-030-91560-5_2

low-quality person images regarding to the target poses. The early works [13,14] usually adopt two-stage networks to generate coarse-grained images in target poses in the first stage, and then refine the images in an adversarial way in the second stage. However, two-stage methods are susceptible to quality of the coarse-grained images generated in the first stage. Recently, end-to-end models have become popular in human pose transfer due to their convenience. For instance, Siarohin et al. [18] employed deformable skip connections in the generator to spatially transform the textures, which decomposed the overall deformation by a set of local affine transformations with limited performance. Zhu et al. [31] proposed a progressive pose attention transfer network, and its performance may decrease due to the uncertainty of intermediate poses when transferring to the intermediate poses.

In this paper, we propose an end-to-end interactive pose attention network (IPAN) to generate the person image in a target pose. More specifically, IPAN comprises a sequence of interactive pose attention (IPA) blocks, which interact with each other via image pathway and pose pathway. On one hand, we update the image in each IPA block by using conditional pose and target pose to calculate an attention mask, which focuses on the image regions of interest that are expected to be transferred. On the other hand, the unattended image regions will be used as additional information to update the generation of pose. In particular, the cascaded IPA blocks in IPAN aim to transfer the person images to the target poses by generating the person images in a sequence of intermedia poses. Furthermore, we devise a long-distance residual connection in IPAN to inject the low-level features of person image into the IPA blocks to keep its appearance characteristics.

The main contributions of this work are summarized below:

- We devise an interactive pose attention (IPA) block by interacting with image and pose pathways to progressively transfer the regions of interest based on human pose, and capture the uninterested regions for subsequent pose transfer.
- We design a generative adversarial network by cascading IPA blocks with long-distance residual, where IPA blocks transfer person images in a sequence of intermedia poses progressively, and long-distance residual keeps the appearance characteristics in person images.
- We conduct extensive experiments on the DeepFashion and Market-1501 datasets, which manifests the effectiveness of IPA block, outperforming the state-of-the-art baselines.

The rest of the paper is organized as follows. Section 2 reviews the related works. Section 3 introduces the details of the proposed IPAN. Experiments and discussions are conducted in Sect. 4, followed by conclusions in Sect. 5.

2   Related Work

2.1   Human Pose Transfer

Human pose transfer is derived from the task of GANs-based [3] image-to-image translation from different domains, e.g., photo-to-map, sketch-to-image, and night-to-day, etc. For instance, Ma et al. [13] firstly proposed a two-stage GAN architecture to generate person images, in which the person in the target pose is coarsely synthesized in the first stage and would be refined in the second stage. Siarohin et al. [18] introduced a special U-Net [17] structure with deformable skip connections as a generator to synthesize person images from decomposed and deformable images, which alleviated crucial shape and appearance misalignments. Tang et al. [21] adopted a cascade strategy to divide the generation procedures into two stages, where stage I aimed to capture the semantic structure of the scene, and Stage II focused on appearance details via the proposed multi-channel attention selection module. Tang et al. [19] proposed a graph generator, which consisted of two blocks to model the pose-to-pose and pose-to-image relations, respectively. Tang et al. [20] proposed a generator consisting of two generation branches to model the person's appearance and shape information, respectively. Men et al. [16] used human parsing maps for attribute-controllable person image synthesis. However, the aforementioned methods ignore the intermediate information in pose transfer, and the gap between the condition pose and the target pose. Zhu et al. [31] transferred image information from the condition pose to the target pose progressively with the local attention mechanism. However, it suffers from the uncertainty while transferring to the intermediate poses.

2.2 Attention Mechanism

Attention mechanism [23,28] builds long-range dependencies among the layers in deep networks and enables the network to make use of non-local information, aiming to refine perceived information and building long-term correlations. It has been proved to be effective in quite a few tasks, e.g., natural language processing [23], image recognition [5,24], and image generation [28], etc. Generally, the attention mechanism can be categorized into global and local ones. Global attention takes all the hidden states to calculate the attention weights, which is time-consuming. In contrast, local attention [12] focuses only on a subset of the source positions per target word. The local attention mechanism draws inspiration from the soft and hard attentional models proposed by Xu et al. [26] to tackle the task of generating image captions. In their work, soft attention refers to the global attention approach in which weights are placed "softly" over all patches in the source image, while the hard attention focuses on one patch at a time. Zhang et al. [28] introduced the self-attention mechanism into the convolution layer of the GANs and obtained impressive performance in image generation tasks. Attention mechanism is able to make the model focus on the critical regions of interest, while the unattended ones may be neglected. In the context of transferring image to a target domain, there may exist a sequence of uncertain intermedia target states [31], while the unattended regions cannot be neglected directly as attention may not always capture meaningful information.

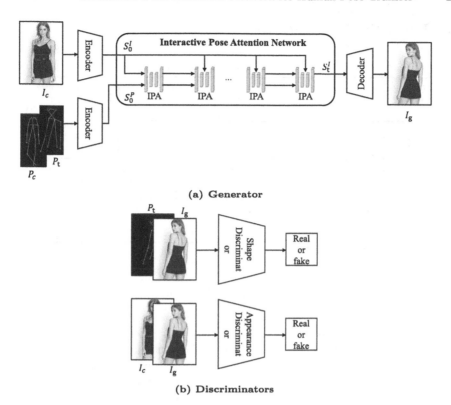

(a) Generator

(b) Discriminators

Fig. 1. Overview of the proposed IPAN.

3 Methodology

3.1 Overview of the Framework

The architecture of the proposed interactive pose attention network (IPAN) is shown in Fig. 1, which consists of one generator and two discriminators. In terms of the generator, it consists of an encoder, an interactive pose attention network (IPAN), and a decoder. Given the raw features extracted from the input image with its pose and target pose extracted via an existing human pose estimator (HPE) [1] encoded by encoders, IPAN constructs a sequence of interactive pose attention (IPA) blocks to transfer the attended regions progressively corresponding to the target pose and retain the texture details of the unattended regions. Finally, the decoder generates the person image in the target pose from the output of IPA blocks via deconvolutional layers. In terms of the discriminators, we adapt shape discriminator to align the pose of the generated image with the target pose, and appearance discriminator to keep the appearance textures of generated image with the input image.

3.2   Interactive Pose Attention Network (IPAN)

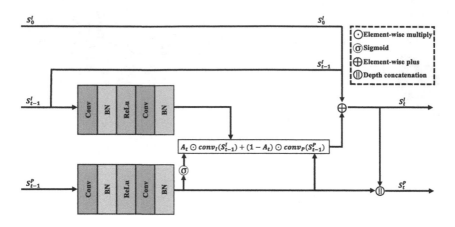

Fig. 2. Details of the t-th interactive pose attention (IPA) block.

1) Interactive pose attention (IPA) block. As mentioned previously, the existing works [20,31] using attention mechanism usually focus on the critical regions of interest, while the unattended ones may be neglected totally. In the context of human pose transfer, a sequence of uncertain intermedia poses may be expected to be transferred progressively instead of transferring from one pose to another pose in a one-step manner, as the attended regions can hardly correspond to the intermedia poses. As a result, the unattended regions cannot be neglected completely as meaningful information still has not been captured fully. These aspects motivate us to transfer poses progressively and pay attention to the regions of interested being attended to and the ones being temporally neglected jointly.

As shown in Fig. 2, IPA block consists of two paths, i.e., image path and pose path, in addition to a skip connection to pass the raw features with the network goes deep. In IPAN with a number of IPA blocks, given an image code S_{t-1}^{I}, pose code S_{t-1}^{P} from the previous IPA block, the t-th block conducts a series of convolution, normalization layers, and ReLU to obtain path-specific feature maps. The activated feature map from the pose path will be used as attention mask A_t to localize the attended image feature map R_a below,

$$R_a = A_t \odot conv_I \left(S_{t-1}^{I} \right) \tag{1}$$

$$A_t = \sigma \left(conv_P \left(S_{t-1}^{P} \right) \right) \tag{2}$$

where σ denotes a sigmoid activation function, and $\odot$ denotes element-wise product.

Furthermore, the attention mask A_t will be used to localize the temporally unattended image feature map R_u below,

$$R_u = (1 - A_t) \odot conv_P \left(S_{t-1}^{P} \right) \tag{3}$$

In terms of image pathway, given A_t, the image code S_t^C is updated as follows,

$$S_t^C = R_a + (1 - A_t) \odot conv_P \left(S_{t-1}^P\right) + S_{t-1}^I \tag{4}$$

where $\odot$ denotes element-wise product. In this way, the IPA block does not need to render static elements and focuses on the pixels defining the pose movements without ignoring the useful information of the part that is not transferred.

In terms of pose pathway, as the image code gets updated through the IPA blocks, the pose code should also be updated to synchronize the changes, i.e., update the location of the pixels to capture and place according to the new image code. Therefore, the pose code update is expected to incorporate the new image code. Specifically, the previous pose code S_{t-1}^P firstly goes through convolutional blocks (two convolution layers, two batch normalization layers, and ReLU). Furthermore, the transformed code is mixed with the updated image code by concatenation. Mathematically, the update is performed below,

$$S_t^P = conv_P \left(S_{t-1}^P\right) \| S_t^C \tag{5}$$

where $\|$ denotes the concatenation of two maps along the depth axis.

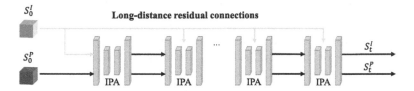

Fig. 3. Details of the long-distance residual connections.

2) Long-distance residual connections. With IPAN goes deep, the network may forget the low-level information corresponding to appearance characteristics of person image conveyed by the former layers, we devise a long-distance residual connection to alleviate intermediate information being lost in IPAN. As shown in Fig. 3, long-distance residual connections constructed in the image path inject the low-level features to each cascaded IPA blocks, respectively. More specifically, given the initial image code S_0^I and the computed image code S_t^C, the update of the t-th IPA block is performed below,

$$S_t^I = S_t^C + S_0^I \tag{6}$$

Finally, the decoder generates the output image I_g from S_t^I via deconvolutional layers.

3.3   Objective Function

In terms of loss terms, IPAN exploits reconstruction loss, perceptual loss, contextual loss and adversarial loss in the generator, and adversarial loss for discriminators. More specifically, given the input image I_c, target image I_t, condition pose P_c, target pose P_t and the generated image I_g, the objective function for the generator is defined below,

$$\mathcal{L}_{gen} = \lambda_{rec}\mathcal{L}_{rec} + \lambda_{per}\mathcal{L}_{per} + \lambda_{cx}\mathcal{L}_{CX} + \mathcal{L}_{GAN}^{G} \tag{7}$$

where $\mathcal{L}_{rec}$, $\mathcal{L}_{per}$, $\mathcal{L}_{cx}$, and $\mathcal{L}_{GAN}^{G}$ denote reconstruction loss, perceptual loss, contextual loss, and adversarial loss for the generator, respectively. The reconstruction loss forces the output to be consistent with the visual appearance of the target image. We adopt perceptual loss [7] to generate more realistic images. The perceptual loss calculates the ℓ_2 distance between activation maps of the pretrained VGG network. Simultaneously, we also adopt contextual loss [15] to ensure less texture distortion and reasonable outputs. The $\mathcal{L}_{GAN}^{G}$ is used to train the generator. The parameters λ_{per}, λ_{rec} and λ_{cx} denote the weights of the loss terms, respectively.

1) Reconstruction loss. The reconstruction loss is used to directly encourage the visual appearance of the generated image I_g to be similar with the target image I_t as follows,

$$\mathcal{L}_{rec} = \|I_g - I_t\|_1 \tag{8}$$

which can reduce visible color distortions and accelerate the convergence process to acquire satisfactory results.

2) Perceptual loss. For texture matching, we exploit deep features extracted from certain layers of the pretrained VGG network, which is effective in image synthesis [7] tasks. As feature correlations can well represent visual style statistics in the high-level features, we utilize the Gram matrix of features to measure the texture similarity. The Gram matrix $\mathcal{G}\left(\mathcal{F}^l\left(I_t\right)\right)$ for the target image I_t over its feature map $\mathcal{F}^l\left(I_t\right)$ can be computed below,

$$\mathcal{G}\left(\mathcal{F}^l\left(I_t\right)\right) = \left[\mathcal{F}^l\left(I_t\right)\right]\left[\mathcal{F}^l\left(I_t\right)\right]^T \tag{9}$$

Furthermore, the perceptual loss $\mathcal{L}_{per}$ is defined below,

$$\mathcal{L}_{per} = \left\|\mathcal{G}\left(\mathcal{F}^l\left(I_g\right)\right) - \mathcal{G}\left(\mathcal{F}^l\left(I_t\right)\right)\right\|_1^2 \tag{10}$$

3) Contextual loss. The contextual loss proposed in [15] is designed to measure how similar two non-aligned images are for image transformation. Unlike the pixel-level loss, which requires pixel-to-pixel alignment, the contextual loss allows for spatial deformations with respect to the target, achieving less texture distortion and more reasonable outputs. We compute the contextual loss $\mathcal{L}_{CX}$ below,

$$\mathcal{L}_{CX} = -\log\left(CX\left(\mathcal{F}^l\left(I_g\right), \mathcal{F}^l\left(I_t\right)\right)\right) \tag{11}$$

where $\mathcal{F}^l\left(I_g\right)$ and $\mathcal{F}^l\left(I_t\right)$ denote the feature maps for image I_g and I_t, respectively, extracted from the certain layer of the pretrained VGG network. $CX(\cdot)$ denotes the similarity metric of features between the generated images and the target images, taking into account both the context of the entire image and the semantic meaning of pixels [15].

4) Adversarial loss. For generator G, we define the following loss function,

$$\mathcal{L}_{GAN}^{G} = \mathbb{E}_{I_c \sim I, P_t \sim P, I_{t'} \sim I} \{ log [D_p (I_c, I_g) \cdot D_t (P_t, I_g)] \} \tag{12}$$

where I and P denote the distribution of real person images and person poses, respectively. D_p and D_t denote the appearance discriminator and shape discriminator, respectively.

In terms of the discriminators, IPAN adapts the shape discriminator to align the pose of the generated image with the target pose, and the appearance discriminator to keep the appearance textures of the generated image with the input image. More specifically, the objective function for the discriminator is defined below,

$$\begin{aligned} \mathcal{L}_{GAN}^{D} = &\mathbb{E}_{I_c \sim I, P_t \sim P, I_t \sim I} \{ log [D_p (I_c, I_t) \cdot D_t (P_t, I_t)] \} \\ &+ \mathbb{E}_{I_c \sim I, P_t \sim P, I_g \sim F} \{ log [(1 - D_p (I_c, I_g)) \cdot (1 - D_t (P_t, I_g))] \} \end{aligned} \tag{13}$$

where I, P and F denote the distribution of real person images, person poses and fake person images, respectively. The adversarial loss with appearance discriminator D_p and shape discriminator D_t is employed to penalize the distribution difference between real pairs $((I_c, P_t), I_t)$ and fake pairs $((I_c, P_t), I_g)$.

4 Experiments

4.1 Datasets

1) In-shop Clothes Retrieval Benchmark DeepFashion [11] contains 52,712 images with a resolution of 256×256. Following the same data configuration [31], we randomly picked 101,966 pairs of images for training and 8,750 pairs for testing.

2) Market-1501 [30] contains 263,632 training pairs and 12,000 testing pairs with a resolution of 128×64. Pose transfer on this dataset is more challenging as the images are low-resolution and quite different in terms of pose, viewpoint, background, and illumination, etc.

For quantitative evaluations, Fréchet inception distance (FID) [4], Structural Similarity (SSIM) [25], Peak Signal-to-Noise Ratio (PSNR), and Learned Perceptual Image Patch Similarity (LPIPS) [29] are adopted in the experimental test. Besides, we follow the previous work [13] to calculate the Mask-SSIM (MS) to alleviate the impact of background by masking it out in the Market-1501 dataset. Especially, Amazon Mechanical Turk (AMT) is used for qualitative evaluations.

4.2 Baselines

The baselines include quite a few recent works on human pose transfer, such as PG2 [13], BGGAN [19], SecGAN [21], PATN [31], XingGAN [20] and ADGAN [16]. The details are as follows.

1) PG2 [13] proposed an architecture to synthesize novel person images in arbitrary poses given as input an image of that person and a new pose.

2) BGGAN [19] proposed a graph generator mainly consisting of two blocks to model the pose-to-pose and pose-to-image relations, respectively.

3) SecGAN [21] adopted a cascade strategy to divide the generation procedure into two stages. Stage I aims to capture the semantic structure of the scene, and Stage II focuses on appearance details via the proposed multi-channel attention selection module.

4) PATN [31] proposed to transfer image information from the condition pose to the target pose progressively with a local attention mechanism.

5) XingGAN [20] modeled appearance information and shaped information with two respective blocks.

6) ADGAN [16] used human parsing maps to extract the style encoding of the target image and inject it into the target pose for attribute-controllable person image synthesis.

4.3 Implementation Details

Our method is implemented in Pytorch framework using an NVIDIA RTX 3090 GPU with 24 GB memory. The weights for the loss terms are set as $\lambda_{rec} = 1$, $\lambda_{per} = 1$, and $\lambda_{cx} = 0.02$. We adopt Adam optimizer [9] with the momentum set to 0.5 to train our model for around 180k iterations. The initial learning rate is set as 0.001 and linearly decayed to 0 after 60k iterations. Instance normalization [22] is applied for DeepFashion, and batch normalization [6] is used for Market-1501. We train the generator and two discriminators alternatively.

4.4 Results

1) Quantitative comparisons. Table 1 summarizes the quantitative results of the approaches on two datasets, from which we have some observations. 1) Compared with the two-stage methods, i.e., PG2 and SecGAN, end-to-end methods, e.g., BGGAN, XingGAN, PATN, ADGAN and our method, can generate images with low FID and LPIPS values, demonstrating that these images are more realistic in human perception and their appearance are more consistent with the condition images. The reason is that the end-to-end methods are not susceptible to the quality of the coarse-grained images generated in the first stage of the two-stage method. 2) In terms of the end-to-end human pose transfer methods, our method achieves the best performance in terms of both FID and LPIPS on the two datasets, benefiting from that the interactive pose attention (IPA) blocks capture a large number of texture details, and long-distance residual connections inject the low-level features of the person image into the IPA blocks. 3) The performance of the approaches on PSNR and SSIM are quite close to each other, as the metrics focus on the pixel-level difference between the groundtruth and generated images. Especially, our method outperforms all baselines on the DeepFashion dataset. In addition, the performance of PSNR and SSIM may be

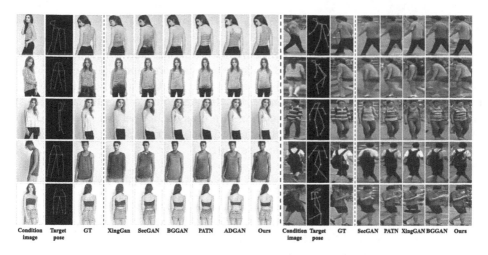

Fig. 4. The qualitative comparisons with several state-of-the-art methods on the Deep-Fashion (left) and Market-1501 datasets (right).

susceptible to the impact of poor quality of the Market-1501 dataset. Nevertheless, our method still outputs comparable results on SSIM, PSNR, and MS mertrics.

Table 1. Quantitative comparisons of the approaches. $^-$ indicates lower is better and $^+$ indicates higher is better. Since ADGAN uses human parsing maps for human pose transfer, it is unsuitable for the low-quality Market-1501 dataset.

	DeepFashion				Market-1501				
	FID$^-$	SSIM$^+$	LPIPS$^-$	PSNR$^+$	FID$^-$	SSIM$^+$	LPIPS$^-$	PSNR$^+$	MS$^+$
PG2	63.192	0.762	0.3046	31.291	142.628	0.261	0.3274	22.177	0.791
SecGAN	32.274	0.765	0.2747	31.306	103.065	**0.331**	0.3489	**22.264**	0.816
BGGAN	24.276	0.777	0.2434	31.369	36.708	0.325	0.3041	22.253	**0.818**
XingGAN	44.294	0.764	0.2921	31.072	37.51	0.304	0.3050	22.254	0.806
PATN	21.884	0.773	0.2527	31.149	38.576	0.311	0.3188	22.193	0.799
ADGAN	15.757	0.776	0.2248	31.288	/	/	/	/	/
Ours	**12.835**	**0.780**	**0.2102**	**31.376**	**22.568**	0.322	**0.2854**	22.258	0.811

2) Qualitative comparisons. The typical results of different methods are provided in Fig. 4, from which we have some observations. 1) From the examples on DeepFashion, we can observe that the XingGAN, SecGAN, and BGGAN fail to generate sharp images and cannot keep the consistency of shape and texture. The reason is that those methods ignore critical regions of interest during the pose transfer process. PATN and ADGAN succeed in keeping shape consistency, but there are still some texture and color deviation (e.g., missing texture on the back of the first row, color mismatch on the fourth row). Nevertheless, our results are more realistic and consistent with the condition images, benefiting from that interactive pose attention (IPA) blocks retain texture details. 2) In terms of the examples on Market-1501, we can observe that our method can generate consistent color with condition image (e.g., in the first row), benefiting from the long-distance residual connections. In addition, our method generates the natural leg layout even when the legs are crossed in the target pose (e.g., legs in the fourth row). 3) Overall, our method can generate realistic images and keep the best consistency of shape and texture.

Table 2. Ranking scores of AMT results.

	Realness	Appearance consistency
XingGAN	5.161	5.044
SecGAN	4.702	4.782
BGGAN	4.466	4.474
PATN	4.080	4.093
ADGAN	4.282	4.251
Ours	**2.795**	**2.821**
Groundtruth	2.513	2.534

3) Qualitative comparisons with AMT. More specifically, we randomly selected 120 generated images of the DeepFashion dataset and recruited 20 volunteers to rank them according to realistic and appearance consistent with the condition images on the Amazon Mechanical Turk (AMT) platform. The ranking scores against the five baselines are summarized respectively in Table 2 and Fig. 6, from which we can observe that our method outperforms the baselines, achieving the best ranking scores on both aspects. In addition, As shown in Fig. 6, our number of the best scores rated by the volunteers are more than twice the best baseline on the measure of realness. The experimental results manifest that our generated images are more realistic and achieve better consistency of appearance.

4) Person image synthesis in arbitrary poses. As shown in Fig. 5, given the same condition person image and several poses extracted from person images in the test set, our method can output natural and realistic results even when the target poses differ significantly from the condition person image in scale, viewpoints, etc.

4.5 Ablation Studies

The proposed IPAN contains two components, i.e., a long-distance residual connection, and an attention mechanism by interacting with image and pose pathways. Let IPAN_R and IPAN_A denote IPAN without long-distance residual

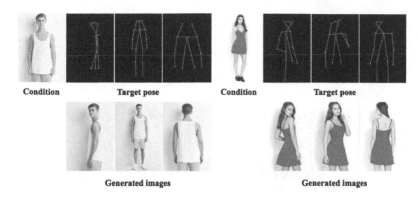

Fig. 5. Results of synthesizing person images in arbitrary poses.

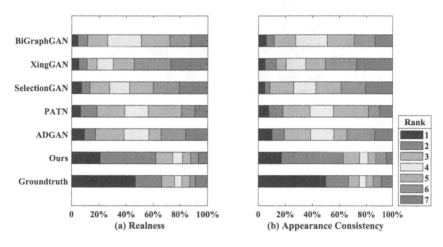

Fig. 6. Ranking score distribution of AMT.

Table 3. Performance of the variations of IPAN on DeepFashion and Market-1501 datasets.

	DeepFashion				Market-1501				
	FID⁻	SSIM⁺	LPIPS⁻	PSNR⁺	FID⁻	SSIM⁺	LPIPS⁻	PSNR⁺	MS⁺
IPAN_A	17.633	0.773	0.2384	31.132	32.049	0.303	0.2972	22.217	0.806
IPAN_R	16.736	0.776	0.2166	31.290	23.216	0.301	0.2881	22.226	0.805
IPAN	**12.835**	**0.780**	**0.2102**	**31.376**	**22.568**	**0.322**	**0.2854**	**22.258**	**0.811**

connection and attention mechanism, respectively. Table 3 summarizes the performance of the variations of IPAN on DeepFashion and Market-1501 datasets, from which we have some observations. 1) IPAN with both long-distance residual connections and attention mechanism tends to generate realistic images with consistent appearance and textures, achieving best FID, LPIPS, SSIM, and PSNR metrics. 2) IPAN without the attention mechanism generate images losing texture details as shown in the visual examples in Fig. 7. 3) IPAN without the long-distance residual connections leads to inconsistent appearance to the condition image as shown in Fig. 7.

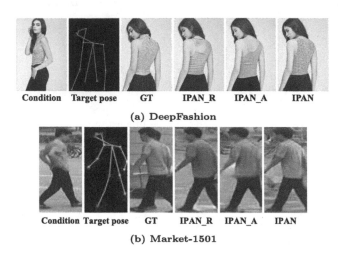

Fig. 7. Visual examples of the variations of IPAN on both datasets.

4.6   Failure Examples

Some failure examples of IPAN shown in Fig. 8 can demonstrate that the poses of the generated images are quite consistent with target poses. However, the patterns fail to be generated in T-shirts. As mentioned above, valid results captured and placed by the pixels can be obtained in the training data. Some irregular patterns, e.g., letters and cartoons shown in Fig. 8, may not be effectively generated from the seen pixels.

Fig. 8. Failure examples of IPAN.

5   Conclusions

In this paper, we propose an end-to-end interactive pose attention network (IPAN) by cascading interactive pose attention (IPA) blocks with long-distance residual for human pose transfer. The network comprises a sequence of interactive pose attention (IPA) blocks which interact with image pathway and pose pathway. Inside the block, we design an attention mechanism to transfer the regions of interest based on the human pose and capture the uninterested regions for subsequent pose transfer. In addition, the long-distance residual keeps the appearance characteristics in person image. In terms of both the effectiveness in appearance consistency and texture consistency, the proposed IPAN exhibits superior performance in both subjective visual realness and objective quantitative scores on two public datasets compared to the state-of-the-art approaches.

Acknowledgement. This work is supported by the National Natural Science Foundation of China (No. 62076073), the Guangdong Basic and Applied Basic Research Foundation (No. 2020A1515010616), Science and Technology Program of Guangzhou (No. 202 102020524), the Guangdong Innovative Research Team Program (No. 2014ZT05G 157), the Key-Area Research and Development Program of Guangdong Province (2019B010136001), and the Science and Technology Planning Project of Guangdong Province (LZC0023).

References

1. Cao, Z., Simon, T., Wei, S.E., Sheikh, Y.: Realtime multi-person 2d pose estimation using part affinity fields. In: Proceedings of the IEEE Conference on Computer Vision and Pattern Recognition, pp. 7291–7299 (2017)
2. Cui, Q., Sun, H., Yang, F.: Learning dynamic relationships for 3d human motion prediction. In: Proceedings of the IEEE/CVF Conference on Computer Vision and Pattern Recognition, pp. 6519–6527 (2020)
3. Goodfellow, I.J., et al.: Generative adversarial networks (2014). arXiv preprint arXiv:1406.2661
4. Heusel, M., Ramsauer, H., Unterthiner, T., Nessler, B., Hochreiter, S.: Gans trained by a two time-scale update rule converge to a local nash equilibrium (2017). arXiv preprint arXiv:1706.08500
5. Hu, H., Zhang, Z., Xie, Z., Lin, S.: Local relation networks for image recognition. In: Proceedings of the IEEE/CVF International Conference on Computer Vision, pp. 3464–3473 (2019)
6. Ioffe, S., Szegedy, C.: Batch normalization: accelerating deep network training by reducing internal covariate shift. In: International Conference on Machine Learning, pp. 448–456. PMLR (2015)
7. Johnson, J., Alahi, A., Fei-Fei, L.: Perceptual losses for real-time style transfer and super-resolution. In: Leibe, B., Matas, J., Sebe, N., Welling, M. (eds.) ECCV 2016. LNCS, vol. 9906, pp. 694–711. Springer, Cham (2016). https://doi.org/10.1007/978-3-319-46475-6_43
8. Kanazawa, A., Zhang, J.Y., Felsen, P., Malik, J.: Learning 3d human dynamics from video. In: Proceedings of the IEEE/CVF Conference on Computer Vision and Pattern Recognition, pp. 5614–5623 (2019)

9. Kingma, D.P., Ba, J.: Adam: a method for stochastic optimization **180** (2014). arXiv preprint arXiv:1412.6980
10. Li, X.: Human-robot interaction based on gesture and movement recognition. Signal Process. Image Commun. **81**, 115686 (2020)
11. Liu, Z., Luo, P., Qiu, S., Wang, X., Tang, X.: Deepfashion: powering robust clothes recognition and retrieval with rich annotations. In: Proceedings of the IEEE Conference on Computer Vision and Pattern Recognition, pp. 1096–1104 (2016)
12. Luong, M.T., Pham, H., Manning, C.D.: Effective approaches to attention-based neural machine translation (2015). arXiv preprint arXiv:1508.04025
13. Ma, L., Jia, X., Sun, Q., Schiele, B., Tuytelaars, T., Van Gool, L.: Pose guided person image generation (2017). arXiv preprint arXiv:1705.09368
14. Ma, L., Sun, Q., Georgoulis, S., Van Gool, L., Schiele, B., Fritz, M.: Disentangled person image generation. In: Proceedings of the IEEE Conference on Computer Vision and Pattern Recognition, pp. 99–108 (2018)
15. Mechrez, R., Talmi, I., Zelnik-Manor, L.: The contextual loss for image transformation with non-aligned data. In: Proceedings of the European conference on computer vision (ECCV), pp. 768–783 (2018)
16. Men, Y., Mao, Y., Jiang, Y., Ma, W.Y., Lian, Z.: Controllable person image synthesis with attribute-decomposed gan. In: Proceedings of the IEEE/CVF Conference on Computer Vision and Pattern Recognition, pp. 5084–5093 (2020)
17. Ronneberger, O., Fischer, P., Brox, T.: U-Net: convolutional networks for biomedical image segmentation. In: Navab, N., Hornegger, J., Wells, W.M., Frangi, A.F. (eds.) MICCAI 2015. LNCS, vol. 9351, pp. 234–241. Springer, Cham (2015). https://doi.org/10.1007/978-3-319-24574-4_28
18. Siarohin, A., Sangineto, E., Lathuiliere, S., Sebe, N.: Deformable gans for pose-based human image generation. In: Proceedings of the IEEE Conference on Computer Vision and Pattern Recognition, pp. 3408–3416 (2018)
19. Tang, H., Bai, S., Torr, P.H., Sebe, N.: Bipartite graph reasoning gans for person image generation (2020). arXiv preprint arXiv:2008.04381
20. Tang, H., Bai, S., Zhang, L., Torr, P.H.S., Sebe, N.: XingGAN for person image generation. In: Vedaldi, A., Bischof, H., Brox, T., Frahm, J.-M. (eds.) ECCV 2020. LNCS, vol. 12370, pp. 717–734. Springer, Cham (2020). https://doi.org/10.1007/978-3-030-58595-2_43
21. Tang, H., Xu, D., Sebe, N., Wang, Y., Corso, J.J., Yan, Y.: Multi-channel attention selection gan with cascaded semantic guidance for cross-view image translation. In: Proceedings of the IEEE/CVF Conference on Computer Vision and Pattern Recognition, pp. 2417–2426 (2019)
22. Ulyanov, D., Vedaldi, A., Lempitsky, V.: Instance normalization: the missing ingredient for fast stylization (2016). arXiv preprint arXiv:1607.08022
23. Vaswani, A., et al.: Attention is all you need (2017). arXiv preprint arXiv:1706.03762
24. Wang, X., Girshick, R., Gupta, A., He, K.: Non-local neural networks. In: Proceedings of the IEEE Conference on Computer Vision and Pattern Recognition, pp. 7794–7803 (2018)
25. Wang, Z., Bovik, A.C., Sheikh, H.R., Simoncelli, E.P.: Image quality assessment: from error visibility to structural similarity. IEEE Trans. Image Process. **13**(4), 600–612 (2004)
26. Xu, K., et al.: Show, attend and tell: neural image caption generation with visual attention. In: International Conference on Machine Learning, pp. 2048–2057. PMLR (2015)

27. Yang, C., Wang, Z., Zhu, X., Huang, C., Shi, J., Lin, D.: Pose guided human video generation. In: Proceedings of the European Conference on Computer Vision (ECCV), pp. 201–216 (2018)
28. Zhang, H., Goodfellow, I., Metaxas, D., Odena, A.: Self-attention generative adversarial networks. In: International Conference on Machine Learning, pp. 7354–7363. PMLR (2019)
29. Zhang, R., Isola, P., Efros, A.A., Shechtman, E., Wang, O.: The unreasonable effectiveness of deep features as a perceptual metric. In: Proceedings of the IEEE Conference on Computer Vision and Pattern Recognition, pp. 586–595 (2018)
30. Zheng, L., Shen, L., Tian, L., Wang, S., Wang, J., Tian, Q.: Scalable person re-identification: a benchmark. In: Proceedings of the IEEE International Conference on Computer Vision, pp. 1116–1124 (2015)
31. Zhu, Z., Huang, T., Shi, B., Yu, M., Wang, B., Bai, X.: Progressive pose attention transfer for person image generation. In: Proceedings of the IEEE/CVF Conference on Computer Vision and Pattern Recognition, pp. 2347–2356 (2019)

Exploiting Intra and Inter-field Feature Interaction with Self-Attentive Network for CTR Prediction

Shenghao Zheng[1], Xuefeng Xian[2(✉)], Yongjing Hao[1], Victor S. Sheng[4], Zhiming Cui[3], and Pengpeng Zhao[1(✉)]

[1] Institute of AI, Soochow University, Suzhou, China
`ppzhao@suda.edu.cn`
[2] Suzhou Vocational University, Suzhou, China
`xianxuefeng@jssvc.edu.cn`
[3] Suzhou University of Science and Technology, Suzhou, China
[4] Texas Tech University, Lubbock, TX, USA

Abstract. Click-Through Rate (CTR) prediction models have achieved huge success mainly due to the ability to model arbitrary-order feature interactions. Recently, Self-Attention Network (SAN) has achieved significant success in CTR prediction. However, most of the existing SAN-based methods directly perform feature interaction operations on raw features. We argue that such operations, which ignore the intra-field information and inter-field affinity, are not designed to model richer feature interactions. In this paper, we propose an **Intra** and **Inter**-field **Self-Attentive Network** (IISAN) model for CTR prediction. Specifically, we first design an effective embedding block named Gated Fusion Layer (GFL) to refine raw features. Then, we utilize self-attention to model the feature interactions for each item to form meaningful high-order features via a multi-head attention mechanism. Next, we use the attention mechanism to aggregate all interactive embeddings. Finally, we assign DNNs in the prediction layer to generate the final output. Extensive experiments on three real public datasets show that IISAN achieves better performance than existing state-of-the-art approaches for CTR prediction.

Keywords: CTR prediction · Recommender system · Self-attentive network · Feature interaction

1 Introduction

CTR, the key to success in computational advertising and recommender system, is defined as the probability that a user clicks through a specific recommended item or advertisement on a web page. With the increasing complexity of real-world data, especially in CTR prediction, effectively process input features has become a widely concerned issue. Machine learning has played a vital role in CTR prediction, which is usually formulated as supervised learning

© Springer Nature Switzerland AG 2021
W. Zhang et al. (Eds.): WISE 2021, LNCS 13081, pp. 34–49, 2021.
https://doi.org/10.1007/978-3-030-91560-5_3

with user profiles and item attributes as input features. Many shallow models such as Logistic Regression (LR) and Factorization Machines (FM) [16] have been widely applied to the industry's CTR prediction task. However, these shallow methods all have a common problem: they are limited to simple low-order feature interactions, which may not capture the implicit information completely in raw features. As shown in recent works [3,6,13,17], high-order feature combinations are beneficial for a good performance. For example, it is reasonable to recommend the NBA Basketball game, a famous sports event, to u_1, a twenty-year-old man. In this case, the third-order combinational feature ⟨*gender* = *male*, *age* = 20, *hobby* = *basketball*⟩ is very informative for prediction. Besides, the data in the real world is often non-linear and has a complex internal structure. Although FM can handle sparse data well, it is still a linear model, limited to capture second-order interactions. High-order FM [16] is also proposed to model arbitrary-order feature interactions. It still belong to the linear level and is claimed to be difficult to estimate, so that the expressive ability is insufficient.

As we mentioned above, there are two main challenges currently facing. First, the raw input data structure is very sparse and high-dimensional. In real-world applications, whether the user clicks on an item depends on many factors. Each factor can be a feature of a click instance. A large number of features are usually continuous or categorical. To make the supervised learning method applicable, we convert features into one-hot vectors, which causes the feature dimension space to increase sharply and inevitably become very sparse. In the famous CTR prediction dataset Criteo, the feature dimension is approximately 30 million, and the sparsity of the feature space after one-hot encoding is greater than 99.99%. A sharp increment in dimensionality will bring unforeseen damages to machine learning models and make machine learning models overfit easily. Simultaneously, the waste of computing resources is unacceptable. Second, we need to capture high-order combinational information in the raw feature space. As shown in some outstanding works of literature [13,18,21], obtaining representations of higher-order features will produce positive performance improvements for CTR prediction.

The development of deep learning open a new path to guide higher-order feature interactions for CTR prediction. In DNN-based methods, a deeper neural network structure capture more complex interactive information from raw input features to learn the combinational influence. For example, xDeepFM [13] can automatically learn high-order feature interactions explicitly and implicitly through multi-layer feedforward neural networks and multi-block compressed interaction networks. DeepFM [6] combined FM and DNN methods in parallel, which can learn low-order and high-order feature interactions at the same time without manual feature combinations. Both of them achieve a remarkable improvement in CTR prediction. However, most of the existing methods conduct feature interactions directly in the raw features embeddings without considering the affinity between fields. For example, AutoInt [18] explicitly constructed high-order features automatically to model the dependencies between features and discovered high-quality feature combinations, which effectively improved CTR prediction accuracy yet focuses on the raw features. NON [15] extracted only the information from features in the same field, which did not consider the relationship between the features in different related fields. AFN [4] proposed to learn any order feature interactive information adaptively, but it still lack consideration of the information existing in the field. In this paper, we design a novel model by exploiting the Intra and Inter-field feature interaction, which aims to refine the raw features before the feature interaction. By this way, the feature interaction layer can obtain richer field-wise feature information. For the intra-field information, we assign a separate fully-connected layer to each field as a local encoder to capture the intra-field information. We use another fully connected layer as a global encoder to capture the information in all fields for the inter-field affinity. Then we will fusion the intra-field information and inter-field information as additional features. After that, a gating mechanism is applied to control information flow to preserve the original feature information. Next, we adopt the multi-head self-attention network that explicitly constructs arbitrary-order feature interactions. In the last step, we adopt the attention mechanism to aggregate the generated interaction vectors and utilize a non-linear projection to generate the final output.

The main contributions of this work are summarized as follows:

- We introduce a GFL that facilitates more valuable combination features by generating multiple refined embeddings in multi-channel based on raw feature inputs. The refined embeddings contain inter-field and intra-field information.
- We propose a novel model named IISAN for CTR prediction, which adopts the state-of-the-art self-attention network structure to construct higher-order combinational feature automatically and extra latent information from intra-field and inter-field.
- We conduct extensive experiments on three public real-world datasets. Our experimental results show that our proposed method outperforms existing state-of-the-art methods.

2 Related Work

2.1 CTR Prediction

CTR prediction have a great significance for computational advertisement, recommender system, and information retrieval [26]. High-dimensional, sparse, and multi-field are typical characteristics of the real-world CTR data. Therefore, effectively modeling high-order feature interactions and obtaining richer feature information are vital for CTR prediction. Much research has been conducted in the previous work.

FM utilizes factorized parameters to model all pairwise interactions. Specifically, it uses polynomials to represent the correlation between features and applies factorization to make up LR shortcomings, which cannot handle high-dimensional sparse data. Although the ability to process large-scale sparse data has been demonstrated, it is limited by the exponential computational cost and the expressive abilities of the shallow model structure. Subsequently, various variants of FM are proposed one after another. For example, Juan et al. [10] assumed that features have different meanings in different fields, so each feature require a low-dimensional vector when interacting with features in different fields. By this way, it obtains a considerable performance improvement. Unfortunately, it sometimes causes the explosion of parameters, and overfitting also occurs easily. Cheng et al. [2] introduced a feature selection algorithm into FM, which helped FM reduce noise and select beneficial feature interactions.

Recently, many methods are inspired by the powerful deep learning methods to model high-order interactive features. For example, He et al. [8] used stacked neural network to simulate non-linear feature combinations, which enhanced high-level representation capabilities compared to FM. Similarly, many other methods use multi-layer feedforward neural networks to capture high-order feature interactions [3, 6, 21]. Since these methods implicitly model feature interactions, interpretability is a common problem for them. Wang et al. [21] and Lian et al. [13], which took the outer-product of feature embedding at the bit- and vector-wise level, performed feature interactions explicitly yet still difficultly explain which feature combination was helpful.

In addition to directly interacting with the raw features, another perspective is to study the raw feature embedding before the feature interaction. Luo et al. [15] used the field-aware network to extract intra-field information for enriching the input features. Huang et al. [9] dynamically learned the importance of different features through the Squeeze-Excitation network (SENET), weighted important features, and weakened features that contain little information. However, none of these works consider intra-field and inter-field information simultaneously before feature interactions, which is an essential difference between these models and IISAN.

2.2 Attentive Neural Network

The attention mechanism has shown powerful capabilities in various domains, such as natural language process [14, 19], recommender system [22, 24], computer vision [23], etc. The attention technique controls the information visible to other machine learning modules, improving the model's performance and making the model more interpretative [12, 18, 25, 27]. Song et al. [18] proposed to use the multi-head self-attentive neural network with residual connections to model high-order feature interactions for CTR prediction and improve the interpretability. Zhao et al. [27] proposed the disentangled self-attentive neural network, which is independently model pairwise and unary terms to reduce the coupling degree and facilitate learning features. Li et al. [12] employed hierarchical attention networks in an efficient manner, which can improve the interpretability of the model and alleviate the ambiguity of features in different semantic subspaces. In summary, all approaches mentioned above conducted feature interactions just on the raw feature embeddings, ignoring additional information in fields.

3   Our Approach

3.1   Problem Definition

In this subsection, we formally define the problem as follows:

Definition 1: CTR Prediction. We use $\mathbf{x} \in \mathbb{R}^n$ to represent the features for each input instance, where scalars represent numerical features and categorical features are encoded by one-hot. Note that each input instance contains multiple numerical features and categorical features, and n denotes the dimension of the embedding layer. Therefore, we define the CTR problem as predicting the probability that input instance be clicked according to its features $\mathbf{x}$.

Definition 2: h-order Feature Combination. Formally, given the feature vector of an input instance $\mathbf{x} \in \mathbb{R}^n$, we define an h-order combinational feature as $\langle x_{i_1}, x_{i_2}, ..., x_{i_h} \rangle = f(x_{i_1}, ..., x_{i_h})$, where h denotes how many fields are contained in $\langle \cdot \rangle$. We can adopt multiplication [16] or outer-product to serve as the combined function $f(\cdot)$. Note that a field may contain multiple features.

Definition 3: Problem Statement. The problem of CTR prediction aims to predict the probability of user u clicking item v according to the feature vector $\mathbf{x}$. For the feature vector of an input instance $\mathbf{x} \in \mathbb{R}^n$, we aim to refine the input feature vector $\mathbf{x} \in \mathbb{R}^n$. After that, we need to discover more beneficial high-order feature combinations to predict the CTR.

3.2   Intra-field and Inter-field Self-Attentive Network

As we mentioned above, directly perform feature interaction operations on top of raw features without considering their extra meaningful field information may lead to suboptimal performance. Thus, we propose IISAN to integrate Intra-field information and Inter-field affinity into an embedding layer through GFL. As shown in Fig. 1, we first design an effective embedding block named GFL to refine raw features. Then we utilize the self-attention mechanism to model the feature dependencies for each field, which can help form meaningful high-order combinations. Next, we make use of the attention mechanism to aggregate all interactive embeddings. At last, we generate the final output by using DNNs to predict CTR with a more significant probability score.

Embedding Layer. Customarily, each feature in fields is categorical or numerical. For example, assuming *user* and *item* are two fields, an input instance usually consists of many field-aware values, such as $\langle user{-}gender = male, user{-}clicks = 375, user{-}like{-}color = black, item{-}size = 23' \rangle$. It can be seen from this example that a field can contain multiple features, and the embedding representation can be enriched by combining the field information based on the feature. Therefore, a general method is to transform the field-aware features into a low-dimensional vector space. Specifically, we represent the categorical features by

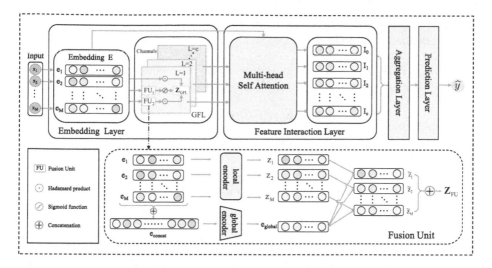

Fig. 1. The overall architecture of our proposed IISAN. GFL denotes the introduced gated fusion layer, in which L denotes the serial number of a channel and c is the number of channels. Each GFL contains two FUs (namely Fusion Unit). To highlight our work, we show the details of FU in the lower part of the figure. The figure is best to view in color. (Color figure online)

one-hot encoding and use a scalar value to represent each numerical features as follows:

$$\mathbf{x} = [\mathbf{x}_1, ..., \mathbf{x}_i, ..., \mathbf{x}_M], \tag{1}$$

where M denotes the number of fields involved in $\mathbf{x}$, and $\mathbf{x}_i$ represents the i-th feature field. Then we parameterize $\mathbf{x}_i$ as a vector representation[1] $\mathbf{e}_i \in \mathbb{R}^d$, where d is the embedding size. In view of the previous work [18], we take a low-dimensional vector to indicate each categorical feature. Likewise, we use a low-dimensional vector of the same size to represent each numerical feature. Formally, an input instance is denoted as:

$$\mathbf{E} = [\mathbf{e}_1, ..., \mathbf{e}_i, ..., \mathbf{e}_M], \tag{2}$$

$$\mathbf{e}_i = \mathbf{V}_i \mathbf{x}_i, \tag{3}$$

where $\mathbf{E} \in \mathbb{R}^{d \times M}$ is the embedding matrix of each input instance, $\mathbf{V}_i$ is a transformation matrix for field i, and $\mathbf{e}_i$ is the latent embedding of the i-th field. Note that $\mathbf{V}_i$ is an embedding vector when $\mathbf{x}_i$ is a scalar value, and when $\mathbf{x}_i$ is a one-hot vector, it is an embedding matrix.

Gated Fusion Layer. In order to capture extra meaningful information between different fields, given the original embedding $\mathbf{e}_i$, we encode intra- and

---

[1] The field embedding $\mathbf{e}_i$ is also known as the feature embedding. If the field is multivalent, the sum of feature embedding is used as the field embedding.

inter-field information by a Fusion Unit (FU), which is shown in Fig. 1. Specifically, we define FU as follows:

Local Encoder: We first adopt an independent fully connected (FC) layer to encode the intra-information of each field, as follows:

$$\mathbf{z}_i = FC_i(\mathbf{e}_i), \tag{4}$$

where z_i is intra-field information representation.

Global Encoder: In this step, we capture the global inter-field information. Given all the field-aware embeddings, we concatenate and feed them to another fully connected layer:

$$\mathbf{e}_{concat} = concat([\mathbf{e}_1, ..., \mathbf{e}_i, ..., \mathbf{e}_M]) \in \mathbb{R}^{Md}, \tag{5}$$

$$\mathbf{e}_{global} = FC_{global}(\mathbf{E}) = \mathbf{W}_{global}\mathbf{e}_{concat} + \mathbf{b}_g \in \mathbb{R}^d, \tag{6}$$

where $\mathbf{W}_{global} \in \mathbb{R}^{Md \times d}$ is the weights of FC_{global}, $\mathbf{b}_g \in \mathbb{R}^1$ is the bias term, and e_{global} contains the inter-field information.

Gated Information Fusion: After the original embeddings pass through the above two encoders, the corresponding new feature representations contain intra-field information and intra-field affinity, respectively. Next, we consider combining them through the fusion function F as follows:

$$\tilde{\mathbf{z}}_i = F(\mathbf{z}_i, \mathbf{e}_{global}) = \mathbf{z}_i \odot \mathbf{e}_{global} \in \mathbb{R}^d, \tag{7}$$

where F can be any fusion function. To ensure model efficiency, we employ the Hadamard product denoted by $\odot$. Since e_{global} contains the information of all fields, a single field representation $\mathbf{z}_i$ can automatically pinpoint the corresponding position in e_{global} via F. The embedding matrix is generated as:

$$\mathbf{Z}_{FU} = [\tilde{\mathbf{z}}_1, ..., \tilde{\mathbf{z}}_i, ..., \tilde{\mathbf{z}}_M] \in \mathbb{R}^{M \times d}, \tag{8}$$

where $\mathbf{Z}_{FU}$ includes intra-field information and inter-field affinity and *FU* indicates *Fusion Unit*. Note that the diversity of information contained in the raw features, we use the gating mechanism [5] to control the ratio of raw information and additional information. Specifically, we first obtain two feature embedding matrices $\mathbf{Z}_{FU_1}$ and $\mathbf{Z}_{FU_2}$, and then we apply the sigmoid function to convert $\mathbf{Z}_{FU_1}$ to $[0, 1]$ as a scaling factor of $\mathbf{Z}_{FU_2}$. We formally express the GFL layer as follows:

$$\mathbf{Z}_{FU_1} = FU_1(\mathbf{E}) \in \mathbb{R}^{M \times d}, \tag{9}$$

$$\mathbf{Z}_{FU_2} = FU_2(\mathbf{E}) \in \mathbb{R}^{M \times d}, \tag{10}$$

$$\mathbf{Z}_{GFL} = \mathbf{Z}_{FU_2} \otimes \sigma(\mathbf{Z}_{FU_1}) + \mathbf{E} \otimes (1 - \sigma(\mathbf{Z}_{FU_1})) \in \mathbb{R}^{M \times d}, \tag{11}$$

where σ denotes a sigmoid function and $\mathbf{Z}_{GFL}$ represents the extra features.

Feature Interaction Layer. Inspired by the ability of the multi-head self-attentive network, which can model the dependency relationship between arbitrary words [20], it is natural to associate it with complex relationships in high-order feature interactions. We first automatically learn high-order combination features based on the multi-head self-attentive network.

From the embedding layer, we can obtain the embedding matrix $\mathbf{E}$ and $\mathbf{Z}_{GFL}$ as the input of the feature interaction layer. We follow the scaled dot-product attention scheme [20] to formulate each feature interaction as a key-value pair and learn the importance of each combinational feature by multiplying each feature embedding. We take $\mathbf{E} = [\mathbf{e_1}, ..., \mathbf{e_M}]$ as an example to formalize the process as follows:

$$\mathbf{h}_i = softmax_i \left(\frac{\mathbf{QK^T}}{\sqrt{\mathbf{d}}} \right), \tag{12}$$

$$\mathbf{Q} = \mathbf{W}_{(i)}^{(\mathbf{Q})}\mathbf{E}, \quad \mathbf{K} = \mathbf{W}_{(i)}^{(\mathbf{K})}\mathbf{E}, \quad \mathbf{V} = \mathbf{W}_{(i)}^{(\mathbf{V})}\mathbf{E}, \tag{13}$$

where the scaling factor $\frac{1}{\sqrt{d}}$ is used to compensate for large dot-product values, $\mathbf{Q}$, $\mathbf{K}$ and $\mathbf{V}$ transformation of $\mathbf{i}$-th head are parameterized by three linear transformation matrices $\mathbf{W}_{(i)}^{(\mathbf{Q})}, \mathbf{W}_{(i)}^{(\mathbf{K})}, \mathbf{W}_{(i)}^{(\mathbf{V})} \in \mathbb{R}^{d' \times d}$, respectively. Through the attention function Eq. 12, we project the original embedding space $\mathbb{R}^{d \times M}$ to the new space $\mathbb{R}^{d' \times M}$. By doing this, feature vectors are updated by aggregating with attention scores. Besides, using multi-head in the self-attentive network enables learning arbitrary feature interactions in different subspaces. Formally, we concatenate combinational features learned in all subspaces and define them as follows:

$$\mathbf{H} = [\mathbf{h}_1; ...; \mathbf{h}_i; ...; \mathbf{h}_{n_h}], \tag{14}$$

where n_h is the number of heads. Note that the "Mask" operation in other sequential models is not selected here. Since there is no sequence between features, each feature can affect all the others.

In some respects, the expressive ability of the model becomes stronger as the network deepens. However, blindly deepening the depth also brings many problems, such as vanishing gradient and exploding gradient, so simply stacking more layers does not directly match better performance. In order to propagate raw feature information to interactive vectors, we add standard residual connections [7] to the network as follows:

$$\mathbf{I} = \text{ReLU}(\mathbf{W}_{\mathbf{res}}^{\mathbf{H}}\mathbf{E} + \mathbf{H}), \tag{15}$$

where $\mathbf{W}_{\mathbf{res}}^{\mathbf{H}} \in \mathbb{R}^{d'n_h \times d}$ is the weights matrix, ReLU is the activation function to learn the non-linearity of the combined information, and $\mathbf{I} \in \mathbb{R}^{d' \times M}$ is the output of the multi-head self attention module. It is necessary to know that we may input multiple $\mathbf{Z}_{GFL}$ in a multi-channel manner into the feature interaction layer, which helps us capture much more intra-field and inter-field information. Moreover, we set the number of channels as a hyperparameter, and one channel outputs one $\mathbf{Z}_{GFL}$. The multi-channel output of the previous layer can be formalized as:

$$\mathbb{I} = \begin{bmatrix} \mathrm{SAN}(\mathbf{E}), \\ \mathrm{SAN}(\mathbf{Z}_{GFL_1}), \\ \mathrm{SAN}(\mathbf{Z}_{GFL_2}), \\ ..., \\ \mathrm{SAN}(\mathbf{Z}_{GFL_c}) \end{bmatrix} = \begin{bmatrix} \mathbf{I_0}, \\ \mathbf{I_1}, \\ \mathbf{I_2}, \\ ... \\ \mathbf{I_c} \end{bmatrix}, \tag{16}$$

where SAN denotes the whole multi-head self-attention mechanism for simplicity, $\mathbf{I}_i$ indicates the result from Eq. 15, and $\mathbb{I}$ is all the outputs of the feature interaction layer.

Aggregation Layer. Next, inspired by AFM [22], we perform weighted aggregation by

$$\mathbb{I}_{agg} = Attentionagg(\mathbb{I}) = \sum_{i=0}^{c} \alpha_i \mathbf{I}_i, \tag{17}$$

where α_i denotes the i-th attention value, which is calculated by

$$\alpha_i = \frac{\exp(\alpha_i')}{\sum_{i=0}^{c} \exp(\alpha_i')}, \tag{18}$$

$$\alpha_i' = \mathbf{h}^T \mathrm{ReLU}(\mathbf{W}_i \mathbf{I}_i), \tag{19}$$

where $\mathbf{W}_i \in \mathbb{R}^{s \times d' M}$, $\mathbf{h}^T \in \mathbb{R}^s$ is model parameters, and s denotes the hidden layer size, namely *attention factor* in the attention network.

Prediction Layer. Finally, the prediction score is calculated by

$$\hat{y} = \sigma(\mathbf{W^T}(\mathbb{I}_{agg}) + \mathbf{b}), \tag{20}$$

where $\mathbf{W^T} \in \mathbb{R}^{n_h d' M}$ and bias $\mathbf{b}$ refer to non-linear projection parameters. In the training phase, we set binary cross-entropy loss function as the objective function:

$$\mathcal{L}(y, \hat{y}) = -y \log \hat{y} - (1 - y) \log (1 - \hat{y}), \tag{21}$$

where $y, \hat{y}$ indicates the ground-truth and the final model prediction, respectively. Following previous state-of-the-art methods, we optimize $\mathcal{L}$ by the Adam gradient descent optimizer [11].

4   Experiments

In this section, we first set up the experiments. Then we evaluate the performance of IISAN compared with state-of-the-art baseline methods and analyze their experimental results. Finally, we explore the influence of hyper-parameters.

Table 1. Statistics of the datasets

Dataset	#Samples	#Fields	#Features
Criteo	$45,840,617$	39	998960
Avazu	$40,428,967$	23	1544488
MovieLens-1M	$739,012$	7	3529

4.1 Experiment Setup

Datasets. We conduct comparative experiments on three commonly used real-world datasets, i.e. *Criteo*[2], *Avazu*[3], and *MovieLens-1M*[4]. The statistical information of the three datasets is shown in Table 1. *Criteo* and *Avazu* are benchmarks released by two online advertising companies, which are the standard CTR prediction datasets for Kaggle competitions, including chronologically ordered click-through records. *Criteo* contains one month of ad click logs. There are 13 continuous features and 26 categorical ones. *Avazu* contains click logs with 40 million data instances. For each click data, 23 fields indicate elements of a single advertisement track. *MovieLens-1M* is a well-known e-commerce recommendation dataset including user ratings, user profiles, and item attributes. For *MovieLens-1M*, we perform data preprocessing on whether a user will click a movie when the user clicks through the movie according to user ratings to adapt our CTR prediction task. Specifically, when the rating is more than 3, we consider it a positive sample (click); when the score is less than 3, we consider it a negative sample (not click), and the rating of 3 will be eliminated.

For all datasets above, we first randomly divide an original dataset into a training set, a validation set, and a test set according to the ratio of [0.8, 0.1, 0.1]. We further remove infrequent feature categories. Specifically, in *Criteo*, we regard features that appear no more than ten times as an "⟨*unknown*⟩" category, while in *Avazu*, the threshold is set to 5. Particularly, considering *MovieLens-1M* is small and contains fewer infrequent features than the other two datasets, we do not set a threshold to filter the dataset. Secondly, since numerical features may not capture non-linearity of the data, we follow the method proposed by the winner of the *Criteo* competition[5] and use the function $normalize(x) = \lfloor \log^2(x) \rfloor$ to convert them into discrete values.

Evaluation Metrics. Referring to previous methods, we adopt two popular metrics to evaluate the prediction performance for IISAN, i.e., Area Under the ROC Curve (AUC) and Logloss. A smaller Logloss or a larger AUC represents reliable performance.

---

[2] https://www.kaggle.com/c/criteo-display-ad-challenge.
[3] https://www.kaggle.com/c/avazu-ctr-prediction.
[4] https://grouplens.org/datasets/movielens/.
[5] https://www.csie.ntu.edu.tw/~r01922136/kaggle-2014-criteo.pdf.

4.2   Baselines

We select previous representative baselines for comparison, which are briefly described as follows:

- **LR**. Logistic Regression is a simple yet authoritative approach for CTR prediction. It simply sums up all raw features via linear combination.
- **FM** [16]. FM is a typical method for CTR prediction tasks, aiming to solve feature combination problems under sparse data.
- **AFM** [22]. Attentional Factorization Machine extends the vanilla FM via an attentive network to model the importance of second-order feature interactions.
- **DeepCrossing** [17]. Deep Crossing Network can automatically combine features to discover important crossing features implicitly.
- **NFM** [8]. Neural Factorization Machine (NFM) adopts a bi-interaction layer based on the element-wise product of feature embeddings to capture the higher-order interactions in an implicit manner.
- **CrossNet(Deep&Cross)** [21]. CrossNet is extracted from the Deep&Cross model, which attempts to model cross features of bounded degree by carrying the outer product of concatenated feature representation at the bit-wise level.
- **CIN(xDeepFM)** [13]. Compressed Interaction Network (CIN) comes from the xDeepFM model, which constructs combinational features by means of the outer product of latent embeddings and then compresses the feature projections derived from the outer product to reconstruct each feature representation.
- **HOFM** [1]. Higher-Order Factorization Machine (HOFM) is the first training algorithm to implement arbitrary-order FM.
- **FiBiNET** [9]. FiBiNET utilizes the Squeeze Excitation network (SENET) to capture the importance of features and generate SENET-like embedding for further feature interactions.
- **AutoInt** [18]. AutoInt conducts each feature to interact with the others and model the dependency between features using a multi-head self-attentive network, which can automatically learn high-order feature interactions and provide high-grade explainability of combinational features.
- **AFN** [4]. AFN learns adaptive-order cross features via a logarithmic transformation layer, which verifies that learning adaptive-order cross features can bring better predictive performance than modeling fixed-order feature interactions.

4.3   Parameter Settings

All baseline methods are implemented with Tensorflow version 1.14 and follow the best parameters provided in corresponding papers. In IISAN, we set the batch size to 1024, the self-attention heads to 3, the stacking blocks to 3, and the embedding size to 24 to achieve the best performance. In particular, we made the dropout set for the MovieLens-1M dataset because we found that the model is easy to overfit on a small dataset. On the contrary, we do not set dropout in the other two large datasets because it does not affect the final performance.

Table 2. Experimental results of IISAN and baselines. We mark the best performance in boldface.

Type	Method	Criteo		Avazu		ML-1M	
		AUC	Logloss	AUC	Logloss	AUC	Logloss
First-order	LR	0.7819	0.4697	0.7561	0.3961	0.7712	0.4427
Second-order	FM	0.7836	0.4700	0.7706	0.3856	0.8252	0.3998
	AFM	0.7938	0.4584	0.7718	0.3854	0.8227	0.4048
High-order	DeepCrossing	0.8012	0.4513	0.7643	0.3889	0.8453	0.3814
	NFM	0.7957	0.4562	0.7708	0.3864	0.8357	0.3883
	CrossNet	0.7907	0.4591	0.7667	0.3868	0.7968	0.4266
	CIN	0.8009	0.4517	0.7758	0.3829	0.8286	0.4108
	HOFM	0.8005	0.4508	0.7701	0.3854	0.8304	0.4013
	FiBiNET	0.8060	0.4451	0.7757	0.3829	0.8477	0.3751
	AutoInt	0.8061	0.4454	0.7752	0.3823	0.8460	0.3784
	AFN	0.8061	0.4453	0.7754	0.3830	0.8477	0.3762
	IISAN(ours)	**0.8078**	**0.4439**	**0.7769**	**0.3813**	**0.8543**	**0.3703**

4.4 Comparisons of Performance

As shown in Table 2, our proposed neural model IISAN is better than all baselines on three datasets in terms of each metric, including the first-order (i.e., LR), second-order (i.e., FM, AFM), and high-order feature interaction models (i.e., the rest shown in Table 2). Specifically, we analyzed that because we modeled high-order features and obtained more fine-grained combinational features, the model performs better than low-order methods. In addition, we enrich the raw feature embedding based on the self-attention mechanism via introducing intra- and inter-field information, so IISAN is better than AFN and other high-order methods. In MovieLens-1M, we can notice that IISAN is 0.78% higher in terms of AUC than the best baseline, and 1.59% lower in Logloss. Compared with *MovieLens-1M*, *Criteo* and *Avazu* datasets are too lagers to train completely. Recent researchers [3, 6, 21] have pointed out that a **0.001-level** change of Logloss or AUC is powerful for CTR prediction tasks. Given that our improvement on these two datasets is more than 0.001, it is the best proof of our model's effectiveness.

4.5 Influence of Hyper-Parameters

In this subsection, we discuss the impact of hyperparameters on IISAN.

Impact of the Embedding Size. We first investigate how the performance change w.r.t. the dimension d of field embedding vectors. We set the embedding dimension in $\{8, 16, 24, 32\}$, and show the performance of IISAN in different

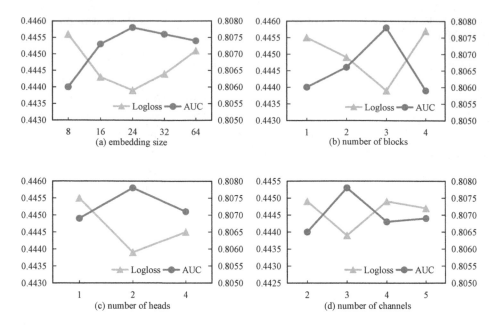

Fig. 2. The figure shows the impact of hyperparameters, i.e., the embedding size (a), the number of blocks (b), the number of heads (c), the number of channels (d). Considering space constraints, we only show the impact on the Criteo dataset. On the other two datasets, we will give the best parameter settings below. This figure is best viewed in color. (Color figure online)

embedding sizes in terms of AUC and Logloss. The results on the Criteo dataset are shown in Fig. 2(a). We conclude that setting the embedding size to 24 is the best choice for Criteo. The AUC metric gradually decreases after 24, and Logloss gradually increases. Note that the larger embedding size may not be equivalent to better performance because overfitting may occur if the embedding size is too large.

Impact of the Number of Self-attention Blocks. Stacking self-attention blocks b on raw feature embeddings and supplementary feature embeddings helps learn more complex transition patterns, while higher-order feature interactions can be learned by stacking multiple blocks in the feature interaction layer. The effect of the number of self-attention blocks on Criteo is shown in Fig. 2(b). We can observe that stacking the appropriate number of blocks can boost the performance of IISAN. Nevertheless, the performance is significantly reduced when b is too large. Especially as b = 3, the AUC and Logloss result achieves the best and then decreases to 0.8059 and 0.4457, respectively. We analyzed that stacking too many blocks makes it difficult to transmit adequate information.

Impact of the Number of Self-attention Heads. The multi-head mechanism allows a model to learn relevant information in different representation subspaces using different weight matrices. As shown in Fig. 2(c), the best performance occurs when head = 2. As increasing the number of heads continuously, performance will decrease sharply. We argue that it will lead to overfitting if the subspaces are divided too much.

Impact of the Number of Channels. The number of channels is a vital hyper-parameter in our model. Stacking channels on the embedding layer helps learn richer transition patterns. Here, we change the number of channels according to $\{2, 3, 4, 5\}$ to demonstrate its advantage over zero-channel methods (the baselines). From the results on Criteo shown in Fig. 2(d), our model achieves the best performance when the number of channels is 3. Since the number of features and the number of samples on different datasets are inconsistent, we set different numbers of channels to achieve the best performance, which we state in the Sect. 4.3. To summarize, IISAN generates multiple supplementary embeddings by employing GFL in different channels and captures richer interactions.

5   Conclusion

In this paper, we proposed a novel model named IISAN for CTR prediction. Our model considers not only the preference of arbitrary high-order combinational features but also the intra- and inter-field information. Experimental results demonstrated that IISAN outperforms the state-of-the-art methods on three public real-world datasets in terms of the standard metrics. At the same time, it further verified that it is a feasible idea to consider richer embedding input before feature interaction. In the future, we will make efforts to obtain supplementary embeddings such as inter-field and intra-field information in the knowledge graph or graph structure to extract the rich potential structure information contained in the raw features.

Acknowledgements. This research was partially supported by NSFC (No.61876117, 61876217, 61872258, 61728205), Open Program of Key Lab of IIP of CAS (No. IIP2019-1) and the Priority Academic Program Development of Jiangsu Higher Education Institutions (PAPD).

References

1. Blondel, M., Fujino, A., Ueda, N., Ishihata, M.: Higher-order factorization machines. In: NIPS, pp. 3351–3359 (2016)
2. Cheng, C., Xia, F., Zhang, T., King, I., Lyu, M.R.: Gradient boosting factorization machines. In: RecSys, pp. 265–272. ACM (2014)
3. Cheng, H., et al.: Wide & deep learning for recommender systems. In: DLRS@RecSys, pp. 7–10. ACM (2016)

4. Cheng, W., Shen, Y., Huang, L.: Adaptive factorization network: Learning adaptive-order feature interactions. In: AAAI, pp. 3609–3616. AAAI Press (2020)
5. Gehring, J., Auli, M., Grangier, D., Yarats, D., Dauphin, Y.N.: Convolutional sequence to sequence learning. In: ICML, Proceedings of Machine Learning Research, vol. 70, pp. 1243–1252. PMLR (2017)
6. Guo, H., Tang, R., Ye, Y., Li, Z., He, X.: Deepfm: a factorization-machine based neural network for CTR prediction. In: IJCAI, pp. 1725–1731. ijcai.org (2017)
7. He, K., Zhang, X., Ren, S., Sun, J.: Deep residual learning for image recognition. In: CVPR, pp. 770–778. IEEE Computer Society (2016)
8. He, X., Chua, T.: Neural factorization machines for sparse predictive analytics. In: SIGIR, pp. 355–364. ACM (2017)
9. Huang, T., Zhang, Z., Zhang, J.: Fibinet: combining feature importance and bilinear feature interaction for click-through rate prediction. In: RecSys, pp. 169–177. ACM (2019)
10. Juan, Y., Zhuang, Y., Chin, W., Lin, C.: Field-aware factorization machines for CTR prediction. In: RecSys, pp. 43–50. ACM (2016)
11. Kingma, D.P., Ba, J.: Adam: a method for stochastic optimization. In: ICLR (Poster) (2015)
12. Li, Z., Cheng, W., Chen, Y., Chen, H., Wang, W.: Interpretable click-through rate prediction through hierarchical attention. In: WSDM, pp. 313–321. ACM (2020)
13. Lian, J., Zhou, X., Zhang, F., Chen, Z., Xie, X., Sun, G.: xdeepfm: Combining explicit and implicit feature interactions for recommender systems. In: KDD, pp. 1754–1763. ACM (2018)
14. Lu, J., Yang, J., Batra, D., Parikh, D.: Hierarchical question-image co-attention for visual question answering. In: NIPS, pp. 289–297 (2016)
15. Luo, Y., Zhou, H., Tu, W., Chen, Y., Dai, W., Yang, Q.: Network on network for tabular data classification in real-world applications. In: SIGIR, pp. 2317–2326. ACM (2020)
16. Rendle, S.: Factorization machines with LIBFM. ACM Trans. Intell. Syst. Technol. **3**(3), 1–22 (2012)
17. Shan, Y., Hoens, T.R., Jiao, J., Wang, H., Yu, D., Mao, J.C.: Deep crossing: Web-scale modeling without manually crafted combinatorial features. In: KDD, pp. 255–262. ACM (2016)
18. Song, W., et al.: Autoint: automatic feature interaction learning via self-attentive neural networks. In: CIKM, pp. 1161–1170. ACM (2019)
19. Tan, Z., Wang, M., Xie, J., Chen, Y., Shi, X.: Deep semantic role labeling with self-attention. In: AAAI, pp. 4929–4936. AAAI Press (2018)
20. Vaswani, A., et al.: Attention is all you need. In: NIPS, pp. 5998–6008 (2017)
21. Wang, R., Fu, B., Fu, G., Wang, M.: Deep & cross network for ad click predictions. In: ADKDD@KDD, pp. 1–7. ACM (2017)
22. Xiao, J., Ye, H., He, X., Zhang, H., Wu, F., Chua, T.: Attentional factorization machines: Learning the weight of feature interactions via attention networks. In: IJCAI, pp. 3119–3125. ijcai.org (2017)
23. Xiao, T., Xu, Y., Yang, K., Zhang, J., Peng, Y., Zhang, Z.: The application of two-level attention models in deep convolutional neural network for fine-grained image classification. In: CVPR, pp. 842–850. IEEE Computer Society (2015)
24. Xu, C., et al.: Graph contextualized self-attention network for session-based recommendation. In: IJCAI, pp. 3940–3946. ijcai.org (2019)
25. Zhao, Y., Liang, S., Ren, Z., Ma, J., Yilmaz, E., de Rijke, M.: Explainable user clustering in short text streams. In: SIGIR, pp. 155–164. ACM (2016)

26. Zhu, J., Liu, J., Yang, S., Zhang, Q., He, X.: Fuxictr: an open benchmark for click-through rate prediction (2020). CoRR abs/2009.05794
27. Zhu, Y., Xu, Y., Yu, F., Liu, Q., Wu, S., Wang, L.: Disentangled self-attentive neural networks for click-through rate prediction (2021). CoRR abs/2101.03654

AMBD: Attention Based Multi-Block Deep Learning Model for Warehouse Dwell Time Prediction

Xingyi Lv[1], Wei Zhao[1], Jiali Mao[1(✉)], Ye Guo[2], and Aoying Zhou[1]

[1] School of Data Science and Engineering, East China Normal University,
Shanghai, China
52195100008@stu.ecnu.edu.cn, {jlmao,ayzhou}@dase.ecnu.edu.cn
[2] Jing Chuang Zhi Hui (Shanghai) Logistics Technology, Shanghai, China
guoye@jczh56.com

Abstract. Warehouse dwell time consists of working time and waiting time of trucks that have loading tasks in a warehouse. Warehouse dwell time prediction plays a crucial role for improving the truck scheduling strategies as well as the truck drivers' experiences, and further proliferating the efficiency of warehouse logistics. *Queuing theory* based time prediction methods mainly focus on some queuing events with regularity to conform. However, the warehouse queuing system has a more complex structure. Specifically, the dwell time of any truck depends on the dwell time of its preceding trucks and the loading ability of warehouse. While warehouse loading ability keeps dynamically changing due to several factors like the capacity of the production line and loading device failures. This greatly increases the difficulty of warehouse dwell time predicting. In this paper, we first put forward the definition of *block* to represent the loading task statuses of different trucks. On the basis of that, we propose a deep learning based multi-block dwell time prediction model, called *AMBD*. It incorporates the loading ability of warehouse and the execution process of loading tasks of preceding trucks in the queue. Moreover, to proliferate the precision of warehouse dwell time prediction, we introduce attention mechanism to extract strong correlations among the trucks' dwell time. Experimental results on real-world steel logistics data demonstrate the efficacy of our proposed models.

Keywords: Warehouse · Dwell time prediction · Queuing system · Attention

1 Introduction

Fueled by the continuous development of informatization, the bulk logistics industry is going through a rapid transformation. The truck drivers are accustomed to the convenience of *simple-and-immediate* receiving and submitting assignments through online platforms or applications. However, a few issues in

W. Zhang et al. (Eds.): WISE 2021, LNCS 13081, pp. 50–66, 2021.
https://doi.org/10.1007/978-3-030-91560-5_4

actual scenarios have yet to be solved. For instance, when the truck drivers are queuing for the accomplishment of tasks, they are not clear about the rest of dwell time in warehouse, which makes them have bad experiences. Additionally, in order to optimize truck scheduling and proliferate the efficiency of plant logistics, it is necessary for the managers to be aware of time consumption of warehouse operations for the truck drivers. Therefore, it is particularly important to predict warehouse dwell time (or *WDT* for short) consisting of working and waiting time.

Some existing methods in queuing theory have been widely used to solve the issues of customers' dwell time prediction in queuing system [5,12,17]. But they can only be used to deal with the queuing scenarios with regular service time distributions. As more available data is generated from online platforms or applications, machine learning techniques are leveraged to solve the issues of time prediction, e.g., *Estimated Time of Arrival* (or *ETA* for short), prediction of *Order Fulfillment Cycle Time* (or *OFCT* for short) [9,21,25], etc. But the aforementioned methods cannot be directly applied to *WDT* prediction issues due to the particularity and limitations of the scenarios.

To be specific, there are some unique challenges for predicting the dwell time in warehouse queuing system as follows:

- Dynamically change of warehouse loading ability. Warehouse loading ability will fluctuate due to the periodic change in the production and inventory of cargoes. Specifically, as part of the dwell time, the working time of the truck is influenced by the warehouse loading ability, which varies greatly during different time periods or when loading different cargoes.
- Execution time of loading tasks of preceding trucks in the queue has influence on that of the current truck to various degree. The time of performing different loading tasks is not the same due to loading different categories or weights of cargoes. In addition, long dwell time of preceding trucks will have great influence on the time consumption of the current truck. Hence, it necessitates to know the extent of execution process of preceding trucks' loading tasks in the queue affecting on the current truck.

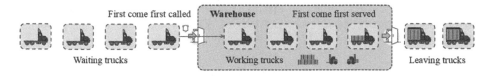

Fig. 1. The entire warehouse loading process

The warehouse queuing process of trucks is illustrated in Fig. 1. When the previous *working truck* finishes the loading task and leaves the warehouse, the next *waiting truck* can enter the warehouse to load cargoes. To extract the execution process of loading tasks of preceding trucks in the queue, we design a deep learning based multi-block transmission architecture according to the characteristics of the warehouse queuing system. Three blocks are leveraged to extract latent information from three kinds of truck sequences respectively. The leaving block generates the warehouse loading ability embedding using task completion information, which includes dwell time in warehouse and loading task information. Combining with the warehouse loading ability, the working block generates the task execution process embedding of working trucks. The waiting block outputs the execution process of waiting trucks' tasks after receiving the final working truck task process embedding. There are various latent features that will impact the target truck's WDT, e.g., historical loading efficiency of a particular cargo category. Inspired by the context-aware property of attention mechanism, we add the masked multi-head attention layer to capture the task execution process of trucks that has greatest impact on the current truck's WDT prediction. Compare one truck of executing 5-min loading task with another one of executing 30-min loading task, the latter one has greater influence on WDT for subsequent trucks.

The major contributions of this paper are listed below:

- We design a multi-block structure based on the loading task statuses of the trucks for extracting the loading ability of warehouse and trucks' task execution process respectively.
- We proposed a multi-block deep learning based prediction model, called $AMBD$, to predict dwell time of the trucks with loading tasks in a warehouse.
- We introduce local attention mechanism to improve the precision of dwell time prediction model, which can mine the influence degree of preceding trucks' task execution process on the current truck's WDT.
- We conduct comparative experimental evaluation to manifest the effectiveness of our proposed prediction model on real logistics datasets, and further deploy it on-line at the application for a steel logistics company successfully.

Outline. The rest part of this paper is organized as follows. Section 2 reviews relevant researches about the time prediction issue. Section 3 defines the problem formally. In Sect. 4 and 5, we introduce the extracted features and elaborate the details of our proposed prediction method. In Sect. 6, we evaluate the predictive performance of our proposal on a real dataset. Finally, we conclude the paper in the last section.

2   Related Work

Recently, the time prediction issue has become one of the research focuses and attracted extensive attention from academia and industry. The existing approaches can be divided into four categories: queuing theory based methods, machine learning based methods, deep learning based methods and attention based methods.

2.1   Queuing Theory Based Methods

Queuing theory could suitably model queuing systems with varieties of structures [2,8,13]. Shajin et al. proposed K time frame model for the advance reservation [16]. Zhan et al. focused on the N-system with many servers and proposed a model to analyze the customers' and servers' behaviors [22]. However, the above mentioned methods had a same limitation, i.e. the arrival time and the service time of customers in queuing system should be subject to a certain probability distribution. In the steel logistics scene, the arrival time and the service time of the trucks have no specific distributions, thus, the queuing theory based methods cannot be used to solve our proposed prediction issue.

2.2   Machine Learning Based Methods

WDT prediction issues for queues can be viewed as a variant of time series prediction problems, which are often solved by combining the knowledge of mathematics. *ARIMA* was a popular linear model with good performance in flexibility and statistics, but it processed inefficiently. Liu et al. proposed an online learning algorithm for increasing the noise tolerance of *ARIMA* [15]. To capture non-linear relationships that *ARIMA* couldn't get, IBM Tokyo came up with Gaussian *DyBM* to make up for the *DyBM*'s shortcoming [7]. Li et al. proposed an improvement to structure prediction of time series data for exponential outputs with an adversarial sequence tagging approach [14]. Nevertheless, the aforementioned models couldn't be suited for queuing scenarios with complex structures because they took no account for latent features.

2.3   Deep Learning Based Methods

Deep learning techniques had ability to extract latent features, and were widely used to solve the time estimation issues like *ETA* and *OFCT* prediction [9,21, 25]. Alibaba group used *Deep Neural Network* (or *DNN* for short) to predict *OFCT* at Eleme platform [25]. Since *DeepFM* is capable of handling sparse features in click-through rate prediction [10], it is applied by Meituan to forecast the delivery time. Most of neural network based methods regarded each data item as an individual, but they ignored the correlations among data items. To address this, *Recurrent Neural Network* (or *RNN* for short) and its derivative models were used for time series prediction [4,19]. Although deep learning based methods could capture the relationships among individuals, they had problems with gradient explosion and vanishing.

2.4   Attention Based Methods

Attention mechanism was originally used in machine translation [20]. Due to it could mitigate the effect of information loss on prediction precision, it had been widely used to tackle the sequential prediction issue. Adhikari et al. proposed the attention based *EpiDeep* model to make the epidemic forecasting [1]. In traffic prediction, $GMAN$ was proposed with temporal-spatial attention mechanism using graphic data [23]. Zhou et al. built a multi-block model using local and global attention for passenger demand prediction [24]. Inspired by them, in this paper, self-attention mechanism is leveraged to mine influence degree of previous tasks' execution process on WDT prediction of the current truck.

3   Problem Statement

3.1   Basic Concept

Initially, we take the queuing system of trucks in steel logistics as the background to introduce the rationality and suitability of our proposed WDT prediction model.

We design a WDT prediction model including three blocks by trucks' different loading task statuses. There is a truck queue in each block, and the trucks in each queue are queued by the check-in time in the warehouse. Three truck queues are connected one by one as shown in Fig. 1. The loading task information and warehouse dwell time of these trucks in the front of the queue have a certain impact on that of the later trucks in the queue. The first block contains the loading task information and warehouse dwell time of the leaving trucks recently, the main aim of which is to mine the warehouse loading ability in the current time instant. The last two blocks try to extract the impact of the loading task information of the preceding trucks on that of the later trucks in the queue, and warehouse dwell time of the working trucks and waiting trucks respectively.

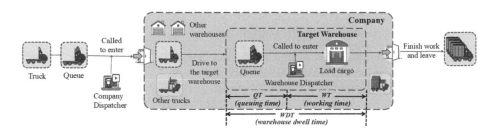

Fig. 2. A complete cargo loading cycle

The steel logistics enterprise needs to arrange varieties of trucks to load different cargoes everyday. The entire cargo loading cycle is shown in Fig. 2, and yellow part represents the main research content of this paper. It contains the operations of trucks waiting inside and working outside the warehouse:

Waiting Outside the Warehouse: The number of trucks that are allowed entering the warehouse is limited in a same period due to limited loading ability of the warehouse. In other words, only when the warehouse berth is available, the other waiting trucks can enter the warehouse for loading.

Working Inside the Warehouse: The order of loading cargoes of the trucks in the warehouse follows the rule of first-come-first-served (or *FCFS* for short). The working process of these trucks includes registration, cargo loading and inspection of cargoes.

Dwell time prediction of the trucks plays an important role in the optimization of plant logistics, for which there are several reasons.

- Some drivers may become irritable when they are unsure about their queuing time in advance. Accurate *WDT* prediction results can provide a better service experience for such drivers and hence avoid loosing drivers for the logistic platform.
- Warehouses cannot be ensure at full capacity due to the uncertainty of berth's vacant time. Accordingly, the resources of the enterprise can't be fully utilized. Therefore, it necessitates an accurate *WDT* predicting result to get a reasonable entry time for each truck.
- To avoid congestion on some roads in the plant, it requires to design appropriate truck scheduling mechanism and warehouse working sequence recommendation strategies. And all of the above scheduling depends on accurate *WDT* prediction.

3.2   Problem Formulation

Based on the concepts introduced above, we then give a formal definition of *WDT* prediction problem in the warehouse queuing system.

Let S denote all the warehouses which can provide loading services and C denote all the trucks in the queuing system. We use t_p to represent the current time instant when the prediction is requested. Let $t_{s,c}^{arrival}$, $t_{s,c}^{call}$, $t_{s,c}^{finish}$ denote the arrived time, called time and finished time of the truck $c \in C$ respectively, l_d^s and l_w^s denote the upper limit number of all trucks and waiting trucks in the queue system $s \in S$. Definitions of these symbols can be viewed in Table 1.

The trucks in waiting, working and leaving statuses at time t are represented by $C_{s,t}^{wait}$, $C_{s,t}^{work}$, $C_{s,t}^{leave}$ respectively. The waiting truck set of warehouse s is defined as follows:

$$C_{s,t}^{wait} = \left\{ c \in C | t_{s,c}^{arrival} \leq t_p \leq t_{s,c}^{call} \right\} \tag{1}$$

The preceding waiting trucks are defined as follows:

$$PC_{\tilde{c},t}^{wait} = \left\{ c \in C_{s,t}^{wait} | t_{s,c}^{arrival} \leq t_{s,\tilde{c}}^{arrival} \right\} \tag{2}$$

The working truck set of warehouse s is defined as follows:

$$C_{s,t}^{work} = \left\{ c \in C | t_{s,c}^{call} \leq t_p \leq t_{s,c}^{finish} \right\} \tag{3}$$

The preceding working trucks are defined as follows:

$$PC_{\tilde{c},t}^{work} = \left\{ c \in C_{s,t}^{work} | t_{s,c}^{arrival} \leq t_{s,\tilde{c}}^{arrival} \right\} \tag{4}$$

The set of the leaving trucks is defined as:

$$C_{s,t}^{leave} = \left\{c \in C | 0 \leq t_p - t_{s,c}^{finish} \leq \alpha\right\} \tag{5}$$

where α is a predefined time interval.

Problem: Given the execution process information of cargoes loading tasks of $C_{s,t}^{wait}$, $C_{s,t}^{work}$, $C_{s,t}^{leave}$ and other information at t_p, the WDT of the truck c is defined as:

$$WDT_{s,c} = t_{s,c}^{finish} - t_{s,c}^{arrival} \tag{6}$$

WDT is divided into *queuing time (QT)* and *working time (WT)*, whose definitions are listed below:

$$QT_{s,c} = t_{s,c}^{call} - t_{s,c}^{arrival} \tag{7}$$

$$WT_{s,c} = t_{s,c}^{finish} - t_{s,c}^{call} \tag{8}$$

Table 1. Table of symbols

Symbol	Definition
S	The set of warehouses
C	The set of trucks
s	One warehouse
c	One truck
l	The upper limit number of trucks for s
$WDT_{s,c}$	The warehouse dwell time of c in the queue of s
$QT_{s,c}$	The waiting time of c in the queue of s
$WT_{s,c}$	The working time of c in the queue of s
$CT_{s,c}$	The consumed time of c in the queue of s

4   Feature Extraction

Appropriate feature extraction is necessary for *AMBD* model. In this section, we will describe various features that are extracted from the queuing system in the steel factory for *WDT* prediction.

4.1   Cargo Features

Different categories of cargoes need different handling equipment for loading, and weights of cargoes will affect the loading efficiency of such equipment. Therefore, we attempt to extract the category and weight as cargo features.

4.2 Temporal Features

Trucks' *WDT* may have a periodic fluctuation in the temporal dimension. For example, the warehouse dwell time may become longer in rush time due to warehouse loading ability limit, and it may also become longer in holidays for fewer workers in the warehouse. Additionally, it necessitates to obtain the consumed dwell time of the preceding trucks, which is defined as:

$$CT_{s,c} = t_p - t_{s,c}^{arrival} \tag{9}$$

The consumed dwell time of trucks reflects the latest loading ability of the warehouse. Imagine if some accidents occurred in the warehouse, the consumed dwell time of preceding trucks in the queue will take longer. Accordingly, the dwell time of these trucks behind in the queue may become longer. Further, the consumed dwell time can also reflect the current truck's remaining warehouse dwell time in some degree.

4.3 Loading Ability Features

Warehouse loading ability is a determining factor for trucks' warehouse dwell time prediction. It is affected by the cargo inventory, the production line capacity and the loading ability of handling equipment. But such information is hard to obtain for model training. Considering the recent loading task completion situation can reflect the future trend of warehouse loading ability, we try to generate warehouse loading ability embedding information based on the task completion information of latest n leaving trucks.

4.4 Warehouse Features

All warehouses have their respective upper limit numbers l_d^s and l_w^s (i.e. the maximum number of trucks and waiting trucks in queue system), which determine the available lengths of queues in the working block and the waiting block. To a certain extent, these limits also reflect overall scale and loading ability of the warehouse.

5 Model

5.1 Overview

The architecture of *AMBD* model is shown in Fig. 3. Latent variables such as the warehouse loading ability and the execution process of preceding trucks' loading tasks have an important impact on *WDT*. To capture these latent variables, a deep learning based model is designed. Firstly, all the relevant trucks are sorted out by their check-in time and then divided into three partitions according to their loading task statuses, i.e. the leaving trucks, the working trucks and the waiting trucks. Secondly, we fusion the temporal and loading task status features into an embedding vector. After that, the task embedding of different loading task statuses is fed into three different blocks to obtain warehouse loading ability embedding and the task execution process embedding of waiting and working trucks. Finally, the self-attention is utilized to capture the impact of preceding trucks' task information on the current truck's *WDT*. Here, *Long Short-Term Memory* (or *LSTM* for short) is regarded as an example to implement blocks of *AMBD* model. True values of warehouse dwell time are used as the labels. We choose *Mean Square Error* (or *MSE* for short) as the objective function to minimize the *WDT* predictive loss of our proposal.

$$loss(WDT_{s,c}^{pred}, WDT_{s,c}) = \left(WDT_{s,c}^{pred} - WDT_{s,c}\right)^2 \tag{10}$$

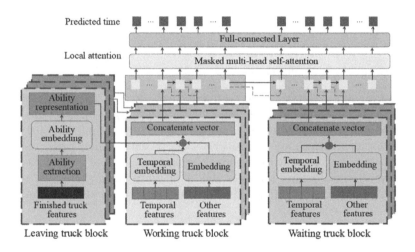

Fig. 3. The structure of *AMBD*

5.2   Leaving Truck Block

For the leaving truck block, the output is the warehouse loading ability representation. The input is the task completion information of trucks that have completed loading, which includes dwell time in warehouse and loading task information. Their latent information can imply the recent warehouse loading ability. Fed these trucks' features into an *LSTM* model and the embedding layer, a vector contains warehouse loading ability information is obtained. It is regarded as a captured latent feature and the input of the working truck block.

5.3   Working Truck Block

For the working truck block, the warehouse loading ability interrelates with their *WDT* tightly and straightly. Therefore, the input features of this block are the fundamental loading task statuses and the extracted warehouse ability. After processed by an embedding layer, the embedding of each working truck is obtained and then put into $LSTM$ model. The outputs of $LSTM$ in the working truck block are fed into the attention module. In order to transmit execution process of tasks, $LSTM$ cell output of the final truck is fed into the waiting truck block.

5.4   Waiting Truck Block

For the waiting truck block, the main difference from working trucks is that waiting trucks don't have direct relationship with the warehouse loading ability. The waiting block receives the final truck output of the working block as the initial hidden state of its $LSTM$. The input feature vectors of $LSTM$ are the embedded task information from waiting trucks. Finally, the outputs of $LSTM$ are put into the attention module.

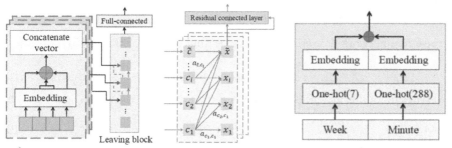

(a) Ability extraction module (b) Attention module (c) Temporal embedding module

Fig. 4. Main modules of $AMBD$

5.5   Warehouse Ability Extraction

Warehouse loading ability is key factor for trucks' working time. To obtain the warehouse ability information, we design an ability extraction structure as shown in Fig. 4(a). Information of the latest n leaving trucks' task completion are used to extract the current warehouse loading ability. The fundamental features of each truck are embedded using several full-connected layers and then merged into one representation vector. Loading tasks that are completed earlier have less representation value for warehouse loading ability representation. As a result, the warehouse ability mining process can be regarded as a series prediction problem. The output of the final cell in the leaving block is transformed to a vector using the full-connected layer, which represents the warehouse loading ability.

5.6   Masked Multi-Head Attention

The preceding trucks with various loading task execution process influence WDT of the current truck to different degrees. The loading task embedding of the preceding trucks in the queue are transmitted backwards through $LSTM$, thus each $LSTM$ neuron contains the task execution process embedding for the relative truck in the queue. In order to further obtain the effect degree on the current truck's warehouse dwell time from the task execution process embedding of the preceding trucks, we introduce the self-attention mechanism into our prediction model.

The attention module is implemented by using scaled dot-product multi-head attention, whose core idea is to capture the impact degree of preceding tasks' execution process for WDT prediction. In addition, WDT won't be influenced by the loading task statuses of subsequent trucks. So we mask out the weight matrices by setting the weights of the irrelevant subsequent trucks to $-\infty$. The $softmax$ function is used to calculate the relevance weights for WDT prediction. Then, the results of heads are concatenated to $mh_{\tilde{c},t}$:

$$mh_{\tilde{c},t} = concat(ha_{\tilde{c},t,1}, ha_{\tilde{c},t,2}, \ldots, ha_{\tilde{c},t,n}) \tag{11}$$

The residual network is used to obtain the final output of the module $op_{\tilde{c},t}$:

$$op_{\tilde{c},t} = ip_{\tilde{c},t} + mh_{\tilde{c},t} \tag{12}$$

The structure of our attention module is shown as Fig. 4(b), where c_i denote the preceding truck:

$$c_i \in PC_{\tilde{c},t}^{wait} \cup PC_{\tilde{c},t}^{work} \tag{13}$$

Take head i at time t as an example, where $ha_{\tilde{c},t,i}$ denotes the target truck $\tilde{c}$'s weighted sum of the correlations and the information from the previous truck set PC, which is the union set of $PC_{\tilde{c},t}^{wait}$ and $PC_{\tilde{c},t}^{work}$:

$$ha_{\tilde{c},t,i} = \sum_{c \in PC} a_{\tilde{c},c,i} \cdot h_{c,t,i} \tag{14}$$

where $h_{c,t,i}$ is the vector of head i for truck c. $a_{\tilde{c},c,i}$ denotes the attention weight indicating the task importance of target truck $\tilde{c}$ to relevant truck c in head i. The matrix $A_{i,t}$ consists of $a_{\tilde{c},c,i}$ and is derived from the following equation:

$$A_{i,t} = softmax(\frac{Q_{i,t}W_i^q \cdot (K_{i,t}W_i^k)^T}{\sqrt{d_k}}) \tag{15}$$

In our self-attention module, $Q_{i,t}$ and $K_{i,t}$ are the same. Both of them consist of $h_{c,t,i}$. The differences are their weight matrix W_i^q and W_i^k, where the parameters are learned by the whole model. d_k in the scaling factor $\frac{1}{\sqrt{d_k}}$ is the dimension of each head vector. The sum of the weights is equal to 1:

$$\sum_{c \in PC} a_{\tilde{c},c,i} = 1 \tag{16}$$

5.7   Temporal Embedding

Temporal embedding is used to present the temporal information of a truck's loading task as shown in Fig. 4(c). Firstly, we use one-hot coding to encode the day-of-week and five-minute-of-day of each truck into vectors $\mathbb{R}^7$ and $\mathbb{R}^{288}$ respectively. Secondly, one-hot vectors are encoded to two vectors $\mathbb{R}^n$ through several full-connected layers of the neural network. Finally, the embedded vectors are transformed to one vector $\mathbb{R}^{10}$ by a full-connected layer to present the temporal features.

6   Experiments

6.1   Dataset

We evaluate the performance of our method on the dataset from a steel logistics enterprise in Shandong province, China. It contains 76,142 loading task records of trucks from January to April, 2021. We use 70% of the dataset for training, 20% for validation, and 10% for testing.

6.2   Evaluation Metrics

We compare the predictive performance of our proposal with that of the other methods by four evaluation metrics: *Mean Absolute Error* (or *MAE* for short), *Root Mean Squared Error* (or *RMSE* for short), *Mean Absolute Percentage Error* (or *MAPE* for short).

$$MAE = \frac{1}{|C_{test}|}\sum_{s,c}\left|W\hat{D}T_{s,c} - WDT_{s,c}\right| \tag{17}$$

$$RMSE = \sqrt{\frac{1}{|C_{test}|}\sum_{s,c}(W\hat{D}T_{s,c} - WDT_{s,c})^2} \tag{18}$$

$$MAPE = \frac{100\%}{|C_{test}|}\sum_{s,c}\left|\frac{W\hat{D}T_{s,c} - WDT_{s,c}}{WDT_{s,c}}\right| \tag{19}$$

6.3   Baselines

We choose three representative methods: machine learning based one, deep learning based one and attention based one. Besides, our proposal is also compared with *AMBD-RNN* and *AMBD-GRU* whose sequential prediction blocks are implemented by *RNN* and *Gated Recurrent Units* (or *GRU* for short), respectively. The comparison results are given in Table 2.

Table 2. Relative performance of various methods

Model	MAE	RMSE	MAPE
LASSO	14.684	22.074	0.579
RFR	14.526	21.639	0.592
DNN	23.225	35.716	0.908
Wide&Deep	15.636	25.601	0.444
DeepFM	19.220	28.994	0.812
Transformer	14.803	22.126	0.585
AMBD-RNN	14.661	21.669	0.587
AMBD-GRU	14.298	21.229	0.571
AMBD-LSTM	**13.732**	**21.278**	**0.503**

- *LASSO* [3,18]: A linear regression method contains L1-regularization.
- *RFR* [11]: An ensemble technique that combines multiple decision trees to deal with regression problems.
- *DNN*: A complex neural network with more than three hidden layers and various activation functions.
- *Wide&Deep* [6]: A deep learning based prediction model consists of a wide generalized linear part to learn feature correlation from historical data and deep feed-forward neural network to reveal the interaction between implicit features.
- *DeepFM* [10]: A deep learning based prediction model that proposed by Huawei is an improved version of *Wide&Deep* model, which changes the *LR* of wide part of *Wide&Deep* into *FM* for solving the problem of feature combination under large-scale sparse data.
- *Transformer* [20]: A series prediction model that proposed by Google contains the attention mechanism. In our comparison experiments, only the encoder component is used.
- *AMBD-RNN*: A variant model of our proposal with *RNN* in block.
- *AMBD-GRU*: A variant model of our proposal with *GRU* in block.

From the comparative results with other approaches, we can find that neural network models have worse prediction performance than the others.

The structure of *AMBD* models are designed according to the characteristics of warehouse queue system. As a result, they show excellent predictive performance compared with other methods for *WDT* prediction. The difference among *AMBD-RNN*, *AMBD-GRU*, *AMBD-LSTM* is the design details involved in blocks. The result shows that the model with *LSTM* has the best performance. The reason is that long-term memory can deal with the internal queue information with long queue length more better. Therefore, we choose *AMBD-LSTM* as the representative model to illustrate the prediction structure in our paper.

6.4   Ablation Study

In this section, we evaluate the availability of each feature. Table 3 shows the variance of prediction performance after removing the relevant feature compared with the original method having all features. The analysis results show that all the features involved in our prediction model provide essential information for warehouse dwell time prediction.

Table 3. Results of ablation study

Ablation feature group	MAE	RMSE	MAPE
Warehouse ability	**13.00%**	**8.71%**	**27.58%**
Cargo features	14.92%	6.41%	31.17%
Consumed time	9.93%	4.80%	22.37%
Temporal features	4.40%	3.61%	5.02%
Attention	3.89%	0.92%	10.88%

Since warehouse ability feature can reflect trucks' working efficiency in current time instant and will influence all trucks' subsequent warehouse dwell time, it has the greatest impact on the model's predictive performance. Cargo feature including the weights and categories of cargoes affects overall loading time mostly. Without consumed time and temporal features, the prediction precision of *AMBD* decrease a lot due to lack of preceding trucks' the loading task status information. What's more, the local attention module contributes to the prediction greatly for its extraction on important preceding trucks' information.

6.5   Attention Module Interpretation

In this part, we interpret how the local attention mechanism reflects the truck correlations of *WDT* prediction. We select three specific queue scenarios whose inner relationships among queuing trucks are different.

As shown in Fig. 5, we present the weight matrices of trucks in three truck blocks. For each graph, the grid with coordinate (i, j) corresponds to the impact degree of i on j. The deeper color means greater correlation on *WDT* prediction. These weight graphs are lower triangular matrices because the later trucks have no effect on the preceding truck's *WDT* prediction. The waiting trucks and working trucks are represented by index 10 to 19 and 0 to 9 respectively. In Fig. 5(c), *WDT* of waiting trucks are influenced more by the closer waiting trucks in the queue. As for the working trucks, the trucks at the head of the queue with longer duration of dwell impact most to the subsequent trucks because they are about to finish the work and leave. In Fig. 5(a)(b), the correlation weight distributions of working trucks are irregular. It indicates that the influence of working trucks on subsequent *WDT* prediction are not simply determined by the trucks' distances.

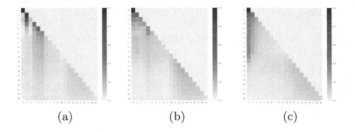

Fig. 5. Three cases of truck attention weight analysis

In conclusion, the masked multi-head attention module of our proposal plays a bigger role than normal time series models. It can capture the inner correlations among trucks with varieties of task execution process, and find out the most important factors which have a great impact on the trucks' dwell time.

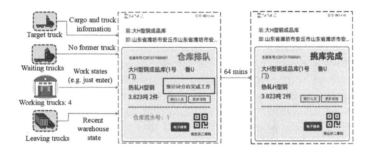

Fig. 6. One case of WDT prediction at the steel logistics enterprise

6.6 Demonstration

We present a real-world case to prove the availability and reliability of $AMBD$ online as shown in Fig. 6. A driver came to the company early at 8:44 in the morning. The number of required cargoes was two. As soon as he arrived at the company, he was called to enter into the factory for loading because of the low morning peak. But when he reached the warehouse, due to the number of working trucks reached the warehouse's working limit number, he had to wait for being called. At the moment, his queue serial number was one. After analyzing the information of working tucks and leaving trucks, we found four trucks in that warehouse had just started working. It meant that the current trucks needed to wait for a long time. As a result, our prediction model can give a predicted warehouse dwell time of 58 min, which is basically in line with WDT of 64 min in actual case.

7   Conclusion

We proposed an attention based multi-block deep learning model ($AMBD$) for WDT prediction. To extract the warehouse loading ability and preceding trucks' warehouse dwell time, we design a multi-block structure to obtain the loading task status information of leaving trucks, working trucks and waiting trucks. Further, we introduce local attention mechanism to mine the influence degree of the preceding trucks' task execution process on the current truck for the improvement of the WDT prediction accuracy. Experiments on the real-world dataset show that $AMBD$ can achieve much better performance than other available ones. In the future, we will further consider the distinctions among the queue systems of different warehouses for improvement of WDT prediction performance.

Acknowledgments. This research was supported by NSFC (Nos. 62072180, U1911203 and U1811264).

References

1. Adhikari, B., Xu, X., Ramakrishnan, N., Prakash, B.A.: Epideep: exploiting embeddings for epidemic forecasting. In: Teredesai, A., Kumar, V., Li, Y., Rosales, R., Terzi, E., Karypis, G. (eds.) KDD, pp. 577–586. ACM (2019)
2. Aveklouris, A., Vlasiou, M., Zwart, B.: Bounds and limit theorems for a layered queueing model in electric vehicle charging. Queueing Syst. Theory Appl. **93**(1–2), 83–137 (2019)
3. Breiman, L.: Better subset regression using the nonnegative garrote. Technometrics **37**(4), 373–384 (1995)
4. Chaki, S., Doshi, P., Patnaik, P., Bhattacharya, S.: Attentive RNNs for continuous-time emotion prediction in music clips. In: Chhaya, N., Jaidka, K., Healey, J., Ungar, L., Sinha, A.R. (eds.) AAAI, CEUR Workshop Proceedings, vol. 2614, pp. 36–46. CEUR-WS.org (2020)
5. Chen, Y., Whitt, W.: Algorithms for the upper bound mean waiting time in the GI/GI/1 queue. Queueing Syst. Theory Appl. **94**(3–4), 327–356 (2020)
6. Cheng, H., et al.: Wide & deep learning for recommender systems. In: Karatzoglou, A., et al. (eds.) DLRS@RecSys, pp. 7–10. ACM (2016)
7. Dasgupta, S., Osogami, T.: Nonlinear dynamic boltzmann machines for time-series prediction. In: Singh, S.P., Markovitch, S. (eds.) AAAI, pp. 1833–1839. AAAI Press (2017)
8. Feinberg, E.A., Yang, F.: Optimal pricing for a gi/m/k/n queue with several customer types and holding costs. Queueing Syst. Theory Appl. **82**(1–2), 103–120 (2016)
9. Fu, K., Meng, F., Ye, J., Wang, Z.: Compacteta: a fast inference system for travel time prediction. In: Gupta, R., Liu, Y., Tang, J., Prakash, B.A. (eds.) KDD, pp. 3337–3345. ACM (2020)
10. Guo, H., Tang, R., Ye, Y., Li, Z., He, X.: Deepfm: a factorization-machine based neural network for CTR prediction. In: Sierra, C. (ed.) IJCAI, pp. 1725–1731. ijcai.org (2017)
11. Ho, T.K.: The random subspace method for constructing decision forests. IEEE Trans. Pattern Anal. Mach. Intell. **20**(8), 832–844 (1998)

12. Horváth, G.: Waiting time and queue length analysis of markov-modulated fluid priority queues. Queueing Syst. Theory Appl. **95**(1), 69–95 (2020)
13. Legros, B.: M/G/1 queue with event-dependent arrival rates. Queueing Syst. Theory Appl. **89**(3–4), 269–301 (2018)
14. Li, J.: Structured prediction in time series data. In: Singh, S.P., Markovitch, S. (eds.) AAAI, pp. 5042–5043. AAAI Press (2017)
15. Liu, C., Hoi, S.C.H., Zhao, P., Sun, J.: Online ARIMA algorithms for time series prediction. In: Schuurmans, D., Wellman, M.P. (eds.) AAAI, pp. 1867–1873. AAAI Press (2016)
16. Shajin, D., Krishnamoorthy, A., Dudin, A.N., Joshua, V.C., Jacob, V.: On a queueing-inventory system with advanced reservation and cancellation for the next K time frames ahead: the case of overbooking. Queueing Syst. Theory Appl. **94**(1–2), 3–37 (2020)
17. Stanford, D.A., Taylor, P., Ziedins, I.: Waiting time distributions in the accumulating priority queue. Queueing Syst. **77**(3), 297–330 (2013). https://doi.org/10.1007/s11134-013-9382-6
18. Tibshirani, R.: Regression shrinkage and selection via the lasso. J. Roy. Stat. Soc. Ser. B (Methodol.) **58**(1), 267–288 (1996)
19. Vassøy, B., Ruocco, M., de Souza da Silva, E., Aune, E.: Time is of the essence: a joint hierarchical RNN and point process model for time and item predictions. In: Culpepper, J.S., Moffat, A., Bennett, P.N., Lerman, K. (eds.) WSDM, pp. 591–599. ACM (2019)
20. Vaswani, A., et al.: Attention is all you need. In: Guyon, I., et al. (eds.) Advances in Neural Information Processing Systems 30: Annual Conference on Neural Information Processing Systems, pp. 5998–6008 (2017)
21. Wang, Z., Fu, K., Ye, J.: Learning to estimate the travel time. In: Guo, Y., Farooq, F. (eds.) KDD, pp. 858–866. ACM (2018)
22. Zhan, D., Weiss, G.: Many-server scaling of the n-system under FCFS-ALIS. Queueing Syst. Theory Appl. **88**(1–2), 27–71 (2018)
23. Zheng, C., Fan, X., Wang, C., Qi, J.: GMAN: a graph multi-attention network for traffic prediction. In: AAAI, pp. 1234–1241. AAAI Press (2020)
24. Zhou, X., Shen, Y., Huang, L., Zang, T., Zhu, Y.: Multi-level attention networks for multi-step citywide passenger demands prediction. IEEE Trans. Knowl. Data Eng. **33**(5), 2096–2108 (2021)
25. Zhu, L., et al.: Order fulfillment cycle time estimation for on-demand food delivery. In: Gupta, R., Liu, Y., Tang, J., Prakash, B.A. (eds.) KDD, pp. 2571–2580. ACM (2020)

Performance Evaluation of Pre-trained Models in Sarcasm Detection Task

Haiyang Wang, Xin Song, Bin Zhou$^{(\boxtimes)}$, Ye Wang, Liqun Gao, and Yan Jia

National University of Defense Technology, ChangSha, China
{wanghaiyang19,songxin,binzhou,ye.wang}@nudt.edu.cn,
jiayanjy@vip.sina.com

Abstract. Sarcasm is a widespread phenomenon in social media such as Twitter or Instagram. As a critical task of Natural Language Processing (NLP), sarcasm detection plays an important role in many domains of semantic analysis, such as stance detection and sentiment analysis. Recently, pre-trained models (PTMs) on large unlabelled corpora have shown excellent performance in various tasks of NLP. PTMs have learned universal language representations and can help researchers avoid training a model from scratch. The goal of our paper is to evaluate the performance of various PTMs in the sarcasm detection task. We evaluate and analyse the performance of several representative PTMs on four well-known sarcasm detection datasets. The experimental results indicate that RoBERTa outperforms other PTMs and it is also better than the best baseline in three datasets. DistilBERT is the best choice for sarcasm detection task when computing resources are limited. However, XLNet may not be suitable for sarcasm detection task. In addition, we implement detailed grid search for four hyperparameters to investigate their impact on PTMs. The results show that learning rate is the most important hyperparameter. Furthermore, we also conduct error analysis by means of several sarcastic sentences to explore the reasons of detection failures, which provides instructive ideas for future research.

Keywords: Sarcasm detection · Pre-trained models · Natural language processing

1 Introduction

With the thriving of social media platforms such as Twitter and Instagram, the exchange of opinions and ideas among people becomes more frequent than ever. The rise of communication among people stimulates the use of figurative and creative language including sarcasm or irony. Sarcasm usually occurs when there is some discrepancy between the literal and the intended meaning of a text. Consider the following text: *yeah! It is so great to be able to work all day on weekends.* This sentence contains *yeah* and *great,* which seems to express the happiness of the author but actually expresses the negative and complaining sentiment of the overworked weekends. Besides, sarcasm can be manifested in

© Springer Nature Switzerland AG 2021
W. Zhang et al. (Eds.): WISE 2021, LNCS 13081, pp. 67–75, 2021.
https://doi.org/10.1007/978-3-030-91560-5_5

many implicit and concise ways so that Oscar Wilde [3] described sarcasm as "the lowest form of wit but the highest form of intelligence."

It is common knowledge that the goal of Natural Language Processing (NLP) is to understand and generate human language. The sarcasm detection task is of great significance in NLP tasks due to sarcasm can express different meanings from the literal. Moreover, sarcasm detection has great potential in the domains of semantic analysis, such as stance detection and sentiment analysis. However, sarcasm detection is a very difficult task. Firstly, the expression of sarcasm is influenced by many factors such as culture, author's background and conversation context etc. Then, the effectiveness of sarcasm detection models depends on the availability and quality of labelled data used for training. However, collecting such data is challenging due to the subjective nature of sarcasm.

Recently, substantial research has shown that pre-trained models (PTMs) can learn universal language representations. Such excellent representations may be beneficial for overcoming the difficulties of sarcasm detection and can avoid training a new model from scratch. Therefore, we consider the following questions: (1) Can PTMs help improve the performance of the sarcasm detection task? (2) How different PTMs perform in sarcasm detection tasks? (3) How do different hyperparameters of PTMs affect the performance of sarcasm detection?

In order to explore the performance of PTMs in sarcasm detection task, we evaluate five models: BERT [2], RoBERTa [8], DistilBERT [11], ALBERT [6] and XLNet [12]. All models are pre-trained on large unlabelled corpora by applying self-supervised objective. We develop a detailed grid search for four hyperparameters to investigate the function and effectiveness of various hyperparameters on four well-known Twitter sarcasm detection datasets which are SemEval [5], iSarcasm [9], Ptacek [10]and Ghosh [3]. SemEval and Ptacek datasets are labelled by distant supervision where texts are considered sarcastic if they meet predefined criteria, such as including specific hashtags (#sarcasm). The datasets of iSarcasm and Ghosh are labelled by tweets authors.

Our contributions are summarized as follows:

- We evaluate five PTMs comparing with the state-of-the-art baselines in sarcasm detection task. According to the results of comprehensive experiments, we analyse the performance difference of PTMs.
- We implement a detailed grid search for four hyperparameters. Based on the results, we give some constructive and meaningful suggestions for the hyperparameter selection of PTMs in the sarcasm detection task.
- We employ four well-known and up-to-date sarcasm detection datasets and introduce the characteristics of different datasets in detail. We also perform error analysis to better explain the power and limitation of PTMs.

2   Related Work

The methods of sarcasm detection in recent research work can be divided into three categories, machine learning methods based on feature engineering [5], conventional deep learning methods [7], and PTMs [1]. Moreover, the models can also be divided into single text-based or context-based according to the information used in the detection [9].

3 Methods and Datasets

3.1 Pre-trained Models

With the development of deep learning, various PTMs have been widely used to solve NLP tasks and yield outstanding performance.

BERT is designed to pretrain deep bidirectional representations from unlabelled text by conditioning simultaneously the possibility of both left and right word context in all layers [2]. It applies the combination of MLM and NSP as a pre-training objective.

RoBERTa is pre-trained with larger batches of longer sequences for longer time on over 160GB unlabelled text corpora [8]. It only uses MLM as the training objective after comprehensive comparative experiments.

DistilBERT is a distilled version of BERT [11]. It reduces the size of BERT model while retaining 97% language understanding capabilities by a leverage distinct knowledge distillation approach.

ALBERT is a lite BERT that has significantly fewer parameters than traditional BERT [6]. It combines two parameter reduction techniques which are the factorized embedding parameterization and the cross-layer parameter sharing technique. It leverages a self-supervised loss for sentence-order prediction to further improve the performance.

XLNet is a generalized autoregressive pre-trained method [12]. It leverages the best of both autoregressive (AR) language modeling and autoencoding (AE) while avoiding their limitations. It employs a PLM objective to capture dependency structures among tokens better and eliminate the independence assumption.

In summary, PTMs shows that a very deep model can significantly improve the performance of NLP tasks and can be pre-trained from unlabelled datasets. In this paper, the PTMs are all implemented in PyTorch using the huggingface transformers library. They are finetuned using the Adam algorithm in Ubuntu 18.04.5 LTS with 72 CPUs and 6 GPUs.

3.2 Datasets

We apply four well-known sarcasm datasets of English tweets.

SemEval: It is an irony detection dataset in English tweets [5]. The ironic tweets were collected using irony-related hashtags and were subsequently manually annotated to minimise the amount of noise in the corpus.

iSarcasm: The iSarcasm [9] is a sarcasm dataset labelled for sarcasm directly by their authors. It is created by asking Twitter users to provide both sarcastic and non-sarcastic tweets that they had posted in the past.

Ptacek: Ptacek et al.[10] created a Czech and English tweets dataset for sarcasm detection. In this paper, we only use English tweets. Same as the dataset of SemEval2018 Task 3, they also collected 780,000 English tweets with *#sarcasm* hashtag.

Ghosh: Ghosh and Veale created a self-annotations tweets dataset for sarcasm detection[3]. They used a Twitterbot named @onlinesarcasm. Twitterbot retweets a chosen unlabelled tweet to its author, appending a yes/no question as a comment to elicit a reply. Then the tweet will be labelled according to the author's reply.

4   Experiments Results

In this section, we conduct comparative experiments, detailed grid search and error analysis research to explain the power and limitation of PTMs in sarcasm detection task.

4.1   Comparative Experiments

The goal of the comparative experiments is to evaluate the general performance of PTMs. We evaluate PTMs on four sarcasm datasets and compare five PTMs with the best baseline of each dataset. Table 1 shows the F1-score results.

Table 1. The results of comparative experiments on four sarcasm datasets

Model	SemEval	iSarcasm	Ptacek	Ghosh
Best-Baseline	0.724 [5]	0.364 [9]	0.874 [9]	**0.900** [3]
BERT	0.700	0.654	0.864	0.828
RoBERTa	**0.731**	**0.668**	**0.882**	0.849
DistilBERT	0.678	0.617	0.849	0.846
ALBERT	0.679	0.541	0.856	0.792
XLNet	0.655	0.455	0.872	0.829

As can be observed from the Table 1, RoBERTa performs best on the four sarcasm detection datasets compared to the other four pre-trained models. That's probably because RoBERTa pre-trained in a larger corpus which may contain more ironic texts. In addition, RoBERTa may learn better representations to capture the implicit semantics of texts. BERT performs better than DistilBERT, ALBERT, and XLNet on the SemEval and iSarcasm datasets. For the two light versions of BERT, DistilBERT and ALBERT, they show competitive performance in SemEval and Ptacek. The XLNet model does not show superior performance. Especially on the iSarcasm dataset, XLNet may fall into the trap of local maximum. More detailed explanation can be found in the Sect. 4.2 .

4.2   Grid Search

The fine-tuning of PTMs is similar to transfer learning but different from training the model from scratch. Thus, the experience accumulated in traditional
hyperparameter optimization research may not be suitable for the settings of
fine-tuning[4]. Our goal is to research the effectiveness and necessity of hyperparameter tuning for fine-tuning in the sarcasm detection task. Thus, we apply a
grid search for the hyperparameters: Batch size, Sequence length, Learning rate
and Learning rate schedule, which are shown in Table 2 .

Table 2. Search space over chosen hyperparameters.

Hyperparameter	Considered Configurations
Batch size	8; 16
Sequence length	128; 256
Learning rate	1e–05; 2e–05; 3e–05; 4e–05
Learning rate schedule	constant (cst);linear (lin);cosine (cos)

As we all know, the learning rate is the most common and important hyperparameter when fine-tuning. Compared to a model trained from scratch, setting
a smaller learning rate during fine-tuning is a better choice. The pre-training
model has learned fairly good representations in large-scale corpora during the
pre-training process. Excellent representations are very helpful for downstream
tasks. A small learning rate can help models make full use of the already gained
semantic knowledge. Therefore, we consider learning rate values from 1e-5 to
4e-5. Moreover, we apply three different learning rate schedules. For batch size,
we choose 8 and 16 after considering memory limitations and evaluate sequence
lengths of 128 and 256 for the grid search.

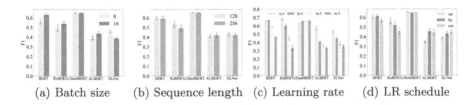

(a) Batch size   (b) Sequence length   (c) Learning rate   (d) LR schedule

Fig. 1. Average of F1-score on SemEval dataset.

Given the defined hyperparameters in Table 2. There are 48 combinations
evaluated for each model and dataset. We analyse the performance of each combination vary with learning rate and the results are shown in Table 3. According

to the F1-score, we can get two different types of configurations which are winning and optimum. (1) **Winning configurations:** Given the learning rate, winning configurations are the hyperparameter combinations in which each model achieves the highest F1-score under each dataset. (2) **Optimum configurations:** Given the dataset and pre-trained model, optimum configurations are the hyperparameter combinations which achieve the highest F1-score. In total, there are 80 winning configurations and 20 optimum configurations in Table 3. Moreover, in order to visually observe the impact of different hyperparameters on performance, we calculate the average F1-score and variance on the dataset of SemEval, as shown in the Fig. 1. We select configurations that have specified hyperparameters and calculate the average of the F1-score.

Table 3. The results of the grid search vary with the Learning Rate (LR). Winner is the winning configuration out of the 12 possible configurations per LR. The format is (Batch size, Sequence length, Learning rate schedule). The F1-score of optimum configurations are indicated in bold. The highest F1-score on each dataset is shown in red.

	LR	BERT Winner	F1	RoBERTa Winner	F1	DistilBERT Winner	F1	ALBERT Winner	F1	XLNet Winner	F1
SemEval	1e-5	(8,256,cos)	0.6784	(8,256,cst)	0.7314	(16,256,cst)	0.6586	(8,256,lin)	**0.6794**	(16,256,cos)	0.6178
	2e-5	(16,256,cst)	**0.6999**	(8,256,lin)	0.7204	(8,128,cst)	**0.6783**	(16,128,lin)	0.6331	(8,128,cst)	**0.6545**
	3e-5	(16,128,lin)	0.6928	(16,128,lin)	0.7104	(16,128,lin)	0.6629	(16,128,cos)	0.6677	(16,128,cos)	0.6066
	4e-5	(8,256, cos)	0.6745	(16,128,lin)	0.3763	(16,128,lin)	0.6758	(16,128,cos)	0.3763	(8,128,cst)	0.3763
iSarcasm	1e-5	(8,128,cst)	0.6334	(8,128,lin)	0.6539	(16,128,cst)	0.5114	(8,128,cos)	0.5265	(8,128,cst)	**0.4546**
	2e-5	(8,256,lin)	**0.6539**	(16,256,lin)	0.6682	(16,256,cst)	0.5959	(16,128,cos)	**0.5410**	(8,128,cst)	0.4546
	3e-5	(8,128,cst)	0.6364	(8,128,cst)	0.4546	(16,256,cst)	0.5815	(16,128,cos)	0.4714	(8,128,cst)	0.4546
	4e-5	(16,256,lin)	0.6104	(8,128,cst)	0.4546	(16,128,lin)	**0.6166**	(16,256,lin)	0.5234	(8,128,cst)	0.4546
Ptacek	1e-5	(8,256,cos)	0.8630	(16,128,cos)	0.8819	(8,128,cos)	0.8462	(16,128,lin)	**0.8559**	(16,256,lin)	**0.8724**
	2e-5	(16,128,cos)	**0.8636**	(16,256,lin)	0.8813	(8,128,lin)	**0.8486**	(16,128,cos)	0.8517	(16,256,cos)	0.8706
	3e-5	(16,128,cos)	0.8617	(16,128,cos)	0.8752	(16,128,lin)	0.8470	(16,128,cst)	0.7999	(16,256,lin)	0.8699
	4e-5	(16,128,lin)	0.8611	(16,128,cos)	0.8685	(8,256,lin)	0.8451	(8,128,cos)	0.7759	(16,128,lin)	0.8598
Ghosh	1e-5	(8,256,cos)	0.8231	(16,256,cst)	0.8353	(16,256,cos)	0.8161	(8,256,cos)	**0.7921**	(16,128,cst)	**0.8294**
	2e-5	(16,128,lin)	**0.8281**	(16,128,cst)	0.8485	(16,128,cst)	0.8407	(16,256,lin)	0.7644	(16,256,cos)	0.8051
	3e-5	(16,128,lin)	0.8226	(16,256,cos)	0.8198	(16,256,cos)	0.8164	(8,128,cst)	0.3305	(16,128,cos)	0.7700
	4e-5	(16,128,lin)	0.7939	(16,256,cos)	0.7946	(16,128,cst)	**0.8455**	(16,256,lin)	0.7539	(8,128,cst)	0.3361

Next, we discuss the results of grid search according to the types of hyper-parameters.

Batch Size. The best-performing batch is 16 whether it is considered globally or from the 20 highest F1-scores. Related research also shows that larger batches could help achieve more accurate gradient estimation to some extent. On the contrary, the small batch may cause additional noise during model training. However, we also find that the model with a lower learning rate and smaller batch is more likely to achieve better performance.

Sequence Length. Intuitively from Fig. 1(b), a sequence length of 256 does not lead to better performance. For the RoBERTa, the longer sequence length makes the average F1-score decrease greatly. According to the results of grid search, there is an indication that short sequence length may help PTMs achieve better performance on tweets datasets.

Learning Rate. In general, the best performing learning rate is 1e–5 and 2e–5. From Table 3, we can see that updating the pre-trained parameters by a small learning rate is important to maximize the performance of PTMs. Further more, we can see that from Fig. 1(c). Only DistilBERT model has a slight ascent with the increase of learning rate. BERT, RoBERTa, ALBERT and XLNet all have huge performance declines when the learning rates are 3e-5 and 4e-5. Considering the reason, DistilBERT use knowledge distillation approach, which may help it become more robust.

Learning Rate Schedule. Evaluating the learning rate schedule, the number of winning configuration with the constant, the linear or the cosine schedule is very close which are 25, 27 and 28 respectively. As for optimum configurations, the linear and constant schedule are chosen frequently. As shown in Fig. 1(d), different PTMs have various learning rate schedule preferences.

Comprehensive consideration of overall performance, RoBERTa achieves excellent performance in the sarcasm detection task. Surprisingly, DistilBERT outperforms BERT on the Ghosh dataset. It also outperforms XLNet on the SemEval, iSarcasm and Ghosh datasets. However, ALBERT and XLNet do not perform as well as expected. The performance of ALBERT still lags behind that of BERT despite the use of large variants. As shown in Table 3 , XLNet falls into the trap of local maxima on iSarcasm dataset. The first reason is the unbalanced distribution of labels in test set. Besides, XLNet uses the PLM as the training objective which can help XLNet capture dependencies between tokens. However, sarcasm detection belongs to segment-level task [4] while PLM may not help XLNet achieve better performance.

4.3   Error Analysis

In this section, we analyse two typical error instances. These instances are all sarcastic tweets while the model judges them to be non-sarcastic tweets. However, it's easy for humans to recognize them as sarcasm.

Sometimes I feel maths is the only place in this world that really accepts me. Numbers are my life.

The author of this tweet expresses his love for mathematics and his disappointment with the real world. The irony of this sentence is aimed at the author himself. The reason why this tweet is wrongly classified may be that there are no obvious positive and negative words in the text at the same time. Sarcastic statements are often characterised by a form of opposition or contrast. In this tweet, the form of opposition is very cryptic.

I love when my parents yell and scream at me and then get pissed at me for crying.

This is an ironic tweet. The author expressed dissatisfaction with his/her parents. The misclassification of tweets may be due to the corpora of PTMs. In large unlabelled corpora, the text when *love* and *parents* appear together often express a positive emotion and true love to parents. A pre-trained model fine-tuned with a little sarcasm data cannot capture the implicit meaning of the tweet. Therefore, the pre-trained model is likely to make some preconceived judgements due to the impact of the corpora.

5 Conclusion

In this paper, we evaluate the performance of various PTMs on the four well-known sarcasm detection datasets. Experimental results show that the RoBERTa model performs strongly on several sarcasm detection datasets and exceeds the best baseline. Furthermore, the BERT-based approaches outperform XLNet on several datasets. Then we summarize some rules about the hyperparameter selection of PTMs for the sarcasm detection task. Smaller learning rate, shorter sequence length and larger batch size can help PTMs achieve better performance in detecting sarcastic tweets task. In the future, we plan to investigate sarcasm detection with contextual information and explore whether pre-training on large unlabelled corpora cause PTMs to make some biased judgments.

Acknowledgements. This work was supported by the National Key Research and Development Program of China No. 2018YFC0831703.

References

1. Baruah, A., Das, K., Barbhuiya, F.A., Dey, K.: Context-aware sarcasm detection using bert. In: Fig-Lang@ACL (2020)
2. Devlin, J., Chang, M.W., Lee, K., Toutanova, K.: Bert: pre-training of deep bidirectional transformers for language understanding. In: NAACL-HLT (2019)
3. Ghosh, A., Veale, T.: Magnets for sarcasm: making sarcasm detection timely, contextual and very personal. In: EMNLP (2017)
4. Guderlei, M., Aßenmacher, M.: Evaluating unsupervised representation learning for detecting stances of fake news. In: COLING (2020)
5. Hee, C.V., Lefever, E., Hoste, V.: Semeval-2018 task 3: irony detection in english tweets. In: SemEvalNAACL-HLT (2018)
6. Lan, Z., Chen, M., Goodman, S., Gimpel, K., Sharma, P., Soricut, R.: Albert: a lite bert for self-supervised learning of language representations. ArXiv (2020)
7. Lemmens, J., Burtenshaw, B., Lotfi, E., Markov, I., Daelemans, W.: Sarcasm detection using an ensemble approach. In: Fig-Lang@ACL (2020)
8. Liu, Y., et al.: Roberta: a robustly optimized bert pretraining approach (2019). ArXiv abs/1907.11692
9. Oprea, S., Magdy, W.: isarcasm: a dataset of intended sarcasm (2020). ArXiv abs/1911.03123
10. Ptácek, T., Habernal, I., Hong, J.: Sarcasm detection on Czech and English twitter. In: COLING (2014)

11. Sanh, V., Debut, L., Chaumond, J., Wolf, T.: Distilbert, a distilled version of bert: smaller, faster, cheaper and lighter. ArXiv (2019)
12. Yang, Z., Dai, Z., Yang, Y., Carbonell, J., Salakhutdinov, R., Le, Q.V.: Xlnet: generalized autoregressive pretraining for language understanding. In: NeurIPS (2019)

Deep Learning (2)

News Popularity Prediction with Local-Global Long-Short-Term Embedding

Shuai Fan[1], Chen Lin[1(✉)] (iD), Hui Li[1], and Quan Zou[2]

[1] School of Informatics, Xiamen University, Xiamen, China
chenlin@xmu.edu.cn
[2] Institute of Fundamental and Frontier Sciences, University of Electronic Science and Technology of China, Chengdu, China

Abstract. Predicting news popularity is an essential topic in the news industry. It is challenging because numerous factors influence public response to the news. This paper presents F^4, a neural model to predict news popularity by learning news embedding from global, local, long-term and short-term factors. F^4 integrates a sentence encoding module to represent the local context of each news story; a heterogeneous graph-based module to capture the short-term information propagation from current buzz words to each news story; a sequential module to extract long-term popularity features in entity sequence; and an attention module to learn global news-entity correlations. Extensive experiments on real-world Chinese and English news datasets demonstrated that F^4 outperforms state-of-the-art baselines in predicting and ranking news popularity.

1 Introduction

The development of the Internet has revolutionized the news industry. We have seen news agencies starting online news portal services and online news apps persistently boosting audiences. Predicting news popularity (i.e., estimating how many people will read a particular piece of news) becomes an important topic, and it builds the foundation for a broad spectrum of downstream tasks. For example, based on the predicted news popularity, online news portals can optimize the page layout and resource management; advertisers can tailor ads and save costs; news recommender systems can improve recommendation quality for news users, to name a few.

Recent years have witnessed numerous research in predicting the popularity of online news, which is measured by various user behaviors, including the number of likes [12,28], number of clicks [4], number of comments [21], number of instant messages [15], and so on. Conventional approaches are built upon hand-crafted features, including context features and news content features [4]. They apply shallow learning models, such as Support Vector Machines [4] and Logistic Regression [2]; topic models, such as named entity topic model [1]; and

© Springer Nature Switzerland AG 2021
W. Zhang et al. (Eds.): WISE 2021, LNCS 13081, pp. 79–93, 2021.
https://doi.org/10.1007/978-3-030-91560-5_6

statistical sequence models, such as the neural hawks process [13]. As features are costly to obtain and sometimes not accessible, modern distributed learning, which represents each news as a numerical vector called news embedding, has been applied in related problems such as news recommendation [26] and stock price movement prediction [11].

However, news popularity is not easy to predict since many factors (i.e., global, local, long-term, and short-term factors) influence how online users respond to a piece of news. We illustrate two example news in Fig. 1. We can see that, *news1* is more popular than *news2*, because (1) the content of *news1* (i.e., **local** factors) contains terms that are current buzz words (i.e., **short-term** factors) (2) *news1* is related, with different association strengths (i.e., **global** factors), to named entities that are consistently popular over time (i.e., **long-term** factors). In this paper, we propose F^4: a news embedding learning model from local, global, long-term, and short-term factors to predict news popularity. (1) For long-term factors, F^4 adopts a Gated Recurrent Unit (GRU) module on entity sequence to learn the sequential popularity embedding. (2) For the short-term factors, F^4 adopts a heterogeneous graph-based module to propagate information from current buzz words to the news. (3) For the local factors, F^4 adopts a sentence encoder to represent the local context of each news story. (4) For the global factors, F^4 adopts the attention mechanism to associate the popularity of each news story with different entities. (5) F^4 incorporates the above modules in a unified framework to make predictions by integrating all factors.

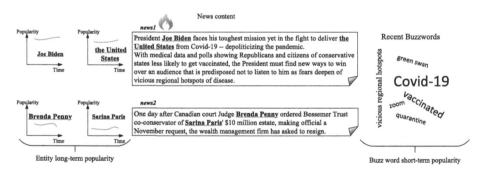

Fig. 1. Global, local, long-term and short-term factors affect news popularity. A popular news is related to long-term popular entities and the news content contains current buzz words.

In summary, our contributions are: (1) We propose to predict news popularity by integrating local, global, long-term, and short-term factors. (2) We develop neural models to learn local contextual representation, short-term popularity propagation, long-term sequential entity popularity, and global news-entity associations. (3) We demonstrate the effectiveness of our proposed method through extensive experiments on real-world Chinese and English datasets.

2   Related Work

With the development of online social networks, predicting the popularity of online news receives much attention and has been an active research area. Existing research generally falls into two categories. The first type forecasts news popularity based on textual contents. Clustering method [14], regression method [8,9], and classification method [22], have been proposed. Tatar et al. [22] used three models, including a linear model, a linear model on a logarithmic scale, and a constant scaling model to predict online news popularity. Stoddard et al. [20] proposed a simple Poisson regression model to estimate the quality of articles and found that news with higher popularity on Reddit and Hacker News is, to a large extent, articles with higher quality. The second type focuses on exploiting context features [4] and multi modal features [12].

The majority of previous research used the non-neural network models such as SVM classifiers [4], logistic regression [2], Tree Regression [6], etc. Recently, deep neural network which automatically extract vectorized numerical feature representations have shown promising results, including Convolutional Neural Network (CNN) [19], Recurrent Neural Network (RNN) [10], Long Short-Term Memory network(LSTM) [29], Graph Neural Network (GNN) [5], Autoencoder [18], and the Attention Mechanism [16]. We have seen applications of neural network models in similar tasks such as dwell time prediction [3] and news recommendation [27].

However, previous studies ignore the complex factors that influence the popularity of news. As we show in the following sections, purely relying on local content and missing the long-term, global entity information leads to low prediction accuracy. On the contrary, F^4 obtains high prediction accuracy by capturing different factors and their influence on news popularity.

3   Model

3.1   Preliminaries

Problem Definition. Suppose we have a training set $\mathcal{D}$, where each training instance is a tuple $< \mathbf{x}, y > \in \mathcal{D}$. The input for each news story is $\mathbf{x} = \{\mathbf{s}_1, \mathbf{s}_2, \ldots, \mathbf{s}_{N(\mathbf{x})}\}$, which contains $N(\mathbf{x})$ sentences. Each sentence is a sequence of words, i.e. $\mathbf{s}_i = \{\mathbf{w}_1, \mathbf{w}_2, \ldots \mathbf{w}_{M(\mathbf{s}_i)}\}$, where $M(\mathbf{s}_i)$ is the length of $\mathbf{s}_i$. The output label y is the normalized public response to $\mathbf{x}$, i.e., $y \in (0,1)$ (more details in Sect. 4). Our goal is to forecast $\hat{y}$ for any test instance $\hat{\mathbf{x}}$.

Model Overview. Figure 2 presents the framework of F^4, which mainly consists of four parts. (1) The **sentence encoding module** encodes the context within each sentence of a news story to represent the local factor of popularity prediction. (2) The **heterogeneous graph-based news encoding module** encodes the information propagation from current buzz words to each news story, based on a heterogeneous graph of words and sentences, to represent the

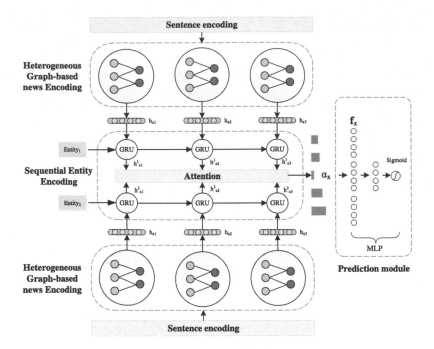

Fig. 2. Framework overview of F^4

short-term factor of popularity prediction. (3) The **sequential entity encod-ing module** encodes the sequence of news stories of an entity to represent the long-term factor of popularity prediction. (4) The **attention module** captures associations among news stories and entities to represent the global factor of popularity prediction. (5) The **prediction module** makes the final prediction by integrating all factors. We will introduce each module in the following sub-sections.

3.2 Sentence Encoding

First, each sentence $\mathbf{s}_i$ is represented as a word embedding matrix $X^{\mathbf{s}_i} \in \mathcal{R}^{M(\mathbf{s}_i) \times D_W}$, where $M(\mathbf{s}_i)$ is the length of sentence $\mathbf{s}_i$, and D_W is the dimen-sion size of the word embedding vectors.

As shown in Fig. 3, the input word embedding matrix is first initialized by pre-trained word embedding. Then, it flows through a CNN layer, which adopts convolutional operations to capture the local n-gram features for each sentence $\mathbf{s}_i$. The output of the CNN component then flows through a BiLSTM layer, which captures dependency between sentences.

3.3 Heterogeneous Graph-Based News Encoding

Inspired by heterogeneous graph [25], we construct a heterogeneous graph for each news story, which consists of two types of nodes: word nodes and sentence

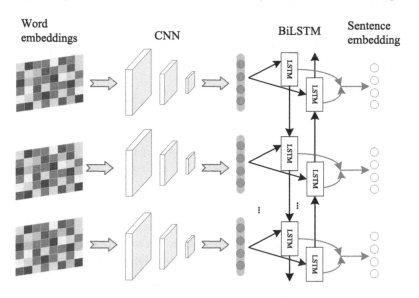

Fig. 3. Sentence encoding

nodes. We initialize each sentence node, i.e., $\mathbf{e}_i^{S,0} \in \mathcal{R}^{D_S}$, for sentence $\mathbf{s}_i$, by the output of sentence encoder in Sect. 3.2. We initialize the word node, i.e., $\mathbf{e}_j^{W,0} \in \mathcal{R}^{D_W}$, for the word $\mathbf{w}_j$, by the pre-trained word embedding vectors. We construct an edge for each pair of word node and sentence node. We initialize the edge vector $\mathbf{e}_{i \leftrightarrow j} \in \mathcal{R}^{D_E}$ for the link between sentence $\mathbf{s}_i$ and word $\mathbf{w}_j$. To distinguish important words from common words, we first compute the TF-IDF weight of each word in each sentence, formally, $tfidf_{i,j} = tf(i,j)/df(j)$, where $tf(i,j)$ is the number of occurrences of word j in sentence i, and $df(j)$ is the number of sentences that contains word j. Then we divide the TF-IDF values to 10 bins, and map each $tfidf(i,j)$ to one vector, corresponding to one bin. The edge vector is not updated during the encoding process.

In the l−th layer of graph neural network, we update the hidden states related to sentence nodes by aggregating adjacent word nodes. In order to propagate the popularity information of current buzz words to the corresponding sentence, we first adopt graph attention[24], in Eq. 1:

$$z_{i,j}^{l+1} = \text{LeakyReLU} \left(\mathbf{W}_a \left[\mathbf{W}_q \mathbf{e}_i^{S,l} \| \mathbf{W}_k \mathbf{e}_j^{W,l} \| e_{i \leftrightarrow j} \right] \right) \tag{1}$$

$$\alpha_{i,j}^{l+1} = \frac{\exp \left(\mathbf{z}_{i,j}^{l+1} \right)}{\sum_{j \in \mathcal{N}_i} \exp \left(\mathbf{z}_{i,j}^{l+1} \right)}$$

$$\mathbf{u}_i^{l+1} = \sigma \left(\sum_{j \in \mathcal{N}_s} \alpha_{i,j}^{l+1} \mathbf{W}_v \mathbf{e}_j^{W,l} \right)$$

where $\|$ is the concatenation operation, σ is the sigmoid function, $\mathbf{W}_a, \mathbf{W}_q, \mathbf{W}_k, \mathbf{W}_v$ are trainable weights, $\mathcal{N}_i$ is the set of adjacent word nodes of sentence $\mathbf{s}_i$, and $\alpha_{i,j}^{l+1}$ is the attention weight between sentence node $\mathbf{s}_i$ to word node $\mathbf{w}_j$. We also adopt multi-head attention. As shown in Eq. 2, suppose $\alpha_{i,j}^{l+1,k}$ denotes the $k-$th attention head,

$$\mathbf{u}_i^{l+1} = \|_{k=1}^{K} \sigma \left(\sum_{j \in \mathcal{N}_i} \alpha_{i,j}^{k,l+1} \mathbf{W}_v^k \mathbf{e}_i^{W,l} \right) \tag{2}$$

We also add a residual connection to avoid gradient vanishing after several iterations. Therefore, the hidden state of sentence node can be represented as:

$$\mathbf{e}_i^{S,l+1} = \mathbf{u}_i^{l+1} + \mathbf{e}_i^{S,0} \tag{3}$$

We apply similar computation to update the hidden states of word nodes in each iteration. Finally, we apply a position-wise feed-forward (FFN) layer consisting of two linear transformations just as Transformer[23] on all sentences to output the news encoding.

$$\mathbf{h_x} = \text{FFN}((\mathbf{e}_1^{S,L} \| \cdots \| \mathbf{e}_{N(\mathbf{x})}^{S,L})) \tag{4}$$

where $N(\mathbf{x})$ is the number of sentences in news $\mathbf{x}$, and L is the number of layers in the graph neural network.

3.4   Sequential Entity Encoding

For each news story, we also extract the entities. Suppose for each news story $\mathbf{x}$, the set of entities extracted from $\mathbf{x}$ is $\mathcal{O}(\mathbf{x})$. We construct an *entity sequence* for each entity o, i.e., $c_o = < \mathbf{x}_{o,1}, \cdots, \mathbf{x}_{o,T(o)} >$, which is the chronically ordered sequence of news stories containing the entity, i.e., $o \in O(\mathbf{x}_{o,t})$, where t is the released time of $\mathbf{x}_{o,t}$ and $T(o)$ is the number of news stories which contain entity o.

As shown in Fig. 2, the sequential entity encoding module operates the Gated Recurrent Unit (GRU) upon the entity sequence. We feed the GRU layer with the hidden state by news encoding from Sect. 3.3. Then GRU adopts a reset gate, an update gate, and a current memory gate to update the hidden state of entity from previously released news, as in Eq. 5

$$
\begin{aligned}
z^{o,t} &= \sigma \left(\mathbf{W}_z \cdot [\mathbf{h}_{\mathbf{x}_{o,t}} \| \mathbf{o}^{t-1}] \right) \\
r^{o,t} &= \sigma \left(\mathbf{W}_r \cdot [\mathbf{h}_{\mathbf{x}_{o,t}} \| \mathbf{o}^{t-1}] \right) \\
\tilde{\mathbf{o}}^t &= \tanh \left(\mathbf{W}_o \cdot [r^{o,t} \mathbf{o}^{t-1} \| \mathbf{h}_{\mathbf{x}_{o,t-1}}] \right) \\
\mathbf{o}^t &= \left(1 - z^{o,t} \right) \times \mathbf{o}^t + z^{o,t-1} \times \tilde{\mathbf{o}^{t-1}}
\end{aligned}
\tag{5}
$$

where $\mathbf{W}_z$, $\mathbf{W}_r$, $\mathbf{W}_o$ are learnable GRU weight matrices, and $\sigma(\cdot), tanh(\cdot)$ are the sigmoid and tanh activation functions, respectively.

3.5   Attention

Since each news story is related to several entities, and each entity is related to numerous news stories, we further adopt an attention layer to capture the global interactions among entities and news stories. Suppose o, p are entities extracted from news story, i.e., $o \in \mathcal{O}(\mathbf{x}), p \in \mathcal{O}(\mathbf{x})$, and o is the entity with the longest sequence, i.e., $o = \arg\max_{T(o)} O(\mathbf{x})$, the attention layer is designed as follows:

$$z_{\mathbf{x}}^{o,p} = \text{LeakyReLU}\left(\mathbf{W}_M\left(\mathbf{W}_q\mathbf{o}^t \| \mathbf{W}_k\mathbf{p}^t\right)\right)$$

$$\alpha_{\mathbf{x}}^{o,p} = \frac{\exp\left(z_{\mathbf{x}}^{o,p}\right)}{\sum_{p\in\mathcal{O}(\mathbf{x})}\exp\left(z_{\mathbf{x}}^{o,p}\right)} \tag{6}$$

$$\mathbf{f_x} = \sum_{p\in\mathcal{O}(\mathbf{x})} \sigma\left(\alpha_{\mathbf{x}}^{o,p}\mathbf{W}_v\mathbf{o}^t\right)$$

where $\mathbf{W}_M, \mathbf{W}_q, \mathbf{W}_k, \mathbf{W}_v$ are trainable weights.

3.6   Prediction

Finally, we can use the news representation obtained in Sect. 3.5 in Eq. 6 to feed a towered MultiLayer Perceptron (MLP) component.

$$\hat{y} = \sigma\left(f_N\left(\cdots f_2\left(f_1(\mathbf{f_x})\right)\cdots\right)\right) \tag{7}$$

where $f_l(\cdot)$ with $l = 1, 2, \cdots, N$ denotes the mapping function for the l-th hidden layer in MLP. $f_l(\mathbf{x}) = \sigma(\mathbf{W}_l\mathbf{x}+\mathbf{b}_l)$, where $\mathbf{W}_l$ and $\mathbf{b}_l$ are learnable weight matrix and bias vector for layer l. The activation function σ for each layer is sigmoid. We set the size of layers (i.e., the dimensionality of $\mathbf{x}$) as one-third of the previous layers. The output layer $f_{out}(\cdot)$ is similar to $f_l(\cdot)$ and its size is the number of selected items.

The Loss function is Mean Absolute Error(MAE), which is given by:

$$\mathrm{L} = \frac{1}{|\mathcal{D}|} \sum_{<\mathbf{x},y>\in\mathcal{D}} \left(|\hat{y} - y|\right) \tag{8}$$

4   Experiment

4.1   Dataset

We used three datasets for evaluation. The statistics of the datasets are shown in Table 1.

Disaster. We crawled $50,000$ Chinese news related to catastrophic events published by the famous news outlet "toutiao" on the Weibo platform during 2015 to 2020 . We labeled each story (i.e., y equals to the sum of numbers of likes, comments, and thumb-ups) at crawl time (January 2021) on Weibo for each story. Catastrophic events mainly include earthquakes, tsunamis, floods, strong winds, sandstorms, landslides, and typhoons occurred in China.

Table 1. Statistics of datasets

Data	#News	#Words	#Entity
Disaster datasets	50,000	100,000	15,000
Entertainment datasets	48,000	68,000	9,000
MIND datasets	50,000	60,000	12,000

Entertainment. We also crawled 48,000 news, published by "toutiao" on the Weibo platform during the year 2018 to 2020, to build a dataset on entertainment-related topics, including movies, shows, music, celebrities, gossip, and so on.

MIND[1] is a commonly used benchmark dataset in news recommendation. We extracted 50,000 news which have more interactions with users. Then, we label each story by the total number of users who browsed and clicked the news.

4.2 Experimental Setup

Data Pre-Processing. For two Chinese datasets, we used the Chinese named entity recognition tool LAC[2] to extract the factual entities. In the MIND dataset, we retained the content and entity information of each news. In text pre-processing, we removed emoji expressions, HTTP links, and mentions (@somebody) in the news content. For each news, we divided the news into a set of sentences, and we used the jieba Natural Language Processing tool[3] for segmentation. In Chinese datasets, we initialized word embedding vectors with 128-dimensional pre-trained embedding from AI Tencent[4]. In the MIND dataset, we initialized with 300-dimensional GloVe embedding [17]. We filtered stop words. Sentence segmentation was conducted based on punctuation marks ".". The maximum number of sentences in each news story is set to 5. To eliminate the common noise words, we further removed 10% of the vocabulary with low TF-IDF values over the whole dataset. We used random $80\% - 10\% - 10\%$ training/valid/test split. All the codes and data are publically available in GitHub[5].

Parameters. For BiLSTM in sentence encoding, we used 2 layers, with $128-$dimensional hidden states and $128-$dimensional output sentence encoding. For heterogeneous graph based news encoding, we used 8 attention heads, $50-$dimensional edge vector, $64-$dimensional hidden states in graph neural network, $512-$dimensional hidden states in the $2-$layer FFN to output a $50-$dimensional news encoding. For sequential entity encoding, we used

---

[1] https://msnews.github.io/.
[2] https://github.com/baidu/lac.
[3] https://github.com/fxsjy/jieba.
[4] https://ai.tencent.com/ailab/nlp/data/Tencent_AILab_ChineseEmbedding.tar.gz.
[5] https://github.com/XMUDM/NewsPopularityPrediction.

128−dimensional hidden states. For the news popularity prediction module, we set the number of MLP layers to 3. During training, we used a batch size of 256 and apply Adam optimizer [7] with an initial learning rate of 5e–3. We set the decay of learning rate until the lowest was 5e–5, and an early stop was performed when validation loss did not decrease for three continuous epochs.

Baselines. We compared F^4 with the following baselines. (1) **SVM** Support Vector Machine (SVM) was used in [2,21] on traditional bag of stemmed words (BOW) vectors. (2) **MLP+CNN** Similar as [3], we implemented two CNN layers, followed by a dense layer, on a concatenation of W2V vectors and TF-IDF vectors. (3) **LSTM** We learned news representation with LSTM, where for each news, we used a CNN component to extract local sentence encoding vectors to feed a 3-layer LSTM. (4) **BiLSTM** We used a PCNN on words and 3-layer BiLSTM on sentences to get the characteristics of the news. (5) **GCN** was adopted on the word-news graph.

4.3 Comparative Regression Study

To study the accuracy of predicted news popularity, we adopted regression evaluation metrics, including Mean Absolute Error (MAE) in Eq. 8, Root Mean Squared Error (RMSE), and Median Absolute Error(MedAE).

$$\text{RMSE} = \sqrt{\frac{1}{|\mathcal{D}|} \sum_{<\mathbf{x},y> \in \mathcal{D}} (\hat{y} - y)^2}, \quad \text{MedAE} = median\{|y - \hat{y}|\} \tag{9}$$

where $|\mathcal{D}|$ represents the number of news stories in the dataset, y is the true popularity, $\hat{y}$ is the predicted popularity, $median\{|y - \hat{y}|\}$ returns the median value in the set $\{|y - \hat{y}|\}$.

From Table 2, we have the following observations. (1) F^4 *consistently outperformed all baselines in terms of RMSE, MedAE, and MAE.* (2) F^4 *produced robust regression performance over three datasets.* (3) Besides F^4, GCN was the second-best method in terms of RMSE, while BiLSTM produced the lowest MedAE and MAE. This suggests that adopting a graph neural network or sequential model per se can not guarantee a robust performance.

Table 2. Comparative Regression results of different baselines

Model	Disaster dataset			Entertainment dataset			MIND dataset		
	RMSE	MedAE	MAE	RMSE	MedAE	MAE	RMSE	MedAE	MAE
SVM	0.0821	0.0246	0.0261	0.0184	0.0078	0.0103	0.1654	0.0524	0.1034
MLP+CNN	0.0704	0.0143	0.0235	0.0156	0.0047	0.0071	0.1281	0.0441	0.0725
LSTM	0.0682	0.0112	0.0167	0.0134	0.0018	0.0034	0.1238	0.0256	0.0595
BiLSTM	0.0669	0.0105	0.0159	0.0135	0.0016	0.0056	0.1338	0.0363	0.0699
GCN	0.0581	0.0121	0.0187	0.0131	0.0022	0.0035	0.1254	0.0125	0.0745
F^4	**0.0424**	**0.0083**	**0.0146**	**0.0087**	**0.0006**	**0.0025**	**0.1076**	**0.0021**	**0.0359**

4.4   Comparative Ranking Study

We were also concerned about whether the order of news popularity is predicted precisely, and we evaluated the ranking performance in terms of two evaluation metrics. For each entity, we derived a gold-standard list of the news based on the actual popularity. Then, we computed $HR@K, NDCG@K$ by comparing each method's ranking list of predicted popularity with the gold-standard list:

$$HR@K = \frac{\sum_{j=1}^{K} IK_j}{K}, \quad NDCG@K = \frac{1}{Z_K} \sum_{j=1}^{K} \frac{2^{rel_j} - 1}{\log_2(1 + j)}, \tag{10}$$

where K refers to the topK ranking result, IK_j returns 1 if the news story at position j is within the topK most popular stories in the gold-standard list. Z_K is a normalizer which ensures that perfect ranking has a value of 1, rel_j is the relevance score at position j. We set $rel_j = 1$ if the jth result is within the top 5 items in the gold-standard list, and $rel_j = 0$ otherwise.

In Fig. 4, We reported the average HR@5 and NDCG@5 results in three datasets. We have three remarks. (1)F^4 *consistently outperformed all competitors in terms of both HR@5 and NDCG@5.* F^4 did not only retrieve popularity news (i.e., high HR@5) but also distinguished most and more popular news(i.e., high NDCG@5). (2) F^4 *produced robust ranking performance over different datasets.* (3) The performance of GCN is worse than that of the sequential model in the Entertainment dataset and MIND dataset. The underlying reason is that graph-based encoding can not capture the long-term influence of news popularity appropriately. This observation validated our assumption of modeling the entity sequences for news popularity prediction.

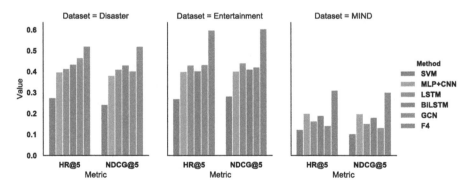

Fig. 4. Comparative ranking performance of different baselines

5   Ablation Study

In order to better understand the contributions of different modules, we conducted an ablation study. (1) F^4 **-entity** removes the sequential entity encoding module in Sect. 3.4; (2) F^4 -**graph** removes the heterogeneous graph based news encoding module in Sect. 3.3; (3) F^4 -**both** removes both the sequential entity encoding module and the heterogeneous graph based news encoding module.

As shown in Table 3, we have the following observations. (1) F^4-entity and F^4-graph were comparable. F^4-entity produced lower RMSE and MedAE, higher MAE in Disaster and Entertainment Datasets, and higher RMSE and lower MedAE and MAE in MIND dataset. This shows that the sequential entity encoding module and the graph-based news encoding module compensate each other, with the former capturing the long-term popularity factor while the latter capturing the short-term popularity factor. (2) F^4-both performed significantly worse than F^4-entity and F^4-graph in terms of most evaluation metrics. This observation verifies the importance of the sequential entity encoding module and the graph-based news encoding module. (3) Comparing F^4 with the rest baselines, F^4 produced the best results in terms of all evaluation metrics across different datasets.

Table 3. Regression performance of different modules

Model	Disaster dataset			Entertainment dataset			MIND dataset		
	RMSE	MedAE	MAE	RMSE	MedAE	MAE	RMSE	MedAE	MAE
F^4-entity	0.0593	0.0126	0.0139	0.0124	0.0013	0.0029	0.1155	0.0164	0.0564
F^4-graph	0.0512	0.0098	0.0146	0.0105	0.0009	0.0031	0.1228	0.0065	0.0481
F^4-both	0.0651	0.0103	0.0156	0.0152	0.0017	0.0048	0.1272	0.0325	0.0652
F^4	**0.0424**	**0.0083**	**0.0146**	**0.0087**	**0.0006**	**0.0025**	**0.1076**	**0.0021**	**0.0359**

The ranking results were shown in Fig. 5, We have the following remarks. (1) F^4-both performed the worst, while F^4 performed the best, in terms of all ranking evaluation metrics, across different datasets. (2) F^4-entity performed significantly worse than F^4-graph, which shows that capturing the long-term factor in entity sequence is especially important for accurate popularity ranking.

6   Performance of Attention in Merging Entities

Finally, we evaluated the impact of the attention mechanism. We implemented several different baselines to merge news embedding. (1) F^4-nomerge fed the prediction module with the news encoding output by a single entity sequence. First, we computed the TF-IDF weighting of each entity in each news story. Then, we used one entity with the maximal TF-IDF weight to represent each news story. Next, we implemented the sequential entity encoding module in Sect. 3.4 and used the output of the corresponding unit of the representative entity sequence. (2) F^4-avg used average pooling to merge the output of news embedding in several entity sequences. (3) F^4-max used max-pooling to merge

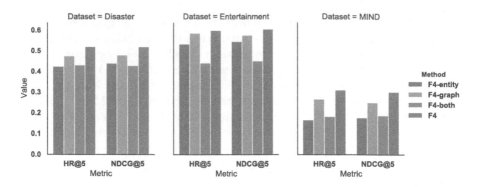

Fig. 5. Ranking performance of different modules

the news embedding. (4) F^4-entity used the pre-trained embedding of entities to calculate attention weights.

As shown in Table 4, F^4 consistently performed the best in terms of regression evaluation metrics. This indicates the effectiveness of the attention mechanism used in merging the news embedding from various relevant entities. In addition, F^4-entity generated the second-best RMSE results in the Disaster and MIND dataset. However, the performance was not stable. One possible reason is that F^4-entity calculated the attention weights based on static entity profiles and did not reflect the dynamic nature of entity sequence.

Table 4. Regression performance of different merging strategies

Model	Disaster dataset			Entertainment dataset			MIND dataset		
	RMSE	MedAE	MAE	RMSE	MedAE	MAE	RMSE	MedAE	MAE
F^4-nomerge	0.0451	0.0142	0.0149	0.0106	0.0009	0.0025	0.1134	0.0152	0.0521
F^4-avg	0.0438	0.0122	0.0149	0.0112	0.0012	0.0031	0.1326	0.0084	0.0674
F^4-max	0.0539	0.0144	0.0191	0.0091	0.0010	0.0028	0.1259	0.0249	0.0612
F^4-entity	0.0445	0.0095	0.0167	0.010	0.0011	0.0027	0.1103	0.0095	0.0498
F^4	**0.0424**	**0.0083**	**0.0146**	**0.0087**	**0.0006**	**0.0025**	**0.1076**	**0.0021**	**0.0359**

The ranking results are shown in Fig. 6. Again, F^4 consistently gained significant performance improvements in terms of both HR@5 and NDCG@5. Interestingly, the ranking performance of F^4-nomerge was better than naive merge, e.g., either F^4-avg or F^4-max. The underlying reason is that, though many entities are involved in a news event, they play different roles, and how to integrate the different entities in predicting news popularity is a non-trivial task. Many factors must be considered, including the relevance of the entity, the temporal impact of the entity, and correlations among entities.

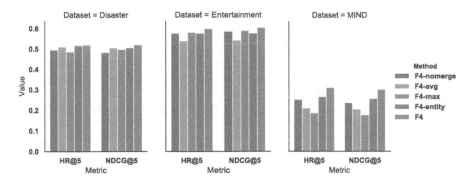

Fig. 6. Ranking performance of different merging strategies

7 Conclusion

In this paper, we study the problem of news popularity prediction. We discuss the local, global, short-term, and long-term factors that affect news popularity. We present F^4, which adopts a CNN based sentence encoding module to represent the local context of each news story; a heterogeneous graph-based module to capture the short-term information propagation from current buzz words to each news story; a sequential module to extract long-term popularity features in entity sequence; and finally, an attention module to learn global news-entity correlations. Experiments on real-world English and Chinese datasets have demonstrated the effectiveness of F^4. In the future, we plan to investigate the problem of forecasting public response to news events via multi-tasking, e.g., incorporating popularity prediction, sentiment classification, and so on.

Acknowledgements. Chen Lin is the corresponding author. Chen Lin is supported by the Natural Science Foundation of China (No. 61972328), Joint Innovation Research Program of Fujian Province China (No.2020R0130). Hui Li is supported by the Natural Science Foundation of China (No. 62002303), Natural Science Foundation of Fujian Province China (No. 2020J05001). Quan Zou is supported by Natural Science Foundation of China (No. 61922020).

References

1. Yang, Y., Liu, Y., Lu, X., Xu, J., Wang, F.: A named entity topic model for news popularity prediction. Knowl.-Based Syst. **208**, 106430 (2020)
2. Ambroselli, C., Risch, J., Krestel, R., Loos, A.: Prediction for the newsroom: Which articles will get the most comments? In: Proceedings of the 2018 Conference of the North American Chapter of the Association for Computational Linguistics: Human Language Technologies, New Orleans - Louisiana, vol. 3 (Industry Papers), pp. 193–199 (2018)
3. Davoudi, H., An, A., Edall, G.: Content-based dwell time engagement prediction model for news articles. In: Proceedings of the 2019 Conference of the North American Chapter of the Association for Computational Linguistics: Human Language Technologies, Minneapolis, Minnesota, vol. 2 (Industry Papers), pp. 226–233 (2019)

4. Gupta, R.K., Yang, Y.: Predicting and understanding news social popularity with emotional salience features. In: Proceedings of the 27th ACM International Conference on Multimedia, New York, NY, USA, pp. 139–147 (2019)
5. Hamid, A., et al.: Fake news detection in social media using graph neural networks and NLP techniques: a COVID-19 use-case. In: MediaEval. CEUR Workshop Proceedings, vol. 2882 (2020)
6. Keneshloo, Y., Wang, S., Han, E.S., Ramakrishnan, N.: Predicting the popularity of news articles. In: Venkatasubramanian, S.C., Jr., W.M. (eds.) Proceedings of the 2016 SIAM International Conference on Data Mining, Miami, Florida, USA, 5–7 May 2016, pp. 441–449 (2016)
7. Kingma, D.P., Ba, J.: Adam: a method for stochastic optimization. In: ICLR (Poster) (2015)
8. Lee, J.G., Moon, S.B., Salamatian, K.: An approach to model and predict the popularity of online contents with explanatory factors. In: Web Intelligence, pp. 623–630 (2010)
9. Lee, J.G., Moon, S.B., Salamatian, K.: Modeling and predicting the popularity of online contents with cox proportional hazard regression model. Neurocomputing 76(1), 134–145 (2012)
10. Lin, P., Mo, X., Lin, G., Ling, L., Wei, T., Luo, W.: A news-driven recurrent neural network for market volatility prediction. In: ACPR, pp. 776–781 (2017)
11. Liu, Q., Cheng, X., Su, S., Zhu, S.: Hierarchical complementary attention network for predicting stock price movements with news. In: Proceedings of the 27th ACM International Conference on Information and Knowledge Management, CIKM '18, pp. 1603–1606. Association for Computing Machinery, New York (2018)
12. Mazloom, M., Rietveld, R., Rudinac, S., Worring, M., van Dolen, W.: Multimodal popularity prediction of brand-related social media posts. In: Proceedings of the 24th ACM International Conference on Multimedia, MM '16, New York, NY, USA, pp. 197–201 (2016)
13. Mei, H., Eisner, J.: The neural hawkes process: a neurally self-modulating multivariate point process. In: NIPS, pp. 6754–6764 (2017)
14. Mukherjee, S.K., Bandyopadhyay, S.: Clustering to determine predictive model for news reports analysis and econometric modeling. In: ReTIS, pp. 302–309 (2015)
15. Naseri, M., Zamani, H.: Analyzing and predicting news popularity in an instant messaging service. In: Proceedings of the 42nd International ACM SIGIR Conference on Research and Development in Information Retrieval, SIGIR'19, New York, NY, USA, pp. 1053–1056 (2019)
16. Okano, E.Y., Liu, Z., Ji, D., Ruiz, E.E.S.: Fake news detection on fake.br using hierarchical attention networks. In: Quaresma, P., Vieira, R., Aluísio, S., Moniz, H., Batista, F., Gonçalves, T. (eds.) PROPOR 2020. LNCS (LNAI), vol. 12037, pp. 143–152. Springer, Cham (2020). https://doi.org/10.1007/978-3-030-41505-1_14
17. Pennington, J., Socher, R., Manning, C.D.: Glove: global vectors for word representation. In: EMNLP, pp. 1532–1543. ACL (2014)
18. Sadiq, S., Wagner, N., Shyu, M., Feaster, D.: High dimensional latent space variational autoencoders for fake news detection. In: MIPR, pp. 437–442 (2019)
19. Shang, Y., Wang, Y.: Study of CNN-based news-driven stock price movement prediction in the a-share market. In: Qin, P., Wang, H., Sun, G., Lu, Z. (eds.) ICPCSEE 2020. CCIS, vol. 1258, pp. 467–474. Springer, Singapore (2020). https://doi.org/10.1007/978-981-15-7984-4_35
20. Stoddard, G.: Popularity dynamics and intrinsic quality in reddit and hacker news. In: ICWSM, pp. 416–425 (2015)

21. Tsagkias, M., Weerkamp, W., de Rijke, M.: Predicting the volume of comments on online news stories. In: Proceedings of the 18th ACM Conference on Information and Knowledge Management, CIKM '09, New York, NY, USA, pp. 1765–1768 (2009)
22. Tsagkias, M., Weerkamp, W., de Rijke, M.: Predicting the volume of comments on online news stories. In: CIKM, pp. 1765–1768 (2009)
23. Vaswani, A., et al.: Attention is all you need. In: NIPS, pp. 5998–6008 (2017)
24. Velickovic, P., Cucurull, G., Casanova, A., Romero, A., Liò, P., Bengio, Y.: Graph attention networks (2017). CoRR arXiv:1710.10903
25. Wang, D., Liu, P., Zheng, Y., Qiu, X., Huang, X.: Heterogeneous graph neural networks for extractive document summarization. In: ACL, pp. 6209–6219 (2020)
26. Wu, C., Wu, F., An, M., Huang, J., Huang, Y., Xie, X.: NPA: neural news recommendation with personalized attention. In: Proceedings of the 25th ACM SIGKDD International Conference on Knowledge Discovery and Data Mining, KDD '19, pp. 2576–2584. Association for Computing Machinery, New York (2019)
27. Wu, C., Wu, F., An, M., Huang, Y., Xie, X.: Neural news recommendation with topic-aware news representation. In: Proceedings of the 57th Annual Meeting of the Association for Computational Linguistics, Florence, Italy, pp. 1154–1159 (2019)
28. Zaman, T., Fox, E.B., Bradlow, E.T.: A bayesian approach for predicting the popularity of tweets (2013). CoRR arXiv:1304.6777
29. Zhao, X., Wang, C., Yang, Z., Zhang, Y., Yuan, X.: Online news emotion prediction with bidirectional LSTM. In: Cui, B., Zhang, N., Xu, J., Lian, X., Liu, D. (eds.) WAIM 2016. LNCS, vol. 9659, pp. 238–250. Springer, Cham (2016). https://doi.org/10.1007/978-3-319-39958-4_19

An Efficient Method for Indoor Layout Estimation with FPN

Aopeng Wang[1], Shiting Wen[2(✉)], Yunjun Gao[2,3], Qing Li[4], Ke Deng[5],
and Chaoyi Pang[2]

[1] Polytechnic Institute, Zhejiang University, Zhejiang, China
[2] School of Computer and Data Engineering, NingboTech University, Ningbo, China
wensht@nit.zju.edu.cn
[3] School of Computer Science and Technology, Zhejiang University, Hangzhou, China
[4] Department of Computing, The Hong Kong Polytechnic University,
Hong Kong, China
[5] School of Science, RMIT University, Melbourne, Australia

Abstract. As a fundamental part of indoor scene understanding, the research of indoor room layout estimation has attracted much attention recently. The task is to predict the structure of a room from a single image. In this article, we illustrate that this task can be well solved even without sophisticated post-processing program, by adopting Feature Pyramid Networks (FPN) to solve this problem with adaptive changes. Besides, an optimization step is devised to keep the order of key points unchanged, which is an essential part for improving the model's performance but has been ignored from the beginning. Our method has demonstrated great performance on the benchmark LSUN dataset on both processing efficiency and accuracy. Compared with the state-of-the-art end-to-end method, our method is two times faster at processing speed (32 ms) than its speed (86 ms), with 0.71% lower key point error and 0.2% higher pixel error respectively. Besides, the advanced two-step method is only 0.02% better than our result on key point error. Both the high efficiency and accuracy make our method a good choice for some real-time room layout estimation tasks.

Keywords: Layout estimation · Scene understanding · Feature pyramid network

1 Introduction

Indoor room layout estimation is a fundamental task for many applications, including in indoor navigation, augmented reality, and scene reconstruction. This study focuses on the indoor layout estimation to predict the structure of a room through a single RGB image.

Figure 1 shows the definition of different room layout types and an example in LSUN dataset. From this figure we can see that a room has three parts: ceiling, wall and floor. The wall can be further divided into the left wall, middle wall and right wall in general, from different perspectives. These parts are labeled with

© Springer Nature Switzerland AG 2021
W. Zhang et al. (Eds.): WISE 2021, LNCS 13081, pp. 94–106, 2021.
https://doi.org/10.1007/978-3-030-91560-5_7

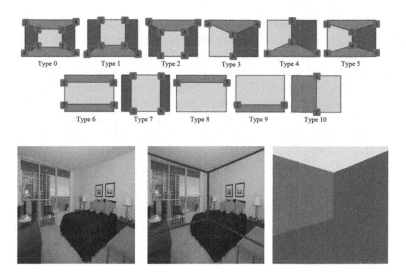

Fig. 1. Definition and an example of room layout in LSUN dataset.

different semantic for semantic segmentation. The junction points of different parts are considered as the key points in connections and each point is labeled with a number, representing their order for a given room type. The example at the bottom of Fig. 1 illustrates the segmentation map and room layout for an input image, which is the expected output of this task.

Intuitively, as long as these key points are correctly detected and connected in a proper order, the correct room layout can be obtained. Therefore, the major task in room layout estimation is to correctly detect the key points in the right order. However, it is challenging to accomplish this task through a single RGB image, because rooms in a photo can be cluttered and shadowy in different ways.

In this paper, we see the key points as small objects and adopt the Feature Pyramid Networks (FPN) [7,11], which is one of the most effective approaches to deal with both the small objects detection and segmentation task, to predict the room layout in end-to-end style. With proper modification, we demonstrate that FPN could solve this task well even without the commonly used post-processing step in prior works and we record the modified FPN as M-FPN. Besides, by keeping the order between key points stable during the data augmentation stage, the proposed model can get a better performance. Moreover, this step is a low cost operation and can be applied in other similar order related work, but have not been discussed in previous works.

The contribution of this paper can be summarized as follows:

1. M-FPN is the first approach that uses FPN to solve room layout estimation in end-to-end style. It shortens the processing time and can be used for real time processing tasks.
2. We propose a new technique called "order-preserving" to enhance model's performance. This step makes each point consistent with its activation area

and can basically eliminate the wrong matching problem by preserving the order between key points in the process of data augmentation. We are the first to devise and deploy them to indoor layout estimation.

3. The proposed model demonstrates great performance on the challenging benchmark LSUN dataset [17]. Our method is comparable to the advanced end-to-end methods on the pixel error metric, while it outperforms the end-to-end methods on the key point error metric. In particular, it achieves a similar level of key point error as that of the state-of-the-art two-step methods. To the best of our knowledge, it is the fastest method on indoor layout estimation and reaches 31fps at processing speed.

The rest of this article is organized as follows. Section 2 reviews the related work of indoor layout estimation. Section 3 describes our proposed method and framework. Section 4 reports the experimental results. Section 5 concludes this article.

2   Related Work

The methods of indoor layout estimation can be divided into two styles according to whether an independent post-processing program is used.

Two-Step Methods: The two-step methods usually follow the proposal-ranking scheme, in which the whole process, from feature extraction to proposal ranking, is processed by a human-designed program [3,4,10,15,21]. As the rise of deep learning, some two-step methods use neural networks for feature extraction with great improvements in accuracy [2,8,13,16,22]. In general, these methods use an additional post-processing step to elevate the results' precision but require a much longer processing time and more manually participating than that of the end-to-end methods.

End-to-End Methods: The end-to-end methods [5,9,25] have faster processing speed, but these methods usually lack of a well-designed post-processing mechanism resulting in less sufficient performance on accuracy than two-step methods. RoomNet [9] is the first end-to-end method to solve the layout estimation task. It follows the SegNet [1] structure, which is featured with encoder-decoder architecture. RoomNet adds an extra classifier composed of three fully connected layers at the bottleneck of SegNet to predict the room type. Since there are 48 points in total in the 11 types of rooms in the LSUN dataset, this network outputs a feature map of 48 channels with each channel corresponding to a point. To avoid using an extra classifier, Roth, et al. [5] proposed a 'Smart Hypothesis Generation' (SHG) method to divide the training data into three groups according to the number of walls in the input images, where each group is used to train a network respectively. Each network has two parts, the first of which outputs semantic segmentation maps while the second outputs key point heatmaps from the segmentation maps. After three sets of semantic segmentation and key point heatmaps are obtained, a scoring function is used to calculate the consistency between the segmentation maps and the key point heatmaps. The algorithm then chooses the highest score as the final result. This method gets rid of an extra classifier by using 3 groups of networks and a scoring function to select the best matches. Although this method avoids an explicit room type classifier, the disadvantage is that it exploits 3 groups of networks, which makes the training process more complicated and takes more time in inference.

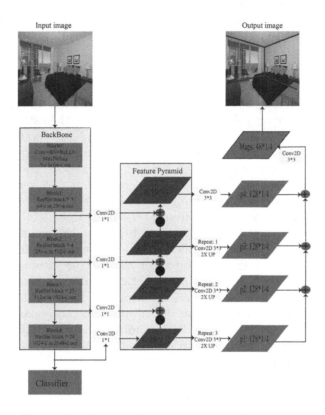

Fig. 2. Pipeline and the proposed network structure.

In this article, we use the same output setting of RoomNet but we increase the accuracy of the classifier by using a stronger feature extraction backbone. Our proposed method is also different from SHG [5] by employing a single network which can be more efficient than SHG method. The output size of our method is only 1/4 of the input size, which is very similar to that of model compression. To enhance the accuracy of M-FPN, we also make each point consistent to its activation area. From this, our method achieves the state-of-the-art performances for both the processing speed and accuracy on the tested data sets.

3   M-FPN Method

In this section, we first introduce the proposed end-to-end method M-FPN in Sect. 3.1. We then propose the "order-preserving" strategy in Sect. 3.2 to optimize the proposed network for better outputs in accuracy.

3.1   M-FPN Network

Although SegNet has been used in both RoomNet and SHG method, we see another choice in basic network architecture: Feature Pyramid Network. As shown in Fig. 2, M-FPN is a single-network and basically modified from the network architecture of [7] and has the following characteristics:

Characteristics. Compared to the encoder-decoder structure of SegNet, FPN sets most parameters and performs most computations on the encoder (backbone) part to get stronger features, which ensures that the extra classifier will have higher accuracy and will not bottleneck the model's overall performance. Meanwhile, it merges the multi-scale features with just a few operations in the decoder part. The proposed network can be seen as an asymmetric, lightweighted encoder-decoder, which is very efficient in computing and very appropriate for saving GPU memory. By adopting this architecture, our proposed model improves the classification accuracy from 81.5% in [9] to 83.5%.

Besides, considering that the key points belong to the small details in the picture, we first upsample the four different scale features in the feature pyramid to the same scale and then integrate them together as shown in Fig. 2. The integrated feature retains both strong semantic information and richer details. The model's performance is then effectively improved. By comparison, RoomNet only uses the pooling index recorded at the subsampling process to produce dense feature maps of a larger size during the upsampling process. This network will miss the information in the former extracted features, a common problem caused by the symmetrical encoder-decoder network.

Our proposed network works as follows:

- It takes a 320*320 image as input and outputs a feature pyramid with the feature size ranging from 1/32 to 1/4 of the original size.
- We add an classifier, which includes an average pooling layer, a dropout layer [18] and a fully connected layer, after the Block4 to predict the room type. The dropout layer can prevent the model from being overfitting. A single fully connected layer can decrease the number of parameters and provides relatively high accuracy.
- As shown in Fig. 2, f1~f4 in the feature pyramid are upsampled different times to get the p1~p4 at 1/4 of the original input size, and then we add them in pixel-wise and perform a 3*3 convolution to get the final result.

Here, we choose 1/4 input resolution (80*80) as the final output size to save memory and reduce the network parameters and as demonstrated from many experimental tests of previous methods, it is sufficient enough to use 80*80 resolution feature maps for the layout estimation task.

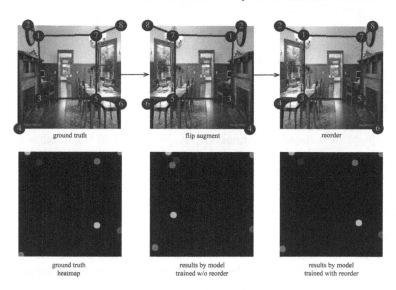

Fig. 3. Top row: process from ground truth to reordered result. Bottom row: impact of the reordering process on the prediction result through a control experiment.

Table 1. Notations

Symbol	Meaning
G	Gaussian function
(I, y, t)	Training sample of input image I: y is the ground truth coordinates of k key points; t is the room type
ψ	Room type classifier (output from the classifier in Fig. 2)
φ	Key point heatmap regressor (final 48-channel maps in Fig. 2)

Loss Function. The ground truth of each key point is a 2D Gaussian centred heatmap at 80*80 size with a standard deviation of 10 pixels. In this article, we follow the notation used in [9] as M-FPN shares the same output settings (Table 1). Different from the paper of [9], we use the focal loss [12] as the cost function for the room type prediction to alleviate the data imbalance problem in the LSUN dataset. The Euclidean loss is used as the cost function for key point heatmaps regression. The total loss function can be written as follows:

$$L(I, y, t) = \sum_{k} I_{k,t} \|G_k(y) - \varphi_k(I)\|^2 - \lambda \sum_{c} II_{c,t} FL(\psi_c(I)) \tag{1}$$

The regressor φ produces an ordered set of 48-channel heatmaps that are ordered according to the room type order. Figure 1 shows the point order defined in the LSUN dataset. For example, the first eight channels correspond to the eight ordered points in room type 0 and the next six channels correspond to the six ordered points in room type 1 and so on. In Eq. (1), $I_{k,t}$ is the Euclidean loss between the predicted heatmaps and the ground truth heatmaps if and only if key point k appears in ground truth room type t. Similarly, $\Pi_{c,t}$ is the focal loss if and only if room type index c is equal to the ground truth room type t. λ is used to balance the two loss items and is set to 1 in our model. Based on many conducted experimental tests, we set the hyper-parameter of Focal Loss $\alpha = 1$, $\gamma = 2$ respectively.

At the test stage, we use the predicted room type to select the corresponding set of key point heatmaps in the final output. We then find the maximum value's coordinates for each heatmap and scale the obtained coordinates back to their original resolution to get the final coordinates.

3.2   Preserving Points Order

Horizontal flipping is the commonly used data augmentation technique in this task. However, after this procedure the order between key points will be messed and no prior work has discussed the result of this. Here, after the augmentation, we preserve the order of flipped points consistent to the order of original points (Fig. 3). By doing this, each point in the flipped image stays at the relatively identical area to its original counterpart and model trained by these rectified images performs better. The reason can be explained as follows:

1. There are 11 different room types defined in the LSUN dataset. As depicted in Fig. 1(top), each room type has a specific points ordering and structure. Since each point belongs to a relatively fixed location in the image, a coarser position of each point can be learned through the network.
2. Preserving the order of flipped points being consistent with that of the original points can greatly reduce the training complexity of the network, enables the network to converge faster and achieves much better outputs.
3. The order-preserving process reduces the semantic ambiguity on points. For example, the activation area of points at the symmetrical position can overlap a lot if we do not execute the order-preserving step. In this case, the network will learn two symmetrical activation areas in the image making false prediction rate doubled for each point.

We demonstrates the effectiveness of the order-preserving step in Fig. 3. The general practice of horizontal flipping is to simply flip the input image (including the points in it) as the way shown in the top middle of Fig. 3. This can disrupt the order among the points and enlarge the activation area of each point, resulting that the network model is more likely to produce bad predictions. For example, if we train our model as the conventional practices, our model will learn two symmetrical activation areas for point 1, the upper left and the upper right part of the ground truth image in Fig. 3, which are overlapped with the activation areas for point 7. This greatly increases the possibility of making wrong predictions. As can be seen in Fig. 3, the area of point 7 exists a bright light in the ground truth so that when we predict it using the model trained without the order-preserving step, we get the result (bottom middle image in Fig. 3) where point 7 is near point 1, both at the upper left position of the image, and a similar situation occurs between point 3 and point 5. In contrast, when we use the model trained with the order-preserving step, we get a good result (bottom right image in Fig. 3). The reason is that during the training process we make the network learn a relatively individual activation area for each point so that the model can give the accurate predictions.

Overall, the order-preserving step is an essential step for improving the model's final accuracy and has not been used in the previous work.

4   Experiments

4.1   Dataset

The current mainstream dataset includes the Hedau [3] dataset and LSUN (Large-scale Scene Understanding Challenge room layout) [23] dataset. The Hedau dataset, released in 2009, was sampled from the internet and LabelMe [17]. It contains 209 training pictures, 53 validation pictures and 105 test pictures. This dataset is relatively small and usually only used to compare pixel error. However, we could not find public resources from the internet. Therefore, we test our model on the LSUN dataset in this article.

The LSUN dataset, sampled from the SUN [19] dataset, includes 4000 training pictures, 394 validation pictures and 1000 test pictures. Since the challenge has been closed, we cannot get the ground truth of the test set, so we use the validation set to evaluate our model, like many other methods [5,9,16,25]. In all the experimental tests, we scale the model's outputs to the original resolution and evaluate the result by the LSUN toolkit as many other methods did.

4.2   Implementation Details

Our experimental environment is a single 2080Ti GPU, CPU i9-9900K, and the operating system is ubuntu16.04. Training our model from scratch takes about 20 h on our machine. All input RGB images are resized to 320*320 and the outputs are the 48-channel key point heatmaps of resolution 80*80 and the labeled room types. The Adam optimizer [6] is employed to train the network for 150 epochs with batch size 8, weight decay 0.0005 and learning rate 0.00001. In addition, we use the ReduceLROnPlateau function to monitor the loss value in order to dynamically adjust the learning rate. The attenuation coefficient is set to 0.5, the patience value is 4, and the minimum learning rate is set to 1e-8. The backbone ResNeXt-101 [20] is set as the default with a group number of 32, group width of 8, and we initialize the backbone with weights pre-trained on ImageNet. The classifier consists of an average pooling layer, a dropout [18] layer and a fully connected layer, with a dropout rate of 0.5. In terms of data augmentation, apart from the horizontal flipping, we also use the ColorJitter function to produce small variations in image brightness, saturation, contrast and hue. We weaken the gradients of the background pixels in the key point heatmaps by multiplying the l2 loss at the background with a factor of 0.2 as did in [9] as the output of the network tends to converge to zero due to the imbalance between the foreground and background pixels. Our model is implemented in the open source deep learning framework PyTorch [14], version 1.1.0.

4.3   Testing Results

The metrics for the room layout estimation are pixel and key point errors (PE and KE) as defined in the LSUN toolkit. PE calculates the pixel-wise accuracy of the segmentation maps between the ground truth and the predicted segmentation. KE calculates the Euclidean distance between the predicted key points and ground truth points, normalized by the image diagonal length. Both PE and KE are averaged over all tested images.

Table 2. Comparison of quantitative results and inference time

Method	Year	Post process	PE (%)/KE (%)	Time
Hedau, et al. [3]	2009	Yes	24.23/15.48	–
Mallya, et al. [13]	2015	Yes	16.71/11.02	–
Dasgupta, et al. [2]	2016	Yes	10.63/8.20	–
Ren, et al. [16]	2016	Yes	9.31/7.95	–
Zhao, et al. [24]	2017	Yes	~~5.29/3.84~~	–
RoomNet [9]	2017	**No**	9.86/6.30	166 ms
LayoutNet [25]	2018	**No**	11.96/7.63	39 ms
Edge semantic [22]	2019	Yes	6.94/5.16	–
Double refinement [8]	2019	Yes	**6.72/5.11**	–
Smart hypothesis [5]	2020	**No**	**7.79**/5.84	86 ms
Our method	2020	**No**	**7.99**/5.13	32 ms

In Table 2, we show the comparison of inference time and quantitative results on the LSUN dataset. Compared to other end-to-end methods, our method clearly outperforms all other methods on both error metrics except the 'Smart Hypothesis', which advantages our method slightly 0.2% in PE. In contrast, our method outperforms it by 0.71% in KE. In fact, our method has reached the level of state-of-the-art two-step method of Double Refinement [8] in KE.

It should be noted that [24] is a transfer learning method which requires a large amount of training data far beyond the LSUN dataset for pre-training and then fine-tunes the model on the LSUN dataset. Therefore, it is incomparable with other methods in our designed environment.

In terms of inference time among the end-to-end algorithms, our model takes only 32 ms, the fastest processing speed among the methods in a single NVIDIA Titan X GPU. It is about 31 fps and suitable for real-time application. The time costs of two-step methods are not listed in Table 2 as these methods usually require from several seconds to several minutes for predictions, which are too high to compare with our method.

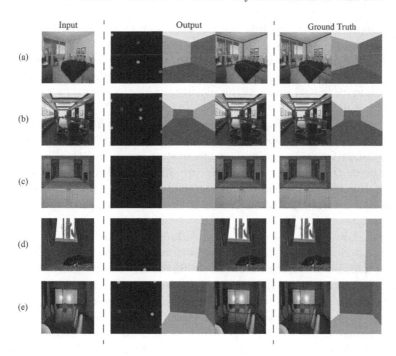

Fig. 4. Some qualitative results. The first column is the input image, the middle 3 columns are the key point heatmaps, semantic segmentation map, and room layout results in sequence produced by the network, and the last two columns are ground truth.

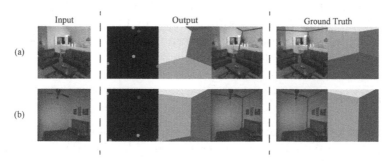

Fig. 5. Some failure cases. In case (a), the oblique line structure on the upper left of the house is mistaken by the network as the room layout. In case (b), the network uses its "imagination" to predict two key points from the occlusion, which turns case (b) from type 3 to type 5.

Figure 4 shows the qualitative results of our model on the LSUN validation set. It is not hard to see that in a daily indoor environment, our network can effectively estimate the layout for different types of rooms. It can be seen from samples (a), (b) and (e) that our network has strong robustness to occlusion for any input room types. In sample (a), the key points are covered by the bed. In samples (b) and (e), the key points are obscured by chairs. Both sample (c) and sample (d) have only two key points that can be identified easily from a human perspective, but the hard thing is how to select the specific tiny key point area from the many potential areas in the picture. From the results, we can see that our model performs well.

Here we want to specifically mention sample (e). This sample belongs to room type 1, but two points are missed in its ground truth label, which can be distinguished from the semantic segmentation map, and the line at the top slightly exceeds the boundary of the picture and beyond the field of view. Even though, in this situation, our model can still accurately predict the layout of the room, demonstrating that our model is very capable of learning the patterns from the dataset.

Figure 5 shows some representative failure cases. In case (a), there is a misleading oblique line in the house structure, which does not fit most cube-structured houses and makes the network mistakenly regard this oblique line as the final prediction result. Especially when there are multiple such structural lines in the house, it will be difficult for the model to make correct judgments. We call the representative situation in case (b) the network using "imagination". Since some areas in the picture, such as corners, have a higher probability of being key points, when these areas are blocked and the classifier predicts a wrong room type, the network intends to use "imagination" to predict possible key points in these areas matching the room type that accounts for a higher proportion in the dataset.

We also conduct a control experiment to illustrate the effectiveness of order-preserving step. The result shows that without this step, KE increases from 5.13% to 6.72%, and PE increases from 7.99% to 10.98%. The reason is that the activation area can overlap a lot between the symmetrical key points if the order-preserving step is not performed, as indicated in Fig. 3.

5   Conclusion

In this paper, we have proposed M-FPN as an end-to-end model for indoor layout estimation. With a processing speed up to 31 fps, M-FPN has reached the level of the state-of-the-art two-step method on the KE metric with the fastest speed. Compared with the most advanced end-to-end method, there is a slight 0.2% gap in the PE metric. This is due to our network's superior capability of integrating multi-scale features and the order-preserving strategy. The order-preserving strategy keeps the consistency of each key point's activation area. The performance on the LSUN dataset has demonstrated the superiority of our model. In the future, we intend to use a single network to predict key point maps and semantic segmentation maps at the same time, so as to further achieve better results.

Acknowledgments. The authors would like to thank the data providers of [23] for the testing data sets. This work was partially supported by the Natural Science Foundation of China (No. 61802344), the Ningbo Science and Technology Special Project(No. 2021Z019), the Hebei "One Hundred Plan" Project (No. E2012100006) and National Talent Program (No. G20200218015).

References

1. Badrinarayanan, V., Kendall, A., Cipolla, R.: Segnet: a deep convolutional encoder-decoder architecture for image segmentation. IEEE Trans. Pattern Anal. Mach. Intell. **39**(12), 2481–2495 (2017)

2. Dasgupta, S., Fang, K., Chen, K., Savarese, S.: Delay: robust spatial layout estimation for cluttered indoor scenes. In: 2016 IEEE Conference on Computer Vision and Pattern Recognition, CVPR 2016, pp. 616–624 (2016)

3. Hedau, V., Hoiem, D., Forsyth, D.A.: Recovering the spatial layout of cluttered rooms. In: IEEE 12th International Conference on Computer Vision, ICCV 2009, pp. 1849–1856 (2009)

4. Hedau, V., Hoiem, D., Forsyth, D.A.: Recovering free space of indoor scenes from a single image. In: 2012 IEEE Conference on Computer Vision and Pattern Recognition, pp. 2807–2814 (2012)

5. Hirzer, M., Roth, P.M., Lepetit, V.: Smart hypothesis generation for efficient and robust room layout estimation. In: IEEE Winter Conference on Applications of Computer Vision, WACV 2020, pp. 2901–2909 (2020)

6. Kingma, D.P., Ba, J.: Adam: a method for stochastic optimization. In: Bengio, Y., LeCun, Y. (eds.) Proceedings of 3rd International Conference on Learning Representations, ICLR 2015 (2015)

7. Kirillov, A., Girshick, R.B., He, K., Dollár, P.: Panoptic feature pyramid networks. In: IEEE Conference on Computer Vision and Pattern Recognition, CVPR 2019, pp. 6399–6408 (2019)

8. Kruzhilov, I., Romanov, M., Babichev, D., Konushin, A.: Double refinement network for room layout estimation. In: Palaiahnakote, S., Sanniti di Baja, G., Wang, L., Yan, W.Q. (eds.) ACPR 2019. LNCS, vol. 12046, pp. 557–568. Springer, Cham (2020). https://doi.org/10.1007/978-3-030-41404-7_39

9. Lee, C., Badrinarayanan, V., Malisiewicz, T., Rabinovich, A.: Roomnet: end-to-end room layout estimation. In: IEEE International Conference on Computer Vision, ICCV 2017, pp. 4875–4884 (2017)

10. Lee, D.C., Gupta, A., Hebert, M., Kanade, T.: Estimating spatial layout of rooms using volumetric reasoning about objects and surfaces. In: Proceedings of the 23rd International Conference on Neural Information Processing Systems, vol. 1, pp. 1288–1296 (2010)

11. Lin, T., Dollár, P., Girshick, R.B., He, K., Hariharan, B., Belongie, S.J.: Feature pyramid networks for object detection. In: 2017 IEEE Conference on Computer Vision and Pattern Recognition, CVPR 2017, pp. 936–944 (2017)

12. Lin, T., Goyal, P., Girshick, R.B., He, K., Dollár, P.: Focal loss for dense object detection. In: IEEE International Conference on Computer Vision, ICCV 2017, pp. 2999–3007 (2017)

13. Mallya, A., Lazebnik, S.: Learning informative edge maps for indoor scene layout prediction. In: 2015 IEEE International Conference on Computer Vision, ICCV 2015, pp. 936–944 (2015)

14. Paszke, A., et al.: Pytorch: an imperative style, high-performance deep learning library. In: Advances in Neural Information Processing Systems 32: Annual Conference on Neural Information Processing Systems 2019, NeurIPS 2019, pp. 8024–8035 (2019)
15. Ramalingam, S., Pillai, J.K., Jain, A., Taguchi, Y.: Manhattan junction catalogue for spatial reasoning of indoor scenes. In: 2013 IEEE Conference on Computer Vision and Pattern Recognition, pp. 3065–3072 (2013)
16. Ren, Y., Li, S., Chen, C., Kuo, C.-C.J.: A coarse-to-fine indoor layout estimation (CFILE) method. In: Lai, S.-H., Lepetit, V., Nishino, K., Sato, Y. (eds.) ACCV 2016. LNCS, vol. 10115, pp. 36–51. Springer, Cham (2017). https://doi.org/10.1007/978-3-319-54193-8_3
17. Russell, B.C., Torralba, A., Murphy, K.P., Freeman, W.T.: Labelme: a database and web-based tool for image annotation. Int. J. Comput. Vis. **77**(1–3), 157–173 (2008)
18. Srivastava, N., Hinton, G.E., Krizhevsky, A., Sutskever, I., Salakhutdinov, R.: Dropout: a simple way to prevent neural networks from overfitting. J. Mach. Learn. Res. **15**(1), 1929–1958 (2014)
19. Xiao, J., Hays, J., Ehinger, K.A., Oliva, A., Torralba, A.: SUN database: large-scale scene recognition from abbey to zoo. In: The Twenty-Third IEEE Conference on Computer Vision and Pattern Recognition, CVPR 2010, pp. 3485–3492 (2010)
20. Xie, S., Girshick, R.B., Dollár, P., Tu, Z., He, K.: Aggregated residual transformations for deep neural networks. In: 2017 IEEE Conference on Computer Vision and Pattern Recognition, CVPR 2017, pp. 5987–5995 (2017)
21. Zhang, J., Kan, C., Schwing, A.G., Urtasun, R.: Estimating the 3d layout of indoor scenes and its clutter from depth sensors. In: IEEE International Conference on Computer Vision, ICCV 2013, pp. 1273–1280 (2013)
22. Zhang, W., Zhang, W., Gu, J.: Edge-semantic learning strategy for layout estimation in indoor environment. CoRR abs/1901.00621 (2019). arXiv:1901.00621
23. Zhang, Y., Yu, F., Song, S., Xu, P., Seff, A., Xiao, J.: Large-scale scene understanding challenge: room layout estimation. In: CVPR Workshop (2015)
24. Zhao, H., Lu, M., Yao, A., Guo, Y., Chen, Y., Zhang, L.: Physics inspired optimization on semantic transfer features: an alternative method for room layout estimation. In: 2017 IEEE Conference on Computer Vision and Pattern Recognition, CVPR 2017, pp. 870–878 (2017)
25. Zou, C., Colburn, A., Shan, Q., Hoiem, D.: Layoutnet: reconstructing the 3d room layout from a single RGB image. In: 2018 IEEE Conference on Computer Vision and Pattern Recognition, CVPR 2018, pp. 2051–2059 (2018)

Lightweight Network Traffic Classification Model Based on Knowledge Distillation

Yanhui Wu and Meng Zhang[✉]

College of Computer Science and Technology, Jilin University,
Changchun 130012, China
zhangmeng@jlu.edu.cn

Abstract. Deep learning have been extensively applied to network traffic classification. This technology reduce much manual design and archive high accuracy in complex and highly variable networks. The existing deep learning approaches usually require abundant space and computing resources to improve the accuracy. However, for on-line network traffic classifications, increments in latency and instability incurred by the high costs of the model make it unsuitable for the case. A promising tool to deal with the challenge is knowledge distillation, which produces space-and-time efficient models from high-accurate large models. In this paper, we propose a light-weight encrypted traffic classification model based on knowledge distillation. We adopt the LSTM structure and apply knowledge distillation to it. The distilling loss is focal loss, which can effectively solve the imbalance in the number of samples and different degrees of difficulty in classification. To enhance the learning efficiency, we design an adaptive temperature function to soften the labels at each training stage. Experiments show that compared with the teacher model, the recognition speed of the student model is increased by 72%, the accuracy of the student model decreased by only 0.45% to 99.52%. Our model achieves both high accuracy and low latency for on-line encrypted traffic classification compared with the state of the art.

Keywords: Network traffic classification · Knowledge distillation · Deep learning · Long short-term memory

1 Introduction

Traffic classification classifies the network traffic into different categories according to policies. As the encrypted traffic is prevalent, the classification of encrypted traffic becomes crucial. In the past few decades, much work focuses on encrypted traffic classification. There are three classic methods: port-based method, payload-based method, and traffic-statistics-based method [2,3]. The accuracy of the port-based approach has been declining in recent years since new applications either use well-known port numbers to disguise their traffic or avoid using standard registered port numbers [4]. The payload-base method is

© Springer Nature Switzerland AG 2021
W. Zhang et al. (Eds.): WISE 2021, LNCS 13081, pp. 107–121, 2021.
https://doi.org/10.1007/978-3-030-91560-5_8

also called the deep packet inspection, which searches for patterns or keywords in the payload of packets. As the encrypted payload does not have stable features to extract [5], payload-base classification is difficult. The statistics-based method classifies encrypted traffic by manually designed features and classical machine learning algorithms, such as random forest or KNN. This method highly depends on the manual-designed features and cannot adapt to the complex and ever-changing network environment [6].

Inspired by the success of deep learning in CV and NLP, researchers applied CNN, LSTM, etc., to traffic classification. These approaches have shown a high accuracy. Compared with the traditional methods, these methods can adapt to the complex and ever-changing network environment without manual-designed features [7]. For highly redundant data sets, the existing deep learning approaches usually require great storage and computing resources [8] for high accuracy. This problem poses a challenge to the online deployment of the model for high-speed network traffic. When deploying the classification model online, we have strict restrictions on latency and computing resources to ensure the quality of service [9]. Therefore, these high-accuracy models are more suitable for the offline scenario other than online.

In recent years, the booming research field of knowledge distillation provides a powerful tool to improve the space and speed of the model. Knowledge distillation has two stages. First, training a complex and large teacher model that has high accuracy. Then, using the logits generated by the teacher model to train a student model, which has fewer number of parameters and a simpler structure. The student model can learn quite an excellent generalization ability directly [8].

In this paper, we propose a lightweight network traffic classification model based on knowledge distillation. We choose LSTM as our underlying model, since the LSTM networks learn the spatial-temporal characteristics of network traffic [5] well. We design an adaptive temperature function so that the temperature can get the appropriate value at different stages of training. Compared with the teacher model, the size of the student model is only 33KB, and the number of parameters is 0.2% of that in the teacher model. The prediction time of the student model is 28% of the teacher model. Though the model size is greatly reduced, the accuracy of the student model reaches 99.55% of the teacher model. Experiments show that this method effectively transfers the knowledge of the teacher model to the student model. Compared with existing work, our model achieves the state of the art in both accuracy and speed.

In summary, we make the following contributions:

1. We propose a lightweight network traffic classification model based on knowledge distillation. As far as we know, this is the first attempt to apply knowledge distillation to the compression of network traffic classification models.
2. We design an adaptive temperature function that can soften the labels to varying degrees at different stages of training so that the student model can learn the knowledge of the teacher model more quickly and accurately.
3. We introduce focal loss as distilling loss to better learn different kinds of information in the process of knowledge distillation. It can effectively solve

the problems of imbalance in the number of samples and different degrees of difficulty in classification.

4. We conduct extensive evaluations and compare our results with other advanced work, our model not only has the highest accuracy but also has the shortest prediction time.

2   Related Work

2.1   Encrypted Traffic Classification

Rezaei et al. [4] proposed a general framework of traffic classification based on deep learning, which includes traffic data collection, traffic data preprocessing, feature selection and extraction, model selection, training, verification, and model evaluation. Shen [10] presented a systematic method for feature selection in encrypted traffic classification. An one-dimensional CNN for image classification is used for encrypted traffic classification [11]. Authors showed that effective classification can be achieved by intercepting only the first 784 bytes of traffic, and the best classification model is based on session information of all layers [1]. Liu et al. [12] employed GRU as the encoder and decoder of SAE. The model can deeply mine the potential sequence characteristics of the flow. Zeng et al. [13] systematically compared three network structures (1D-CNN, LSTM, and SAE), where L1 regularization and L2 regularization are applied respectively. Experiments showed that L1 regularization + 1D-CNN has the best performance.

To investigate the spatial-temporal characteristics of encrypted traffic, researchers [14,15] combined the CNN and the LSTM. Based on this approach, Ran et al. [16] further took advantage of three-dimensional convolutions to learn more effective spatial-temporal characteristics of network traffic.

Due to the great differences in the number of different types of traffic, many researchers try to solve the imbalance of data sets to improve the accuracy. Bu et al. [3] designed a parallel decision depth neural network with multiple MLP convolution modules and proved that the NIN model has better performance than the CNNs model. CGAN is applied to solve the problem of class imbalance in traffic classification at the packet level [17]. Compared with ROS, SMOTE, and GAN, it is proved that CGAN is superior in synthesizing small samples. In [18], experiments show that AC-GAN can effectively improve the accuracy of traffic classification. Https traffic is classified by two-way attention GRU network, and knowledge transfer is used to speed up the learning speed [19]. Because the malicious traffic is less than the normal traffic, the Siamese convolution network is used for few-shot learning [20]. In addition, Multi-Output DNN is designed to study the multi-label and one-shot of traffic classification [21].

We note that researchers have a strong interest in using deep learning to solve the practical problems of traffic classification, but unfortunately, researchers do not pay attention to the size and structure of the model. In the practical application of network traffic classification, speed is as important as accuracy.

2.2   Knowledge Distillation

Knowledge distillation is a popular model compression method that transfers knowledge from a large model to a small model. The idea was first introduced by Bucila et al. [22], and many developments were proposed since then. Ba and Caruana [23] use logits of the deep net as labels and L2 Loss to training a shallow net. The idea is inspired by the fact that logits contain more information than hard labels. In [8], Hinton et al. introduced the concept of knowledge distillation. The authors pointed out the student model can learn complex knowledge faster and have higher generalization ability from teacher model. In [24], K classes with the highest confidence score are selected. The loss function is taken by using the gap between the class with the highest confidence score and the other $K - 1$ classes with the highest score. Two convolution modules, called paraphraser and translator, are introduced to interpret and reorganize the information of the teacher model, so as to better guide the training of the target model [25]. The paraphraser is trained in an unsupervised way to extract effective information from the teacher model. The translator is located in the target model and is used to transform the corresponding information in the target model to fit the information output by the paraphraser. The outputs of multi-teacher models are used as the structural unit to replace the traditional distillation learning with the output of a single teacher model [26]. Through enhance the performance of knowledge transfer by better characterizing the features of the middle layer of the neural network, the student model learns deep information [27–29]. In addition to training large models as teachers, there are some studies that focus on self-learning strategies that optimize themselves as their own teachers [30–32].

3   Methodology

To deployment the model online, we use the end-to-end traffic classification model to improve the size and speed of the model. An overview of the framework is shown in Fig. 1. The workflow of our approach consists of two parts, the setup of the data set and the knowledge distillation.

3.1   Dataset

Selection of Dataset. The ISCX-VPN-nonVPN dataset has become one of the majority datasets for its openness and rich in types and quantity of data [1,14]. We chose ISCX-VPN-nonVPN as the dataset of experiments. This dataset includes seven conventional encryption services and seven encryption services using a VPN [33]. Since the original traffic is not marked, we re-label the traffic according to their type. The media network traffic from browsers may be classified as either browsers or corresponding media tasks. Authors in [1,11] reported this problem. We adopt the approach in [1,11] and exclude Browsing and VPN-Browsing from experiments. The details of the dataset are listed in Table 1.

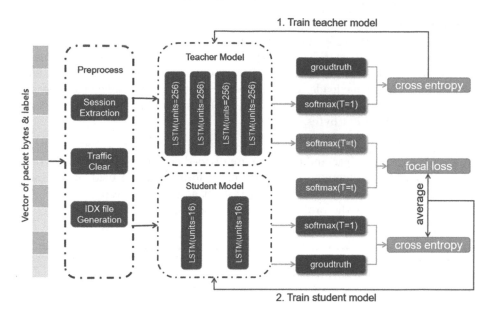

Fig. 1. The overview of lightweight encrypted traffic classification model.

Table 1. ISCX VPN-nonVPN DATASET

Categories	Applications
Chat	ICQ, AIM, Skype, Facebook, Hangouts
Email	SMPT, POP3, IMAP
File transfer	Skype, FTPS, SFTP
VoIP	Facebook, Skype, Hangouts, VoIP buster
P2P	Torrent
Streaming	Vimeo, YouTube, Netflix, Spotify
VPN-Chat	ICQ, AIM, Skype, Facebook, Hangouts
VPN-Email	SMPT, POP3, IMAP
VPN-File transfer	Skype, FTPS, SFTP
VPN-VoIP	Facebook, Skype, Hangouts, VoIP buster
VPN-P2P	BitTorrent
VPN-Streaming	Vimeo, YouTube, Netflix, Spotify

Data Preprocessing. According to existing work [1, 7, 11], data preprocessing is divided into three parts: session extraction, traffic clear and IDX file generation.

Session Extraction. Our work is based on sessions and all layers, so we need to extract the sessions for each pcap file separately.

Traffic Clear. Since the data frame header information (such as MAC address and IP address) may cause the classifier to overfit, traffic clear anonymizes the data frame header information. The address of the datalink layer, IP addresses, and the port number of the TCP/UDP are randomized. Duplicate packets and packets with empty payloads are removed. Packets are normalized to have a consistent length by truncating or padding by 0's. In our experiments, we specify the packet length to 784 bytes.

IDX File Generation. The traffic is classified according to categories in Table 1. The training data and test data are divided according to the proportion of 9:1. IDX file is a common data storage file format in the field of machine learning, so we save the training set and test set as IDX files respectively.

Algorithm 1 shows the process of our preprocessing.

Algorithm 1. Data Preprocess

Input: ISCX-VPN-nonVPN data set.
Output: the IDX files for training and testing.
 1: *sessions* ← sessionExtractor(pacpFiles);
 2: **for** each *session* **in** *sessions* **do**
 3:     Randomize the MAC/IP of the session;
 4:     Remove the duplicate and empty packets;
 5:     *sess* ← Extract data frame from session to sess;
 6:     **if** *sess*.length $\geq$ 784 **then**
 7:         Select the first 784 bytes of *sess*;
 8:     **else**
 9:         Fill *sess* with 0's for less than 784 bytes;
10:     **end if**
11:     SaveFile(*sess*);
12: **end for**
13: Organize the generated files according to Table 1;
14: Split training and testing data according to the portion 9:1 (TrainingData, TestingData);
15: Generate IDX files;
16: **return** IDX files;

3.2   Knowledge Distillation

We present the details of the teacher model and student model in our approach. In the student model training, we evolve the temperature adaptively according to the accuracy and learn from focal loss to solve the problem of learning difficulties in some samples. Algorithm 2 shows the process of knowledge distillation. These two parts are described in detail respectively.

Algorithm 2. Knowledge Distillation

Input: Training Data.
Output: Lightweight encrypted Traffic Classification Model.
 1: Normalize each byte of the traing data to $[0, 1]$;
 2: Split the input with a proportion of 8:2 as the training set and verification set;
 3: $student\_accuracy \leftarrow 0$;
   Step 1: train teacher model.
 4: Add 1st LSTM of 256 units with LeakyReLU activation(alpha=0.3) and L2 regularization;
 5: Add BatchNormalization;
 6: Add 2st LSTM of 256 units with LeakyReLU activation(alpha=0.2) and L2 regularization;
 7: Add BatchNormalization;
 8: Add 3st LSTM of 256 units with LeakyReLU activation(alpha=0.1);
 9: Add BatchNormalization;
10: Add a dense layer with 12 units;
11: Train teacher model with batch size is 128 and optimizer is Adam;
   Step 2: train student model.
12: Add 1st LSTM of 16 units with LeakyReLU activation(alpha=0.1);
13: Add BatchNormalization;
14: Add a dense layer with 12 units;
15: $teacher\_logits \leftarrow$ The training set logits of teacher model;
16: $student\_logits \leftarrow$ The training set logits of student model;
17: $loss \leftarrow$ Get losses using formula 1 with groudth labels, teacher_logits, student_logits, and student_accuracy;
18: Update the weight and bias;
19: $student\_accuracy \leftarrow$ Get the accuracy of student model on verification set;
20: Update T using formula 6;
21: Next epoch;

Teacher Model. The recurrent Neural Network(RNN) is commonly used in the processing of sequence data. Long and short-term memory(LSTM) is a special kind of RNN, which solves the problem of gradient disappearance and gradient explosion in the process of long-sequence training. Considering that the traffic has a temporal correlation not only between packets but also within a packet, we choose LSTM as the teacher model. Specifically, we use three-layers LSTM as the teacher model. Each layer of LSTM consists of 256 units, and an L2 regularization is added. L2 regularization can better control the complexity of the model and prevent overfitting, so better prediction results can be obtained. The activation functions of the three layers are all LeakyReLU, and the alpha of LeakyReLU is 0.3, 0.2, and 0.1 respectively. To speed up the training and make the model learn knowledge more easily, BatchNormalization is added to each layer. Finally, there is a fully connected layer of 12 units. In training the teacher model, the cross-entropy is used as the loss function.

Student Model. In knowledge distillation, the student model can learn the teacher's knowledge well when the two parties share a similar structure. So the student model also uses the LSTM model. Different from the teacher model, the student model has only one layer of LSTM with 16 units. As the student model is relatively small, we do not use L2 regularization. The student model uses the LeakyReLU function as the activation function, where alpha is 0.1. Batch-Normalization is also used in the student model, ending with a fully connected layer of 12 units. The loss function of the student model consists of two parts: $focal\_loss$ and ce_loss.

$$total\_loss = 0.5 \times (focal\_loss + ce\_loss) \tag{1}$$

We define the variables used in the model as follows.
c_i: the groudth label of class i.
v_i: the logits of the teacher model.
z_i: the logits of the student model.
q_i^T: the probability of class i of the teacher model at temperature T.

$$q_i^T = \frac{\exp(v_i/T)}{\sum_j^N \exp v_j} \tag{2}$$

p_i^T: the probability of class i of student model at temperature T.

$$p_i^T = \frac{\exp(z_i/T)}{\sum_j^N \exp z_j} \tag{3}$$

$focal\_loss$: the loss of soft targets with student logits.

$$focal\_loss = -\sum_i^N \alpha q_i^T (1 - p_i^T)^\gamma \log p_i^T \tag{4}$$

ce_loss: the loss of groudth labels with student logits.

$$ce\_loss = -\sum_i^N c_i \log p_i^1 \tag{5}$$

By adjusting T, the labels can be softened to varying degrees. The larger the T, the smoother the label and the more information between categories retained, but the more serious the noise. In the early stage of training, the accuracy of the student model is very low, and the T value is relatively large so that the student model can learn knowledge more quickly from information between categories. However, student model needs less noise and more focus on learning samples difficult to classify with the improvement of the accuracy, so it needs a smaller T. In this experiment, the following functions are used to make T change dynamically:

$$T = 10 - (2 \times accuracy)^3 \tag{6}$$

$focal\_loss$ refers to knowledge distillation loss. With the help of soft targets, it can better solve the problem of unbalanced sample classes and different degrees of difficulty in encrypted traffic classification. According to [34], for $\alpha = 0.25$ and $\gamma = 2$ works best.

4   Evaluation and Experimental Results

4.1   Experimental Setup

We implement our model by Tensorflow and Keras. The experimental platform is Windows 10 with an Intel I7-9750H CPU at 2.6 GHz, 16 GB memory and an external GPU (Nvidia GeForce GTX1650).The mini-batch size of the teacher model is 128, and the mini-batch size of the student model is 64. Adam optimizer built-in Keras is used as the optimizer. To get better results, we use the early stopping on the basis of 1000 epochs. Specifically, stop training when the accuracy has stopped improving, and the patience is 100.

4.2   Evaluation Metrics

In our experiment, we measure the predicting ability of the model by *Accuracy*, *Precision*, *Recall*, and *F*1. We first give some definitions as follows : a) TP: the number of predict a positive class as a positive class; b) FN: the number of predict a positive class as a negative class; c) FP: the number of predict a negative class as a positive class; d) TN: the number of predict a negative class as a negative class. We have the following.

$$Accuracy = \frac{TP + TN}{TP + TN + FP + FN} \qquad (7)$$

$$Precision = \frac{TP}{TP + FP} \qquad (8)$$

$$Recall = \frac{TP}{TP + FN} \qquad (9)$$

$$F1 = \frac{2 \times Precision \times Recall}{Precision \times Recall} \qquad (10)$$

In addition to ensure the speed and accuracy of the model, we also focus on the size of the model($ModelSize$), the count of parameters($ParamCounts$), train time(TT), and the prediction time(PT).

4.3   Experiment Result

First, we compare the model storage and computing resources. We give the experimental results in Table 2. Compared with the teacher model, the number of parameters used in the student model is 0.2% of that of the teacher model, resulting in a student model of the size 33KB. The succinct model greatly saves the storage and computing resources. Stu_NoKD model is a student model trained directly without distillation, Stu_KD model is obtained through traditional knowledge distillation, and Stu_ATKD model is trained by the method described in this paper. The training speed of the Stu_NoKD model is much faster than Stu_KD and Stu_ATKD models. This is because the distillation model needs to calculate the loss between soft targets and student logits. The prediction speed of the student models is about three times faster than the teacher model, which is a great improvement for the online deployment of the model (Table 2).

Table 2. Comparison between model storage and computing resources

	ModelSize	ParamCounts	TT(ms/step)	PT(s)
Teacher	15.5 MB	1,348,620	30	1.64
Stu_NoKD	33KB	3,148	11	0.47
Stu_KD	33KB	3,148	13	0.49
Stu_ATKD	**33KB**	**3,148**	**13**	**0.47**

The performance of the Stu_ATKD is stupendously similar to the teacher model. The experimental results are shown in Table 3, where MA stands for Micro-average, and WA stands for Weighted-average. Compared with the teacher model, the accuracy of the Stu_ATKD decreased by only 0.45%. The decline in *recall* (macro avg), about 1.12%, is the most serious. The Stu_NoKD model is not enough in the ability to learn knowledge, and the final model performance is the worst. After ordinary knowledge distillation, the Stu_KD model has learned some of the knowledge of the teacher model. The accuracy and recall rate have been significantly improved. In the latter stage of training, the noise introduced by similar information between classes causes great interference to the learning ability of the model. The Stu_ATKD model can overcome this problem and achieve higher accuracy.

Table 3. Overall experimental results of four neural networks on network classification.

		Teacher	Stu_NoKD	Stu_KD	Stu_ATKD
Accuracy		0.9997	0.9716	0.9888	**0.9952**
Precision	MA	0.9999	0.9635	0.9838	**0.9965**
	WA	0.9997	0.9721	0.9889	**0.9952**
Recall	MA	0.9999	0.9601	0.9815	**0.9887**
	WA	0.9997	0.9716	0.9888	**0.9952**
F1-score	MA	0.9999	0.9612	0.9826	**0.9925**
	WA	0.9997	0.9716	0.9889	**0.9952**

To obtain a better performance and prevent the model from overfitting, we employ the early stopping in our model. Early stopping ends the training when the performance of the student model stops improving. Therefore, the epoch numbers of different models are not the same. Besides, we use different loss functions for different models. This makes it impossible for us to directly compare the learning abilities of different models. We normalize the loss values of different models to $[0, 1]$, and compare the loss values of each model at the 1, $x * N/10(x \in \{x \mid x \leq 10, x \in N+\})$ epoch. The experimental results are shown in Fig. 2. The Stu_KD model learns similar information between categories with soft targets, so it is faster than the Stu_NoKD model. However, it cannot automatically adjust temperature and filter the noise contained in soft targets in time. In the latter stage of training, noise becomes an obstacle to further improvement of the model. To solve this problem, the temperature of Stu_ATKD decreases with the improvement of accuracy to reduce the noise in similar information. At the same time, the focal loss can adjust the learning direction of the model in time and pay more attention to the samples difficult to classify. Therefore, the Stu_ATKD model has the better learning speed and the highest accuracy.

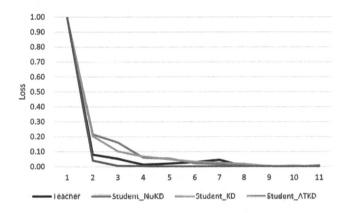

Fig. 2. Comparison of the loss

Our model uses focal loss to solve the imbalance of the difficulty and counts of sample classification. As shown in Fig. 3, Stu_ATKD is better than Stu_KD in terms of prediction, recall, and F1-score. To measure the effectiveness of focal loss, Fig. 3 also shows the performance of two models in terms of Standard Deviation. The Stu_KD model performs poorly in File, VPN-Email, and VPN-Streaming classification. The prediction, recall, and F1-score of Stu_KD in VPN-Email are only 93.33%, indicating that Stu_KD pays the least attention to VPN-Email. This leads to the decline of the overall classification performance. By imposing a higher penalty on the samples that are difficult to classify, Stu_ATKD with focal loss has a much smaller Standard Deviation. It shows that Stu_ATKD pays similar attention to each category in the process of classification.

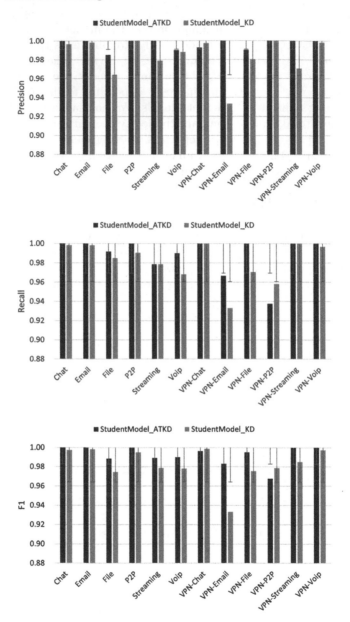

Fig. 3. Comparison of the classification efficiency between Stu_ATKD model and Stu_KD model

We compared our model with that of [35]. As the computer configuration is not the same, we do not compare the predict time of the model. We compared the size, the number of parameters, and the F1-score of the models. The comparison results are shown in Table 4. We can see that our method not only has the smallest model size but also has the highest accuracy. Our method outperforms the state-of-the-art in all the measurements.

Table 4. Comparison between model storage and computing resources

	ModelSize	ParamCounts	F1-Score	Increased
MLP	178KB	12,943	0.9653	2.82%
SAE	1,2681KB	1,359,463	0.9882	0.44%
CNN	1,467KB	182,927	0.9843	0.83%
Stu_ATKD	**33KB**	**3,148**	**0.9925**	–

5   Conclusion and Future Work

In this paper, we propose a lightweight network traffic classification model based on knowledge distillation named Stu_ATKD. The model uses adaptive temperature function and focal loss to better classify network traffic. Compared with the traditional large-scale model, the size of the model and parameters can be greatly reduced and the required computing resources can be greatly reduced under the condition of ensuring accuracy. The Stu_ATKD model provides a new method for the deep learning model to classify network traffic online. Compared with the traditional knowledge distillation, the Stu_ATKD model can further improve the accuracy due to the use of adaptive T and $focal\_loss$. While the recognition speed is increased by 72%, the accuracy of the student model is 99.52% and can reach 99.55% of the teacher model. The proposed method outperforms the state-of-the-art in all the measurements. In the future, we will take consideration of the imbalance of the number of samples of different classes that seriously affects the performance of the model.

References

1. Wang, W., Zhu, M., Wang, J., Zeng, X., Yang, Z.: End-To-end encrypted traffic classification with one-dimensional convolution neural networks. In: 2017 IEEE International Conference on Intelligence and Security Informatics: Security and Big Data, ISI 2017, pp. 43–48 (2017)
2. Yamansavcilar, B., Guvensan, M.A., Yavuz, A.G., Karsligil, M.E.: Application identification via network traffic classification. In: 2017 International Conference on Computing, Networking and Communications, ICNC 2017, pp. 843–848 (2017)
3. Bu, Z., Zhou, B., Cheng, P., Zhang, K., Ling, Z.H.: Encrypted network traffic classification using deep and parallel network-in-network models. IEEE Access **8**, 132 950–132 959 (2020)
4. Rezaei, S., Liu, X.: Deep learning for encrypted traffic classification: an overview. IEEE Commun. Mag. **57**(5), 76–81 (2019)

5. Zou, Z., Ge, J., Zheng, H., Wu, Y., Han, C., Yao, Z.: Encrypted traffic classification with a convolutional long short-term memory neural network. In: Proceedings - 20th International Conference on High Performance Computing and Communications, 16th International Conference on Smart City and 4th International Conference on Data Science and Systems, HPCC/SmartCity/DSS 2018, pp. 329–334 (2019)
6. Aceto, G., Ciuonzo, D., Montieri, A., Pescapè, A.: MIMETIC: mobile encrypted traffic classification using multimodal deep learning. Comput. Netw. **165**, 106944 (2019). https://doi.org/10.1016/j.comnet.2019.106944
7. Wang, W., Zhu, M., Zeng, X., Ye, X., Sheng, Y.: Malware traffic classification using convolutional neural network for representation learning. In: International Conference on Information Networking, pp. 712–717 (2017)
8. Hinton, G., Vinyals, O., Dean, J.: Distilling the Knowledge in a Neural Network, pp. 1–9 (2015). http://arxiv.org/abs/1503.02531
9. Li, H.: Exploring knowledge distillation of deep neural networks for efficient hardware solutions (2018)
10. Shen, M., Liu, Y., Zhu, L., Xu, K., Du, X., Guizani, N.: Optimizing feature selection for efficient encrypted traffic classification: a systematic approach. IEEE Network **34**(4), 20–27 (2020)
11. He, Y., Li, W.: Image-based encrypted traffic classification with convolution neural networks. In: Proceedings - 2020 IEEE 5th International Conference on Data Science in Cyberspace, DSC 2020, pp. 271–278 (2020)
12. Liu, C., He, L., Xiong, G., Cao, Z., Li, Z.: FS-Net: a flow sequence network for encrypted traffic classification. In: Proceedings - IEEE INFOCOM, vol. 2019-April, pp. 1171–1179 (2019)
13. Zeng, Y., Gu, H., Wei, W., Guo, Y.: Deep-full-range: a deep learning based network encrypted traffic classification and intrusion detection framework. IEEE Access **7**, 45 182–45 190 (2019)
14. Wang, W., et al.: HAST-IDS: learning hierarchical spatial-temporal features using deep neural networks to improve intrusion detection. IEEE Access **6**, 1792–1806 (2017)
15. Lopez-Martin, M., Carro, B., Sanchez-Esguevillas, A., Lloret, J.: Network traffic classifier with convolutional and recurrent neural networks for internet of things. IEEE Access **5**, 18042–18050 (2017)
16. Ran, J., Chen, Y., Li, S.: Three-dimensional convolutional neural network based traffic classification for wireless communications. In: 2018 IEEE Global Conference on Signal and Information Processing, GlobalSIP 2018 - Proceedings, pp. 624–627 (2019)
17. Wang, P., Li, S., Ye, F., Wang, Z., Zhang, M.: PacketCGAN: exploratory study of class imbalance for encrypted traffic classification using CGAN. In: IEEE International Conference on Communications, vol. 2020-June (2020)
18. Vu, L., Bui, C.T., Nguyen, Q.U.: A deep learning based method for handling imbalanced problem in network traffic classification. In: ACM International Conference Proceeding Series, vol. 2017, pp. 333–339 (December 2017)
19. Liu, X., et al.: Attention-based bidirectional GRU networks for efficient HTTPS traffic classification. Inf. Sci. **541**, 297–315 (2020). https://doi.org/10.1016/j.ins.2020.05.035
20. Xu, L., Dou, D., Chao, H.J.: ETCNet: encrypted traffic classification using Siamese convolutional networks. In: Proceedings of the Workshop on Network Application Integration/CoDesign, NAI 2020, pp. 51–53 (2020)

21. Sun, H., et al.: Common knowledge based and one-shot learning enabled multi-task traffic classification. IEEE Access **7**, 39485–39495 (2019)
22. Bucila, C., Caruana, R., Niculescu-Mizil, A.: Model compression. In: SIGKDD, pp. 535–541. ACM (2006)
23. Ba, L.J., Caruana, R.: Do deep nets really need to be deep? Adv. Neural Inf. Process. Syst. **3**(January), 2654–2662 (2014)
24. Yang, C., Xie, L., Qiao, S., Yuille, A.: Knowledge distillation in generations: more tolerant teachers educate better students (2018). http://arxiv.org/abs/1805.05551
25. Kim, J., Park, S., Kwak, N.: Paraphrasing complex network: network compression via factor transfer. In: Proceedings of the 32nd International Conference on Neural Information Processing Systems, pp. 2765–2774 (2018)
26. Park, W., Kim, D., Lu, Y., Cho, M.: Relational knowledge distillation. In: Proceedings of the IEEE Computer Society Conference on Computer Vision and Pattern Recognition, vol. 2019, pp. 3962–3971 (2019)
27. Zagoruyko, S., Komodakis, N.: Paying more attention to attention: improving the performance of convolutional neural networks via attention transfer. In: 5th International Conference on Learning Representations, ICLR 2017 - Conference Track Proceedings, pp. 1–13 (2017)
28. Ahn, S., Hu, S.X., Damianou, A., Lawrence, N.D., Dai, Z.: Variational information distillation for knowledge transfer. In: Proceedings of the IEEE Computer Society Conference on Computer Vision and Pattern Recognition, vol. 2019-June, pp. 9155–9163 (2019)
29. Tung, F., Mori, G.: Similarity-preserving knowledge distillation. In: Proceedings of the IEEE International Conference on Computer Vision, vol. 2019-October, pp. 1365–1374 (2019)
30. Xu, T.B., Liu, C.L.: Data-distortion guided self-distillation for deep neural networks. In: 33rd AAAI Conference on Artificial Intelligence, AAAI 2019, 31st Innovative Applications of Artificial Intelligence Conference, IAAI 2019 and the 9th AAAI Symposium on Educational Advances in Artificial Intelligence, EAAI 2019, no. March, pp. 5565–5572 (2019)
31. Zhang, L., Song, J., Gao, A., Chen, J., Bao, C., Ma, K.: Be your own teacher: improve the performance of convolutional neural networks via self distillation. In: Proceedings of the IEEE International Conference on Computer Vision, vol. 2019-October, pp. 3712–3721 (2019)
32. Yun, S., Park, J., Lee, K., Shin, J.: Regularizing class-wise predictions via self-knowledge distillation. In: Proceedings of the IEEE Computer Society Conference on Computer Vision and Pattern Recognition, vol. 1, pp. 13873–13882 (2020)
33. Draper-Gil, G., Lashkari, A.H., Mamun, M.S.I., Ghorbani, A.A.: Characterization of encrypted and vpn traffic using time-related features. In: The International Conference on Information Systems Security and Privacy (ICISSP) (2016)
34. Lin, T.Y., Goyal, P., Girshick, R., He, K., Dollar, P.: Focal loss for dense object detection. IEEE Trans. Pattern Anal. Mach. Intell. **42**(2), 318–327 (2020)
35. Wang, P., Ye, F., Chen, X., Qian, Y.: Datanet: Deep learning based encrypted network traffic classification in SDN home gateway. IEEE Access **6**, 55380–55391 (2018)

RAU: an Interpretable Automatic Infection Diagnosis of COVID-19 Pneumonia with Residual Attention U-Net

Xiaocong Chen[1(✉)], Lina Yao[1], and Yu Zhang[2]

[1] The University of New South Wales, Sydney, NSW 2052, Australia
{xiaocong.chen,lina.yao}@unsw.edu.au
[2] Leihigh University, Bethlehem, PA 18015, USA
yuzi20@lehigh.edu

Abstract. The novel coronavirus disease 2019 (COVID-19) has been spreading rapidly around the world and caused a significant impact on public health and economy. However, there is still lack of studies on effectively quantifying the different lung infection areas caused by COVID-19. As a basic but challenging task of the diagnostic framework, distinguish infection areas in computed tomography (CT) images and help radiologists to determine the severity of the infection rapidly. To this end, we proposed a novel deep learning algorithm for automated infection diagnosis of multiple COVID-19 Pneumonia. Specifically, we use the aggregated residual network to learn a robust and expressive feature representation and apply the soft attention mechanism to improve the capability of the model to distinguish a variety of symptoms of the COVID-19. With a public CT image dataset, the proposed method achieves 0.91 DSC which is 14.6% higher than selected baselines. Experimental results demonstrate the outstanding performance of our proposed model for the automated segmentation of COVID-19 Chest CT images. Our study provides a promising deep learning-based segmentation tool to lay a foundation to facilitate the quantitative diagnosis of COVID-19 lung infection in CT images.

Keywords: Automated segmentation · COVID-19 · Computed tomography · Deep learning

1 Introduction

The novel coronavirus disease 2019, also known as COVID-19 outbreak first noted in Wuhan at the end of 2019, has been spreading rapidly worldwide [32]. As an infectious disease, COVID-19 is caused by severe acute respiratory syndrome coronavirus and presents with symptoms including fever, dry cough, shortness of breath, tiredness and so on. As the Jan 7th, over 87 million people around the world have been confirmed as COVID-19 infection with a case fatality rate of about 5.7% according to the statistic of World Health Organization[1].

---

[1] https://www.who.int/emergencies/diseases/novel-coronavirus-2019/situation-reports.

© Springer Nature Switzerland AG 2021
W. Zhang et al. (Eds.): WISE 2021, LNCS 13081, pp. 122–136, 2021.
https://doi.org/10.1007/978-3-030-91560-5_9

So far, no specific treatment has proven effective for COVID-19. Therefore, accurate and rapid testing is extremely crucial for timely prevention of COVID-19 spread. Real-time reverse transcriptase polymerase chain reaction (RT-PCR) has been referred as the standard approach for testing COVID-19. However, RT-PCR testing is time-consuming and limited by the lack of supply test kits [17,22]. Moreover, RT-PCR has been reported to suffer from low sensitivity and repeated checking is typically required for accurate confirmation of a COVID-19 case. This indicates that many patients will not be confirmed timely [1,15], thereby resulting in a high risk of infecting a larger population.

In recent years, imaging technology has emerged as a promising tool for automatic quantification and diagnosis of various diseases. As a routine diagnostic tool for pneumonia, chest computed tomography (CT) imaging has been strongly recommended in suspected COVID-19 cases for both initial evaluation and follow-up. Chest CT scans play an indispensable role in detecting typical COVID-19 infections [11,13,19]. A systematic review [21] concluded that CT imaging of the chest was found to be sensitive when checking for COVID-19 cases even before some clinical symptoms were observed. Specifically, the typical radiographic features indicating ground class opacification, consolidation and pleural effusion have been frequently observed in the chest CT images scanned from COVID-19 patients [9,24,28].

Accurate segmentation of these important radiographic features is crucial for reliable quantification of COVID-19 infection in chest CT images. Segmentation of medical imaging needs to be manually annotated by well-trained expert radiologists. The rapidly increasing number of infected patients has caused a tremendous burden for radiologists and slowed down the labelling of ground-truth mask. Thus, there is an urgent need for automated segmentation of infection regions, which is a basic but arduous task in the pipeline of computer-aided disease diagnosis [6]. However, automatically delineating the infection regions from the chest CT scans is considerably challenging because of the large variation in both position and shape across different patients and low contrast of the infection regions in CT images [22].

Machine learning-based artificial intelligence provides a powerful technique for the design of data-driven methods in medical imaging analysis [24]. Developing advanced deep learning models would bring unique benefits to the rapid and automated segmentation of medical images [23]. So far, fully convolutional networks have proven superiority over other widely used registration-based approaches for segmentation [6]. In particular, U-Net models work decently well for most segmentation tasks in medical images [2,3,20,22]. However, several potential limitations of U-Net have not been effectively addressed yet. For example, it is difficult for the U-net model to capture complex features such as multi-class image segmentation and recover the complex features into the segmentation image [18]. There are also a few successful applications that adopt U-Net or its variants to implement the CT image segmentation, including heart segmentation [30], liver segmentation [14], or multi-organ segmentation [5]. However, segmentation of COVID-19 infection regions with deep learning remains underexplored. The COVID-19 is a new disease but very similar to common pneumonia in the medical imaging side, which makes its accurate quantification considerably challenging. Recent advancement of the deep learning method provides heaps of insightful ideas about improving the U-Net architecture. The most popular one is the deep residual network (ResNet) [8]. ResNet provided an elegant way to stack CNN layers and demonstrate the strength when combined with U-Net [10]. On the other hand, attention was also applied to improve the U-Net and other deep learning models to boost the performance [18].

Accordingly, we propose a novel deep learning model for rapid and accurate segmentation of COVID-19 infection regions in chest CT scans. Our developed model is based on the U-Net architecture, inspired with recent advancement in the deep learning field. We exploit both the residual network and attention mechanism to improve the efficacy of the U-Net. Experimental analysis is conducted with a public CT image dataset collected from patients infected with COVID-19 to assess the efficacy of the developed model. The outstanding performance demonstrates that our study provides a promising segmentation tool for the timely and reliable quantification of lung infection, toward developing an effective pipeline for precious COVID-19 diagnosis.

Our aim is to develop a plausible segmentation model for automatically identifying the typical COVID-19 infection areas of lungs from chest CT images in order to facilitate COVID-19 diagnosis as the following: (i) our model provides a proper tool for determining the salient regions of CT images, and thus speed up the COVID-19 screening and diagnosis; (ii) comparing with existing approaches, our model is capable of producing fine-grained region of interest relating to COVID-19 infections. It would be useful to identify the different progression stages and therefore offer the groundings for further treatment plan.

The rest of the paper is summarized as follows. Our proposed new deep learning model is detailedly described in Sect. 2 Methodology, including the U-Net structure, the methods used to improve the encoder and decoder. The experimental study and performance assessment are described in Sect. 3, followed by a discussion and summary of our study.

2   Methodology

This section will introduce our proposed Residual Attention U-Net for the lung CT image segmentation in detail. We start by describing the overall structure of the developed deep learning model followed by explaining the two improved components including aggregated residual block and locality sensitive hashing attention as well as the training strategy. The overall flowchart is illustrated in Fig. 1.

2.1   Overview

U-Net was first proposed by Ronneberger et al. [20], which was basically a variant of fully convolutional networks (FCN) [16]. The traditional U-Net is a type

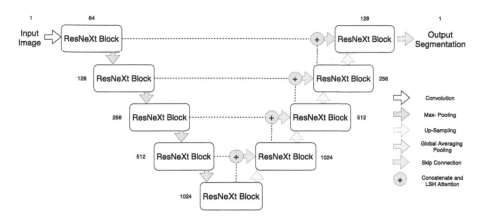

Fig. 1. Illustration of our developed residual attention U-Net model. The aggregated ResNeXt blocks are used to capture the complex feature from the original images. The left side of the U-Shape serves as encoder and the right side as decoder. Each block of the decoder receives the feature representation learned from the encoder, and concatenates them with the output of deconvolutional layer followed by LSH attention mechanism. The filtered feature representation after the attention mechanism is propagated through the skip connection.

of artificial neural network (ANN) containing a set of convolutional layers and deconvolutional layers to perform the task of biomedical image segmentation. The structure of U-Net is symmetric with two parts: encoder and decoder. The encoder is designed to extract the spatial features from the original medical image. The decoder is to construct the segmentation map from the extracted spatial features. The encoder follows a similar style like FCN with the combination of several convolutional layers. To be specific, the encoder consists of a sequence of blocks for down-sampling operations, with each block including two 3×3 convolution layers followed by a 2×2 max-pooling layers with stride of 2. The number of filters in the convolutional layers is doubled after each down-sampling operation. In the end, the encoder adopts two 3×3 convolutional layers as the bridge to connect with the decoder.

Differently, the decoder is designed for up-sampling and constructing the segmentation image. The decoder first utilizes a 2×2 deconvolutional layer to up-sample the feature map generated by the encoder. The deconvolutional layer contains the transposed convolution operation and will half the number of filters in the output. It is followed by a sequence of up-sampling blocks which consists of two 3×3 convolution layers and a deconvolutional layer. Then, a 1×1 convolutional layer is used as the final layer to generate the segmentation result. The final layer adopted Sigmoid function as the activation function while all other layers used ReLU function. In addition, the U-Net concatenates part of the encoder features with the decoder. For each block in encoder, the result of the convolution before the max-pooling is transferred to decoder symmetrically. In decoder, each block receives the feature representation learned from encoder, and concatenates them with the output of deconvolutional layer. The concatenated result is then forwardly propagated to the consecutive block. This concatenation operation is useful for the decoder to capture the possible lost features by the max-pooling.

2.2 Aggregated Residual Block

As mentioned in the previous section, the U-Net only have four blocks of convolution layers to conduct the feature extraction. The conventional structure may not be sufficient for the complex medical image analysis such as multi-class image segmentation in the lung, which is the goal of this study. Although U-Net can easily separate the lung in a CT image, it may have limited ability to distinguish the difference infection regions of the lung which infected by COVID-19. Based on this case, the deeper network is needed with more layers, especially for the encoding process. However, when deeper network converges, a problem will be exposed: with increasing of the network depth, accuracy gets very high and then decreases rapidly. This problem is defined as the degradation problem [7]. He et al. proposed the ResNet [8] to mitigate the effect of network degradation on model learning. ResNet utilizes a skip connection with residual learning to overcome the degradation and avoid estimating a large number of parameters generated by the convolutional layer. The typical ResNet block can be defined as $F(i) = \sum_{j=1}^{D} w_j i_j$ where $i = [i_1, i_2, \cdots, i_D]$ and $W = [w_1, w_2, \cdots, w_D]$ is the trainable weight for the weight layer. Different from the U-Net that concatenates the features map into the decoding process, ResNet adopts the shortcut to add the identity into the output of each block. The stacked residual block can better learn the latent representation of the input CT image. However, the ResNet normally have millions of parameters and may lead to under-fitting or over-fitting due to the model's complexity. Regarding this, Xie et al. proposed the Aggregated Residual Network(ResNeXt) and showed that increasing the cardinality was more useful than increasing the depth or width [29]. The cardinality is defined as the set of the Aggregated Residual transformations with the formulation as $F(i) = \sum_{j=1}^{C} \mathcal{T}_j(i)$ where C is the number of residual transformation to be aggregated and $\mathcal{T}_j(i)$ can be any function. Considering a simple neuron, $\mathcal{T}_j$ should be a transformation which projects i into a low-dimensional embedding ideally and then transforming it. Accordingly, we can extend it into the residual function $y = \sum_{j=1}^{C} \mathcal{T}_j(i) + i$ where the y is the output. The ResNeXt block is visualized in Fig. 2. The weight layer's size is smaller than ResNet as ResNeXt uses the cardinality to reduce the number of layers but keep the performance. One thing is worth to mention that the three small blocks inside the ResNeXt block need to have the same topology, in other words, they should be topologically equivalent.

Similar to the ResNet, after a sequence of blocks, the learned features are fed into a global averaging pooling layer to generate the final feature map. Different from the convolutional layers and normal pooling layers, the global averaging pooling layers take the average of feature maps derived by all blocks. It can sum up all the spatial information captured by each step and is generally more robust

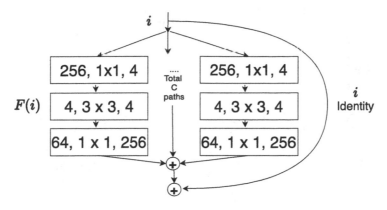

Fig. 2. An example ResNeXt Block. The variable i is the 256-dimension representation of the input image or features map. C represents the cardinality, which indicates that number of blocks inside. $256, 1 \times 1, 4$ represents the size of input image, filter and output's channel. The dimension is determined by the input.

than directly making the spatial transformation to the input. Mathematically, we can treat the global averaging pooling layer as a structural regularizer that is helpful for driving the desired feature maps [31].

Importantly, instead of using the encoder in the U-Net, our proposed deep learning model adopts the ResNeXt block (see Fig. 2) to conduct the features extraction. The ResNeXt provides a solution which can prevent the network from going very deeper but remain the performance. In addition, the training cost of ResNeXt is better than ResNet.

2.3 Locality Sensitive Hashing Attention

The decoder in U-Net is used to up-sampling the extracted feature map to generate the segmentation image. However, due to the capability of the convolutional neural network, it may not able to capture the complex features if the network structure is not deep enough. In recent years, transformers [26] have gained increasingly interest. The key to success is the attention mechanism [27]. Attention includes two different mechanisms: soft attention and hard attention. We adopt soft attention to improve model learning. Different from hard attention, soft attention can let model focus on each pixel's relative position, but hard attention only can focus on the absolute position. There are two different types of soft attention: Scaled Dot-Product Attention and Multi-Head Attention as shown in Fig. 3. The scaled dot-product attention takes the inputs including a query Q, a key K_n of the n-dimension and value V_m of the m-dimension. The dot-product attention is defined as follows:

$$\text{Attention}(Q, K_n, V_m) = \text{softmax}(\frac{QK_n^T}{\sqrt{n}})V_m \tag{1}$$

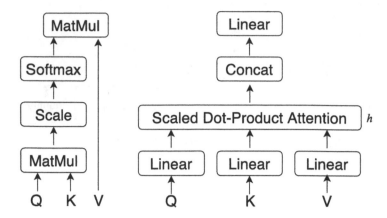

Fig. 3. Attention Mechanism. The left figure shows the simple scaled dot-prodct attention. The right figure depicts the multi-hand attention with the h head.

where K_n^T represents to the transpose of the matrix K_n and $\sqrt{n}$ is a scaling factor. The softmax function $\sigma(\mathbf{z})$ with $\mathbf{z} = [z_1, \cdots, z_n] \in \mathbb{R}^n$ is given by:

$$\sigma(\mathbf{z})_i = \frac{\exp(z_i)}{\sum_{j=1}^n \exp(z_j)} \text{ for } i = 1, \cdots, n \tag{2}$$

Vaswani et al. [27] mentioned that, performing the different linear project of the queries Q, keys K and values V in parallel h layers will benefit the attention score calculation. We can assume that Q, K and V have been linearly projected to d_k, d_k, d_v dimensions, respectively. It is worth noting that these linear projections are different and learnable. On each projection p, we have a pair of the query, key and value Q_p, K_p, V_p to conduct the attention calculation in parallel, which results in a d_v-dimensional output. The calculation can be formulated as:

$$\text{MultiHead}(Q, K, V) = \text{Concatenate}(\text{head}_1, \cdots, \text{head}_h)W^O$$
$$\text{where head}_i = \text{Attention}(QW_i^Q, KW_i^K, VW_i^V)$$

where the the projections $W_i^Q \in \mathbb{R}^{d_{model} \times d_k}$, $W_i^K \in \mathbb{R}^{d_{model} \times d_k}$, $W_i^V \in \mathbb{R}^{d_{model} \times d_v}$ are parameter matrices and $W^O \in \mathbb{R}^{d_{model} \times hd_v}$ is the weight matrix used to balance the results of h layers.

However, the multi-head attention is memory inefficient due to the size of Q, K and V. Assume that the Q, K, V have the shape $[|batch|, length, d_{model}]$ where $| \cdot |$ represents the size of the variable. The term QK^T will produce a tensor in shape $[length, length, d_{model}]$. Given the standard image size, the length $\times$ length will take most of the memory. Kitaev et al. [12] proposed a Locality Sensitive Hashing(LSH) based Attention to address this issue. Firstly, we rewire the basic attention formula into each query position i, j in the partition form:

$$a_i = \sum_{j \in P_i} \frac{\exp(q_i \cdot k_j - z(i, P_i))v_j}{\sqrt{d_k}} \text{ where } P_i = \{j : i \geq j\} \tag{3}$$

where the function z is the partition function, P_i is the set which query position i attends to and k, q, v are elements of the matrix K, Q, V. During model training, we normally conduct the batching and assume that there is a larger set $P_i^L = \{0, 1, \cdots, l\} \supseteq P_i$ without considering elements not in P_i:

$$a_i = \sum_{j \in P_i^L} \frac{\exp(q_i \cdot k_j - N(j, P_i) - z(i, P_i))v_j}{\sqrt{d_k}} \tag{4}$$

$$\text{where } N(j, P_i) = \begin{cases} 0 & j \in P_i \\ \infty & j \notin P_i \end{cases} \tag{5}$$

Then, with a hash function $h(\cdot)$: $h(q_i) = h(k_j)$, we can get P_i as:

$$P_i = \{j : h(q_i) = h(k_j)\} \tag{6}$$

In order to guarantee that the number of keys can uniquely match with the number of quires, we need to ensure that $h(q_i) = h(k_i)$ where $k_i = \frac{q_i}{\|q_i\|}$. During the hashing process, some similar items may fall in different buckets because of the hashing. The multi-round hashing provides an effective way to overcome this issue. Suppose there is n_r round, and each round has different hash functions $\{h_1, \cdots, h_{n_r}\}$, so we have:

$$P_i = \bigcup_{g=1}^{n_r} P_i^g \text{ where } P_i^g = \{j : h^g(q_i) = h^g(q_j)\} \tag{7}$$

Considering the batching case, we need to get the P_i^L for each round g:

$$\widehat{P_i^L} = \left\{ j : \left\lfloor \frac{i}{m} \right\rfloor - 1 \leq \left\lfloor \frac{j}{m} \right\rfloor \leq \left\lfloor \frac{i}{m} \right\rfloor \right\} \tag{8}$$

where $m = \frac{2l}{n_r}$. The last step is to calculate the LSH attention score in parallel. With the above formula , we can derive:

$$a_i = \sum_{g=1}^{n_r} \frac{\exp(z(i, P_i^g) - z(i, P_i))a_i^g}{\sqrt{d_k}} \tag{9}$$

$$\text{where } a_i^g = \sum_{j \in \widehat{P_i^L}} \frac{\exp(q_i \cdot k_j - m_{i,j}^g - z(i, P_i^g))v_j}{\sqrt{d_k}} \tag{10}$$

$$\text{with } m_{i,j}^g = \begin{cases} \infty & j \notin P_i^g \\ 10^5 & i = j \\ \log |\{g' : j \in P_i^{g'}\}| & otherwise \end{cases} \tag{11}$$

2.4   Training Strategy

The task of the lung CT image segmentation is to predict if each pixel of the given image belongs to a predefined class or the background. Therefore, the traditional medical image segmentation problem comes to a binary pixel-wise classification problem. However, in this study, we are focusing on the multi-class image segmentation, which can be concluded as a multi-classes pixel-wise classification. Hence, we choose the multi-class cross-entropy as the loss function:

$$\mathcal{L} = -\sum_{c=1}^{M} y_{o,c} \log(p_{o,c}) \tag{12}$$

where $y_{o,c}$ is a binary value which is used to compare the correct class c and observation class o, $p_{o,c}$ is a probability of the observation o to correct class c and M is the number of classes.

3   Experiment and Evolution Results

3.1   Data Description

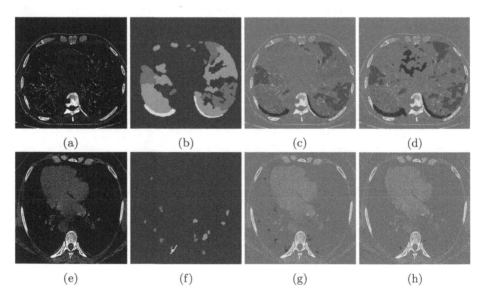

(a)   (b)   (c)   (d)

(e)   (f)   (g)   (h)

Fig. 4. Visualization of segmentation results. The images (a) and (e) show the pre-processed chest CT images of two scans. The images (b) and (f) are the ground-truth masks for these two scans, where the yellow represents the consolidation, blue represents pleural effusion and green corresponds to ground-glass opacities. The images (c) and (g) are the segmentation results generated by our model where the blue represents the consolidation and brown represents the pleural effusion and sky-blue for the ground-glass opacities. The images (d) and (h) are the outputs of the U-Net. In order to make the visualization clear, we choose the light grey as the colour for the background segment. (Color figure online)

We used COVID-19 CT images collected by the Italian Society of Medical and Interventional Radiology (SIRM)[2] for our experimental study. The dataset included 110 axial CT images collected from 43 patients. These images were reversely intensity-normalized by taking RGB-values from the JPG-images from areas of air (either externally from the patient or in the trachea) and fat (subcutaneous fat from the chest wall or pericardial fat) and used to establish the unified Houndsfield Unit-scale (the air was normalized to –1000, fat to –100). The ground-truth segmentation was done by a trained radiologist using Med-Seg[3] with three labels: 1 = ground class opacification, 2 = consolidations, and 3 = pleural effusions. We split the dataset in both patient level and CT image levels to demonstrate the superior of our method. These data are publicly available[4].

3.2   Data Preprocessing and Augmentation

The original CT images have a size of 512 × 512 in matrix form. We use the opencv2[5] to transfer the matrix into gray-scale image to remove some random noises.

As our model is based on deep learning, the number of samples will affect the performance significantly. Consider the size of the dataset, data augmentation is necessary for training the neural network to achieve high generalizability. Our study implements parameterized transformations to realize data augmentation in the training set in this study. We rotate the existing images 90°, 180° and 270° to generate another sets of examples. We can easily generate the corresponding mask by rotating with the same degrees. Scaling has some property with the rotation, so we just scale the image to 0.5 and 1.5 separately to generate another sets of images and its corresponding masks.

3.3   Experiments Setting and Measure Metrics

For the model training, we use the Adma as the optimizer. For a fair comparison, we trained our model and the U-Net with the default parameter in 100 epochs, with learning rate 0.0001 and 3 as the kernel size. Both models are trained under data augmentation and non-augmentation cases. We conducted the experimental analyses on our server consisting of two 12-core/ 24-thread Intel(R) Xeon(R) CPU E5-2697 v2 CPUs, 6 NVIDIA TITAN X Pascal GPUs, 2 NVIDIA TITAN RTX, a total 768 GiB memory. In a segmentation task, especially for the multi-class image segmentation, the target area of interest may take a trivial part of the whole image. Thus, we adopt the Dice Score, accuracy, precision, recall, F1 score and hausdorff distance(HD) as the measure metrics. The dice score is defined as:

---

[2] https://www.sirm.org/category/senza-categoria/covid-19/.

[3] http://medicalsegmentation.com/.

[4] http://medicalsegmentation.com/covid19/.

[5] https://opencv.org/opencv-2-4-8/.

$$DSC(X,Y) = \frac{2|X \cap Y|}{|X| + |Y|} \qquad (13)$$

where X, Y are two sets, and $|\cdot|$ calculates the number of elements in a set. Assume Y is the correct result of the test and X is the predicted result. We conduct the experimental comparison based on 10-fold cross-validation for performance assessment in patient level and image level. And the measure metric for multi-class classification can be calculated by averaging several binary classifications in the same task.

3.4   Results

The Fig. 4 provides two examples about the result images which have data augmentation. The Table 1 shows the measure metric for our proposed model and the U-Net in with data augmentation case and no data augmentation case. Based on this table, we can easily find that our proposed method is out-performed than U-Net that the improvement is at least 10% in all three measure metrics. As shown in Fig. 4(h), we find that the original U-Net almost failed to do the segmentation. The most possible reason is that the range of interest is very small, and the U-Net does not have enough capability to distinguish those trivial difference.

Table 1. Comparison of segmentation performance between our proposed model and U-Net on patient level. All the values are the average value based on the 10-fold cross-validation.

Model	With augmentation					
	DSC	Acc	Precision	Recall	F1	HD(mm)
Ours	0.91	0.85	0.90	0.83	0.86	5.11
U-Net	0.80	0.77	0.81	0.72	0.76	37.27
Improvement	13.8%	10.4%	11.1%	12.2%	13.2%	–
Model	No augmentation					
	DSC	Acc	Precision	Recall	F1	HD(mm)
Ours	0.82	0.77	0.80	0.74	0.78	16.24
U-Net	0.73	0.69	0.71	0.69	0.65	54.22
Improvement	12.3%	11.6%	12.7%	13.2%	20.0%	–

3.5   Ablation Study

In addition to the above-mentioned results, we are also interested in the effectiveness of each component in the proposed model. Accordingly, we conduct the ablation study about the ResNeXt and Attention separately to investigate how these components would affect the segmentation performance. To ensure a fair experimental comparison, we conduct the ablation study in the same experiment

Table 2. Comparison of segmentation performance between our proposed model and U-Net on image level. All the values are the average value based on the 10-fold cross-validation.

Model	With augmentation					
	DSC	Acc	Precision	Recall	F1	HD(mm)
Ours	0.94	0.89	0.95	0.85	0.90	4.85
U-Net	0.82	0.79	0.83	0.74	0.78	34.23
Improvement	14.6%	12.7%	14.5%	14.9%	15.4%	–
Model	No augmentation					
	DSC	Acc	Precision	Recall	F1	HD(mm)
Ours	0.83	0.79	0.82	0.76	0.79	13.82
U-Net	0.75	0.70	0.72	0.62	0.67	44.23
Improvement	10.7%	12.9%	13.9%	22.6%	17.9%	–

environment with our main experiments presented in Sect. 3.3. We implement the ablation study on two variants of our model: Model without Attention and Model without ResNeXt. Our model without ResNeXt is similar with literature [18]. We just use the M-R to represent it. The results are summarized in Table 3, where M-A represents the model without attention and M-R represents the model without ResNeXt block. We can observe that both the attention and ResNeXt blocks play important roles in our model and contribute to derive improved segmentation performance in comparison with U-Net (Tables 3 and 4).

Table 3. Comparison result of ablation study. All the values are the average value based on the 10-fold cross-validation on patient level.

Model	With augmentation					
	DSC	Acc	Precision	Recall	F1	HD(mm)
Ours(M)	0.91	0.85	0.90	0.83	0.86	5.11
M - A	0.84	0.80	0.83	0.77	0.80	15.22
M - R	0.82	0.79	0.81	0.75	0.78	20.42
M-A-R	0.80	0.77	0.81	0.72	0.76	37.27
Model	No augmentation					
	DSC	Acc	Precision	Recall	F1	HD(mm)
Ours(M)	0.82	0.77	0.80	0.74	0.78	16.24
M - A	0.77	0.74	0.73	0.67	0.70	27.51
M - R	0.76	0.73	0.73	0.63	0.68	40.51
M-A-R	0.73	0.69	0.71	0.60	0.65	54.22

Table 4. Comparison result of ablation study. All the values are the average value based on the 10-fold cross-validation on image level.

Model	With augmentation					
	DSC	Acc	Precision	Recall	F1	HD(mm)
Ours(M)	0.94	0.89	0.95	0.85	0.90	4.85
M - A	0.85	0.82	0.84	0.79	0.81	13.66
M - R	0.84	0.81	0.83	0.76	0.79	19.41
M-A-R	0.82	0.79	0.83	0.74	0.78	34.23
Model	No augmentation					
	DSC	Acc	Precision	Recall	F1	HD(mm)
Ours(M)	0.83	0.79	0.82	0.76	0.79	13.82
M - A	0.79	0.74	0.77	0.70	0.73	24.24
M - R	0.77	0.76	0.77	0.67	0.72	32.42
M-A-R	0.75	0.70	0.72	0.62	0.67	44.23

4   Discussion and Conclusions

Up to now, the most common screening tool for COVID-19 is CT imaging. It can help the community to accelerate the speed of diagnosing and accurately evaluate the severity of COVID-19 [22]. In this paper, we presented a novel deep learning-based algorithm for automated segmentation of COVID-19 CT images, and its proved that such algorithm is plausible and superior comparing to a series of baselines. We proposed a modified U-Net model by exploiting the residual network to enhance the feature extraction. An efficient attention mechanism was further embedded in the decoding process to generate high-quality multi-class segmentation results. Our method gained more than 10% improvement in multi-class segmentation when comparing against U-Net and a set of baselines.

A recent study shows that the early detection of COVID-19 is very important [4]. If the infection in chest CT image can be detected at an early stage, the patients would have a higher chance to survive [25]. Our study provides an effective tool for the radiologist to precisely determine the lung's infection percentage and diagnose the progression of COVID-19. It also shed some light on how deep learning can revolutionize the diagnosis and treatment in the midst of COVID-19.

Our future work would be generalizing the proposed model into a wider range of practical scenarios, such as facilitating with diagnosing more types of diseases from CT images. In particular, in the case of a new disease, such as the coronavirus, the amount of ground truth data is usually limited given the difficulty of data acquisition and annotation. The model is capable of generalizing and adapting itself using only a few available ground-truth samples. Another line of future work lies in the interpretability, which is especially critical for the medical domain applications. Although deep learning is widely accepted to its limitation in interpretability, the attention mechanism we proposed in this work can produce the interpretation of internal decision process at some levels. To gain deeper scientific insights, we will keep working along with this direction and explore the hybrid attention model for generating meaningfully semantic explanations.

References

1. Ai, T., et al.: Correlation of chest ct and rt-pcr testing in coronavirus disease 2019 (covid-19) in China: a report of 1014 cases. Radiology 200642 (2020)
2. Alom, M.Z., Hasan, M., Yakopcic, C., Taha, T.M., Asari, V.K.: Recurrent residual convolutional neural network based on u-net (r2u-net) for medical image segmentation (2018). arXiv:1802.06955
3. Baumgartner, C.F., Koch, L.M., Pollefeys, M., Konukoglu, E., et al.: An exploration of 2D and 3D deep learning techniques for cardiac MR image segmentation. In: Pop, M. (ed.) STACOM 2017. LNCS, vol. 10663, pp. 111–119. Springer, Cham (2018). https://doi.org/10.1007/978-3-319-75541-0_12
4. Bernheim, A., et al.: Chest ct findings in coronavirus disease-19 (covid-19): relationship to duration of infection. Radiology, 200463 (2020)
5. Dong, X., et al.: Automatic multiorgan segmentation in thorax ct images using u-net-gan. Med. Phys. **46**(5), 2157–2168 (2019)
6. Gaál, G., Maga, B., Lukács, A.: Attention u-net based adversarial architectures for chest x-ray lung segmentation (2020). arXiv:2003.10304
7. He, K., Sun, J.: Convolutional neural networks at constrained time cost. In: IEEE CVPR, pp. 5353–5360 (2015)
8. He, K., Zhang, X., Ren, S., Sun, J.: Deep residual learning for image recognition. In: Proceedings of the IEEE CVPR, pp. 770–778 (2016)
9. Huang, C., et al.: Clinical features of patients infected with 2019 novel coronavirus in wuhan, china. The Lancet **395**(10223), 497–506 (2020)
10. Ibtehaz, N., Rahman, M.S.: Multiresunet: rethinking the u-net architecture for multimodal biomedical image segmentation. Neural Netw. **121**, 74–87 (2020)
11. Jiang, X., et al.: Towards an artificial intelligence framework for data driven prediction of coronavirus clinical severity. Comput. Mater. Continua **62**(3), 537–551 (2020)
12. Kitaev, N., Kaiser, L., Levskaya, A.: Reformer: the efficient transformer. In: International Conference on Learning Representations (2020). https://openreview.net/forum?id=rkgNKkHtvB
13. Li, Y., Xia, L.: Coronavirus disease 2019 (covid-19): role of chest ct in diagnosis and management. Am. J. Roentgenol **214**, 1–7 (2020)
14. Liu, Z., et al.: Liver ct sequence segmentation based with improved u-net and graph cut. Expert Syst. Appl. **126**, 54–63 (2019)
15. Long, C., et al.: Diagnosis of the coronavirus disease (covid-19): rrt-pcr or ct? Eur. J. Radiol. **126**, 108961 (2020)
16. Long, J., Shelhamer, E., Darrell, T.: Fully convolutional networks for semantic segmentation. In: Proceedings of the IEEE CVPR, pp. 3431–3440 (2015)
17. Narin, A., Kaya, C., Pamuk, Z.: Automatic detection of coronavirus disease (covid-19) using x-ray images and deep convolutional neural networks (2020). arXiv:2003.10849

18. Oktay, O., et al.: Attention u-net: Learning where to look for the pancreas (2018). arXiv:1804.03999
19. Raptis, C.A., et al.: Chest ct and coronavirus disease (covid-19): a critical review of the literature to date. Am. J. Roentgenol. **215**(1), 1–4 (2020)
20. Ronneberger, O., Fischer, P., Brox, T.: U-Net: convolutional networks for biomedical image segmentation. In: Navab, N., Hornegger, J., Wells, W.M., Frangi, A.F. (eds.) MICCAI 2015. LNCS, vol. 9351, pp. 234–241. Springer, Cham (2015). https://doi.org/10.1007/978-3-319-24574-4_28
21. Salehi, S., Abedi, A., Balakrishnan, S., Gholamrezanezhad, A.: Coronavirus disease 2019 (COVID-19): a systematic review of imaging findings in 919 patients. Am. J. Roentgenol **215**, 1–7 (2020)
22. Shan, F., et al.: Lung infection quantification of covid-19 in ct images with deep learning (2020). arXiv:2003.04655
23. Shen, D., Wu, G., Suk, H.I.: Deep learning in medical image analysis. Ann. Rev. Biomed. Eng. **19**, 221–248 (2017)
24. Shi, F., et al.: Review of artificial intelligence techniques in imaging data acquisition, segmentation and diagnosis for covid-19 (2020). arXiv:2004.02731
25. Song, F., et al.: Emerging 2019 novel coronavirus (2019-ncov) pneumonia. Radiology, 200274 (2020)
26. Sutskever, I., Vinyals, O., Le, Q.V.: Sequence to sequence learning with neural networks. In: NIPS, pp. 3104–3112 (2014)
27. Vaswani, A., et al.: Attention is all you need. In: Advances in Neural Information Processing Systems, pp. 5998–6008 (2017)
28. Wang, L.S., Wang, Y.R., Ye, D.W., Liu, Q.Q.: A review of the 2019 novel coronavirus (covid-19) based on current evidence. Int. J. Antimicrob. Agents **55**, 105948 (2020)
29. Xie, S., Girshick, R., Dollár, P., Tu, Z., He, K.: Aggregated residual transformations for deep neural networks. In: Proceedings of the IEEE CVPR, pp. 1492–1500 (2017)
30. Ye, C., Wang, W., Zhang, S., Wang, K.: Multi-depth fusion network for whole-heart ct image segmentation. IEEE Access **7**, 23421–23429 (2019)
31. Zhou, B., Khosla, A., Lapedriza, A., Oliva, A., Torralba, A.: Learning deep features for discriminative localization. In: Proceedings of the IEEE CVPR, pp. 2921–2929 (2016)
32. Zhu, N., et al.: A novel coronavirus from patients with pneumonia in China, 2019. New Engl. J. Med. **382**, 727–733 (2020)

Comparison the Performance of Classification Methods for Diagnosis of Heart Disease and Chronic Conditions

Jiarui Si[1,2](✉), Haohan Zou[1], Chuanyi Huang[1], Huan Feng[3], Honglin Liu[1], Guangyu Li[4], Shuaijun Hu[1], Hong Zhang[6](✉), and Xin Wang[5](✉)

[1] School of Basic Medical Sciences, Tianjin Medical University, Tianjin, China
`sijiarui@tmu.edu.cn`
[2] School of Public Health, Tianjin Medical University, Tianjin, China
[3] School of Medical Humanities, Tianjin Medical University, Tianjin, China
[4] Tianjin Medical University Cancer Institute and Hospital, Tianjin, China
[5] College of Intelligence and Computing, Tianjin University, Tianjin, China
[6] Department of Radiology, Tianjin Chest Hospital, Tianjin, China

Abstract. In this paper, we aim to identify the best ratio of training set, and to evaluate the diagnostic performance of eight classical machine learning methods for chronic diseases. The five categories of chronic disease datasets were collected from UCI and GitHub database include heart disease, breast cancer, diabetic retinopathy, Parkinson's disease and diabetes. Machine learning (ML) methods including six individual learners (logistic regression (LR), BP neural network (BP), learning vector quantization (LVQ), extreme learning machine(ELM), support vector machine (SVM), decision trees (DTs)), and two ensemble learning methods were implemented using BP_Adaboost (Ada) and random forest (RF). We used five indicators, including AUC value, accuracy, sensitivity, specificity and running time to compare the behaviors of each algorithm. The ensemble learning methods were the most suited for the diagnosis of chronic diseases and led to a significant enhancement of performance compared to individual learners.

Keywords: Machine learning · Individual learning · Ensemble learning · Chronic disease

1 Introduction

Chronic diseases usually refer to a group of non-communicable diseases (NCD) with long onset time and complicated etiology [1]. Heart disease (HD) is a common circulatory disease [12]. Diabetes mellitus (DM) and its complications, cancers, cardiopathy and encephalopathy, are several kinds of frequently-occurred chronic diseases reported by the World Health Organization [2–5]. Studies have shown that if chronic diseases can be detected, diagnosed, and treated in early stage, it would significantly improve the therapeutic effect and reduce the economic burden [6–8]. Early detection and control of DM can delay the occurrence of various complications and conducive to treatment

© Springer Nature Switzerland AG 2021
W. Zhang et al. (Eds.): WISE 2021, LNCS 13081, pp. 137–144, 2021.
https://doi.org/10.1007/978-3-030-91560-5_10

[10]. Diabetic retinopathy (DR) is a disease characterized by a progressive damage of the retina, and is a main cause of vision loss among people with diabetes. Screening for DR can reduce the chance of blindness and decrease the overall load on the health care systems. Parkinson's disease (PD) is an encephalopathy characterized by several fundamental motor symptoms, patients with PD are often irreparable, so it is crucial to predict the risk of it [11]. Breast cancer (BC) is the most common life-threatening cancer for women worldwide and the incidence is still increasing particularly in developing countries where many patients are not diagnosed until in late stages [13]. Therefore, finding accurate and inexpensive technology is crucial to reduce the social burden of breast cancer.

As the chronic diseases are continually increasing, manual screening has become a time-consuming task [14]. Fortunately, Machine learning (ML) algorithm such as logistic regression (LR), various kinds of neural networks (NNs), support vector machines (SVM), decision trees (DTs), and ensemble classifiers random forest (RF) and BP_Adaboost (Ada) have been successfully applied in medical diagnosis [15, 16]. Various machine learning techniques have been used to help clinicians analyze problems, but the researchers have not come to a conclusion as to which learning technique is more suitable for chronic disease diagnosis [17–21]. In a previous research about prognosis of breast cancer survival analysis, Jose et al. had concluded that the machine learning method was better than the traditional statistical method. And the best predictions could be obtained by KNN method of six machine learning method [17]. Andre et al. [18] created a model to predict type II diabetes with LR, artificial neural network (ANN), naive Bayes, KNN and RF. The models involving LR and ANN had achieved satisfying results. Tsao et al. used SVM, DT, ANN algorithms and LR to build a prediction model for the DR, the results demonstrated that appropriate algorithms combined with discriminative clinical features are more effective than others [19]. Gao et al. applied LR, RF, SVM, and XGboost constructed predictive models to discriminate fallers and non-fallers of PD patients [20]. For ensemble classifier composed of multiple DTs has appeared the highest accuracy in heart disease detection [21]. Ada is another ensemble learning method in which BP neural network is used as a weak classifier, but its application in chronic disease diagnosis had rarely been reported. To measure the behavior of each algorithm comprehensively, we then compared the average classification efficiency of five different datasets.

2   The Dataset of Chronic Diseases

The dataset of frequently-occurred chronic diseases came from the UCI (http://archive. ics.uci.edu/ml/index.php) and GitHub (https://github.com/) machine learning repository. The Diabetes mellitus dataset was collected from National Institute of Diabetes and Digestive and Kidney Disease [22]. The Diabetic Retinopathy dataset was provided by the department of computer graphics and image processing faculty of informatics in the Debrecen University [23].The SPECT Heart dataset was donated by university of Colorado at Denver in 2001 [24]. The Parkinson dataset was created by Max Little of the University of Oxford, in collaboration with the National Centre for Voice and Speech, Denver, Colorado [26]. The breast cancer dataset was collected by Dr. William

H. Wolberg, W. Nick Street and Olvi L. Mangasarian of University of Wisconsin System [25].

3 Method

3.1 Logistic Regression

LR is a linear model that attempts to model the conditional probability of binary results.

3.2 Neural Networks

Neural Network is developed from simulations of the human brain. It learns statistical law of data in a way like the mind's memory and induces a data model that can describe the features of a sample.

3.3 BP Neural Network

The BP neural network is a multi-layer pre-feedback neural network consisting of input, implicit, and output layers. The information is forwarded and the error is reversed.

3.4 LVQ Neural Network

The LVQ neural network evolved from the Kohonen competition algorithm. It consists of input layer, competition layer, and output layer.

3.5 ELM Neural Network

The ELM neural network consists of input, implicit, and output layers. The weights and thresholds between the input layer and the hidden layer are randomly generated.

3.6 SVM

Similar to multi-layer perception networks and radial basis function networks, SVM can be used for pattern classification and nonlinear regression.

3.7 Decision Trees

DTs is an instance-based inductive learning algorithm based on the idea of increasing information entropy. The algorithm infers the classification rules of the decision tree representation from a set of meta-orders and irregular cases.

4 Ensemble Learning

4.1 Adaboost

Ada is one of great Boosting algorithms. The Ada algorithm is used to combine the outputs of multiple "weak" classifiers to produce a valid classification.

4.2 Random Forest

RF is belonged to ensemble learning algorithm. Actually, it is a classifier containing multiple DTs. These DTs are formed by random methods so they are also called random DTs. The Bootstrap method is re-randomly sampled and it generates a training set without correlation between the DTs.

4.3 Statistical Approach and Evaluating Metrics

Analysis of Variance (ANOVA) was used to compare the differences of accuracy, sensitivity, and specificity in each algorithm. Dunnett's t test was used to test the differences between the best performing algorithm and others. All test were two tailed, and statistical was defined as $P < 0.05$. For evaluating the performance of these methods, the metrics including AUC-value, accuracy, sensitivity, specificity, and running time were applied.

5 Results

Table 1 shows the baseline of each dataset. In order to achieve more reliable results, we prefer to seek the authenticity of the results without processing and deleting the real-world data.

Table 1. Data sets description

Dataset	No. of attributes	No. of instances	Positive cases	Negative cases
HD	23	267	212	55
DM	9	768	500	268
DR	20	1151	1067	94
PD	23	195	171	24
BC	32	569	212	357

* DM: Pima Indians diabetes dataset; DR: Diabetic retinopathy dataset; PD: Parkinson dataset; HD: SPECT Heart dataset; BC: Breast Cancer Wisconsin dataset.

As shown in Fig. 1, Eight machine learning methods were used to classify five chronic diseases, and the average diagnostic performance under different training set ratios were compared, different color lines represent different methods, the points on the line represent the average value of AUC (AUC_value) under the corresponding training sample ratio, then the AUC_value of each method were averaged to get the Average_AUC marked in red dotted line. The higher the Average_AUC, the better the effect. With the decrease of the proportion of training set, the AUC values corresponding to the SVM, BP and Ada algorithm are stable and have little volatility. The AUC value corresponds to ELM, RF and LR is decreasing with fluctuation; However, LVQ is slightly increasing with fluctuation. Based on Average_AUC values the proper percentage of

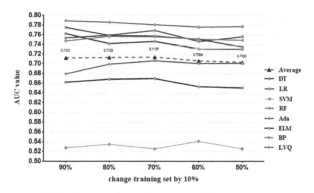

Fig. 1. The Average AUC value corresponds to every 10% change in training set proportion

training samples is 70%, the better results are two ensemble learning algorithms RF and Ada, especially RF is better than any others in any proportion.

After using the AUC value to determine the average optimal ratio of the data set (7:3), we ran each method ten times to calculate average accuracy, sensitivity, specificity, AUC value and running time of each method, in order to measure the pros and cons of various methods more objectively from various aspects. Training sets and testing sets of each data were randomly selected under the optimal ratio. The result was showed in Table 2.

Table 2. Average results from running models ten times

Algorithm	Running time(s)	Accuracy	Sensitivity	Specificity	AUC
LR	1.999	0.799	0.772	0.728	0.757
BP	61.034	0.788	0.737	0.767	0.756
LVQ	12682.516	0.762	0.699	0.721	0.706
ELM	1.382	0.747	0.613	0.712	0.670
SVM	1.351	0.674	0.472	0.590	0.526
DT	1.014	0.781	0.709	0.779	0.746
Ada	49.309	0.834	0.824	0.850	0.769
RF	10.284	0.823	0.757	0.805	0.781

In Table 2, the average results showed the differences of the algorithm. The average accuracy, sensitivity and specificity of Ada algorithm are all optimal, and the average running time of the DT is the shortest. These eight algorithms have significant differences in accuracy ($F = 95.57$, $p < 0.0001$), sensitivity ($F = 96.62$, $p < 0.0001$) and specificity ($F = 71.18$, $p < 0.0001$). The running time of LVQ is longer than other algorithms obviously, the sensitivity and specificity of SVM are relatively lower than others.

In terms of AUC, we compared the best ensemble learner (RF) with the worst (Ada) in accuracy, sensitivity and specificity, the Dunnett's multiple comparisons test was used. In Table 3 there are no difference between Ada and RF for accuracy. Furthermore, compared the best individual learner (LR) with the worst ensemble learner (Ada), the latter is significantly better than the former.

Table 3. Comparison of effect between Ada and LR learners

Indicator	Dunnett's t test		Mean Dif.	95% CI of Dif.	Sig.	P value
Accuracy	Ada	vs. RF	0.0119	(−0.0065, 0.0302)	NS	0.3754
		vs. LR	0.0357	(0.0173, 0.0540)	****	<0.0001
Sensitivity	Ada	vs. RF	0.0447	(0.0121, 0.0774)	**	0.0024
		vs. LR	0.1221	(0.0894, 0.1548)	****	<0.0001
Specificity	Ada	vs. RF	0.0669	(0.0213, 0.1126)	***	0.0009
		vs. LR	0.0515	(0.0058, 0.0971)	*	0.0194

* CI: confidence interval; Dif.: Difference; Sig.: Significant at the 0.05 level; NS: No statistical difference; Significant differences are expressed as * P < 0.05, ** P < 0.01, *** P < 0.001 and *** *P < 0.0001.

In order to eliminate the data dependence of a single type of disease as much as possible, we calculated the average efficacy of each method for the data diagnosis of 5 chronic diseases, as shown in Fig. 2. So far, the research results showed that the classification task based on the ensemble learning algorithm (Ran and Ada) is superior to the single learner algorithm in the prediction and diagnosis of the incidence risk of patients with chronic diseases.

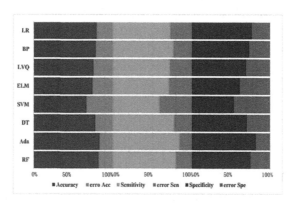

Fig. 2. The average accuracy, sensitivity and specificity of eight methods

6   Conclusion and Discussion

We applied eight machine learning methods to achieve the early diagnosis of five kinds of frequently-occurred chronic diseases based on the datasets from UCI and GitHub database. First, we introduced the harm of chronic diseases, the significance of early prevention and control. Second, we elaborated the characteristics of several chronic diseases selected in our research and the importance of early diagnosis respectively. After that, we reviewed the previous research of several machine learning methods related to this article for the diagnosis of chronic diseases. Next, we introduced the principle of the eight machine learning methods including LR, three kinds of neural networks BP, LVQ and ELM, SVM, DTs, ensemble classifiers RF and Ada. Then these methods were applied to handle diseases data using software MATLAB 2018b. Finally, the results obtained from different machine learning methods to diagnose various kinds of data sets in these two databases were summarized.

We concluded that machine learning may have excellent performance in the diagnosis of chronic diseases because of their significant increase in accuracy. Clinical decision support depends on the available data and the prediction model, thus, the present results might be applied to different chronic disease datasets. Up to this point, the results indicate that clinical decision support based on ensemble learning algorithms outperformed the single learner in the diagnosis of the frequently-occurred chronic diseases and prioritize the Ada algorithm. The proportion of training samples of data has a great impact on the results of machine learning. When selecting specific data sets and samples, the best training sample ratio should be used for prediction.

Acknowledgements. The authors acknowledge support from National Natural Science Foundation of China (91746205, 71673199), and Tianjin Health Science and technology project (ms20015).

References

1. Hunter, D.J., Srinath Reddy, K.: Noncommunicable diseases. N. Engl. J. Med. **369**(14), 1336–1343 (2013)
2. WHO: Global status report on noncommunicable diseases 2014. Women **47**(26), 2562–2563 (2011)
3. Mendis, S.: Combating chronic diseases: the role of the World Health Organization. Glob. Heart **11**(4), 413 (2016)
4. Oni, T., et al.: Chronic diseases and multi-morbidity - a conceptual modification to the WHO ICCC model for countries in health transition. BMC Pub. Health **14**(1), 575 (2014)
5. Yang, G., et al.: Emergence of chronic non-communicable diseases in China. The Lancet **372**(9650), 1697–1705 (2008)
6. Capizzi, S., de Waure, C., Boccia, S.: Global burden and health trends of non-communicable diseases. In: Boccia, S., Villari, P., Ricciardi, W. (eds.) A Systematic Review of Key Issues in Public Health, pp. 19–32. Springer, Cham (2015). https://doi.org/10.1007/978-3-319-136 20-2_3
7. Ambady, R., Chamukuttan, S.: Early diagnosis and prevention of diabetes in developing countries. Rev. Endocr. Metab. Disord. **9**(3), 193–201 (2008)
8. Kuhlman, G.D., Flanigan, J.L., Sperling, S.A., et al.: Predictors of health-related quality of life in Parkinson's disease. Parkinsonism Relat. Disord. **65**, 86–90 (2019)
9. Thomas, R.L., Halim, S., Gurudas, S., Sivaprasad, S., Owens, D.R.: IDF diabetes Atlas: a review of studies utilising retinal photography on the global prevalence of diabetes related retinopathy between 2015 and 2018. Diab. Res. Clin. Pract. **157**, 107840 (2019)

10. Bashir, S., Qamar, U., Khan, F.H.: WebMAC: a web based clinical expert system. Inf. Syst. Front. **20**(5), 1135–1151 (2016)
11. Leddy, A.L., Crowner, B.E., Earhart, G.M.: Utility of the Mini-BESTest, BESTest, and BESTest sections for balance assessments in individuals With Parkinson disease. J. Neurol. Phys. Ther. **35**(2), 90–97 (2011)
12. Mohaghegh, P., Rockall, A.G.: Imaging strategy for early ovarian cancer: characterization of Adnexal masses with conventional and advanced imaging techniques. Radio Graph. **32**(6), 1751–1773 (2012)
13. Azamjah, N., Soltan-Zadeh, Y., Zayeri, F.: Global trend of breast cancer mortality rate: a 25-year study. Asian Pac. J. Cancer Prev. **20**(7), 2015–2020 (2019)
14. Romero, Y., et al.: National cancer control plans: a global analysis. Lancet Oncol. **19**(10), e546–e555 (2018)
15. Tripoliti, E.E., Tzallas, A.T., Tsipouras, M.G., Rigas, G., Bougia, P., Leontiou, M., et al.: Automatic detection of freezing of gait events in patients with Parkinson's disease. Comput. Meth. Progr. Biomed. **110**, 12–26 (2013). https://doi.org/10.1016/j.cmpb.2012.10.016
16. Holzinger, A.: Interactive machine learning for health informatics: when do we need the human-in-the-loop? Brain Inf. **3**(2), 119–131 (2016). https://doi.org/10.1007/s40708-016-0042-6
17. Jerez, J.M., et al.: Missing data imputation using statistical and machine learning methods in a real breast cancer problem. Artif. Intel. Med. **50**(2), 105–115 (2010)
18. Olivera, A.R., et al.: Comparison of machine-learning algorithms to build a predictive model for detecting undiagnosed diabetes - ELSA-Brasil: accuracy study. Sao Paulo Med. J. **135**(3), 234–246 (2017)
19. Tsao, H.-Y., Chan, P.-Y., Emily Chia-Yu, S.: Predicting diabetic retinopathy and identifying interpretable biomedical features using machine learning algorithms. BMC Bioinf. **19**(S9), 283 (2018)
20. Gao, C., Sun, H., Wang, T., et al.: Model-based and model-free machine learning techniques for diagnostic prediction and classification of clinical outcomes in Parkinson's disease. Sci. Rep. **8**(1), 7129 (2018)
21. Tao, R., et al.: Magnetocardiography-based ischemic heart disease detection and localization using machine learning methods. IEEE Trans. Biomed. Eng. **66**(6), 1658–1667 (2019)
22. Smith, J.W., Everhart, J.E., Dickson, W.C., et al.: Using the ADAP learning algorithm to forecast the onset of Diabetes Mellitus. J. Hopkins APL Tech. Dig. **10**, 261–265 (1988)
23. Antal, B., Hajdu, A.: An ensemble-based system for automatic screening of diabetic retinopathy. Knowl. Based Syst. **60**, 20–27 (2014)
24. Kurgan, L.A., Cios, K.J., Tadeusiewicz, R., et al.: Knowledge discovery approach to automated cardiac SPECT diagnosis. Artif. Intell. Med. **23**(2), 149–169 (2001)
25. Street, W.N., Wolberg, W.H., Mangasarian, O.L.: Nuclear feature extraction for breast tumor diagnosis (1993)
26. Little, M.A., Mcsharry, P.E., Roberts, S.J., et al.: Exploiting nonlinear recurrence and fractal scaling properties for voice disorder detection. Biomed. Eng. Online **6**(1), 23 (2007)

Recommender Systems (1)

Capturing Multi-granularity Interests with Capsule Attentive Network for Sequential Recommendation

Zihan Song[1], Jiahao Yuan[2], Xiaoling Wang[1(✉)], and Wendi Ji[1(✉)]

[1] School of Computer Science and Technology, East China Normal University,
Shanghai, China
zhsong@stu.ecnu.edu.cn, xlwang@cs.ecnu.edu.cn
[2] Software Engineering Institute, East China Normal University, Shanghai, China
jhyuan@stu.ecnu.edu.cn

Abstract. The sequential recommender system attempts to predict the next interaction based on user's historical behaviors, which is a challenging problem due to intricate sequential dependencies and user's various interests underneath the interactions. Existing works regard each item that the user interacts with as an interest unit and apply advanced deep learning techniques to learn a unified interest representation. However, user's interests vary in multiple granularities. An item mirrors preferences for a specific item, while a set of items reflect general user interests, which are barely captured by a unified representation at the same granularity level. Furthermore, since the unrelated items are treated at the same granularity level as these decisive items, the model cannot focus on the items that help make the accurate recommendation. In this paper, we propose a novel Capsule Attentive Network (**CAN**) for sequential recommendation, which integrates the dynamic routing algorithm to capture diverse user interests at coarse-grained levels with a transformer module to learn more informative embeddings at fined-grained levels. Experimental results on three datasets demonstrate that CAN achieves substantial improvement over state-of-the-art methods.

Keywords: Sequential recommendation · Capsule network · Attention network

1 Introduction

Capturing user's potential interests is fundamental to sequential recommendation. In many real-world applications, user's diverse interests are embodied by different combinations of items at multiple granularity levels. For example, a user clicks an iPhone out of her desire to buy a mobile phone, while the other user may click iPhone, iPad and Mac because his preference for Apple's products. The user's interests are different, even if both users click the iPhone, due to the multiple granularities that interactions are treated.

© Springer Nature Switzerland AG 2021
W. Zhang et al. (Eds.): WISE 2021, LNCS 13081, pp. 147–161, 2021.
https://doi.org/10.1007/978-3-030-91560-5_11

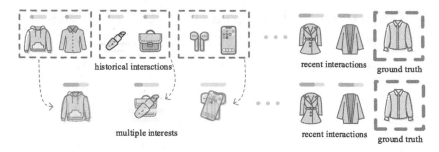

Fig. 1. The differences between general sequential recommender systems and our proposed method which models the sequences at two different granularity levels. The bars above the items represent the possible attention weight corresponding to the item.

To model diverse interests in interaction sequences, traditional methods [7] investigate nearest neighbors to find interactions that obtain similar interests. Recently, approaches based on deep learning have achieved revolutionary improvements in the field of sequential recommendation. RNN based methods [4,5] treat each item as an interest unit and capture the dynamic interest transition in the sequences. SASRec [8] applies attention mechanism to extract the correlation between items and regards the last interaction as the user's interest. SHAN [21] obtains a general interest by fusing user's long-term interests and short-term interests by a hierarchical attention network.

Although all of the methods mentioned above have achieved promising results, there are still limitations. Firstly, these approaches regard each item as an interest unit and obtain a unified interest representation, which cannot captures diverse user's preferences at different granularity levels. Some researchers try to represent user's interests with different vectors [1,2,10]. However, these methods neglect the positional information in the sequences, barely explore user's intentions and recommend items only based on each interest, which are insufficient to infer what users are interested in at present. Secondly, the unified interest learned by model will be dominated by the most recent interactions. As a result, items that should be genuinely considered contribute less to the final recommendations as they are far from the last interaction, leading to inadequate use of sequence information. Lastly, due to the uncertainty of user behaviors, some of the user-item interactions may be caused by mistouches or guided by the e-commerce platform, meaning they may contain some irrelevant interactions that interfere with the next interaction prediction [18]. Despite that the arrival of the attention mechanism [8,17] allows models to focus more on the relevant items, these irrelevant items still hold a few attention weights, which introduce noises.

As illustrated in Fig. 1, the picture above is about the traditional sequential recommender system, which models the whole sequences based on each unit. Intuitively, from the user-item interactions, we can find that the user's intention is likely to buy a piece of clothing. Items such as the lipsticks and the bag may contributes less from the point of view of a single unit, but they may still reflect specific user preferences as a group of items because cosmetics, accessories and clothing are all about fashion. Furthermore, the model can not focus on the clothing item in the historical interactions for the long distance from the last interaction, and the phone and earphone become noises.

To this end, we propose a capsule attentive network for sequential recommendation. As shown below in Fig. 1, we split the sequences into two parts according to sequential order and model the sequence at two different granularity levels in order to better learn and utilize user's multiple interests. Specifically, the historical interactions converge to multiple user's interests at coarse-grained levels by the dynamic routing algorithm. Subsequently, the selected user's interests will be further analyzed with the recent interactions in the following modules at fine-grained levels, which retains favorable information in the historical interactions and eliminates the influence of the unrelated items to some extent. Finally, we leverage an attention module to distinguish the importance of recent interactions and user's interests and obtain the final output. The contributions of our work are summarized as follows.

– We propose a novel framework that extracts the user's interests at both coarse-grained levels and fine-grained levels.
– We introduce the idea of capsule network into sequential recommendation and employ the dynamic routing algorithm to learn user's multiple interests.
– We conduct extensive experiments on three real-world datasets to verify the effectiveness of our proposed model and provide insight into how CAN works under the hood.

2 Related Work

2.1 Sequential Recommendation

Sequential recommender system has become a popular branch in the field of recommender systems, owing to its capability to capture complex dependencies between items. GRU4Rec [5] firstly introduces the recurrent neural network with gated recurrent units into the sequential recommendation. It regards user-item interactions as sequences and predicts the user's next interaction by user's historical interactions, which significantly improves the performance compared to the previous works [12]. Caser [16] treats user-item interactions as an image and applies the convolutional neural networks to extract user interests by continuous items. RUM [2] designs a memory-augmented neural network that dynamically stores and manipulates user's interactions and interests, which enhances the expressiveness of the model. SASRec [8] leverages a self-attention module to capture long-term semantics using the attention mechanism. Similar to SASRec, BERT4Rec [14] further improves the performance by a bidirectional architecture. GC-SAN [20] adopts graph neural networks and a self-attention module to learn the intricate transition of user-item interactions in sequences, reaching competitive results. SDM [11] applies a multi-head self-attention module to capture user's multiple preferences by combining short-term sessions and long-term behaviors with side information. However, these approaches merely consider the relationship between items by a unified representation and neglect the diverse user's interests reflect by a set of items at different granularity levels.

2.2 Multi-Interest Network for Recommendation

One line of works is based on capsule networks. The concept of capsules is first proposed by [6] and employed in the handwritten digit recognition tasks since the dynamic routing algorithm [13] is proposed. MIND [10] introduces the capsule network into recommender systems and designs a multi-interest extractor layer based on the dynamic routing algorithm, which is applicable for clustering historical behaviors and extracting diverse interests. ComiRec [1] further refines the dynamic routing algorithm to make it more suitable for the recommender system scenario and proposes a self-attentive method based approach, which achieves encouraging performance in real-world applications. Another line of works seeks to classify items into several groups by model itself. MCPRN [19] proposes a mixture-channel model to accommodate the multi-purpose item subsets for more precisely representation of a session, which recommends more diverse items to satisfy different purposes. MIIM [3] proposes a multiple interleaving interests modeling framework to model the user behavior sessions. SINE [15] integrates large-scale item clustering and sparse-interest extraction jointly for recommender system. With the help of the sparse-interest module, it adaptively infers a sparse set of concepts for each user from the large concept pool and outputs multiple embeddings accordingly. However, although these methods capture multiple user interests, they ignore the crucial positional information in the sequences and cannot recognize the user's current intention.

3 Proposed Method

3.1 Problem Formulation

Let $\mathcal{U} = \{u_1, u_2, ...u_{|\mathcal{U}|}\}$ denote a set of users, $\mathcal{V} = \{v_1, v_2, ...v_{|\mathcal{V}|}\}$ indicate a set of items and $\mathcal{S}_u = \{v_1^u, v_2^u, ..., v_{|\mathcal{S}_u|}^u\}$ be the user interaction sequence in chronological order for user $u \in \mathcal{U}$, where $v_i^u \in \mathcal{V}$ represents the ith interacted item of the user u. Given a user's interaction truncated at timestep t, $\mathcal{S}_{u,t} = \{v_1^u, v_2^u, ..., v_t^u\}$, a sequential recommender system aims to predict the next item v_{t+1}^u that the user u is most likely to interact. More precisely, our proposed model $\mathcal{M}$ learns to output a score $y_v = \mathcal{M}(\mathcal{S}_{u,t})$ for each item v in the item set $\mathcal{V}$, then the top-K items with the highest scores will be recommended to the user.

3.2 Capsule Attentive Network

The proposed model, Capsule Attentive Network(CAN), consists of four components: (1) an embedding layer that transforms the user-item interaction sequences into embeddings, (2) a coarse-grained interest capsule network to capture user's diverse interests in the sequence, (3) a fine-grained transformer layer that exploits item-level collaborative information among the recent interactions and interest-level user preference obtained by the interest capsule network, and (4) a prediction layer to gather all the information for the final result. The architecture of CAN is shown in Fig. 2.

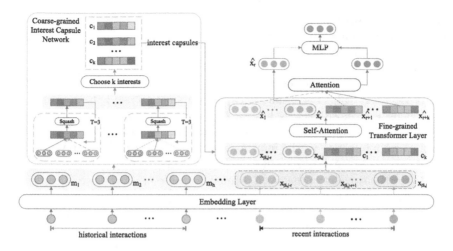

Fig. 2. The overall workflow of the CAN model.

Embedding Layer. The embedding layer maps the items to dense vectors endowing the model with more powerful presentation capabilities. Since the positional information of an item is proved to be rather crucial in sequential recommendation [8,14], in addition to the item embedding matrix $M \in \mathbb{R}^{|\mathcal{V}| \times d}$, we incorporate a position embedding matrix $P \in \mathbb{R}^{n \times d}$ to capture the positional information based on the relative distance from the last interaction, where d denotes the latent dimensionality and n is the maximum sequence length. Following [8,14,15], we ensure that the dimensionality of position embeddings is the same as that of item embeddings so that they can be added together. Formally, the representation for each item $v_i \in \mathcal{S}_u$ can be formulated as

$$x_i = m_i + p_i, \tag{1}$$

where $m_i \in \mathbb{R}^d$, $p_i \in \mathbb{R}^d$ denotes the item embedding and the positional embedding, respectively.

Coarse-Grained Interest Capsule Network. The core of the proposed model CAN is to capture user's multiple interests at two different granularity levels: processing the user's historical interactions at coarse-grained levels while modeling the user's recent interactions and the underlying interests at fine-grained levels. In this layer, we first leverage the dynamic routing algorithm[13] to mine the user's multiple interests buried in the historical interactions instead of compressing all the information into one unified vector.

Let $\mathcal{S}_{u,h} = \{m_1, m_2, ..., m_h\}$ denote user's historical interaction embeddings, where h represents the length of historical interactions truncated in the interaction sequence from the front. In the real world, some user-item interactions are strictly ordered while others may not be, namely, not all adjacent interactions are sequentially dependent in a sequence [18]. Thus, we do not inject any positional information here because we intend to capture user's interests at coarse-grained levels. The item embeddings of the historical interactions can be regarded as primary capsules $m_i \in \mathbb{R}^{d \times 1}$, and the user's interests can be viewed as interest capsules $c_j \in \mathbb{R}^{d \times 1}$. A capsule is a group of neurons whose activity vectors represent the instantiation parameters of a specific type of entity such as an object or an object part [13]. The primary capsule prefers to send its output to the interest capsules whose vectors have a big scalar product with the prediction coming from the primary capsule, which explicitly allows the model to abstract away the user's interests as the information propagates to the interest capsules. The calculation of interest capsules is iterative. The routing logits b_{ij}, which standing for the correlation between primary capsule m_i and interest capsule c_j, are initially sampled from Gaussian distribution $\mathcal{N}(0, \sigma^2)$ to guarantee that each interest capsule is different. Then, b_{ij} will be updated with the cosine similarity between interest capsules c_j and primary capsules m_i during the iteration. Specifically,

$$b_{ij} = b_{ij} + cosine\_similarity(c_j, W_c m_i), \tag{2}$$

where $W_c \in \mathbb{R}^{d \times d}$ denotes the transformation matrix. With routing logits calculated, the interest capsule is computed by the weighted sum of all primary capsules

$$e_j = \sum_{i=1}^{h} w_{ij} W_c m_i, \tag{3}$$

where w_{ij} are the coupling coefficients calculated by softmax function

$$w_{ij} = \frac{\exp b_{ij}}{\sum_{k=1}^{h} \exp b_{ik}}, \tag{4}$$

The length of the capsule denotes the probability of the existence of the entity (i.e., the user's interests), hence a non-linear squashing function is applied to ensure that short interest capsules get shrunk to zero length while long ones get shrunk to a length close to one. Formally, the interest capsule is computed by

$$c_j = squash(e_j) = \frac{\|e_j\|^2}{1 + \|e_j\|^2} \frac{e_j}{\|e_j\|}, \tag{5}$$

Finally, we attain e interest capsules from historical interactions of user u. e is fix as 10 in our experiments. According to the capsule length, the top-k capsules with the longest length $C_u = \{c_1, ..., c_k\}$ are selected, which means that these interests are more likely to appear in the historical interactions. The whole procedure is listed in Algorithm 1.

Fine-Grained Transformer Layer. Compared with the historical interactions, the small amount of interest capsules we obtain from the last layer is more concentrated, reducing the impact of the unrelated interactions and allowing the model to focus more on relevant interactions and user's interests. In this layer, we utilize a transformer module to extract information and obtain more

---

Algorithm 1: Coarse-grained Interest Capsule Network

Input: user historical interactions $\mathcal{S}_{u,h} = \{m_1, m_2, ..., m_h\}$, iteration times r,
  number of interest capsules e, number of selected interest capsules k

Output: user interest capsules $C_u = \{c_1, ..., c_k\}$

1   for all primary capsules i and interest capsules j: initialize $b_{ij} \sim \mathcal{N}(0, \sigma^2)$

2   **for** $iter = 1, ..., r$ **do**

3   for all primary capsule i: $w_i = softmax(b_i)$

4   for all interest capsule j: $e_j = \sum_i w_{ij} W_c m_i$

5   for all interest capsule j: $c_j = squash(e_j)$

6   for all primary capsules i and interest capsules j:
       $b_{ij} = b_{ij} + cosine\_similarity(c_j, W_c m_i)$

7   **end**

8   select top-k interest capsules based on capsule length $\|c_j\|^2$

9   **return** $\{c_j, j = 1, ..., k\}$

---

informative representations for items and interest capsules at fine-grained levels. Let $\mathcal{S}_{u,r} = \{x_{|\mathcal{S}_u|-r}, ..., x_{|\mathcal{S}_u|-1}, x_{|\mathcal{S}_u|}\}$ denote the item embeddings in the user's recent interactions, where r represents the length of recent interactions. We put the interest capsules $C_u = \{c_1, ..., c_k\}$ together as special items with recent item embeddings $\mathcal{S}_{u,r}$ to form a new sequence $X = \{x_{|\mathcal{S}_u|-r}, ..., x_{|\mathcal{S}_u|}, c_1, ..., c_k\}$ and adopt the *Scaled Dot-Product Attention* to capture the correlation between items and interests. Formally,

$$A = Attention(Q, K, V) = softmax(\frac{QK^T}{\sqrt{d}}V),$$

$$Q = \sigma(W_Q X + b_Q), K = \sigma(W_K X + b_K), V = X, \qquad (6)$$

where $W_Q, W_K \in \mathbb{R}^{d \times d}$ are the weight matrices, $b_Q, b_K \in \mathbb{R}^d$ are the bias terms, and $\sigma(\cdot)$ is the selu activation function. Note that we add the selu activation function for Q and K to endow the model with non-linearity.

Referring to the original paper [17], we apply a *Position-wise Feed-Forward Network* to further improve the capability of the model. Also, layer normalization and residual network are employed to improve the model stability. Specifically,

$$X^{(1)} = LayerNorm(AX + \sigma(W_1 AX + b_1)W_2 + b_2), \qquad (7)$$

where $W_1, W_2 \in \mathbb{R}^{d \times d}$ are the weight matrices, $b_1, b_2 \in \mathbb{R}^d$ are the bias terms and A is the attention weight matrix. We stack multiple transformer layers to learn higher-order sequential dependencies and more informative item embeddings. The fine-grained transformer layer above can be defined as follows for simplicity:

$$X^{(1)} = Transformer(X), \qquad (8)$$

The l-th ($l >= 1$) layer is defined as:

$$X^{(l)} = Transformer(X^{(l-1)}), \qquad (9)$$

where $X^{(0)} = X$ and $\hat{X} = X^{(l)} = \{\hat{x_1}, ..., \hat{x_r}, ..., \hat{x_{r+k}}\}$ is the output of this layer, including both item embeddings and interest capsules.

Prediction Layer. Various models [5,8,20] follow the assumption that the last interaction has a strong influence upon the next user interaction, resulting in only the last interaction is used in the prediction procedure. However, the last interaction merely involves the user's recent interests, and the user's overall preference is ignored. Therefore, we leverage the attention mechanism to learn the user's general interest, and merge it with the last interaction to obtain an intact user interest as our model's eventual output. To be more specific, the attention weights are given by

$$\alpha_i = softmax(W_4\sigma(W_3concat(\hat{x}_i, \hat{x}_r) + b_3)), \tag{10}$$

where $W_3 \in \mathbb{R}^{2d \times d}, W_4 \in \mathbb{R}^{d \times 1}$ are the weight matrices, $b_3 \in \mathbb{R}^{2d}$ is the bias term. $\hat{x}_i$ and $\hat{x}_r$ are embeddings and last interaction embedding obtained from the fine-grained transformer layer, respectively. The general preference is calculated by the weighted sum of the $\hat{x}_i$. Finally, a feed forward network is applied to attain the overall user interest vector q which can be formulated as:

$$q = W_5(concat(\hat{x}_r, \sum_{i=1}^{r+k} \alpha_i\hat{x}_i)) + b_5, \tag{11}$$

where $W_5 \in \mathbb{R}^{2d \times d}$ is a learnable weight matrix and $b_5 \in \mathbb{R}^d$ is the bias vector. The score of each candidate item $v_i \in \mathcal{V}$ is calculated by

$$\hat{y}_i = softmax(q^T v_i), \tag{12}$$

where $\hat{y}_i$ denotes the recommendation probability of the item v_i to be the next user interactions. We adopt the cross-entropy loss as the objective function:

$$L(y, \hat{y}) = -\sum_{i=1}^{|\mathcal{V}|} y_i log(\hat{y}_i), \tag{13}$$

where y is a one-hot encoding vector exclusively activated by the ground truth. The *Adam* optimizer is then performed to optimize the cross-entropy loss.

4   Experiments

This section introduces the experiment setup, including the datasets, baseline methods, implementation details and evaluation metrics used in the experiments. Then, we conduct experiments to prove the effectiveness of the proposed model by studying the following research questions:

- **RQ1**: Does our proposed model outperforms state-of-the-art baseline methods of sequential recommendations?
- **RQ2**: What are the effects of each module in the proposed model through ablation studies?
- **RQ3**: How sensitive are the hyper-parameters, especially the number of chosen interests k?
- **RQ4**: Are we able to derive any insight about what do interest capsules learned from the historical interactions?

Table 1. Statistics of datasets (after preprocessing)

Dataset	#Users	#Items	#Actions	Sparsity	l_i	l_u
MovieLens-1M	6040	3416	1.0M	95.16%	5	5
Amazon Books	18319	168724	2.2M	99.93%	20	50
MovieLens-10M	43600	8940	8.8M	97.74%	20	50

4.1 Experimental Setup

Datasets. We evaluate our proposed model on three real-world datasets, i.e., *MovieLens-1M*[1], *Amazon Books*[2] and *MovieLens-10M*[3]. *MovieLens* is a popular benchmark dataset for evaluating sequential recommendation algorithms. Due to the gap in sparsity and size, we used both *MovieLens-1M* and *MovieLens-10M* versions. *Amazon* is a series of product review datasets crawled from *Amazon.com*. In this work, we adopt the *Books* category, which is notable for its large volume and sparsity.

Following [8,16], we treat the presence of a review or a rating as an interaction between the user and the item. Furthermore, the interaction orders are determined by timestamps. To ensure the dataset's quality, we filter out all items with fewer than l_i occurrences and users with fewer than l_u interactions. For all datasets, we follow [8,15] to split user interaction sequences $\mathcal{S}_u$ into three parts: (1) the most recent action $v^u_{|\mathcal{S}_u|}$ for testing, (2) the second most recent action $v^u_{|\mathcal{S}_u|-1}$ for validation, and (3) all remaining actions for training. Note that during testing, the input sequences involve both training actions and validation actions. The statistics of the processed datasets are summarized in Table 1.

Baseline Methods. To evaluate the effectiveness of the model we proposed, we compare it with the following baseline methods, including the general sequential recommendation methods[2,4,8,16,20,21] and the multi-interest methods [1,10,15]:

- **GRU4Rec+** [4] is a session-based recommendation model based on GRU cells. In the experiments, we treat each user's interaction sequence as a session. GRU4Rec+ [4] is an improved version of GRU4Rec [5], which takes up an advanced loss function. We adopt the latter in our experiments.
- **Caser** [16] is a CNN-based model for sequential recommendation, which regards a sequence of items as an image and applies horizontal and vertical convolutional filters to capture high-order relations between items.
- **RUM** [2] is a sequential recommendation model based on the memory network, which leverages external memory slots to store and update user's preferences explicitly. RUM-I regards each item as a memory unit, while the RUM-F stores diverse user interests.

---

[1] https://grouplens.org/datasets/movielens/1m/.
[2] http://jmcauley.ucsd.edu/data/amazon/.
[3] https://grouplens.org/datasets/movielens/10m/.

- **SHAN** [21] proposes a hierarchical attention network and employs a two-layer attention network to couple user's long-term and short-term interests.
- **SASRec** [8] utilizes self-attention blocks to capture long-term semantics and make predictions based on relatively few actions.
- **BERT4Rec** [14] employs the deep bidirectional self-attention to model user behaviors sequences, which achieves state-of-the-art performance in sequential recommendation.
- **GC-SAN** [20] models user sequences as directed graphs and combines a graph neural network and a multi-layer self-attention network.
- **MIND** [10] represents user's interests by multiple vectors, which encodes different aspects of the user's interests based on the dynamic routing algorithm.
- **ComiRec**[1] proposes two approaches to model user's diverse interests, which is similar to the work MIND. One takes advantage of the dynamic routing algorithm(ComiRec-DR), and the other leverages the self-attentive method(ComiRec-SA). We adopt the ComiRec-SA as our baseline method.
- **SINE** [15] is able to adaptively infer a sparse set of concepts from huge concept pool and develop an interest aggregation module to predict the user's current preference.

Evaluation Metrics. To evaluate the performance of all models, we adopt three metrics, including *Hit Ratio*(HR@K), *Mean Reciprocal Rank*(MRR@K) and *Normalized Discounted Cumulative Gain*(NDCG@K). HR@K is used to evaluate unranked results, while MRR@K and NDCG@K are used for ranked results. Since we only have one ground truth item for each user, HR@K is equivalent to Recall@K and Precision@K. In this work, due to space limitations, we report HR, MRR and NDCG with $K = 10$. For all metrics, the higher the better.

Implementation Details. In this work, all hyper-parameters and initialization strategies follow the original paper's suggestion and are fine-tuned on the validation sets via grid search on each dataset. For GRU4Rec, Caser, SHAN, SASRec, BERT4Rec and GC-SAN, we use code provided by *RecBole* [22], an open source framework for research purpose in recommendation. For RUM[4], MIND, ComiRec[5] and SINE, we refer to the open source code and implement ourselves. It is worth noting that we don't adopt negative sampling strategies because sampled metrics do not persist relative statements [9].

We implement our proposed model with Pytorch. All parameters are drawn from the normal distribution $\mathcal{N}(0, 0.02)$. Through the experiments, we find that the length of historical interactions h has little effect on the results, so we use complete sequences as historical interactions, which is 50. To reflect the contributions of capturing interest at both fine-grained levels and coarse-grained levels, for all datasets, we take the length of the recent interactions $r = 20$. We set the batch size equals 256 and explore the case that latent dimension

---

[4] https://github.com/ch-xu/RUM.
[5] https://github.com/THUDM/ComiRec.

Table 2. Performance comparison on three datasets. The best results are highlighted with bold fold, while the second-best results are underlined. Improvements over baseline methods are statistically significant with $p < 0.001$.

Models	MovieLens-1M			Amazon books			MovieLens-10M		
	H@10	M@10	N@10	H@10	M@10	N@10	H@10	M@10	N@10
GRU4Rec+	0.2586	0.1115	0.1457	0.0417	0.0157	0.0217	0.1753	0.0721	0.0960
Caser	0.2174	0.0824	0.1138	0.0439	0.0162	0.0226	0.1508	0.0565	0.0783
RUM-I	0.1426	0.0352	0.0599	0.0353	0.0103	0.0161	0.0750	0.0209	0.0334
RUM-F	0.1094	0.0263	0.0453	0.0345	0.0089	0.0147	0.0986	0.0242	0.0413
SHAN	0.1578	0.0450	0.0711	0.0463	0.0139	0.0214	0.0868	0.0220	0.0369
SASRec	0.2530	0.1094	0.1430	<u>0.0551</u>	0.0205	0.0286	0.1748	0.0717	0.0956
BERT4Rec	<u>0.2603</u>	<u>0.1147</u>	<u>0.1487</u>	0.0538	<u>0.0237</u>	<u>0.0307</u>	<u>0.1784</u>	<u>0.0757</u>	<u>0.0996</u>
GC-SAN	0.2490	0.1061	0.1394	0.0539	0.0203	0.0282	0.1616	0.0655	0.0878
MIND	0.1219	0.0416	0.0601	0.0228	0.0069	0.0105	0.0693	0.0221	0.0330
ComiRec	0.1710	0.0496	0.0776	0.0380	0.0115	0.0176	0.0988	0.0272	0.0436
SINE	0.1947	0.0732	0.1013	0.0433	0.0159	0.0222	0.1126	0.0428	0.0589
CAN	**0.2735**	**0.1203**	**0.1561**	**0.0577**	**0.0261**	**0.0335**	**0.1923**	**0.0852**	**0.1102**

d equals 100 for all models. According to the experimental results, the optimal hyper-parameters are $\{\eta : 0.001, \varepsilon : 0.25, k : 5\}$ on *MovieLens-1M* dataset, $\{\eta : 0.001, \varepsilon : 0.25, k : 8\}$ on *Amazon Books* dataset, and $\{\eta : 0.001, \varepsilon : 0, k : 5\}$ on *MovieLens-10M* dataset, where η is the learnig rate, ε is the dropout rate and k is the number of selected interests. Our source code is available on Github[6].

4.2   Comparison with Baseline Methods

To illustrate the effectiveness of our proposed model, we compare it with baseline methods as shown in Table 2(**RQ1**). We can find that both GRU4Rec, SASRec and BERT4Rec achieve promising results, proving that the RNN-based model and transformer modules are capable of learning a unified interest representation. Although RUM-I and RUM-F adopt extra memory slots to store and update user's interests explicitly, they discard the delicate dependencies between interests. Similar to our work, SHAN divides the interaction sequences into two parts. Owing to the limited capacity of the attention module, it reaches a general result on all datasets. Moreover, Caser and GC-SAN also achieve competitive results, but still inferior to our proposed model. Compared with general sequential recommendation models, multi-interest methods show slight improvement. MIND and ComiRec extract user's interests but ignore the crucial positional information. SINE utilizes the self-attentive method, which injects position weights, performing better than MIND and ComiRec. Nevertheless, the diverse conceptual prototypes utilized in SINE are learned by the model itself rather than abstracted from the interactions. It is hard for models to learn conceptual prototypes without any guides. According to the results, our model outperforms

---

[6] https://github.com/xiaohanhan1019/CAN.

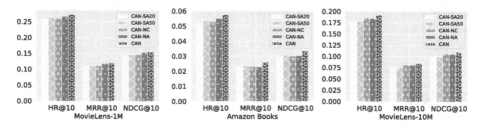

Fig. 3. Effectiveness of each module of CAN on three datasets.

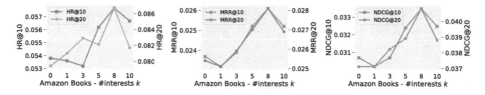

Fig. 4. Performance with different number of interests k on *Amazon Books* dataset.

all baseline methods consistently. This can be attributed to three points: 1) We model the user-item interactions at two different granularity levels, which can better mine user's diverse interests and attain descriptive embeddings. 2) We condense historical interactions into multiple interest capsules, allowing our model to focus more on relevant interactions and user's interests. 3) We do not exaggerate the importance of the last interaction but take full advantage of all the interactions in the sequence.

4.3 Ablation Study

To further verify the contribution of each module, we compare the performance of CAN with four submodules(**RQ2**). Both CAN-SA20 and CAN-SA50 merely apply a transformer module and make predictions by the user's last interaction. To demonstrate the influence of the length of recent interactions on our model, CAN-SA20 set the sequence length to 20 and CAN-SA50 takes the whole sequence, which is 50. CAN-NC refers to CAN without the coarse-grained interest capsule network, and CAN-NA refers to CAN without the prediction layer, which only uses the last interaction for prediction. As shown in Fig. 3, we first find that CAN-SA50 outperforms CAN-SA20, suggesting that longer sequences boost the model performance. However, compared with our proposed model, which only sends the most recent 20 interactions to the fine-grained transformer layer but model all the interactions in the sequence at coarse-grained levels, there is still a big gap, indicating that the coarse-grained interest capsule network makes better use of user's historical interactions. We also try to swap the granularity levels, that is, we leverage the capsule network to extract fine-grained interests and the transformer for coarse-grained interests. Consequently, the model's performance degrades drastically, which demonstrates the rationality of our model

Fig. 5. A case study of a user in *MovieLens-1M* dataset. We generate four interest capsules by coarse-grained interest capsule network. By investigating the nearest neighbour items in the latent space calculated by cosine similarity, we find that the four interest capsules are about adventure, thriller, comedy and sci-fi films. The tags under the movie posters are the film genres.

design. Furthermore, CAN achieves a better performance than CAN-NC, which shows the effectiveness of modeling user's multiple interests, and the module do mine the underlying user's interests buried in the historical interactions. Finally, CAN outperforms CAN-NA, proving that the last interaction is not enough for modeling the whole sequence.

4.4 Impact of Number of Interests k

We investigate the sensitivity of the number of interests k of our proposed model(**RQ3**). Figure 4 illustrates the performance of our model when the number of interests k changes. Generally, as k increases, the model's performance is enhanced, which means user's multi-interests benefit sequential recommendation models to make more accurate predictions. However, the performance drops significantly when k is too large, showing that the number of user's interests in the sequences are limited, and too many interest capsules result in useless information. Moreover, we can find out that the model's performance except HR@20 drops when the number of interests k increases initially, which shows the necessity of multiple interests.

4.5 Case Study

We derive insight about what do interest capsules learned from the historical interactions by a case study(**RQ4**). Figure 5 shows that our proposed model learns four different user's interests from her interaction sequence. On the right side of each interest, three nearest neighbor items found in the latent space according to the cosine similarity between items and interest capsules are displayed. Each interest capsule learned by our model approximately corresponds to one specific film genre, which demonstrates the effectiveness of interest capsules and enhances the interpretability of our model. It is worth noting that we only use the item ID during training, and the model will not obtain any information about film genres. Nevertheless, our model can still learn category information between items.

5   Conclusion

In this paper, we propose a capsule attentive network (CAN) to capture multi-granularity interests for sequential recommendation. Firstly, we leverage the dynamic routing algorithm to learn user's multiple interests at coarse-grained levels. Subsequently, the selected interests will be put into the user's recent interaction sequence, and contribute more informative representations through the fine-grained transformer layer. Finally, we apply an attention module to learn an intact user interest representation instead of picking the last interaction. We have conducted extensive experiments on three benchmark datasets and demonstrates that our proposed model outperforms baseline methods over different evaluation metrics. Furthermore, we have done detailed analyses on each module of our proposed model and found that capturing interests at both coarse-grained and fine-grained levels leads to better use of the user-item interactions.

Acknowledgements. This work was supported by NSFC grants (No. 61972155), the Science and Technology Commission of Shanghai Municipality (20DZ1100300) and the Open Project Fund from Shenzhen Institute of Artificial Intelligence and Robotics for Society.

References

1. Cen, Y., Zhang, J., Zou, X., Zhou, C., Yang, H., Tang, J.: Controllable multi-interest framework for recommendation. In: Proceedings of the 26th ACM SIGKDD International Conference on Knowledge Discovery & Data Mining, pp. 2942–2951 (2020)
2. Chen, X., Xu, H., Zhang, Y., Tang, J., Cao, Y., Qin, Z., Zha, H.: Sequential recommendation with user memory networks. In: Proceedings of the Eleventh ACM International Conference on Web Search and Data Mining, pp. 108–116 (2018)
3. Han, Y., Li, Q., Xiao, Y., Zhou, H., Yang, Z., Wu, J.: Multiple interleaving interests modeling of sequential user behaviors in e-commerce platform. World Wide Web, pp. 1–26 (2021)
4. Hidasi, B., Karatzoglou, A.: Recurrent neural networks with top-k gains for session-based recommendations. In: Proceedings of the 27th ACM International Conference on Information and Knowledge Management, pp. 843–852 (2018)
5. Hidasi, B., Karatzoglou, A., Baltrunas, L., Tikk, D.: Session-based recommendations with recurrent neural networks. arXiv preprint arXiv:1511.06939 (2015)

6. Hinton, G.E., Krizhevsky, A., Wang, S.D.: Transforming auto-encoders. In: Honkela, T., Duch, W., Girolami, M., Kaski, S. (eds.) ICANN 2011. LNCS, vol. 6791, pp. 44 51. Springer, Heidelberg (2011). https://doi.org/10.1007/978-3-642-21735-7_6

7. Jannach, D., Ludewig, M.: When recurrent neural networks meet the neighborhood for session-based recommendation. In: Proceedings of the Eleventh ACM Conference on Recommender Systems, pp. 306–310 (2017)

8. Kang, W.C., McAuley, J.: Self-attentive sequential recommendation. In: 2018 IEEE International Conference on Data Mining (ICDM), pp. 197–206. IEEE (2018)

9. Krichene, W., Rendle, S.: On sampled metrics for item recommendation. In: Proceedings of the 26th ACM SIGKDD International Conference on Knowledge Discovery & Data Mining, pp. 1748–1757 (2020)

10. Li, C., et al.: Multi-interest network with dynamic routing for recommendation at tmall. In: Proceedings of the 28th ACM International Conference on Information and Knowledge Management, pp. 2615–2623 (2019)

11. Lv, F., et al.: SDM: Sequential deep matching model for online large-scale recommender system. In: Proceedings of the 28th ACM International Conference on Information and Knowledge Management, pp. 2635–2643 (2019)

12. Rendle, S., Freudenthaler, C., Schmidt-Thieme, L.: Factorizing personalized markov chains for next-basket recommendation. In: Proceedings of the 19th International Conference on World Wide Web, pp. 811–820 (2010)

13. Sabour, S., Frosst, N., Hinton, G.E.: Dynamic routing between capsules. arXiv preprint arXiv:1710.09829 (2017)

14. Sun, F., et al.: Bert4rec: sequential recommendation with bidirectional encoder representations from transformer. In: Proceedings of the 28th ACM International Conference on Information and Knowledge Management, pp. 1441–1450 (2019)

15. Tan, Q., et al.: Sparse-interest network for sequential recommendation. In: Proceedings of the 14th ACM International Conference on Web Search and Data Mining, pp. 598–606 (2021)

16. Tang, J., Wang, K.: Personalized top-n sequential recommendation via convolutional sequence embedding. In: Proceedings of the Eleventh ACM International Conference on Web Search and Data Mining, pp. 565–573 (2018)

17. Vaswani, A., et al.: Attention is all you need. arXiv preprint arXiv:1706.03762 (2017)

18. Wang, S., Hu, L., Wang, Y., Cao, L., Sheng, Q.Z., Orgun, M.: Sequential recommender systems: challenges, progress and prospects. arXiv preprint arXiv:2001.04830 (2019)

19. Wang, S., Hu, L., Wang, Y., Sheng, Q.Z., Orgun, M.A., Cao, L.: Modeling multipurpose sessions for next-item recommendations via mixture-channel purpose routing networks. In: IJCAI, vol. 19, pp. 3771–3777 (2019)

20. Xu, C., et al.: Graph contextualized self-attention network for session-based recommendation. In: IJCAI, vol. 19, pp. 3940–3946 (2019)

21. Ying, H., et al.: Sequential recommender system based on hierarchical attention network. In: IJCAI International Joint Conference on Artificial Intelligence (2018)

22. Zhao, W.X., et al.: Recbole: Towards a unified, comprehensive and efficient framework for recommendation algorithms. arXiv preprint arXiv:2011.01731 (2020)

Multi-Task Learning with Personalized Transformer for Review Recommendation

Haiming Wang[1,2], Wei Liu[1,2,3(✉)], and Jian Yin[1,2]

[1] School of Artificial Intelligence, Sun Yat-sen University, Guangzhou, China
wanghm39@mail2.sysu.edu.cn, {liuw259,issjyin}@mail.sysu.edu.cn
[2] Guangdong Key Laboratory of Big Data Analysis and Processing,
Guangzhou, China
[3] SKL of Internet of Things for Smart City, University of Macau, Macau, China

Abstract. Drastic increase in item/product review volume has caused serious information overload. Traditionally, product reviews are exhibited in chronological or popularity order without personalization. The review recommendation model provides users with attractive reviews and efficient consumption experience, allowing users to grasp the characteristics of items in seconds. However, the sparsity of interactions between users and reviews appears to be a major challenge. And the multi-relationship context, especially potential semantic feature in reviews, is not fully exploited. To address these problems, **M**ulti-**T**ask **L**earning model incorporating **P**ersonalized **T**ransformer (**MTL-PT**) is proposed to provide users with an interesting review list. It contains three tasks: the main task models user's preference to reviews with the proposed poly aggregator, incorporating the user-item-aware semantic feature. Two auxiliary tasks model the quality of reviews, and user-item interactions, respectively. These tasks collaboratively learn the multi-relationship among user, item, and review. The shared latent features of user/item link the three tasks together. Especially, the personalized semantic features of reviews are also fused into the tasks with the proposed personalized transformer. Two new real-world datasets for personalized review recommendation are collected and constructed. Extensive experiments are conducted on them. Compared with the state-of-the-arts, the results validate the effectiveness of our model for review recommendation.

Keywords: Multi-task learning · Personalized transformer · Review recommendation

1 Introduction

Nowadays, many web services like Amazon, Google PlayStore, Reddit, etc., provide commenting function. Users could describe their experience about the items or products by reviews. Meanwhile, via browsing others' reviews, users could acquire the characteristics of items, and discover reviews that interest themselves, and share subjective opinions with other users. The reviews would be more

W. Zhang et al. (Eds.): WISE 2021, LNCS 13081, pp. 162–176, 2021.
https://doi.org/10.1007/978-3-030-91560-5_12

detailed and authentic than the description of item. The commenting function plays a win-win role in modern web platform, where reviews not only help users make better decisions by providing more various information, but also promote community activity [11,15,23,24].

Since user community becomes more active, reviews for items might increase drastically. And thus, it is laborious for users to find helpful or interesting reviews that meet their preferences. Meanwhile, most current studies are devoted to **non-personalized recommendations** [5,16,23], which neglects user's personalized preferences. In fact, given an item, the preferences of different users for reviews of the item vary greatly. For instance, When a game fanatic browses mobile phones, she would be interested in reviews about the CPU performance and storage, while other customers may prefer different aspects like appearance and camera. Therefore, personalized review recommendation is urgent. Its goal is to recommend useful reviews that satisfy users' unique preference and assist users to make decisions, shown in Fig. 1.

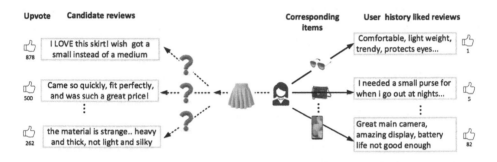

Fig. 1. An example of personalized review recommendation. Right part is a user's history interacted reviews, each of which consists of review text, upvote, and the corresponding item. Left part shows that when the user browses a new item, how should the system recommend interesting reviews to the user from candidate reviews.

There are only a few works on the personalized review recommendations [1,13,14,19,20]. Matrix factorization is utilized to model the interactions between users and reviews [1,13,20]. Further, assuming that review is generated by review writer (a.k.a. reviewer) and the corresponding item, Moghaddam et al. [14] represented review with reviewer and item, and utilized tensor factorization to model the interactions among user, reviewer and items. However, previous works ignore users' preference variation for reviews depending on different items. And merely (or don't) use user/item invariant review semantic representations. Moreover, compared with tremendous reviews, reviews a user actually interacts with (giving feedback, such as reply or like) are very few. Thus, personalized review recommendation faces challenges in two ways. **i)** How to construct a context-aware user preference? Previous works [8,14,20] mainly ignore that user's preference for reviews would change given different items. For instance, shown in Fig. 1, when browsing mobile phones, a user would be interested in reviews about the performance and camera of the mobile phone. While browsing clothes, the user would prefer reviews about the appearance and comfort. **ii)** How to extract latent feature for review with limited interactions, especially considering the corresponding item? Some researchers [1,14,20,25] mainly consider limited feedback data of review, ignoring the semantic information and quality of review itself. Especially, many potentially high-quality reviews may be lack of exposure to the interested users. Most of the related works just utilize traditional hand-crafted textual features from reviews [19]. While, the semantic information in review text, especially user's history interacted reviews, is rich, and could remedy data sparsity efficiently.

Inspired by the above observations, we propose a multi-task learning model to provide users with personalized review lists. The model contains 3 tasks: **1)** A user-review preference modeling task is proposed to make personalized reviews recommendation. **2)** An item-review quality modeling task is proposed to estimate the review's quality with review text given the corresponding item, improving exposure opportunity of potentially high-quality reviews. **3)** A user-item interaction modeling task is proposed to model the interactions between user and item, sharing similar patterns between users and items. Moreover, personalized transformer (PT) is proposed to extract context-aware semantic features from review texts in the first two tasks. Item-aware poly aggregator is built on top of semantic features to model user's preference further.

Our contributions are listed as follows:

- We solve personalized review recommendation problem with a novel multi-task learning model, which collaboratively models the multi-relationship among user, item and review, and learns the user's context-aware preference for history reviews given different items.
- To extract deep semantic feature from review texts and estimate the quality of each review, we propose personalized transformer and utilize item-aware poly aggregator to model user's preference.
- Two new real-world datasets for personalized review recommendation are collected and constructed. Extensive experiments demonstrate the superior performance of our model, especially in the non-e-commerce scenarios.

2   Related Work

The existing researches on review recommendations could be mainly categorized into non-personalized review recommendations (a.k.a. review helpfulness recommendation) and personalized review recommendations. Most of the current works are devoted to review helpfulness recommendation, whereas personalized review recommendations are seldom studied.

2.1   Non-personalized Review Recommendations

Non-personalized review recommendations are mainly based on evaluating the helpfulness of reviews [5]. Mainstream methods [4] generally extract various linguistic features from the text of an online review for helpfulness prediction. Conventional approaches for review helpfulness prediction leverage domain-specific knowledge and use hand-crafted features from the review text, which are fed into linear classifiers such as SVM or Random Forest to predict helpfulness. There are structural features (STR) [27], emotion states [12], etc. However, these features extracted for helpfulness prediction are still shallow, lacking exploration of deep semantic features in reviews. With the advance in deep learning, researchers adopted convolutional neural network (CNN) [2] and LSTM [5,6] to extract textual features from reviews and avoid tedious labor on feature engineering for reviews. Although these deep neural network based methods show promising results in textual feature extraction, they can merely extract local features, ignoring global interactions between words, especially effects of context (e.g., user/item information). Different from them, we would propose a transformer-based method to capture context-aware semantic features from reviews.

2.2 Personalized Review Recommendations

To make personalized review recommendations, researchers [17] integrated individual preference into non-personalized recommendation methods. Moghaddam et al. [13] applied latent factor models by matrix factorization and tensor factorization to make personalized review recommendations, ignoring review semantic information. A follow-up work [14] generates personalized review ranking based on personalized review helpfulness prediction. Further, researchers considered social contexts [20], fine-grained user's preference aspects [8], users similarity [10,19–21] to make recommendation. However, most of the works mentioned above neglect the important multi-interaction among user, item and review, and cannot incorporate review semantic feature. Especially, the item plays an essential role in personalized review recommendation, since user preferences may vary depending on items. Thus, different with existed works, we suggest that personalized review recommendations should be fully aware of both user and item, and model the multi-relationship among them with semantic features.

3   Proposed Model

3.1   Notations and Problem Statement

Let $\mathcal{U} = \{u_1, u_2, ..., u_{|\mathcal{U}|}\}$ denote the user set, $\mathcal{V} = \{v_1, v_2, ..., v_{|\mathcal{V}|}\}$ be the item set, and $\mathcal{R} = \{r_1, r_2, ..., r_{|\mathcal{R}|}\}$ be the review set. Based on users' history feedback to reviews, we define the user-review interaction matrix $\mathbf{Y} \in \mathbb{R}^{|\mathcal{U}| \times |\mathcal{R}|}$, where $y_{ur} = 1$ indicates user u engaged with review r (e.g., behaviors of upvoting, replying), $y_{ur} = 0$ otherwise. Let $\phi(r)$ indicates the corresponding item of review r. Further, the reviews set of item v is defined as $\mathcal{R}_v = \{r | r \in \mathcal{R} \wedge \phi(r) = v\}$, and user u's history reviews set as $\mathcal{H}_u = \{r | r \in \mathcal{R} \wedge y_{ur} = 1\}$. The notations of embeddings/features are denoted by bold letters, and scalars are represented by lower-case letters.

Personalized Review Recommendation: Given user u's history reviews $\mathcal{H}_u$, review recommendation aims to estimate user's preference y_{ur} to each candidate review r in the review set $\mathcal{R}_v$ of item v, and provides a personalized re-ranked review list for the user, when u browses a target item v.

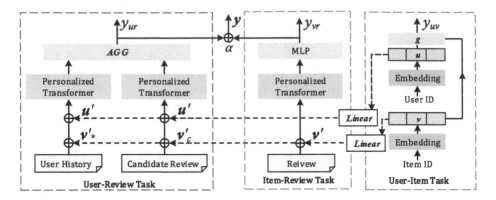

Fig. 2. Model overview. Three tasks are bridged by the shared item and user embeddings (denoted with black dotted lines).

3.2   Model Overview

MTL-PT models multiple interactions among the users, items and reviews. As shown in Fig. 2, MTL-PT consists of three tasks: the user-review recommendation (main) task, the item-review quality modeling (auxiliary) task, and the user-item interaction modeling (auxiliary) task. In the following, three tasks are abbreviated as user-review task, item-review task, and user-item task, respectively. We will introduce the details of each task in the following parts. Since personalized transformer provides deep semantic feature extraction function and appears in multiple tasks, it would be introduced firstly. Then three tasks in our model would be presented successively.

3.3   Personalized Transformer

Personalized transformer is mainly utilized for semantic feature extraction. Though the vanilla transformer uses the multi-head attention mechanism to capture the relative dependency between words in review [22], it is not capable of capturing the personalized semantic feature. Because users usually have personalized interests and different attentions on reviews[26], given the corresponding items. Thus, we propose a personalized transformer, which could extract user/item aware semantic features from reviews in the user-review task and the item-review task.

Based on standard Transformer encoder architecture [3], which contains multi-layer transformer blocks built with multi-head self-attention and position-wise feed-forward network, we construct personalized embeddings as the input, which fuses user's and items' representations with word embeddings in reviews, enabling the encoder to extract personalized semantic representation. The details are introduced as follows.

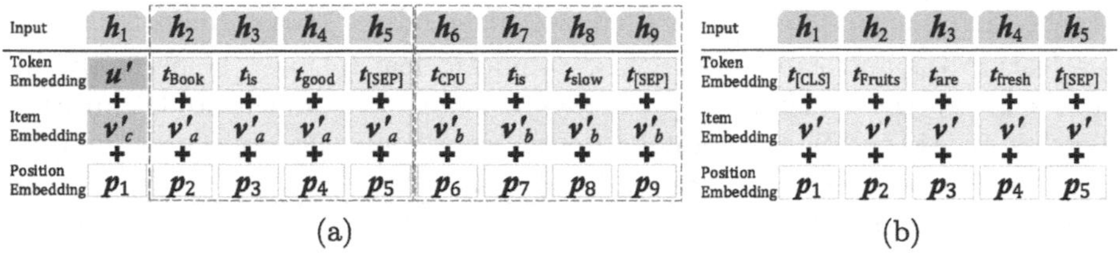

Input	h_1	h_2	h_3	h_4	h_5	h_6	h_7	h_8	h_9
Token Embedding	u'	t_{Book}	t_{is}	t_{good}	$t_{[SEP]}$	t_{CPU}	t_{is}	t_{slow}	$t_{[SEP]}$
	+	+	+	+	+	+	+	+	+
Item Embedding	v'_c	v'_a	v'_a	v'_a	v'_a	v'_b	v'_b	v'_b	v'_b
	+	+	+	+	+	+	+	+	+
Position Embedding	p_1	p_2	p_3	p_4	p_5	p_6	p_7	p_8	p_9

(a)

Input	h_1	h_2	h_3	h_4	h_5
Token Embedding	$t_{[CLS]}$	t_{Fruits}	t_{are}	t_{fresh}	$t_{[SEP]}$
	+	+	+	+	+
Item Embedding	v'	v'	v'	v'	v'
	+	+	+	+	+
Position Embedding	p_1	p_2	p_3	p_4	p_5

(b)

Fig. 3. (a) personalized embeddings in user-review task. (b) personalized embeddings in item-review task. v'_c, v'_a, and v'_b denote the representations of corresponding item of candidate review, respectively. A $[SEP]$ token is added to each review's tail, indicating its ending.

Personalized Embedding as Input. Since a review is not independent and owns rich context information (e.g., the target user, the corresponding item), the context would be helpful to extract semantic feature from reviews. Therefore, we construct personalized word embedding in reviews with additional user and item information. As shown in Fig. 3(a)[1], user embedding $u' = f_1(u)$ and item embedding $v' = f_2(v)$ are both aligned with the dimension D_t of word embedding, and added to word/token embeddings, where $f_1(\cdot)$ and $f_2(\cdot)$ are both linear function. In this way, it allows the transformer to extract user-item-aware feature and grasp more semantic information. We also add positional embedding p_k following [22] to capture the position information of words in review. Therefore, the input for personalized transformer is computed by combining word/token, user, item and token-wise positional embedding:

$$\mathbf{h}_k^{(0)} = \begin{cases} \mathbf{u}' + \mathbf{v}'_c + \mathbf{p}_1, & \text{if } k = 1 \\ \mathbf{t}_k + \mathbf{v}'_* + \mathbf{p}_k, & \text{if } k > 1, \end{cases} \tag{1}$$

where $\mathbf{h}_k^{(0)}$ represents the k-th token in transformer blocks of input layer, $\mathbf{v}'_c$ means the aligned representation of the corresponding item of candidate review r_c. And $\mathbf{v}'_*$ denotes the aligned representation of the corresponding item of the history review r, where r is the review word t_k belongs to.

Finally, we feed personalized embedding into transformer encoders and take the tokens $(\mathbf{h}_1^{(l)}, .., \mathbf{h}_N^{(l)})$ from the last layer as the output, where N is the length of the sequence. For simplicity of notation, in the rest of the paper, the personalized transformer would be abbreviated as function $PT(\cdot)$ with output $(\mathbf{h}_1, .., \mathbf{h}_N)$, where the superscript indicating the number of layer for each token is omitted.

---

[1] Since the inputs of personalized transformers in the user-review task and the item-review task are similar, here we only take the personalized transformer input in the user-review task as an example.

3.4   User-Review Task

The user-review task aims to model the user's preference for reviews given the corresponding item v_c. The task takes the user's history review $\mathcal{H}_u$, the corresponding items v_*, candidate review r_c, its corresponding item v_c and user identification u as inputs. They are fed into personalized transformers to extract user feature $\mathbf{o}_u$ and candidate review semantic feature $\mathbf{o}_r^{(c)}$, respectively. To better capture the semantic dependencies between user and candidate review, following [9], MTL-PT incorporates three types of semantic aggregators built on top of the personalized transformer, detailed as follows:

Dot product aggregator has simple architectures and allows precomputation embedding $\mathbf{o}_r^{(c)}$ for candidate reviews, which is a fast yet effective structure. It is formulated as follow:

$$\mathbf{o}_u = MLP(first(PT_1(\mathcal{H}_u, v_*, v_c, u)))   \quad \mathbf{o}_r^{(c)} = MLP(first(PT_2(r_c, v_c, u))) \tag{2}$$

where $PT_1(\cdot)$ and $PT_2(\cdot)$ are personalized transformers initialized by the same set of pre-trained parameters but fine-tuned separately, $first(\cdot)$ means taking the first vector of the sequence of vectors produced by the personalized transformer. The score of a candidate review r_c is given by the dot-product $\hat{y}_{ur} = \mathbf{o}_u \cdot \mathbf{o}_r^{(c)}$.

Cross aggregator allows full interactions between $\mathcal{H}_u$ and r_c by concatenate them and fed into personalized transformer. With a single encoder, cross aggregator could perform self-attention between user's history and candidate review, which produces a candidate-sensitive input representation as follow:

$$\hat{y}_{ur} = MLP(first(PT(\mathcal{H}_u, v_*, r_c, v_c, u)) \tag{3}$$

However, the cross aggregator doesn't allow precomputation for candidates, since each candidate review has to be concatenated with history review and computed through $PT(\cdot)$ repeatedly, resulting in non-scalable issue.

Poly aggregator utilizes two additional attention modules, which allow richer interactions between history review and candidate review, with precomputation for candidate reviews by $PT(\cdot)$ once. It is formulated as follow:

$$\mathbf{o}_u = \sum_{i=1}^{m} w_i \mathbf{o}_u^{(i)} \ \ where \ (w_1, .., w_m) = softmax(\mathbf{o}_r^{(c)} \cdot \mathbf{o}_u^{(1)}, .., \mathbf{o}_r^{(c)} \cdot \mathbf{o}_u^{(m)}) \tag{4}$$

$$\mathbf{o}_u^{(i)} = \sum_{k=1}^{N} w_k^{q_i} \mathbf{h}_k \ \ where \ (w_1^{q_i}, .., w_N^{q_i}) = softmax(\mathbf{q}_i \cdot \mathbf{h}_1, .., \mathbf{q}_i \cdot \mathbf{h}_N), \tag{5}$$

where $(\mathbf{h}_1, .., \mathbf{h}_N) = PT_1(\mathcal{H}_u, v_*, v_c, u)$ are outputs of the persoanlized transformer for user's history, $\mathbf{q}_i = f_3(\mathbf{v}_i)$ projects corresponding item embeddings $\mathbf{v}_i$ of history review with linear function, m is the number of items in user history. In Eq. 5, the poly aggregator decompose user's history review into representations $\mathbf{o}_u^{(i)}$ corresponding to each items of history review using m query vectors $\mathbf{q}_i$.

Then we use $\mathbf{o}_r^{(c)}$ (Eq. 2) as query and aggregate user's history representations in Eq. 4. The final score is given by the dot-product $\hat{y}_{ur} = \mathbf{o}_u \cdot \mathbf{o}_r^{(c)}$.

The performance of different aggregators are detailed in Sect. 4.4. And the optimization of the parameters of this task will be introduced in Sect. 3.7.

3.5   Item-Review Task

The item-review task aims to model review quality given the corresponding item. Review's quality is important for user to make decisions, and lots of reviews are not rated or lack of sufficient vote scores to represent the actual quality. Moreover, the quality prediction of review should be aware of the corresponding item [6], since the corresponding metric for high-quality review may differ for different items. We propose to estimate the item-aware review's quality through review text.

The item-aware review representation is extracted with personalized transformer, which is fed into MLP to estimate the quality score as follows:

$$\hat{y}_{vr} = MLP(first(PT(r_c, v_c))), \tag{6}$$

The candidate review input for personalized transformer is detailed in Fig. 3(b). To optimize the parameters, we minimize the loss $\mathcal{L}_{IR}$ of Item-Review task, comparing the output with ground truth by Mean Square Error (MSE):

$$\mathcal{L}_{IR} = \frac{1}{N_{vr}} \sum_{v \in \mathcal{V}, r \in \mathcal{R}_v} ||y_{vr} - \hat{y}_{vr}||^2, \tag{7}$$

where N_{vr} is the sample number of the item-review task's training set, review's quality score x, such as helpfulness, upvotes, is normalized by $y_{vr} = x/(1+x)$.

3.6   User-Item Task

This task bridges the above two tasks with shared user/item latent features. We argue that similar users would have similar preference pattern. At the same time, users would also have similar criteria for the reviews in similar items. User-item task learns the user/item latent representation by modeling the interactions between users and items. Here, we suppose that users have interactions with the corresponding items of interacted reviews. We formulate the user-item interaction prediction as $\hat{y}_{uv} = g(\mathbf{u}, \mathbf{v})$, where $\mathbf{u}$ is the user embedding, and $\mathbf{v}$ is the item embedding. $g(\cdot)$ denotes the interaction relation function. For task training, we minimize the log-likelihood loss $\mathcal{L}_{UI}$ of the User-Item task with Binary Cross Entropy (BCE) as follows:

$$\mathcal{L}_{UI} = \frac{1}{N_{uv}} \sum_{u \in \mathcal{U}, v \in \mathcal{V}_u \wedge \mathcal{V}_u^-} -[y_{uv} log(\hat{y}_{uv}) + (1 - y_{uv})log(1 - \hat{y}_{uv})], \tag{8}$$

where N_{uv} is the sample number of the user-item task's training set, and $\mathcal{V}_u$ denotes the item set u has interacted with, otherwise $\mathcal{V}_u^-$.

3.7   Preference Estimation and Parameters Optimization

To consider user's personalized preference and review quality simultaneously, we update the user preference score y_{ur} to the candidate review, with the review quality score y_{ir} as the final preference score formulated as $\hat{y}'_{ur} = \alpha\hat{y}_{ur} + (1-\alpha)\hat{y}_{vr}$, where α denotes the weight hyper-parameter, controlling the ratio between $\hat{y}_{ur}$ and $\hat{y}_{vr}$. For model training, we employ the pairwise loss function [18], which encourages the preference score of an observed entity (review that user u has interacted with, i.e., positive examples) to be higher than its unobserved counterparts (reviews that u has not interacted with, i.e., negative examples),

$$\mathcal{L}_{UR} = -\sum_{u \in \mathcal{U}} \sum_{r^+ \in \mathcal{H}_u} \sum_{r^- \notin \mathcal{H}_u} \ln \sigma(\hat{y}'_{ur+} - \hat{y}'_{ur-}), \tag{9}$$

where $\hat{y}'_{ur+}$ means u's preference to the interacted review, and $\hat{y}'_{ur-}$ otherwise. Θ denotes all the parameters in our model. Since the inputs for each task are different, three tasks are optimized alternately. For each training iteration, we first update the parameters of auxiliary tasks using loss $\mathcal{L}_{IR}$ and $\mathcal{L}_{UI}$ by K and J epochs separately. And then we update parameters for main task with loss $\mathcal{L}_{UR}$.

According to the introduction of three tasks in MTL-PT, we find that the computation is mainly cost on personalized transformer. Therefore, the time complexities of updating parameters in the user-review task and item-review are both $O(N_s^2 D_t)$, where N_s is the max length of word sequence for PT input, D_t is the dimension of word embedding. The time complexity of updating the parameters in user-item task is $O(D_e)$. Since N_s, D_t, D_e are all constants, each epoch of the model optimization could be completed in linear time.

4   Experiments

4.1   Dataset Description

To evaluate the proposed models, we perform experiments on two real-world datasets, Reddit and Taptap[2], collected by web crawler.

Reddit is a website comprising user-generated content. Its primary contents are text-based posts from users. Each post consists of a title, a detailed description (could be empty), and comments from other users. In this paper, posts are treated as items. Top-level comments in the post are regarded as reviews. Due to privacy issues, user's upvote/downvote behavior to comment could not be crawled, we treat user's reply behavior to the review as user's feedback[3]. To the

---

[2] https://www.reddit.com/ and https://www.taptap.com/.
[3] If user replies a review, we assume user is interested in the review, otherwise not.

best of our knowledge, we are the first to carry out personalized review recommendations on the non-e-commerce scenario. Different from typical e-commerce settings, users in Reddit browse comments to discover interesting information about the main post, instead of merely finding helpful information to make purchase decisions.

Taptap is a mobile app store towards games. Each game in the store has a commenting section, where users can comment and communicate with each other. Here, we treat apps in the store as items. Like Reddit, we utilize users' reply behavior to comments as users' feedback due to privacy issue. Taptap dataset is similar to the E-commerce setting, where users would decide whether downloading the App or not by browsing the corresponding reviews[4].

Dataset. The statistical details for both datasets are shown in Table 1. To remove extreme cases, we filter out users with less than 15 interactions and items with less than 15 reviews in Reddit dataset. Since Taptap dataset is relatively small, we filter out users and items less than 5 and 15 interactions, respectively. We could find both datasets are very sparse, and Taptap is much sparser.

Table 1. Statistics of experiment datasets. #reviews ($\geq$1) is the number of reviews with at least one interaction. #inter is the number of user-review interactions.

Dataset	#reviews	#reviews ($\geq$1)	#items	#users	#inter	Sparsity
Reddit	339,038	125,107	8,344	1,730	54,579	99.991%
Taptap	58,569	26,623	1,238	1,559	18,827	99.997%

4.2 Task Setting and Evaluation Metrics

Task Settings. To evaluate the review recommendation models, we adopt the leave-one-out evaluation task, which has been widely used in [7,11]. We randomly select two of the reviews he/she has interacted with as validation and test data for each user, respectively. The remaining reviews are used as the training set. Following [7,11] for easy and fair evaluation, we couple each ground truth with 10 other reviews from the same corresponding item, which the user has not interacted with, as candidate reviews.

Hyperparameter Settings. User/item embedding dimension D_e is set as 50. (1) For personalized transformer settings, the encoder follows the setting of **BERT**$_{BASE}$ [3] with 12 layers, and the parameters are initialized with pretrained English model and Chinese model[5] for Reddit and Taptap datasets, respectively. (2) For user-review task, each positive review that the user interacted with is paired with 4 negative reviews. (3) For user-item task, user-item

---

[4] The datasets and code would be public later.
[5] huggingface.co/bert-base-uncased and huggingface.co/bert-base-chinese.

interactions are constructed in a negative sampling manner with the ratio set as 5. The user-item interaction function $g(\cdot)$ in is set as 2-layer MLP. Adam is used with learning ráte of 0.0001 for fine-tuning the model parameters. Since the influence of the hyperparameters is not our main concerns and the space is limited, we do not show the results of tuning hyperparameters. The models are implemented with Pytorch library. GPU we used is GeForce GTX 2080ti with a memory of 11 GB. The data collection programs are implemented with python.

Evaluation Metrics. To evaluate the ranking list of all the models, we employ a variety of evaluation metrics. Hit Ratio (HR), Normalized Discounted Cumulative Gain (nDCG), and Mean Reciprocal Rank (MRR) are calculated for each test case. Since we only have one ground truth review for each user, HR@k is equivalent to Recall@k and proportional to Precision@k; MRR is equivalent to Mean Average Precision (MAP). And we report HR and nDCG with $k = 1, 3, 5$. While works on personalized review recommendations [13,14,20] usually use the MSE metric to indicate model performance, users' rating towards comments is not prevalent and commonly used. Thus, we argue that ranking metrics can better represent the model performance in practice. For each metric, the higher the value is, the better the performance is. We run each model 10 times, and utilize the average of the results for each model.

4.3   Baselines and Implementation Details

We define two sets of baselines comprising of non-personalized and personalized review recommendation algorithms to evaluate our model's effectiveness. For non-personalized review recommendation, we adopt the baselines following [8]:

- **Rand.** This method randomly sorts the review list.
- **New.** This method ranks the reviews by review's post time.
- **Pop.** This method ranks the reviews by its score, which is calculated by subtracting the downvotes from upvotes[6].

For personalized review recommendation, we consider the following baselines:

- **MF** [13]. This method merely utilizes matrix-factorization to extract users' and reviews' latent representation from user-review interactions.
- **TF** [13]. This method utilizes Tucker decomposition to extract users, reviews, and items latent representation from user-review-item interactions.
- **BETF** [14]. This method extends TF [13] by removing the biases in ratings as well as incorporating an extra task predicting users' rating for the items.
- **CAP** [20]. Review's hand-crafted text features are added into matrix factorization in order to capture more precise review representation.

---

[6] Since other non-personalized review recommendation models [2,6] learn to reproduce the ranking of scores-based recommendation, and SR is the optimal situation of them, we didn't select other non-personalized models as baseline any more.

- **NeuMF** [7]. This method utilizes neural matrix factorization to learn the user/item/review latent representation.
- **A2SPR** [8]. A state-of-the-art review recommendation model, which incorporates review's aspect information for personalized review recommendation.
- **FIM** [25]. A state-of-the-art text recommendation model, which utilizes fine-grained text matching model for recommendation.

4.4   Experiments Results

The experimental results are summarized in Table 2 and Table 3. According to the tables, we have the following observations.

Table 2. Personalized review ranking results in Reddit dataset. The improvement over the state-of-the-arts is statistically significant (p-value < 0.01). Row Imp. vs Per. indicates the MTL-PT (Poly) model improvements over the best personalized method.

Methods	MRR	HR@1	HR@3	HR@5	nDCG@1	nDCG@3	nDCG@5
Rand.	0.2774	0.0968	0.2708	0.4657	0.0943	0.1946	0.2742
New.	0.2084	0.0527	0.1675	0.2973	0.0576	0.1166	0.1695
Pop.	0.3916	0.2283	0.4323	0.5422	**0.2376**	0.3468	0.3921
MF	0.3638	0.1805	0.4297	0.6082	0.1954	0.3221	0.3955
TF	0.3720	0.1981	0.4223	0.6033	0.1993	0.3226	0.3835
BETF	0.3821	0.1982	0.4212	0.5886	0.1926	0.3263	0.3944
CAP	0.3780	0.1977	0.4223	0.5890	0.1827	0.3224	0.3903
NeuMF	0.3672	0.1426	0.4670	0.6593	0.1574	0.3275	0.4066
A2SPR	0.3722	0.1552	0.4582	0.6493	0.2034	0.3432	0.4023
FIM	0.4028	0.2205	0.4802	0.6574	0.2274	0.3667	0.4177
MTL-PT (Dot)	0.4203	0.2289	0.4987	0.6534	0.2235	0.3807	0.4384
MTL-PT (Cross)	**0.4251**	0.2301	**0.5055**	0.6660	0.2301	0.3881	**0.4488**
MTL-PT (Poly)	0.4250	**0.2322**	0.5049	**0.6734**	0.2332	**0.3882**	0.4456
Imp. vs Per.	+5.51%	+5.30%	+ 5.14%	+2.43%	+2.55%	+5.86%	+6.68%

Firstly, reviews' scores-based recommendation method (Pop.) achieves the best performance among non-personalized methods. The prominent result is caused by the website review exhibition mechanism. Most of the web services tend to rank their item's review by scores, which leaves the low ranking reviews' with lower exposure. It accumulates interactions for popular reviews and neglects non-popular ones (Matthew Effect).

Secondly, most of the methods based on personalized review recommendation outperform non-personal ranking methods. The result shows that the variety among users preference does exist, and omitting it would cause a significant performance decrease in recommendation. Though MTL-PT's performance in HR@1 and nDCG@1 is slightly worse than baseline Pop. in Taptap dataset, it may be caused by Matthew Effect mentioned before. Moreover, observing other metrics, it could be proved personalized methods achieve better performance when users have more choices apart from the most popular review, especially in non-e-commerce setting, Reddit.

Table 3. Personalized review ranking result in Taptap dataset.

Methods	MRR	HR@1	HR@3	HR@5	nDCG@1	nDCG@3	nDCG@5
Rand.	0.2932	0.1115	0.2925	0.4863	0.1126	0.2133	0.2929
New.	0.2243	0.0923	0.1724	0.2894	0.0957	0.1545	0.2072
Pop.	0.4651	**0.2852**	0.5430	0.6728	**0.2898**	0.4366	0.4902
MF	0.3732	0.1883	0.4022	0.5850	0.1826	0.3018	0.3869
TF	0.3966	0.1181	0.3740	0.5276	0.1280	0.2671	0.3301
BETF	0.3347	0.1426	0.3583	0.5631	0.1478	0.2665	0.3509
CAP	0.3346	0.1443	0.3590	0.5721	0.1477	0.2771	0.3653
NeuMF	0.3573	0.1463	0.4314	0.6161	0.1428	0.3092	0.3848
A2SPR	0.3423	0.1522	0.4108	0.6100	0.1595	0.3083	0.3353
FIM	<u>0.4183</u>	<u>0.2340</u>	<u>0.4735</u>	<u>0.6252</u>	<u>0.2344</u>	<u>0.3722</u>	<u>0.4346</u>
MTL-PT (Dot)	0.4150	0.2327	0.5197	0.7294	0.2322	0.4382	0.5066
MTL-PT (Cross)	**0.4665**	0.2651	0.5547	**0.7632**	0.2622	**0.4459**	**0.5109**
MTL-PT (Poly)	0.4457	0.2650	**0.5649**	0.7536	0.2633	0.4450	0.5089
Imp. vs Per.	+6.55%	+13.24%	+19.30%	+20.53%	+12.32%	+19.55%	+17.09%

Thirdly, MTL-PT (Poly) outperforms most personalized review recommendation methods by at most 20.53%, with an average improvement of 4.7% in the Reddit dataset and 15.5% in the Taptap dataset. Comparing different aggregators, inner-product aggregator (MTL-PT (Dot)) yields the worst result among all three but armed with simple structure and swift performance. Cross aggregator (MTL-PT (Cross)) and poly aggregator (MTL-PT (Poly)) yield similar results while poly aggregator has far less computation cost compared to its counterpart.

To sum up, our approach achieves excellent performance under all metrics by incorporating the semantic feature from reviews into multi-relationship learning.

4.5 Ablation Analysis

Impact of Personalized Transformer. We compare our personalized transformer with two variations: widely used CNN [26] text encoder and vanilla transformer. The second column in Table 4 shows the result of the CNN variant. The model's performance decreases significantly without the personalized transformer. Compared with the local feature extracted by CNN from text, the global semantic feature extracted by Transformer plays a more critical role. The third column in Table 4 shows the result of the vanilla transformer. The model's MRR drop around 2.5% along with other metrics, which validates the effectiveness of our personalized inputs for the transformer.

Impact of Each Task in MTL-PT. We perform ablation studies by leaving out some crucial tasks in MTL-PT model to investigate their effectiveness. The results are presented in Table 4, where UR is user-review main task, IR is the item-review task and UI is the user-item task. We can find that both item-review task and user-item task play an essential role in contributing useful review representation for the ranking. Specifically, the model without item-review task suffers larger performance decrease. A simple user-review correlation is not enough for representing the user's preference for reviews.

Table 4. The result of ablation study on Reddit dataset.

Metrics	PT ablation			Task ablation		
	MTL-PT	MTL-CNN	MTL-T	UR	UR+IR	UR+UI
MRR	**0.4251**	0.3834	0.4142	0.4080	0.4103	0.4092
HR@1	**0.2322**	0.1936	0.2175	0.2080	0.2304	0.2287
HR@3	**0.5049**	0.4434	0.4922	0.4729	0.4882	0.4789
HR@5	**0.6734**	0.6017	0.6511	0.6436	0.6533	0.6455
nDCG@1	**0.2322**	0.1936	0.2233	0.2080	0.2101	0.2080
nDCG@3	**0.3882**	0.3355	0.3776	0.3621	0.3722	0.3701
nDCG@5	**0.4488**	0.4002	0.4475	0.4323	0.4433	0.4415

5 Conclusions and Future Work

In this paper, we propose a Multi-Task Learning model with Personalized Transformer (MTL-PT) to recommend reviews that meet the user's preferences. Multiple tasks are conducted simultaneously to learn interactions among user, item and review to remedy data sparsity. Considering the varied user's preference to reviews given different items, a personalized transformer is proposed to extract context-aware semantic feature, and item-aware poly aggregator is utilized further for exploiting rich interactions. Meantime, user/item latent features collaboratively learned in multiple tasks could improve the generalization performance of our model. We also construct two real-world datasets, Reddit and Taptap. Reddit dataset is the first non-e-commence dataset for review recommendation. Experiments on the datasets demonstrate the superior performance of our model. In future works, we would conduct experiments on more large datasets. We would also consider temporal information of reviews further and take advantage of graph network for personalized review recommendation.

Acknowledgments. This work is supported by the National Natural Science Foundation of China (61902439, U19112031), Guangdong Basic and Applied Basic Research Foundation (2021A1515011902, 2019A1515011159), National Science Foundation for Post-Doctoral Scientists of China under Grant (2019M663237), Macao Young Scholars Program (UMMTP2020-MYSP-016), the Major Science and Technology Research Programs of Zhongshan City (2019B2006, 2019A40027, 2021A1003).

References

1. Agarwal, D., Chen, B., Pang, B.: Personalized recommendation of user comments via factor models. In: EMNLP, pp. 571–582 (2011)
2. Chen, C., Yang, Y., Zhou, J., Li, X., Bao, F.S.: Cross-domain review helpfulness prediction based on convolutional neural networks with auxiliary domain discriminators. In: NAACL, pp. 602–607 (2018)
3. Devlin, J., Chang, M., Lee, K., Toutanova, K.: BERT: pre-training of deep bidirectional transformers for language understanding. In: NAACL, pp. 4171–4186 (2019)

4. Diaz, G.O., Ng, V.: Modeling and prediction of online product review helpfulness: a survey. In: ACL, pp. 698–708 (2018)
5. Du, J., Zheng, L., He, J., Rong, J., Wang, H., Zhang, Y.: An interactive network for end-to-end review helpfulness modeling. Data Sci. Eng. **5**(3), 261–279 (2020)
6. Fan, M., Feng, C., Guo, L., Sun, M., Li, P.: Product-aware helpfulness prediction of online reviews. In: WWW, pp. 2715–2721 (2019)
7. He, X., Liao, L., Zhang, H., Nie, L., Hu, X., Chua, T.: Neural collaborative filtering. In: WWW, pp. 173–182 (2017)
8. Huang, C., Jiang, W., Wu, J., Wang, G.: Personalized review recommendation based on users' aspect sentiment. TOIT **20**(4), 1–26 (2020)
9. Humeau, S., Shuster, K., Lachaux, M., Weston, J.: Poly-encoders: architectures and pre-training strategies for fast and accurate multi-sentence scoring. In: ICLR (2020)
10. Lu, Y., Tsaparas, P., Ntoulas, A., Polanyi, L.: Exploiting social context for review quality prediction. In: WWW, pp. 691–700 (2010)
11. Ma, J., Wen, J., Zhong, M., Chen, W., Li, X.: MMM: multi-source multi-net micro-video recommendation with clustered hidden item representation learning. Data Sci. Eng. **4**(3), 240–253 (2019)
12. Martin, L., Pu, P.: Prediction of helpful reviews using emotions extraction. In: AAAI, pp. 1551–1557 (2014)
13. Moghaddam, S., Jamali, M., Ester, M.: Review recommendation: personalized prediction of the quality of online reviews. In: CIKM, pp. 2249–2252 (2011)
14. Moghaddam, S., Jamali, M., Ester, M.: ETF: extended tensor factorization model for personalizing prediction of review helpfulness. In: WSDM, pp. 163–172 (2012)
15. Momeni, E., Cardie, C., Diakopoulos, N.: A survey on assessment and ranking methodologies for user-generated content on the web. ACM Comput. Surv. **48**(3), 41:1–41:49 (2016)
16. Paul, D., Sarkar, S., Chelliah, M., Kalyan, C., Nadkarni, P.P.S.: Recommendation of high quality representative reviews in e-commerce. In: RecSys, pp. 311–315 (2017)
17. Prado, T.R.P., Moro, M.M.: Review recommendation for points of interest's owners. In: HT, pp. 295–304 (2017)
18. Rendle, S., Freudenthaler, C., Gantner, Z., Schmidt-Thieme, L.: BPR: Bayesian personalized ranking from implicit feedback. CoRR arXiv:1205.2618 (2012)
19. Suresh, V., Roohi, S., Eirinaki, M., Varlamis, I.: Using social data for personalizing review rankings. In: RecSys (2014)
20. Tang, J., Gao, H., Hu, X., Liu, H.: Context-aware review helpfulness rating prediction. In: RecSys, pp. 1–8 (2013)
21. Tang, J., Gao, H., Liu, H.: mTrust: discerning multi-faceted trust in a connected world. In: WSDM, pp. 93–102 (2012)
22. Vaswani, A., et al.: Attention is all you need. In: NIPS, pp. 5998–6008 (2017)
23. Wang, C., Chen, G., Wei, Q.: A temporal consistency method for online review ranking. Knowl. Based Syst. **143**, 259–270 (2018)
24. Wang, S., Li, C., Zhao, K., Chen, H.: Context-aware recommendations with random partition factorization machines. Data Sci. Eng. **2**(2), 125–135 (2017)
25. Wu, C., Wu, F., An, M., Huang, J., Huang, Y., Xie, X.: Neural news recommendation with attentive multi-view learning. In: IJCAI, pp. 3863–3869 (2019)
26. Wu, C., Wu, F., An, M., Huang, J., Huang, Y., Xie, X.: NPA: neural news recommendation with personalized attention. In: SIGKDD, pp. 2576–2584 (2019)
27. Xiong, W., Litman, D.J.: Empirical analysis of exploiting review helpfulness for extractive summarization of online reviews. In: COLING, pp. 1985–1995 (2014)

ADQ-GNN: Next POI Recommendation by Fusing GNN and Area Division with Quadtree

Yu Wang[1,2], An Liu[1,2(✉)], Junhua Fang[1,2], Jianfeng Qu[1,2], and Lei Zhao[1,2]

[1] School of Computer Science and Technology, Soochow University, Su Zhou, China
[2] Collaborative Innovation Center of Novel Software Technology
and Industrialization, Nanjing, China
20194227025@stu.suda.edu.cn,
{anliu,jhfang,jfqu,zhaol}@suda.edu.cn

Abstract. Point-of-interest (POI) recommendation, i.e., recommending attractive POIs to users, has attracted great attention in both academia and industry. In order to improve recommendation accuracy and rich users' experience, researchers put much effort into modeling users' check-in sequence information as well as the physical distance between POIs. However, the state-of-the-art methods do not take hierarchical relations into consideration when mining spatial information and neglect the impact of different time intervals when capturing temporal correlation. To this end, we propose a novel model by fusing Graph Neural Network (GNN) and area division with quadtree, which takes joint consideration of users' spatial, long-term, and short-term preferences in a unified way. Specifically, the proposed model exploits users' regional transition patterns in different granularity by means of quadtree map division, and adopts attention mechanism to dynamically update weight values of the above three types of preferences. Furthermore, GNN is introduced to capture the complex interaction relationship between different areas and POIs. Finally, we conduct extensive experiments on three real-world datasets, and the results demonstrate that our model significantly outperforms the state-of-the-art methods.

Keywords: POI recommendation · Location-based services · Graph Neural Network

1 Introduction

Benefiting from the popularization of smart mobile devices and the development of GPS technology, location-based social networks (LBSNs) such as Foursquare and TripAdvisor, have becoming pervasive in our daily life. People on these LBSNs are increasingly inclined to share their daily life dynamics, which we call POI check-in records. Industry and academia have been aware of the importance of such information, thus carrying out a lot of research to capture user preferences for enriching user experience and increasing business interests.

© Springer Nature Switzerland AG 2021
W. Zhang et al. (Eds.): WISE 2021, LNCS 13081, pp. 177–193, 2021.
https://doi.org/10.1007/978-3-030-91560-5_13

User's next POI is closely related to the current state, which shows the temporal dependency, so early research introduces the Markov chain to model check-in sequence [13]. These methods model users' general taste and sequential behavior in a unified way via the fusion of common Markov chain and matrix factorization. However, Markov-chain based methods can only model the transition relationship between adjacent POIs and lack the capacity to capture users' long-term preferences in the sequence. More recently, Recurrent Neural Networks (RNN) have been successfully applied in sequence modeling and become pervasive in the field of next-POI recommendation [2, 26]. Equipped with well-designed time or distance gates [10, 17, 28, 29], RNN-based methods can make full use of spatial and temporal information.

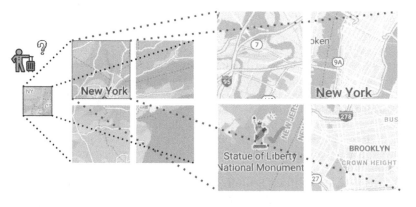

Fig. 1. An illustrative example of user decision-making, which is similar to positioning layer by layer in space.

Although these methods can effectively model spatial-temporal preference, three key issues remain to be resolved. First, beyond POI-level sequential relation, users' check-in records show more regional characteristics because people tend to visit POIs which are close to their current locations. For example, in Gowalla and Foursquare, 90% of users' adjacent check-in records are within the distance less than 50 km [11]. In addition to distance proximity, [24] discovers regional clustering phenomenon: people may be interested in exploring nearby POIs of a preferred POI, even if they are far away from home. Thus, effective use of area-level information can not only solve the sparsity problem, but also better capture user's spatial preferences. Second, before users choose candidate POIs, they may usually first consider which district of the city they are in, then consider the nearby business center, and finally visit some specific POIs in it. Such a decision-making process is similar to layer-by-layer positioning as shown in Fig. 1, which prompts us to hierarchically divide the region. Meanwhile, the coarse-grained regional information is less closely related to users, yet containing more POIs, thus it is possible to recommend long-distance but popular POIs to users, which reflects the idea of exploration. Third, the difference in time intervals mirrors the long-term and short-term preferences of users, and the spatial preferences will also change accordingly. Hence, we need to effectively combine the three preferences to better understand user behavior and make predictions.

To this end, we propose a new model named ADQ-GNN by fusing GNN and area division with quadtree, to effectively model spatial-temporal information and accurately make next POI recommendation. In ADQ-GNN, we first introduce quadtree to divide the area hierarchically and let areas of different granularities contain various spatial information. GNN is then utilized to capture complex user transitions between areas and user-POI interactions in short term. We also employ LSTM to extract users' general interests, denoted as long-term preferences. Furthermore, we propose a two-layer attention-based fusion method to dynamically incorporate spatial, short-term and long-term preferences in a unified way.

In conclusion, our main contributions are listed as follows:

- We propose ADQ-GNN, a spatiotemporal-aware model for next POI recommendation, which can comprehensively exploit user preference in check-in sequences. Specifically, it can adaptively incorporate three types of user preference under an attention-based framework.
- We combine quadtree with GNN to take joint consideration of users' spatial and short-term preference, such that transferring preference in areas of different granularities along with immediate interest change can be exactly modeled. In addition, we utilize LSTM to capture long-term preference.
- Experiments on three real-world datasets are conducted to verify the effectiveness of our method, and the results show that our model outperforms state-of-the-art methods.

2   Related Work

In this section, we divide the related work into two topics: POI recommendation and graph neural networks for recommendation.

2.1   POI Recommendation

Different from other recommendation systems which exhibit digital goods, e.g. music, books and movies, POI recommendation aims to provide users with attractive POIs which should be checked-in in the physical world [18]. As a result of which, data sparsity problem has been a tricky problem for a long time [19]. GeoMF [8] uses weighted matrix factorization to cope with the implicit user-POI matrices. In order to better learn the similarity of behavior between users, CAPRF [3] combines three types of information w.r.t. POI properties, user interests, and sentiment, in a unified way.

Compared to conventional POI recommendation, research on next POI recommendation pays more attention to modeling users' dynamic preferences since users' behavior shows periodicity and time specificity. For example, some users usually go to work on weekdays and go out for entertainment on the weekends; some users tend to go to bars for a drink at night instead of in the morning or noon. In recent years, RNN has shown strong capabilities in sequence modeling, thus widely used in next-POI recommendation. ST-RNN [10] utilizes RNN to capture the periodical temporal contexts, and introduces time-specific and distance-specific transition matrices to better represent the spatial-temporal relationship. DeepMove [2] uses two attention mechanisms to match historical records with the current status and meanwhile preserve the sequential information. Although RNN is good at capturing long-range dependencies of sequences, gradient vanishing/exploding problem is inevitable when the sequence is too long. Thus, some variants of RNN, such as LSTM, GRU, are utilized to better extract the long periodic and short sequential features. Time-LSTM [29] emphasizes the importance of time intervals between users' actions, thus equipping LSTM with specifically designed time gates to better capture users' preferences. Under LSTM architecture, STGN [28] introduces two pairs of time gates and distance gates to capture the spatial-temporal interval information and models users' short-term and long-term interests simultaneously. Considering that the geographical connections between non-consecutive POI are key factors in deciding users' next movements, LSTPM [17] consists of a non-local neural operation and a geo-dilated LSTM to model the long-term and short-term preferences in a unified way. GeoSAN [7] addresses the sparsity issue by emphasizing the use of informative negative samples and captures spatial proximity between nearby locations via a self-attention based geography encoder.

However, only considering the distance relationship is not comprehensive, thus in our work, we additionally consider the interactions between different regions where the POIs are located.

2.2   Graph Neural Network for Recommendation

Nowadays, graph neural network (GNN) has gradually become an important branch in machine learning and shone great power in multiple fields, e.g., image classification [12], object detection [15] and recommendation system [20]. GNN aims to generalize convolutional neural networks to non-Euclidean domains for modeling graph structure data. To obtain accurate item embedding and take complex transitions of items into account, SR-GNN [21] constructs sub-graphs for each session sequence, then uses latent vectors of item in graphs and combines the attention mechanism to make recommendations. GC-SAN [23] integrates the self-attention network (SAN) with graph neural network (GNN), where SAN is designed to obtain contextualized non-local representations and GNN is introduced to model the local graph-structured dependencies of separated session sequences. Despite the compelling success achieved by previous researches, little attention has been paid to POI recommendation with GNNs. GE [22], a generic graph-based embedding model, embeds four relational graphs (POI-POI, POI-Region, POI-Time and POI-Word) in a unified way to capture the sequential effect, geographical influence, temporal cyclic effect and semantic effect. Surpassing GE, STP-UDGAT [9] is able to exploit personalized user preferences and explore new POIs in global spatial-temporal-preference neighborhoods. In view of the above large advantages of GNN on topological structure graphs and sequences, we use it to model users' spatial and short-term preferences.

3   Problem Formulation

We denote M users and N POIs in an LBSN as $U = \{u_1, u_2, \ldots, u_M\}$, $V = \{v_1, v_2, \ldots, v_N\}$ respectively. For each user $u \in U$, his historical check-in record up to current time t_{cur} is defined as $H_u = \{h_1^u, h_2^u, \ldots, h_{|H_u|}^u\}$, where $h_i^u \in H_u$ is a tuple containing four elements, i.e., $h_i^u = \{v_i, lng_{v_i}, lat_{v_i}, t_i^u\}$, which indicate POI, longitude, latitude and check-in time respectively. Since check-in sequences in just one day reflect more dependence, we split the historical records into daily trajectory sequence for each user $u \in U$, represented by $S_u = \{s_1, s_2, \ldots, s_n\}$ where $s_i \in S_u$ contains h_i^u belonging to the same day, and last item $s_n = \{h_1, h_2, \ldots, h_{t_{cur}}\}$ represents check-ins of the current day.

The goal of next POI recommendation is to recommend a list of preferable POIs for a target user, which he/she will visit at next timestamp t_{cur+1}. Specifically, with regard to POI $v_j \in V$ which is more likely to be visited by user $u_i \in U$, the corresponding prediction score $\hat{y}_{ij}$ is higher. Finally, the top-k POIs will be recommended to user according to prediction scores.

4   The Proposed Method

As shown in Fig. 2, we first utilize quadtree to divide the area according to the density of POIs in it, and then use GNN to capture spatial preference and short-term preference. Subsequently, we use LSTM to capture long-term preference. Finally we introduce adaptive fusion approach composed of two fusion units to merge the above three kinds of preferences.

4.1   Spatial Preference Modeling

Most of the existing work models spatial-temporal preferences of users according to the latitude and longitude in check-in data. Based on positioning coordinates, we can effectively describe the distance relationship between two physical points by Haversine formula. However, merely focusing on point-level distance can not effectively describe the spatial characteristics [16,27] and the hierarchical structure of the regions. Therefore, it is necessary to split the geographic space appropriately. Considering that quadtree can capture not only the relationship between regions of the same level, but also the hierarchical relationship among them. So we introduce the idea of quadtree to divide the geographic space as Algorithm 1 summarizes. Note that we set a threshold Θ to control the number of POIs in each area (Line 20), because it makes no sense to divide an

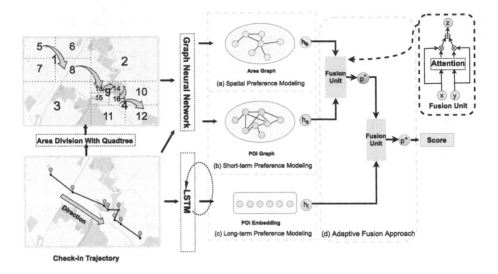

Fig. 2. Framework overview of ADQ-GNN.

area that contains few POIs. For each area, no matter it is a leaf area (cannot be subdivided) or a parent area (containing other children areas), we assign a unique id to it. Since the relationship between regions exhibits a non-Euclidean nature, thus compared to CNN and RNN, using GNN is a more suitable choice to effectively capture the relationship between them. Based on vanilla graph neural network [6,14] further adds gated recurrent units and modern optimization techniques, then proposes Gated Graph Neural Networks (GG-NNs). Inspired by [21] and [23], we construct the graph as follows. The whole quadtree of area can be modeled as a directed graph $\mathcal{G}_a = (\mathcal{V}_a, \mathcal{E})$, where $v \in \mathcal{V}_a$ represents specific area and $(v_i, v_j) \in \mathcal{E}$ indicates how many users flow from area i to area j. We use embedding mechanism to map the unique id of each area to node on the graph to learn its latent representation. Specifically, the node vector $\mathbf{v} \in \mathbb{R}^d$ indicates the latent vector of dimension d. Through the learning strategy of GNN, the final embedding representation of each region contains the connection with other regions, that is, the transfer relationship between adjacent areas and the hierarchical relationship between nested areas.

Next, we will elaborate on how to obtain the latent vector of regions via GNN. For each daily trajectory s_i, we first concatenate area vectors in it to generate a representation $s_a \in \mathbb{R}^{n \times d}$ as follows:

$$s_a = [\mathbf{v}_1, \mathbf{v}_2, \ldots, \mathbf{v}_n], \tag{1}$$

Then we represent the learning process of node vectors and the update functions as follows:

$$a_t = [M_o s_a W_o + b_o, M_i s_a W_i + b_i], \tag{2}$$

$$z_a = \sigma(W_z a_t + U_z \mathbf{v}_{t-1}), \tag{3}$$

Algorithm 1: Split area with quadtree

Input: $\mathcal{L}$: POI sets; Θ: threshold

1 $lats \leftarrow [\,]$, $lngs \leftarrow [\,]$

2 **for** *each POI p in $\mathcal{L}$* **do**

3 | Add lat_p to $lats$

4 | Add lng_p to $lngs$

5 **end**

6 Find $lat_{min} \leftarrow MIN(lats), lat_{max} \leftarrow MAX(lats)$

7 Find $lng_{min} \leftarrow MIN(lngs), lng_{max} \leftarrow MAX(lngs)$

8 $lat_{mid} \leftarrow \frac{lat_{max}+lat_{min}}{2}, lng_{mid} \leftarrow \frac{lng_{max}+lng_{min}}{2}$

9 **if** $lng_{min} \leq lng_p \leq lng_{mid}$ *and* $lat_{min} \leq lat_p \leq lat_{mid}$ **then**

10 | Add POI p to $area_{1st}$

11 **else if** $lng_{mid} \leq lng_p \leq lng_{max}$ *and* $lat_{min} \leq lat_p \leq lat_{mid}$ **then**

12 | Add POI p to $area_{2nd}$

13 **else if** $lng_{min} \leq lng_p \leq lng_{max}$ *and* $lat_{mid} \leq lat_p \leq lat_{max}$ **then**

14 | Add POI p to $area_{3rd}$

15 **else**

16 | Add POI p to $area_{4th}$

17 **end**

18 **for** *each a in $area_{1\sim4}$* **do**

19 | Count the number of POIs $\mathcal{N}_a$ in the area a

20 | **if** $\mathcal{N}_a \geq \Theta$ **then**

21 | | Split area with quadtree(a, Θ)

22 | **end**

23 **end**

$$r_a = \sigma(W_r a_t + U_r \mathbf{v}_{t-1}), \tag{4}$$

$$\widetilde{h}_a = \tanh(W_h a_t + U_h(r_a \odot \mathbf{v}_{t-1})) \tag{5}$$

$$h_a = (1 - z_a) \odot \mathbf{v}_{t-1} + z_a \odot \widetilde{h}_a, \tag{6}$$

where Eq. 2 aims to extract the contextual information of neighborhoods and then propagate information between different nodes. M_i and M_o are two well-designed weight matrix according to inter-regional flow, which we will elaborate in Sect. 4.2. $W_i, W_o \in \mathbb{R}^{d \times d}$ are parameter matrices, $b_i, b_o \in \mathbb{R}^d$ are bias vectors. Intuitively, not all connections between regions are equally important. Some regions have frequent interactions, but some have less exchanges. z_a and r_a are update gate and reset gate, which are designed to preserve and discard information respectively, and connections between regions can be captured more effectively. $W_z, W_r, W_h \in \mathbb{R}^{2d \times d}, U_z, U_r, U_h \in \mathbb{R}^{d \times d}$, are learnable parameters. $\sigma(\cdot)$ is the sigmoid function, which is defined as $\sigma(x) = \frac{1}{1+e^{-x}}$, and $\odot$ denotes Hardamard product operation. Similar to the operation in GRU [1], the candidate state is $\widetilde{h}_a$ is constructed by current state, reset gate and previous state. Finally, the hidden state h_a contains the regional preferences. We aggregate leaf area node vectors to form the final spatial preference h_a.

4.2   Dynamic Preference Modeling

After capturing users' regional preference, we focus on modeling dynamic preference. It is obvious that users' preferences and tastes are not static all the time [5,25], but will change over time, social hot spots or other factors. Therefore, how to effectively capture this kind of time-varying preferences is of great significance to improve the efficiency and accuracy of recommendation.

Short-Term Preference Modeling: Similar to spatial preference modeling in Sect. 4.1, we construct a directed graph $\mathcal{G}_p = (\mathcal{V}_p, \mathcal{E})$ for each trajectory, where each poi $v \in \mathcal{V}_p$ is also embedded into $\mathbf{v} \in \mathbb{R}^d$. The difference lies in the construction of connection matrix. In this scene, the connection weight value between nodes depends on the order and the frequency of nodes. For example, given a user's daily check-in trajectory $\{v_1, v_2, v_3, v_4\}$, where v_1 is user's home, v_2 is his/her work place, v_3 and v_4 are a cinema and a restaurant respectively, the normalized connection weight matrix can be constructed as shown in Fig. 3. After the information propagation of the graph neural network, we can get the users' daily short-term preferences by an average pooling:

$$h_s = \frac{1}{|S_n|} \sum_{t=1}^{|S_n|} h_t, \tag{7}$$

where h_t is the hidden state containing users' short preferences.

Long-Term Preference Modeling: LSTM, a variant of RNN, is proven to be able to effectively capture the user's long-term preferences benefiting from its three gate mechanisms, i.e. input, forget and output gates. As a result, we input all POI sequences before the current moment into the LSTM model, and use the last hidden $h_{t_{cur}}$ state as user's long-term preferences h_l:

$$h_t = LSTM(\mathbf{v}_t, h_{t-1}), \, t \le t_{cur} \tag{8}$$

where h_{t-1} is the hidden state of LSTM, and $\mathbf{v}_t \in \mathbb{R}^d$ is d-dimensional embedding of POI.

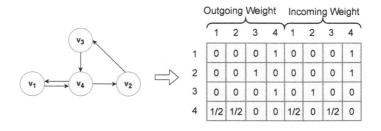

Fig. 3. An example of a trajectory graph and the connection matrix M.

4.3   Adaptive Fusion Approach

In different scenarios, factors that drive users to choose the next POI are different, i.e., how (if a POI in an area with convenient transportation and can be reached quickly, then spatial preference plays a dominating role), when (if the next check-in behavior occurs shortly after the previous action, then short-term preference have greater influence), what (if a POI appears repeatedly in user's historical check-in sequence, user shows a static preference, that is, long-term preference). There is no doubt that effectively fusing users' preferences is of great significance. The intuitive way is to sum and then average them. However, this naive way cannot distinguish the importance of these preference components. Thus we utilize two-layer attention-based dynamic fusion method:

$$\alpha = \sigma(W^\alpha(h_a + h_s) + b^\alpha), \tag{9}$$

$$p^* = \alpha * h_a + (1 - \alpha) * h_s, \tag{10}$$

$$\beta = \sigma(W^\beta(p^* + h_l) + b^\beta), \tag{11}$$

$$p^+ = \beta * p^* + (1 - \beta) * h_l, \tag{12}$$

where W^α, W^β are parameter matrices and b^α, b^β are bias vectors. We dynamically combine user spatial preferences with short-term preferences (Eq. 10), and then integrate them with long-term preferences in the same way (Eq. 12).

4.4   Prediction

After we have obtained the embedding representation of each check-in trajectory, the prediction score $\hat{z}_i$ for each candidate POI $v_i \in V$ is computed via the following:

$$\hat{z}_i = (p^+)^T \mathbf{v}_i, \tag{13}$$

finally, we apply a softmax function to normalize the score to the probability, that is:

$$\hat{y} = softmax(\hat{z}), \tag{14}$$

where $\hat{y} \in \mathbb{R}^N$ denotes the probabilities of POIs to be visited next and the top-k POIs will be recommended.

5   Experiment

5.1   Evaluation Datasets

We conduct experiments on three public LBSNs datasets and the statistic of them are listed in Table 1. NYC (New York City) and Tokyo are both extracted from Foursquare[1] dataset, and their time spans are about 10 month (from 12

---

[1] https://sites.google.com/site/yangdingqi/home/foursquare-dataset.

April 2012 to 16 February 2013). Check-ins in Gowalla[2] dataset are over period of February 2009 to October 2010. For fair comparison, following [17] and [28], we eliminate users with fewer than 10 check-ins and unpopular POIs appearing less than 10 times in all datasets. Then, for each user, we sort the check-in records in chronological order, and treat the first 70% as training set, then 20% for testing and the remaining 10% as the validation set.

Table 1. Statistics of three datasets

Dataset	User	POI	Check-in	Sparsity
NYC	1,083	38,333	227,428	94.52%
Tokyo	2,293	61,858	573,703	99.60%
Gowalla	9,819	44,186	2,455,077	99.43%

5.2 Evaluation Metrics

We employ two widely-used ranking evaluation metrics for next POI recommendation, namely Precision@k and Mean Reciprocal Rank@k (denoted by *Pre@k* and *MRR@k*, respectively). *Pre@k* indicates the accuracy of recommendation. Specifically, it represents the proportion of correctly recommended POIs among the top-k POIs. *MRR@k* is used to assess the quality of ranked list, which considers the order of recommendation ranking. When correct POIs are in the top of the ranking list, the MRR is large.

5.3 Baselines and Settings

To evaluate the performance of the proposed method, we compare ADQ-GNN with the following baselines:

- **LSTM** [4]: This is a variant of RNN model, which can solve the problem of vanishing gradient and has natural advantages in sequence modeling.
- **ST-RNN** [10]: ST-RNN incorporates time-specific and distance-specific transition matrices to model temporal and spatial contexts simultaneously.
- **Time-LSTM** [29]: Time-LSTM equips LSTM with well-designed time gates to model the order information and actions' relations in sequential data.
- **DeepMove** [2]: DeepMove proposes a multi-modal embedding recurrent neural network to capture the pattern of complicated sequential transitions, and introduce attention mechanism to model users' long-term and short-term preferences.

---

[2] https://snap.stanford.edu/data/loc-gowalla.html.

- **STGN** [28]: STGN is the extension of LSTM by introducing spatial and temporal gates. Through the modeling of time and distance intervals between neighbor check-ins, it can effectively learn users' sequential visiting behavior and improve accuracy in the next POI recommendation.
- **LSTPM** [17]: LSTPM consists of a context-aware non-local network and a geo-dilated LSTM, which can explore the spatial-temporal correlations between historical and current trajectories, and exploit the geographical relations among non-consecutive POIs.

In this paper, we set the hidden size of latent vectors $d = 100$. We train our model using the Adam optimizer with a learning rate of 0.001 and set the regulation parameter to 0.0001. Moreover, the number of training epochs and the batch size is set to 20 and 100 respectively. For the parameters of other baselines, we follow the best settings in their papers. To guarantee the accuracy and the fairness the experimental results, we take the average of the results obtained from 10 experiments as the final result.

Table 2. Performance comparison on three datasets evaluated by *Pre@k* and *MRR@K*.

Dataset	Metrics	LSTM	ST-RNN	STGN	Time-LSTM	DeepMove	LSTPM	ADQ-GNN
NYC	*Pre@1*	8.7727	11.2595	14.3882	17.3679	19.0254	21.5696	**23.4252**
	Pre@5	21.3509	25.8728	42.4729	42.6083	43.1740	44.6196	**46.1355**
	Pre@10	28.7639	31.4594	51.5722	52.1098	52.7126	56.7430	**59.1822**
	MRR@1	8.7727	11.2595	14.3882	17.3679	19.0254	21.5696	**23.4252**
	MRR@5	12.7673	16.7002	22.5570	23.8932	29.3204	30.1221	**31.0527**
	MRR@10	13.3672	17.4517	23.7609	24.4527	30.5732	31.4893	**32.4896**
Tokyo	*Pre@1*	5.2305	7.9206	12.4893	18.0182	19.0335	18.9102	**22.4973**
	Pre@5	17.8044	23.0123	35.7561	36.2922	42.7419	37.0550	**44.2093**
	Pre@10	25.9548	31.2391	46.3794	47.9920	51.3673	48.1098	**53.6349**
	MRR@1	5.2305	7.9206	12.4893	18.0182	19.0335	18.9102	**22.4973**
	MRR@5	10.7589	13.2744	21.3227	22.6581	27.2689	25.1098	**27.6958**
	MRR@10	11.2890	14.3724	22.0975	23.2570	27.9758	26.3301	**28.1034**
Gowalla	*Pre@1*	3.2748	5.4528	10.6037	10.0776	9.8689	11.6328	**12.4220**
	Pre@5	5.6740	12.9573	21.7025	21.3890	20.4461	23.8749	**24.4734**
	Pre@10	7.0952	16.4801	31.8639	29.0074	28.5984	33.9202	**34.9648**
	MRR@1	3.2748	5.4528	10.6037	10.0776	9.8689	11.6328	**12.4220**
	MRR@5	6.2293	8.1960	17.0254	16.9896	16.9039	18.7946	**19.6722**
	MRR@10	7.0899	8.6632	17.9982	17.8734	18.7870	19.4558	**20.2871**

5.4 Results

Table 2 shows the results of different methods for next-POI recommendation. In order to highlight the comparison, we bold the best results and underline the second best results. Based on the experimental result data, we can get the following observations:

- Our proposed model ADQ-GNN consistently outperforms all other baselines on three datasets with respect to both *Pre@k* and *MRR@k*. Specifically, compared with six counterparts, the improvement provided by ADQ-GNN are about 4.12%–51.40%, 4.23%–51.61% and 2.99%–79.71% with reference to *Pre@10* metric on NYC, Tokyo and Gowalla respectively. The experimental result indicates the effectiveness of our method for fusing users' spatial preferences with long-term and short-term preferences.

- All methods preform better on NYC and Tokyo. The reason behind it may lie in that POIs in Gowalla are world-wide and show great sparsity. Despite this, ADQ-GNN can also outperform the second best method.
- LSTM shows great advantage in sequence modeling for it can remember the important parts in history and ignore the unimportant, but compared to other baseline methods it has poor performance. The reason is that it only models the user's check-in sequence but ignores the spatial-temporal information contained in it, which confirms that spatial-temporal information plays a vital role in POI recommendation.
- Compared with other baselines, LSTPM and DeepMove have relatively better performances, which may benefit from their modeling of users' long-term and short-preferences. Though STGN takes time and distance intervals into consideration, it doesn't consider the relationship between non-adjacent POIs like LSTPM, thus its performance is not outstanding. However, we utilize GNN to capture connections between POIs, which can embed the POIs more effectively to improve performance.

Table 3. Result of ablation analysis.

| | NYC | | Tokyo | | Gowalla | |
	Pre@10	*MRR@10*	*Pre@10*	*MRR@10*	*Pre@10*	*MRR@10*
ADQ-GNN-S	57.3672	31.4927	50.8561	26.9472	33.2902	19.1377
ADQ-GNN-SA	58.3763	31.9738	51.5819	27.8083	34.0092	19.4793
ADQ-GNN-SL	57.6287	31.7032	51.1433	27.6532	34.3831	19.6409
ADQ-GNN	**59.1822**	**32.4896**	**53.6349**	**28.1034**	**34.9648**	**20.2871**

5.5   Ablation Experiment

To verify the effectiveness of each module in our model, we conduct ablation experiments in this section. Based on original ADQ-GNN, we keep part of components to form variants as follows:

- ADQ-GNN-S: This variant only retains the modeling of users' short-term preference.
- ADQ-GNN-SA: This denotes the version which considers both short-term and spatial preference, removing long-term preference.
- ADQ-GNN-SL: This version preserves short-term and long-term preference.

The result is shown as Table 3, and we can find that the performance of ADQ-GNN is better than all variants, which confirms the effectiveness of each module. Specifically, ADQ-GNN-SA shows the best performance compared with

ADQ-GNN-S and ADQ-GNN-SL on NYC and Tokyo datasets. This indicates that spatial preference plays an important role in these two city datasets. On gowalla where POIs are sparser than NYC and Tokyo, ADQ-GNN-SL achieves a better performance, which confirms the importance of long-term preference modeling for POI recommendation.

5.6   Parameter Sensitivity Analysis

Sensitivity w.r.t. Level of Quadtree. Since the different levels of regional information show various spatial transfer preferences, thus we vary the number of division from 3 to 11 on NYC and Tokyo dataset. Due to POIs in gowalla dataset are worldwide, we set number of division from 5 to 13. Based on experience, we set Θ to 30. As shown in Fig. 4, the performance will gradually reach a stable peak point with level of quadtree increasing, then it will slowly decrease. This is caused by too fine-grained division which is difficult to capture users' general spatial preferences.   We also explored the method of replacing the quadtree

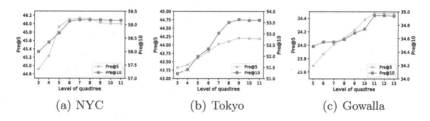

(a) NYC   (b) Tokyo   (c) Gowalla

Fig. 4. Impact of level of quadtree.

division method with equal division to compare the differences. Due to space constraint, we only show the result of NYC in Fig. 5(a). When the size of grid is $200\,\text{m} \times 200\,\text{m}$, the *Pre@10* reaches the maximum vale but still lags behind our method, which confirms the significance of using a quadtree to divide the region hierarchically.

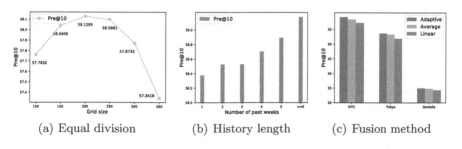

(a) Equal division   (b) History length   (c) Fusion method

Fig. 5. (a) shows the effect of different grid sizes in equal division. (b) and (c) show impact of history length and fusions respectively.

Sensitivity w.r.t. History Length. In our model, users' historical check-ins are utilized to model long-term preferences. We select POI check-in records in the past k weeks, and users having 6 or more ones are in the last group. As Fig. 5(b) shows, the accuracy of recommendation is positively related to the history length on NYC dataset. As a result of which, we use all historical data of users to make predictions.

Sensitivity w.r.t. Fusion Method. Finally, we verify the fusion method of the three types of preferences. We derived two simple variants, namely Average Fusion ($p_{avg} = \frac{h_s + h_s + h_l}{3}$) and Linear Fusion ($p_{lin} = W[h_s, h_s, h_l] + b$). From the Fig. 5(c) we can see that our adaptive fusion approach performs best, and Linear Fusion is the second. On the other two data sets, the results are similar. It proves that introducing the attention mechanism to dynamically combine preferences is of great significance.

6   Conclusion

In this paper, we proposed a novel model named ADQ-GNN for next POI recommendation. We split the area with quadtree to identify spatial preferences. We adopted GNN to model short-term preferences and LSTM to model long-term preferences. To effectively integrate the three preferences, we proposed an attention-based dynamic preference modeling. The experimental results showed that the ADQ-GNN outperforms state-of-the-art methods consistently. In the future, we would take social relationships and POI category into consideration to further improve the performance of next POI recommendation.

Acknowledgement. This work was supported by National Natural Science Foundation of China (Grant Nos. 61802273, 62102276), the Major Program of the Natural Science Foundation of Jiangsu Higher Education Institutions of China (Grant Nos. 18KJ-A520010, 19KJA610002), and the Postdoctoral Science Foundation of China under Grant No. 2020M681529. This work was partially supported by Collaborative Innovation Center of Novel Software Technology and Industrialization.

References

1. Cho, K., et al.: Learning phrase representations using RNN encoder-decoder for statistical machine translation. CoRR abs/1406.1078, pp. 1–15 (2014)
2. Feng, J., et al.: DeepMove: predicting human mobility with attentional recurrent networks. In: WWW, pp. 1459–1468 (2018)
3. Gao, H., Tang, J., Hu, X., Liu, H.: Content-aware point of interest recommendation on location-based social networks. In: AAAI, pp. 1721–1727 (2015)
4. Hochreiter, S., Schmidhuber, J.: Long short-term memory. Neural Comput. **9**, 1735–1780 (1997)

5. Hosseini, S., Yin, H., Zhou, X., Sadiq, S., Kangavari, M.R., Cheung, N.-M.: Leveraging multi-aspect time-related influence in location recommendation. World Wide Web **22**(3), 1001–1028 (2018). https://doi.org/10.1007/s11280-018-0573-2

6. Li, Y., Tarlow, D., Brockschmidt, M., Zemel, R.: Gated graph sequence neural networks. In: ICLR, pp. 1–20 (2015)

7. Lian, D., Wu, Y., Ge, Y., Xie, X., Chen, E.: Geography-aware sequential location recommendation. In: KDD, pp. 2009–2019 (2020)

8. Lian, D., Zhao, C., Xie, X., Sun, G., Chen, E., Rui, Y.: GeoMF: joint geographical modeling and matrix factorization for point-of-interest recommendation. In: KDD, pp. 831–840 (2014)

9. Lim, N., et al.: STP-UDGAT: spatial-temporal-preference user dimensional graph attention network for next POI recommendation. In: CIKM, pp. 845–854 (2020)

10. Liu, Q., Wu, S., Wang, L., Tan, T.: Predicting the next location: a recurrent model with spatial and temporal contexts. In: AAAI, pp. 194–200 (2016)

11. Liu, Y., Pham, T.A.N., Cong, G., Yuan, Q.: An experimental evaluation of point-of-interest recommendation in location-based social networks. Proc. VLDB Endow. **10**, 1010–1021 (2017)

12. Marino, K., Salakhutdinov, R., Gupta, A.: The more you know: using knowledge graphs for image classification. In: CVPR, pp. 20–28 (2017)

13. Rendle, S., Freudenthaler, C., Schmidt-Thieme, L.: Factorizing personalized Markov chains for next-basket recommendation. In: WWW, pp. 811–820 (2010)

14. Scarselli, F., Gori, M., Tsoi, A.C., Hagenbuchner, M., Monfardini, G.: The graph neural network model. Trans. Neural Netw. **20**, 61–80 (2008)

15. Shi, W., Rajkumar, R.: Point-GNN: graph neural network for 3d object detection in a point cloud. In: CVPR, pp. 1708–1716 (2020)

16. Su, C., Wang, J., Xie, X.: Point-of-interest recommendation based on geographical influence and extended pairwise ranking. In: IEEE INFOCOM 2020 Workshops (INFOCOM WKSHPS), pp. 966–971 (2020)

17. Sun, K., Qian, T., Chen, T., Liang, Y., Nguyen, Q.V.H., Yin, H.: Where to go next: modeling long- and short-term user preferences for point-of-interest recommendation. In: AAAI, pp. 214–221 (2020)

18. Wang, H., Shen, H., Ouyang, W., Cheng, X.: Exploiting poi-specific geographical influence for point-of-interest recommendation. In: IJCAI, pp. 3877–3883 (2018)

19. Wang, W., Liu, A., Li, Z., Zhang, X., Li, Q., Zhou, X.: Protecting multi-party privacy in location-aware social point-of-interest recommendation. World Wide Web **22**(2), 863–883 (2018). https://doi.org/10.1007/s11280-018-0550-9

20. Wang, X., He, X., Wang, M., Feng, F., Chua, T.S.: Neural graph collaborative filtering. In: SIGIR, pp. 165–174 (2019)

21. Wu, S., Tang, Y., Zhu, Y., Wang, L., Xie, X., Tan, T.: Session-based recommendation with graph neural networks. In: AAAI, pp. 346–353 (2019)

22. Xie, M., Yin, H., Wang, H., Xu, F., Chen, W., Wang, S.: Learning graph-based poi embedding for location-based recommendation. In: CIKM, pp. 15–24 (2016)

23. Xu, C., et al.: Graph contextualized self-attention network for session-based recommendation. In: IJCAI, pp. 3940–3946 (2019)

24. Ye, M., Yin, P., Lee, W.C., Lee, D.L.: Exploiting geographical influence for collaborative point-of-interest recommendation. In: SIGIR, pp. 325–334 (2011)

25. Ying, H., et al.: Time-aware metric embedding with asymmetric projection for successive POI recommendation. World Wide Web **22**, 2209–2224 (2019)

26. Zhang, L., et al.: An interactive multi-task learning framework for next POI recommendation with uncertain check-ins. CAL **301**, 13954 (2020)

27. Zhang, X., et al.: Personalized geographical influence modeling for POI recommendation. IEEE Intell. Syst. **35**, 18–27 (2020)
28. Zhao, P., et al.: Where to go next: a spatio-temporal gated network for next POI recommendation. In: AAAI, pp. 5877–5884 (2020)
29. Zhu, Y., et al.: What to do next: modeling user behaviors by time-LSTM. In: IJCAI, pp. 3602–3608 (2017)

MGSAN: A Multi-granularity Self-attention Network for Next POI Recommendation

Yepeng Li[1], Xuefeng Xian[2(✉)], Pengpeng Zhao[1(✉)], Yanchi Liu[3],
and Victor S.Sheng[4]

[1] Soochow University, Suzhou, China
[2] Suzhou Vocational University, Suzhou, China
[3] Rutgers University, New Brunswick, NJ, USA
[4] Texas Tech University, Lubbock, TX, USA
xianxuefeng@jssvc.edu.cn, ppzhao@suda.edu.cn

Abstract. Next Point-of-Interest (POI) recommendation has become a vital research trend, helping people find interesting and attractive locations. Existing methods usually exploit the individual-level POI sequences but failed to utilize the information of collective-level POI sequences. Since collective-level POIs, like shopping malls or plazas, are common in the real world, we argue that only the individual-level POI sequences cannot represent more semantic features and cannot express complete transition patterns. To this end, we propose a novel Multi-Granularity Self-Attention Network (MGSAN) for next POI recommendation, which utilizes the multi-granularity representation and the self-attention mechanism to capture the transition patterns of individual-level and collective-level POI sequences on two different levels of granularities. Specifically, individual-level and collective-level POI sequences are first constructed and embeddings of each check-in tuple are normalized. Then, MGSAN incorporates spatio-temporal features by introducing two temporal-aware encoders and two spatial-aware encoders and learns sequential patterns with the self-attention network for two granularities. Finally, we recommended a user's next POI with the help of two sub-tasks, i.e., the activity task to predict the next category and the auxiliary task to predict the next POI type. Extensive experiments on three real-world datasets show that MGSAN outperforms state-of-the-art methods consistently.

Keywords: POI recommendation · Multi-granularity · Self-attention

1 Introduction

A noticeable advancement in social network applications in recent years is the introduction of which has led to the emergence of Location-Based Social Networks (LBSNs), such as Foursquare, Twinkle, and GeoLife. POI recommendation

© Springer Nature Switzerland AG 2021
W. Zhang et al. (Eds.): WISE 2021, LNCS 13081, pp. 193–208, 2021.
https://doi.org/10.1007/978-3-030-91560-5_14

system plays a vital role in the LBSNs as it can help users discover interesting local attractions or facilities. Moreover, it can help shop-keepers attract thousands of customers from the globe via advertisements.

Recently, latent factor models are widely applied in POI recommendation. We can extract the latent relationships between users and POIs by utilizing Matrix Factorization. Lian et al. augment Matrix Factorization by incorporating geographical features for next POI recommendation [9]. Li et al. propose a ranking-based geographical factorization model that exploits the check-in frequency to represent users' visiting preference and learns latent factors by ranking POIs [8]. More Recently, POI recommendation is often modeled as a sequential recommendation to take advantage of the sequential patterns of user check-ins. Recurrent Neural Networks (RNNs) have been widely applied to POI recommendation and achieve state-of-the-art performance. Liu et al. propose a RNN-based method that exploits spatio-temporal contexts for the next POI recommendation [11]. Zhao et al. propose a spatio-temporal gated network to improve LSTM, which utilizes two time gates and two distance gates to simultaneously model user's short-term interest and long-term interest [18]. Huang et al. improve the RNN method by employing an attention mechanism, and the model achieves better performance for next POI recommendation [5]. Above-mentioned methods exploit individual-level POIs, ignoring the existence of collective-level POIs.

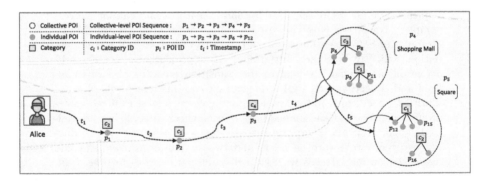

Fig. 1. An example of Alice's visited POIs, where p_4 and p_5 are collective-level POIs containing multiple individual-level POIs; c_i is the particular category id; and t_i is the particular check-in time.

As illustrated in Fig. 1, Alice visited *individual-level POIs* at p_1, p_2 and p_3 successively, namely *certain check-ins*. She then visited individual-level POI p_6 inside the shopping mall p_4 and individual-level POI p_{12} inside the square p_5. Note that p_4 and p_5 are defined as *collective-level POIs* containing several individual-level POIs. Most existing studies are built upon Alice's individual-level POIs: $p_1 \rightarrow p_2 \rightarrow p_3 \rightarrow p_6 \rightarrow p_{12}$. But Alice may leave a rough footprint, e.g., p_4, instead of the precise POI p_6. The accessible sequence of check-ins will become: $p_1 \rightarrow p_2 \rightarrow p_3 \rightarrow p_4 \rightarrow p_5$, where the *uncertain check-ins* at collective-level POIs p_4 and p_5 are involved. Thus, collective-level POIs are common and essential in real life.

Although the recent studies [5,11,18] achieve state-of-the-art performance by utilizing the sequence individual-level POIs only, they fail to consider the sequential patterns of collective-level POIs. Recently, Zhang et al. propose a model that exploits the sequences of collective-level POIs, demonstrating the effectiveness of utilizing the sequence of collective-level POIs[17]. As a higher level of POI granularity, collective-level POIs can help represent latent features of particular POIs (e.g., shopping malls). And collective-level POI sequence can also reveal the latent relationships of POI transitions (e.g., relationships between shopping mall and user office). But Zhang et al. fail to model the sequential patterns of individual-level POIs. However, as both collective-level POIs and individual-level POIs are helpful in modeling transition patterns, we argue that a method that combines the two granularities of POI sequences can facilitate each other to achieve better performance for next POI recommendation.

To achieve more accurate next POI recommendation, we propose a Multi-Granularity Self-Attention Network (MGSAN), which captures the sequential transition patterns of two granularities: collective-level check-ins and individual-level check-ins. Specifically, MGSAN incorporates spatio-temporal features by introducing two temporal-aware encoders and two spatial-aware encoders for both individual-level POI sequences and collective-level POI sequences. These encoders are devised to capture the latent transition patterns of spatio-temporal preferences based on the self-attention mechanism. Next, two sub-tasks, i.e., an activity task and an auxiliary task, are introduced to enhance the performance of next POI recommendation. The activity task is utilized to predict the next POI category, and the auxiliary task is applied to predict the next POI type. Finally, we recommend next POI with the multi-granularity representation and the results of two sub-tasks.

To summarize, we make the following contributions:

1. To the best of our knowledge, we are the first to introduce the multi-granularity representation in next POI recommendation. We combine individual-level and collective-level POI sequences to extract full transition patterns of user check-ins.
2. We propose a novel self-attention network named MGSAN that incorporates the spatio-temporal features of individual-level and collective-level POI sequences. To achieve better recommendation performance, we also adopt a multi-task learning approach.
3. The empirical performance of our proposed model MGSAN on three real-world datasets shows that MGSAN outperforms state-of-the-art models consistently.

2   Related Work

In this section, we review closely related work from two perspectives, which are next POI recommendation and the self-attention mechanism.

2.1 Next POI Recommendation

Next POI recommendation, as an extension of POI recommendation, is recently proposed and becomes a research interest. Compared to POI recommendation, Next POI recommendation is mostly affected by the sequential influence of user check-ins. He et al. propose a category-aware method to model transitions of check-in sequences via listwise bayesian personalized ranking [4]. Xie et al. propose a graph-based POI embedding model named GE, which incorporates sequential influence, spatial effect, temporal effect, and semantic effect by embedding their corresponding bipartite graphs into a lower latent space [16]. Liu et al. propose a RNN-based method that exploits spatio-temporal contexts for next POI recommendation [11]. Compared to previous studies that mainly focus on individual-level POIs, Zhang et al. propose a multi-task learning model that utilizes collective-level POIs for next POI recommendation. However, Zhang et al. only consider the sequential patterns of collective-level POIs, ignoring the sequential patterns of individual-level POIs, which are also beneficial for modeling the user's preferences [17]. Therefore, in this paper, we will propose a multi-granularity network that utilizes both collective-level POIs and individual-level POIs to extract full sequential patterns of user check-ins.

2.2 Self-attention Mechanism for Recommendation

The self-attention mechanism takes advantage of the attention mechanism to calculate the correlations between each element and all other elements. As the self-attention network achieves state-of-the-art performance in machine translation [15], many researchers begin to investigate improving model performance by utilizing the self-attention mechanism. SASRec adopts the self-attention mechanism to model latent transition patterns of users' historical items for sequential recommendation [7]. CSAN applies the self-attention mechanism to capture heterogeneous and contextualized user behaviors for sequential recommendation [6]. Self-attention based approaches [6,7] generally achieve better performance than RNN based methods, indicating the superiority of the self-attention mechanism. As next POI recommendation is often modeled as a sequential recommendation, we will apply the self-attention mechanism in our proposed model to take advantage of user check-ins' sequential patterns

3   Preliminaries

This section first formulates the investigated research problem and then presents the self-attention mechanism that will be later applied in our proposed model.

3.1   Problem Formulation

We start by defining several concepts.

Definition 1 (POI). *A POI is a spatial item related to a geographical location. We use p to represent a POI identifier.*

Definition 2 (collective-level POI). *A collective-level POI is a region of places which usually includes several individual-level POIs.*

Definition 3 (Check-in). *A user's check-in activity is represented by a six-element tuple (u, p, t, c, y, g) which indicates a user u visits POI p at timestamp t. Note that there are three attributes of p, which are its category(c), geo-location (g) and type (y). POI type can tell if p is a collective-level POI (when $y = 1$) or an individual-level POI (when $y = 0$).*

In next POI recommendation, we aim to recommend the next POI based on a user's historical check-ins. We introduce the multi-granularity representation here to improve the performance. Individual-level and collective-level POI sequences are denoted as follows:

In the individual-level POI sequences, we split users' historical check-in records, i.e., (user, POI, time, category, POI type, geo-location) or (u, p, t, c, y, g), into check-in sequences by day, where each record is sorted by timestamps as in Zhao et al. [19]:

- The i-th temporal-aware activity sequences for user u can be denoted as a set of temporal tuples, i.e., $A^{u,i} = \{A^u_{t_1}, A^u_{t_2}, \dots\}$, where $A^u_{t_i} = (c^u_{t_i}, y^u_{t_i}, t^u_i)$, $y^u_{t_i} \in \{0, 1\}$. The category $c^u_{t_i}$ of a specific POI $p^u_{t_i}$ is essential to represent temporal preference, e.g., food-oriented POIs and night-life POIs have different timelines. POI type is also vital to represent the user's temporal preference, as $p^u_{t_i}$ can be a massive place like a shopping mall containing many individual-level POIs with diverse business hours.
- The i-th spatial-aware location sequences for user u can be denoted as a set of spatial tuples, i.e., $S^{u,i} = \{S^u_{t_1}, S^u_{t_2}, \dots\}$, where $S^u_{t_i} = (p^u_{t_i}, g^u_{t_i})$; $g^u_{t_i}$ consists of the corresponding longitude and latitude.

Similarly, in the collective-level POI sequences, we denote the i-th temporal-aware tuple $\tilde{A}^u_{t_i} = (\tilde{c}^u_{t_i}, \tilde{y}^u_{t_i}, t^u_i)$ and the i-th spatial-aware tuple $\tilde{S}^u_{t_i} = (\tilde{p}^u_{t_i}, \tilde{g}^u_{t_i})$.

Given $A^u_{t_n}, S^u_{t_n}, \tilde{A}^u_{t_n}, \tilde{S}^u_{t_n}$, we first conduct two sub-tasks: category ($c^u_{t_{n+1}}$) prediction task and POI type ($y^u_{t_{n+1}}$) prediction task at time t_{n+1} for user u. Our aim is to recommend next POI $p^u_{t_{n+1}}$ with the help of previous task results.

3.2 The Self-attention Network

The self-attention network takes advantage of the attention mechanism to calculate the correlations between each element and all other elements. In order to calculate self-attention scores, we have the following processes: We first calculate the dot product of matrix Q and K and apply $\sqrt{d_k}$ to scale the result. Then, a softmax operation is utilized to normalize the result into a probability distribution. Finally, we multiply the result by matrix V to get the representation of the weighted sum. To summarize, we have the following formulation:

$$Z = Attention(Q, K, V) = softmax(\frac{QK^T}{\sqrt{d_k}})V, \tag{1}$$

where d_k is the dimension of Key K. In self-attention mechanism, Query (Q), Key (K), and Value (V) come from the same input matrix X; we can calculate Q, K, V as:

$$\begin{cases} Q = XW^Q, \\ K = XW^K, \\ V = XW^V, \end{cases} \tag{2}$$

where W^Q, W^K, and W^V are the parameters which can be learned during model training. As previous formulations in the self-attention mechanism are linear, we apply a point-wise feed-forward network (FFN) [14] to model nonlinearity:

$$F = \text{FFN}(Z) = ReLU(ZW_1 + b_1), \tag{3}$$

where $ReLU$ is an activation function to give nonlinearity to the result of self-attention network.

By adopting the self-attention network, we get a weighted representation that can be effective to take context into account.

4   Proposed Model

The proposed model MGSAN is outlined in Fig. 2, comprised of two granularities, i.e., individual-level POIs and collective-level POIs, two temporal-aware encoders, two spatial-aware encoders, and prediction tasks. We apply two types of encoders for the coarse-grained sequential check-ins (collective-level POIs) and fine-grained check-ins (individual-level POIs). These encoders, equipped with embedding, aggregation, and self-attention layers, seek to capture sequential spatio-temporal preferences. We recommend user's next POI after interactively performing two sub-tasks.

4.1   Multi-granularity Representation Learning

In order to extract full transition patterns of sequential data, we adopt individual-level POI sequences and collective-level POI sequences. We divide each sequence into a set of five embeddings, which are category, time, POI type, POI id, and distance. To better learn the multi-granularity representation of two sequences, we introduce two temporal-aware encoders to encode temporal features and two spatial-aware encoders to encode spatial features. We have an aggregation layer in each encoder to aggregate the corresponding features and a self-attention layer to learn sequential transition patterns.

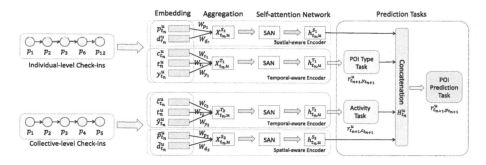

Fig. 2. The architecture of the proposed model MGSAN, comprised of individual-level and collective-level POI sequences. Each sequence is divided into five embeddings including spatial-aware features (distance, POI id) and temporal-aware features(category, time, POI type) which are aggregated and then fed into self-attention networks. Afterwards, we recommended a user's next POI with the help of two subtasks, i.e., the activity task to predict the next category and the auxiliary task to predict the next POI type.

Representation Learning of Individual-Level Check-Ins. We capture the sequential patterns of the fine-grained sequences, i.e., individual-level POI sequences. Given an activity tuple $A_{t_n}^u = (c_{t_n}^u, y_{t_n}^u, t_n^u)$, we aim to encode temporal-aware features. In order to model temporal preference, it's essential to incorporate sequential contexts. The aggregation $x_{t_n,u}^{T_1}$ of three elements from an activity tuple is represented by,

$$x_{t_n,u}^{T_1} = W_{c_1} c_{t_n}^u + W_{t_1} t_n^u + W_{y_1} y_{t_n}^u + b_1, \tag{4}$$

where W is the weight matrix; b_1 is the bias terms; $c_{t_n}^u \in \mathbb{R}^{D_c}$ is the embeddings of category; $t_n \in \mathbb{R}^{D_t}$ is the embeddings of timestamp mapped into 24 h; and $y_{t_n}^u \in \mathbb{R}^{D_y}$ is the embeddings of POI type. The aggregation $x_{t_n,u}^{T_1}$ is then fed into a self-attention layer to capture the fine-grained sequential correlations of user's activity:

$$h_{t_n,u}^{T_1} = SAN(x_{t_n,u}^{T_1}), \tag{5}$$

where $SAN(\cdot)$ represents a self-attention layer as formulated in Sect. 3.2 and $h_{t_n,u}^{T_1}$ is the encoding of temporal representation.

Given a spatial-aware location tuple $S_{t_n}^u = (p_{t_n}^u, g_{t_n}^u)$, we aim to encode spatial-aware features. As a user's check-in is generally affected by the distance between the current location and the next visiting one [2], we can adopt the location distance between two POIs as a new feature. Therefore, the aggregation $x_{t_n,u}^{S_1}$ of distances and geo-locations is represented by,

$$x_{t_n,u}^{S_1} = W_{p_1} p_{t_n}^u + W_{d_1} d_{t_n}^u + b_2, \tag{6}$$

where W is the weight matrix; $p_{t_n}^u \in \mathbb{R}^{D_p}$ is the embeddings of POI $p_{t_n}^u$; and $d_{t_n}^u \in \mathbb{R}^{D_d}$ is the embedding of the distance $d_{t_n}^u$ between $p_{t_{n-1}}^u$ and $p_{t_n}^u$ based on geo-locations, i.e., $g_{t_{n-1}}^u$ and $g_{t_n}^u$. Note that the distance is rounded into integer (e.g., $1.18 \to 1$) to reduce the complexity of parameters [12]. Followed by the aggregation layer, a self-attention layer is adopted:

$$h_{t_n,u}^{S_1} = SAN(x_{t_n,u}^{S_1}), \tag{7}$$

where $SAN(\cdot)$ represents a self-attention layer as formulated in Sect. 3.2 and $h_{t_n,u}^{S_1}$ is the encoding of spatial representation. The spatio-temporal features captured by the fine-grained sequences are represented by,

$$h_{t_n,u}^{T_1,S_1} = [h_{t_n,u}^{T_1}; h_{t_n,u}^{S_1}], \tag{8}$$

where $h_{t_n,u}^{T_1,S_1}$ is the concatenation of temporal-aware representation and spatial-aware representation.

Representation Learning of Collective-Level Check-Ins. Similarly, we model the spatio-temporal preference of the coarse-grained sequences, i.e., collective-level POI sequences. The aggregation $x_{t_n,u}^{T_2}$ of three temporal features is represented by,

$$x_{t_n,u}^{T_2} = W_{c_2}\tilde{c}_{t_n}^u + W_{t_2}t_n^u + W_{y_2}\tilde{y}_{t_n}^u + b_3. \tag{9}$$

The final representation of the temporal-aware encoder is formulated as,

$$h_{t_n,u}^{T_2} = SAN(x_{t_n,u}^{T_2}). \tag{10}$$

The aggregation of two spatial features is calculated as,

$$x_{t_n,u}^{S_2} = W_{p_2}\tilde{p}_{t_n}^u + W_{d_2}\tilde{d}_{t_n}^u + b_4. \tag{11}$$

Followed by the aggregation layer, a self-attention layer is adopted:

$$h_{t_n,u}^{S_2} = SAN(x_{t_n,u}^{S_2}). \tag{12}$$

The spatio-temporal features captured by the coarse-grained sequences are represented by,

$$h_{t_n,u}^{T_2,S_2} = [h_{t_n,u}^{T_2}; h_{t_n,u}^{S_2}], \tag{13}$$

where $h_{t_n,u}^{T_2,S_2}$ is the concatenation of two encoders.

To summarize, we formulate the multi-granularity representation by concatenating two representations from two different granularities:

$$h_{t_n,u}^{T,S} = [h_{t_n,u}^{T_1,S_1}; h_{t_n,u}^{T_2,S_2}]. \tag{14}$$

4.2   The Prediction Tasks

By utilizing the multi-task learning strategy, we interactively perform three prediction tasks (i.e., the next activity, POI type, and POI).

POI Type Prediction. In order to predict the next POI type, we use the latent temporal features from two granularities to formulate as:

$$\begin{cases} h_{t_n,u}^T = [h_{t_n,u}^{T_1}; h_{t_n,u}^{T_2}], \\ r_{t_{n+1},y_{t_{n+1}}}^u = \sigma(W_h^y h_{t_n,u}^T), \end{cases} \tag{15}$$

where $h_{t_n,u}^T$ is the concatenation of the temporal features of two granularity sequences and σ is the sigmoid function.

Activity Prediction with Auxiliary Task. The activity task is utilized to predict the next POI category, and POI type is required for activity prediction. Inspired by Girshick et al. [3], we perform the major task (next activity prediction task) with the help of an auxiliary task (next POI type prediction task). In doing so, jointly learning both tasks facilitates to enhance the model generalization. The probability of next possible activity $c_{t_{n+1}}$ at time t_{n+1} for user u is formulation as:

$$\begin{cases} x_{t_{n+1},u}^{T_1} = W_{c_1} c_{t_{n+1}}^u + W_{t_1} t_{n+1}^u + W_{y_1} y_{t_{n+1}}^u + b_1, \\ x_{t_{n+1},u}^{T_2} = W_{c_2} \tilde{c}_{t_{n+1}}^u + W_{t_2} \tilde{t}_{n+1}^u + W_{y_2} \tilde{y}_{t_{n+1}}^u + b_3, \\ A_{rep} = W^{T_1} x_{t_{n+1},u}^{T_1} + W^{T_2} x_{t_{n+1},u}^{T_2} + b_5, \\ r_{t_{n+1},c_{t_{n+1}}}^u = (W_h^c h_{t_n,u}^T)^T A_{rep}, \end{cases} \tag{16}$$

where $x_{t_{n+1},u}^{T_1}$ and $x_{t_{n+1},u}^{T_2}$ are the aggregations of temporal-aware contexts from two granularities. A_{rep} denotes the temporal-aware activity representation, and $W^c h_{t_n,u}^T$ represents user u's temporal preference.

4.3 POI Recommendation

Inspired by Chen et al. [1], we concatenate the latent representations of two encoders and the results of two prediction tasks:

$$H_{t_n}^u = [h_{t_n,u}^{T,S}; r_{t_{n+1},c_{t_{n+1}}}^u; r_{t_{n+1},y_{t_{n+1}}}^u], \tag{17}$$

where $H_{t_n}^u$ denotes the aggregation of the final latent representations. Therefore, the probability of next possible POI $p_{t_{n+1}}^u$ at time t_{n+1} for user u is calculated as:

$$\begin{cases} x_{t_{n+1},u}^{S_1} = W_{p_1} p_{t_{n+1}}^u + W_{d_1} d_{t_{n+1}}^u + b_2, \\ x_{t_{n+1},u}^{S_2} = W_{p_2} \tilde{p}_{t_{n+1}}^u + W_{d_2} \tilde{d}_{t_{n+1}}^u + b_5, \\ P_{rep} = W^{S_1} x_{t_{n+1},u}^{S_1} + W^{S_2} x_{t_{n+1},u}^{S_2} + b_6, \\ r_{t_{n+1},p_{t_{n+1}}}^u = (W_H H_{t_n}^u)^T P_{rep}, \end{cases} \tag{18}$$

where $x_{t_{n+1},u}^{S_1}$ and $x_{t_{n+1},u}^{S_2}$ are the aggregations of spatial-aware contexts from two granularities. P_{rep} denotes the spatial-aware location representation and $W_H H_{t_n}^u$ represents user's spatio-temporal preference.

4.4   Model Training

Followed by Huang et al. [5], training instances Ω are generated for each sequence, and negative instances are also uniformly sampled. We adopt cross entropy loss function for POI type prediction task.

$$L_y = -\sum_{i=1}^{|\Omega|}[y_{t_i}ln(r^u_{t_i,y_{t_i}}) + (1 - y_{t_i})ln(1 - r^u_{t_i,y_{t_i}})], \tag{19}$$

After the binary classification, i.e., POI type prediction task, we adopt Bayesian Personalized Ranking loss function [13] for activity prediction task and next POI prediction task.

$$\begin{cases} L_c = \sum_{(c,c')\in\Omega} ln(1 + e^{-(r^u_{t,c}-r^u_{t,c'})}), \\ L_p = \sum_{(p,p')\in\Omega} ln(1 + e^{-(r^u_{t,p}-r^u_{t,p'})}), \end{cases} \tag{20}$$

where c' and p' denote the negative samples of activity and POI location respectively. Finally, we can conclude the sum loss function:

$$L = \lambda_y L_y + \lambda_c L_c + \lambda_p L_p + \frac{\lambda}{2}||\Theta||^2, \tag{21}$$

where λ_y, λ_c, and λ_p ($\lambda_y + \lambda_c + \lambda_p = 1$) are weights to adjust the importance of different losses; λ is the regularization coefficient; and $\Theta = (W, c, y, p, t, d, b)$ is the set of learnable parameters. For optimization, we adopt AdaGrad to optimize the network parameters.

5   Experiments

In this section, we conduct experiments to evaluate the performance of our proposed method MGSAN on three real-world datasets. Our experiments are designed to answer the following research questions:

- **RQ1**: How does the performance of our proposed model MGSAN compare with other baselines?
- **RQ2**: How do the parameters affect the performance of MGSAN?
- **RQ3**: Is it practical for the model to adopt the multi-granularity representation and self-attention mechanism?

5.1   Experimental Setup

Datasets. Due to the lack of public datasets with collective-level POIs, we collect three datasets, i.e., Charlotte (CLT), Calgary (CAL), and Phoenix (PHO), from Foursquare[1]. We then manually group individual-level POIs into different collective-level POIs from Google Maps, according to Zhang et al. [17]. For

Table 1. Statistics of three POI datasets

Dataset	# User	# POI	# Check-in	# Category
CLT	1580	1791	20940	239
CAL	301	985	13954	184
PHO	1623	2441	22620	251

instance, individual-level POIs, e.g., p_2, p_3, p_4, p_5, are grouped into a collective-level POI p_1 as shown in Fig. 1. The statistics of the three datasets are shown in Table 1.

We remove users with fewer than 10 check-ins and POIs with fewer than 10 users. We sort historical check-in records by timestamps in ascending order. Then we take the first 80% sequences as the training set, while the validation set and test set occupy 10%, respectively.

Metrics. We adopt two standard metrics, i.e., MAP@K (Mean Average Precision) and Recall@K, to validate the performance of all models. We only choose $K = 10$ here due to space limitation, but similar trends can be observed when $K = 5$ or $K = 15$.

Parameter Settings. In our experiments, all learnable matrices are initialized in a uniform distribution. Embedding sizes are set to 120/140/120, and the iteration numbers are set to 25/25/30 for CLT, CAL, and PHO, respectively. We choose learning rate $\eta = 0.0001$ and regularization coefficient $\lambda = 0.0025$ for all three datasets. For the weights that balance the importance of three different loss functions, we select the best parameter pair $\lambda_c = 0.4, \lambda_y = \lambda_p = 0.3$

Baselines. We compare our proposed model MGSAN with seven representative models for next POI recommendation.

- **Pop**: It recommends next POI location by most popularity.
- **CMF** [12]: It exploits users' transition probability matrices over activity categories to improve POI recommendation accuracy. It can recommend next POI based on next activity (category).
- **ME** [2]: It is a metric embedding model for next POI recommendation.
- **LBPR** [4]: It proposes category-aware method via listwise bayesian personalized ranking for next POI recommendation. Similarly, It can also make recommendations for next activity (category) and next POI.
- **STRNN** [11]: It improves the RNN method by including spatial and temporal features for next POI recommendation.

---

[1] https://sites.google.com/site/yangdingqi/home/foursquare-dataset.

- **ATST** [5]: It incorporates spatio-temporal features via an attention-based RNN for next POI recommendation. By adopting the attention mechanism, ATST can utilize spatio-temporal context to selectively focus on related historical check-in records in a check-in sequence.
- **MCA** [10]: It proposes a multi-task context-aware recurrent neural network to recommend next POI with the help of an auxiliary task, i.e., activity task.
- **iMTL** [17]: It is the recently proposed model that utilizes collective-level POIs via a multi-task learning approach for next POI recommendation.

5.2   Method Comparisons (RQ1)

The performance of our proposed model MGSAN and other baselines are evaluated by Recall and MAP on three datasets. For all the baselines, parameters are tuned for optimal results according to their original papers. We summarize the empirical performance (Rec@K and MAP@K) of all methods across the three datasets for next POI recommendation in Table 2. We show K = 10 due to space limitation, but similar results can be observed when K is smaller or bigger.

Table 2. Performance comparison of POI recommendation (**RQ1**)

Models	CLT		CAL		PHO	
	Recall@10	Map@10	Recall@10	Map@10	Recall@10	Map@10
POP	0.0305	0.0104	0.0317	0.0119	0.0323	0.0125
CMF	0.0323	0.0124	0.0341	0.0155	0.0352	0.0180
ME	0.0401	0.0137	0.0416	0.0205	0.0434	0.0207
LBPR	0.0446	0.0174	0.0483	0.0226	0.0525	0.0218
STRNN	0.0421	0.0162	0.0479	0.0230	0.0506	0.0243
ATST	0.0465	0.0201	0.0522	0.0328	0.0591	0.0260
MCA	0.0458	0.0210	0.0545	0.0364	0.0608	0.0275
iMTL	<u>0.0534</u>	<u>0.0238</u>	<u>0.0691</u>	<u>0.0443</u>	<u>0.0769</u>	<u>0.0352</u>
MGSAN	**0.0582**	**0.0257**	**0.0749**	**0.0473**	**0.0832**	**0.0381**
impr.	8.99%	7.98%	8.39%	6.77%	8.19%	8.24%

Note that in Table 3, boldface represents the best performance; the runner-up performance is underlined; and impr. denotes the improvement ratio of the best method compared to the runner-up method. From Table 3, we can summarize the following analysis:

Firstly, CMF, ME, and LBPR outperform Pop on three datasets, indicating the effectiveness of personalized recommendation methods. We notice that sequential models (STRNN, ATST, MCA, iMTL) usually outperform the non-sequential models (CMF, ME, LBPR), suggesting the importance of modeling sequential patterns for next POI recommendation.

Secondly, compared with STRNN, ATST has a considerable improvement in terms of two metrics. The success of ATST suggests that, by incorporating the attention mechanism, sequential recommendation models can achieve better performance.

Thirdly, MCA usually outperforms STRNN by a large margin, while MCA slightly outperforms ATST. MCA utilizes a multi-task learning strategy, demonstrating the effectiveness of multi-task learning.

Fourthly, iMTL outperforms the previously mentioned methods. iMTL adopts collective-level POIs to model the latent sequential patterns and utilize a novel multi-task learning strategy for better performance.

Finally, the results of our proposed model MGSAN outperform all other state-of-the-art baselines. The improvements can be summarized in a two-fold way: (1) We consider both granularities in one model to retain more significant features. (2) We adopt the self-attention mechanism for better modeling sequential dependencies.

5.3   Sensitivity of Parameters (RQ2)

We investigate the sensitivity of parameters to study how they affect the performance. Parameters are tuned and illustrated on how they affect the performance, as depicted in Fig. 3.

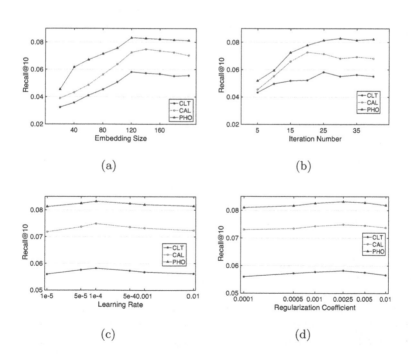

Fig. 3. Sensitivity analysis of the following parameters: embedding size, iteration number, learning rate η, and regularization coefficient (λ). (**RQ2**)

From Fig. 3(a), we can observe that the performance increases rapidly for the early period as embedding size grows. When the performance reaches a summit, it gradually becomes stable. MGSAN can converge within 30 iterations for all three datasets as shown in Fig. 3(b). We further study the influence of learning rate (η) and regularization coefficient (λ) on the performance of MGSAN in Fig. 3(c, d). They show similar trends that they gradually increase to a summit and then decrease slowly.

5.4   Ablation Study (RQ3)

We analyze the effectiveness of different components of MGSAN and illustrate the corresponding performance in Table 2. We have three variants, which are as follows:

Table 3. Performance comparison of model variants (**RQ3**)

Variants	CLT		CAL		PHO	
	Recall@10	Map@10	Recall@10	Map@10	Recall@10	Map@10
MGSAN-i	0.0547	0.0235	0.0712	0.0442	0.0782	0.0341
MGSAN-c	0.0486	0.0219	0.0642	0.0438	0.0753	0.0354
MGSAN-a	0.0562	0.0248	0.0735	0.0461	0.0814	0.0367
MGSAN	**0.0582**	**0.0257**	**0.0749**	**0.0473**	**0.0832**	**0.0381**

(1) MGSAN-i: It removes the sequences of individual-level POIs from MGSAN, therefore no aggregation of both granularities.
(2) MGSAN-c: It considers individual-level POIs only by ignoring collective-level POIs.
(3) MGSAN-a: It replaces the self-attention mechanism adopted in MGSAN with the basic RNN approach.

We notice that, by removing either individual-level POIs or collective-level POIs, variant performs worse than MGSAN, explaining the effectiveness of the multi-granularity representation.

6   Conclusion

In this paper, we proposed a multi-granularity attention network named MGSAN for next POI recommendation. MGSAN incorporated the individual-level and collective-level POI sequences to represent more semantic features and express and express full transition patterns. Specifically, Embeddings of check-in tuples were aggregated and fed into self-attention networks in order to form the multi-granularity representation. Finally, we recommended user's next POI via an effective multi-task learning approach. Extensive experiments on three real-world datasets showed that MGSAN outperforms state-of-the-art methods consistently.

Acknowledgements. This research was partially supported by NSFC (No. 61876117, 61876217, 61872258, 61728205), Open Program of Key Lab of IIP of CAS (No. IIP2019-1) and the Priority Academic Program Development of Jiangsu Higher Education Institutions (PAPD).

References

1. Chen, Z., et al.: Co-attentive multi-task learning for explainable recommendation. In: IJCAI, pp. 2137–2143 (2019)
2. Feng, S., Li, X., Zeng, Y., Cong, G., Chee, Y.M., Yuan, Q.: Personalized ranking metric embedding for next new POI recommendation. In: IJCAI, pp. 2069–2075 (2015)
3. Girshick, R.B.: Fast R-CNN. In: ICCV, pp. 1440–1448 (2015)
4. He, J., Li, X., Liao, L.: Category-aware next point-of-interest recommendation via listwise Bayesian personalized ranking. In: IJCAI, pp. 1837–1843 (2017)
5. Huang, L., Ma, Y., Wang, S., Liu, Y.: An attention-based spatiotemporal LSTM network for next poi recommendation. IEEE Trans. Serv. Comput. (2019). https://doi.org/10.1109/TSC.2019.2918310
6. Huang, X., Qian, S., Fang, Q., Sang, J., Xu, C.: CSAN: contextual self-attention network for user sequential recommendation. In: ACM Multimedia, pp. 447–455 (2018)
7. Kang, W., McAuley, J.J.: Self-attentive sequential recommendation. In: ICDM, pp. 197–206 (2018)
8. Li, X., Cong, G., Li, X., Pham, T.N., Krishnaswamy, S.: Rank-GeoFM: a ranking based geographical factorization method for point of interest recommendation. In: SIGIR, pp. 433–442 (2015)
9. Lian, D., Zhao, C., Xie, X., Sun, G., Chen, E., Rui, Y.: GeoMF: joint geographical modeling and matrix factorization for point-of-interest recommendation. In: KDD, pp. 831–840 (2014)
10. Liao, D., Liu, W., Zhong, Y., Li, J., Wang, G.: Predicting activity and location with multi-task context aware recurrent neural network. In: IJCAI, pp. 3435–3441 (2018)
11. Liu, Q., Wu, S., Wang, L., Tan, T.: Predicting the next location: a recurrent model with spatial and temporal contexts. In: AAAI, pp. 194–200 (2016)
12. Liu, X., Liu, Y., Aberer, K., Miao, C.: Personalized point-of-interest recommendation by mining users' preference transition. In: CIKM, pp. 733–738 (2013)
13. Rendle, S., Freudenthaler, C., Gantner, Z., Schmidt-Thieme, L.: BPR: Bayesian personalized ranking from implicit feedback. In: UAI, pp. 452–461 (2009)
14. Tang, Z., Fishwick, P.A.: Feedforward neural nets as models for time series forecasting. INFORMS J. Comput. **5**(4), 374–385 (2017)
15. Vaswani, A., et al.: Attention is all you need. In: NIPS, pp. 5998–6008 (2017)
16. Xie, M., Yin, H., Wang, H., Xu, F., Chen, W., Wang, S.: Learning graph-based POI embedding for location-based recommendation. In: CIKM, pp. 15–24 (2016)
17. Zhang, L., et al.: An interactive multi-task learning framework for next POI recommendation with uncertain check-ins. In: IJCAI, pp. 3551–3557 (2020)

18. Zhao, P., et al.: Where to go next: a spatio-temporal gated network for next POI recommendation. In: AAAI, pp. 5877–5884 (2019)
19. Zhao, S., Zhao, T., King, I., Lyu, M.R.: Geo-teaser: geo-temporal sequential embedding rank for point-of-interest recommendation. In: WWW, pp. 153–162 (2017)

HRFA: Don't Ignore Strangers with Different Views

Senhui Zhang, Wendi Ji$^{(\boxtimes)}$, Jiahao Yuan, and Xiaoling Wang

School of Computer Science and Technology, East China Normal University,
Shanghai, China
{51194501081,jhyuan}@stu.ecnu.edu.cn, xlwang@cs.ecnu.edu.cn

Abstract. Review-based recommender suffers from the sparsity of reviews: only a few users leave substantial comments in the real world. As a result, some recent methods resort to supplementary reviews written by similar users, which only leverage homogeneous preferences. However, users holding different views could also supply valuable information with heterogeneous preferences. In this paper, we propose a recommendation model for rating prediction, named **H**eterogeneous **R**eview-based Recommendation via **F**our-way **A**ttention (**HRFA**). To take advantage of the heterogeneous preferences, the supplementary reviews in HRFA are redefined as reviews from all users with common purchase history, no matter whether they give similar ratings. Specially, we integrate the heterogeneous preferences into the one semantic space via introducing a similarity projection based on rating difference. Experiments conducted on five datasets demonstrate that our model achieves higher rating prediction accuracy than other baselines.

Keywords: Review-based recommendation · Rating prediction · Data sparsity · Heterogeneous preferences.

1 Introduction

Data sparsity and content ignorance are two major problems of traditional collaborative filtering (CF) recommenders. Many review-based approaches [2,15,21] integrate ratings with user reviews to tackle these issues. Although exploiting textual information from reviews is the most effective way to improve rating prediction accuracy for users and items with few records, it still suffers from review sparsity. Some recent researches [19,20] exploit supplementary reviews from similar users to deal with the sparsity problem of reviews. These methods assume that only reviews written by similar users, who give similar ratings on the same item, can be used as supplementary reviews. The example in Fig. 1 reflects the limitations of this assumption. For user A, a review from a similar user (e.g., user C) only provides homogeneous preferences information, that is, what two users both like or dislike. However, other users with different ratings (e.g., user B) can also provide opposite attitudes on the same item. Their reviews could help to infer user's interest points implicitly.

© Springer Nature Switzerland AG 2021
W. Zhang et al. (Eds.): WISE 2021, LNCS 13081, pp. 209–217, 2021.
https://doi.org/10.1007/978-3-030-91560-5_15

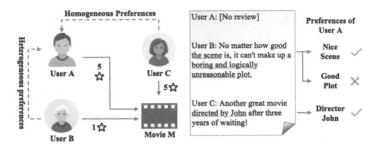

Fig. 1. An example of heterogeneous preferences. Considering user B's review and his opposite attitude to the movie with user A, we could infer that user A probably doesn't care about the movie's plot so much but prefers the movie with nice scenes.

We define these different preferences as heterogeneous preferences. If supplementary reviews are limited to similar users' reviews, heterogeneous preferences will be ignored. To exploit heterogeneous preferences from diverse review sources, we propose a recommendation model for rating prediction named **H**eterogeneous **R**eview-based Recommendation via **F**our-way **A**ttention (**HRFA**). Our model includes heterogeneous reviews into the supplementary reviews and utilizes an attentive network to learn the semantic information of reviews at word-level and review-level. Specifically, heterogeneous preferences from supplementary reviews will be integrated into one semantic space via a similarity projection based on ratting differences. Moreover, four-way attention means collecting supplementary reviews for unpopular items and applying heterogeneous preferences learning on the item-side, which is absent in previous research. In this way, HRFA could supplement features for items lacking reviews. The main contributions of our work can be summarized as follows:

- To relieve the review sparsity, we leverage heterogeneous preferences to extend supplementary reviews and introduce a similarity projection to unify heterogeneous preferences.
- We propose a four-way attention network to learn heterogeneous preferences from both the user and item sides.
- Experiments on five sparse datasets demonstrate that our model achieves higher performance than existing models while facing review sparsity.

2   Related Work

Collaborative Filtering(CF) [5,11,18] technology is a widely used recommendation technique. Its main task is to model relationships between users and items from their past interactions. Furthermore, matrix factorization methods [6,9,14] map users and items to a latent factor space by factorizing the rating or interaction matrix. Although traditional CF-based recommenders have achieved success in many applications, data sparsity and content ignorance are two major problems. To tackle the limitation of collaborative filtering, many review-based recommendations approaches [1,2,12,13,21] using TextCNN [7] are proposed to generate better latent semantic representations for reviews. DeepCoNN [21] introduces two parallel neural convolutional networks to extract text features from the user review document and the item review document, respectively. NARRE [2] and NRPA [13] introduce word-level and review-level attention networks to explore reviews' usefulness. Meanwhile, AHN [4] utilizes an asymmetric attentive module to filter out a subset of contents associated with the current target item from user reviews. Moreover, CARP [10] uses two sentiment capsules to learn positive and negative sentiments in user reviews on multi-aspect of target items. However, the above review-based recommenders still suffer from the sparsity of review data. In the real world, not all users have the habit of writing their reviews. To relieve this problem, PARL [20] and MRMRP [19] pick supplementary reviews by randomly selecting the review from another user who gives a similar rating score as the specified user.

3   Problem Setup

Let $U = \{u_1, u_2, ..., u_j, ...u_P\}$ and $I = \{i_1.i_2, ..., i_j, ...i_Q\}$ denote the user set and item set, $R = \{r_{u,i} | u \in U, i \in I\}$ and $\Delta = \{\delta_{u,i} | u \in U, i \in I\}$ denote the ratings and reviews that user u leaves for item i. The interaction of user $u \in U$ and item $i \in I$ can be formulated as a tuple $(u, i, r_{u,i}, \delta_{u,i})$, which denotes a review $\delta_{u,i}$ written by user u for item i that explains why rating $r_{u,i}$ is given. The input of a review-based recommender system is an unknown user-item pair (u, i), it should output the rating prediction $\hat{r}_{u,i}$ that user i probably gives to item i.

4   Proposed Method

This section will introduce our HRFA model in detail. First, we explain how our model queries supplementary reviews from the review set. Subsequently, we elaborate on the process of extracting features from reviews and supplementary reviews. Finally, we explain how these features are combined and involved in rating prediction. The architecture of HRFA is shown in Fig. 2.

4.1   Supplementary Reviews

We redefine supplementary reviews as the reviews written by other users who have common purchase history, no matter whether the ratings are similar or not. In the meantime, we also collect supplementary reviews for items from items with a common buyer due to the review sparsity problem on the item side. Ultimately, for each record (u, i), the model will receive four sets of reviews, including the user reviews δ_u, the user supplementary reviews δ_u^s, the item review δ_i, and the item supplementary reviews δ_i^s.

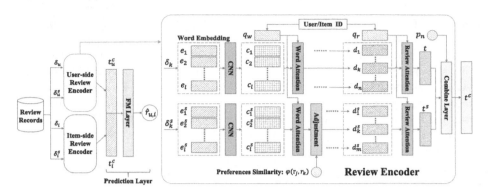

Fig. 2. The architecture of HRFA. User-side and item-side review encoders share the same structure and process user and item reviews in parallel.

4.2   Review Feature Extraction

The review encoder in HRFA is designed to extract segment features of reviews. It consists of a word embedding layer, two attention layers, and a combine layer.

Word Embedding: A pre-trained word embedding is used to map each word into a low-dimensional vector. Given a review $\delta_{u,i}$, which is written by user u for item i. $W = [w_1, w_2, ..., w_l]$ is the sequence of words in $\delta_{u,i}$. We map each word w_i in review $\delta_{u,i}$ to its embedding vector $e_i \in \mathbb{R}^{1 \times d_w}$.

Convolution Layer: Following the word embedding layer, a convolution layer is used to extract semantic information. With the kernel function, each embedding vector and its neighbors are integrated into a local context matrix $C \in \mathbb{R}^{l \times K}$:

$$C = \sigma(W_j * E + b_j), 1 \leqslant j \leqslant K, \tag{1}$$

where σ is the ReLU [16] nonlinear activation function and $*$ is the convolution operator, W_j and b_j are the weight and basis of the j-th kernel filter.

Word-Level Attention: Inspired by NARRE [2] and NRPA [13], we use a word-level attention layer to concentrate on keywords. At first, the semantic feature of words and user/item embedding are used to calculate the attention score ai:

$$a_i = softmax(h^T \sigma(W_q q + W_c c_i + b_1) + b_2) \tag{2}$$

The q is the user/item embedding as a query vector. The final representation of each review is the weighted sum of context vectors: $d = \sum_{i=1}^{l} a_i c_i$.

Adjustment for Heterogeneous Preferences: Heterogeneous preferences will become an obstacle to learning homogeneous preferences if used directly. Therefore, we designed the preference similarity based on the rating difference. Given two ratings r_{u,i^s} and r_{u^s,i^s} that user u and supplementary user u^s leave to the common item i^s, the preference similarity is calculated as:

$$\varphi(r_{u,i^s}, r_{u^s,i^s}) = cos\left(\frac{r_{u,i^s} - r_{u^s,i^s}}{r_{max} - r_{min}}\pi\right) \tag{3}$$

where r_{max} and r_{min} are the maximum and minimum ratings allowed by the system. The semantic feature of words will be multiplied by the rating similarity:

$$\boldsymbol{d}^s = \varphi(r_j, r_k) \odot \sum_{j=1}^{l} a_j \boldsymbol{c}_j^s \tag{4}$$

where $\odot$ denotes the element-wise product operation, and $\boldsymbol{c}_j^s$ is the semantic feature from supplementary reviews.

Review-Level Attention: Similar to word-level attention, review-level attention networks are used to select more important reviews. Given a set of reviews $D = [\boldsymbol{d_1}, \boldsymbol{d_2}, ..., \boldsymbol{d_n}]$, the attention score for each review is calculated as: $t = \sum_{i=1}^{n} o_i \boldsymbol{d_i}$, where o_i is the attention score for each review.

Feature Combination: The task of the combination layer is to merge the text features learned from user/item reviews and supplementary reviews. Following [19], a dimension reduction is performed on review vector t before the combination:

$$t_{low} = \sigma(\boldsymbol{W_l} t + \boldsymbol{b_l}) \tag{5}$$

In general, the less preference information provided by user/item reviews, the more supplementary reviews we need to learn user preferences or item characteristics. So we design a gate value to control how many heterogeneous preferences from supplementary reviews will be included in the final representation:

$$t^c = (1 - g) \odot t_{low} + g \odot t_{low}^s \tag{6}$$

$$g = sigmoid(\boldsymbol{p_n}(\boldsymbol{W_g} t_{low} + b_g)), 0 \le n \le 5 \tag{7}$$

where t^s is the supplementary review vectors. The $\boldsymbol{p} \in \mathbb{R}^{1 \times N}$ is a parameter associate with the count of user/item reviews, and $\boldsymbol{p_n}$ is the n-th element of $\boldsymbol{p}$, where n is the count of user/item reviews.

4.3 Rating Prediction and Optimization

In HRFA, FM layer works as the final rating prediction, which is widely used in many models [4, 13, 20, 21]. We concatenate user features t_u^c and item features t_i^c as $\boldsymbol{x}$, then feed them into FM layer:

$$\hat{r}_{u,i} = w_0 + \sum_{j=1}^{|\boldsymbol{x}|} w_j \boldsymbol{x}_j + \sum_{j=1}^{|\boldsymbol{x}|} \sum_{k=j+1}^{|\boldsymbol{x}|} \langle \boldsymbol{v}_j \boldsymbol{v}_k \rangle \boldsymbol{x}_j \boldsymbol{x}_k \tag{8}$$

where w_0 and w_j are parameters for linear regression, $\boldsymbol{v}_j$ and $\boldsymbol{v}_k$ are latent vectors of Factorization Machine [17]. Once the rating is predicted, we use Mean Square Error (MSE) loss as the objective function for parameter optimization with Adam [8] algorithm.

5   Experiment

5.1   Datasets and Experiments Settings

Four 0-Core datasets of Amazon Review[1] and a random sample of Yelp[2] dataset are used in our experiments. These datasets will be randomly divided into training set, validation set, and test set at a ratio of 8:1:1. For textual pre-processing, we follow the work [7] and only perform the basic tokenization. We pre-train 50-dimensional word embedding vectors by Gensim[3] with a minimum word frequency of 5. To evaluate the performance of all models, we calculate Mean Square Error (MSE) as the metric. Furthermore, a t-test on pair-wise samples will be performed to verify the statistical significance of performance improvements between HRFA and the best baseline method. For the hyperparameters settings, we perform a grid search to tune the parameters on all five datasets individually. The detailed configuration for hyperparameters could be found in source code[4].

5.2   Baseline Methods

- **NMF** [14]: NMF is a basic non-negative matrix factorization-based approach to collaborative filtering for recommender systems.
- **DeepCoNN** [21]: DeepCoNN uses the FM layer to process the review information extracted by TextCNN [7].
- **NRPA** [13]: NRPA utilizes review-level and word-level attention to learn more personalized review features.
- **ANR** [3]: ANR performs aspect-based representation learning for both users and items via an attention-based component.
- **DAML** [12]: DAML uses local and mutual attention of the convolutional neural network to learn the features of reviews jointly.
- **CARP** [10]: CARP model reasons a rating behavior by learns the informative logic units from the reviews with capsule network.
- **PARL** [20]: PARL addresses the issue of review data sparsity by using reviews with similar ratings.
- **MRMRP** [19]: MRMRP seeks supplementary reviews by user similarity rather than single rating.

5.3   Overall Performance

The MSE metrics of all methods are shown in Table 1. HRFA achieves statistically significant improvement on all five datasets, which shows the robust effectiveness of using heterogeneous preference information. Besides, we can observe

---

[1] https://nijianmo.github.io/amazon/index.html.
[2] https://www.yelp.com/dataset.
[3] https://radimrehurek.com/gensim.
[4] https://github.com/KindRoach/HRFA.

Table 1. MSE metrics of all models. The last line denotes our model's improvement. †
denotes the statistical significance for $p < 0.01$ compared to the best baseline method.

Dataset	Appliances	Digital Music	Luxury	Pantry	Yelp
NMF [14]	1.7160	0.7337	1.6572	1.4534	2.7152
DeepCoNN [21]	1.5195	0.6508	1.5615	1.3098	2.1622
NRPA [13]	1.4742	0.6399	1.5159	1.2697	2.1474
ANR [3]	1.4769	0.6511	1.5603	1.2666	2.1687
DAML [12]	1.4721	0.6324	1.5242	1.2605	2.1594
CARP [10]	1.4709	0.6398	1.5518	1.2625	2.1404
PARL [20]	1.4785	0.6348	1.4808	1.2538	2.1473
MRMRP [19]	1.4836	0.6525	1.6523	1.2683	2.1308
HRFA	**1.4542**	**0.6215**	**1.4656**	**1.2369**	**2.1042**
Improvement(%)	1.36†	1.72†	1.41†	1.64†	1.25†

that: 1) Approaches using supplementary reviews alleviate the reviews sparsity
problem gain better performance than the other baseline on three datasets. The
supplementary reviews are beneficial to relieve the problem of review sparsity.
However, the lack of attention limits PARL and MRMRP to achieve better performance. 2) The methods that process reviews with attention networks, such
as CARP and DAML, still show higher prediction accuracy than other baselines
on two datasets. It's necessary to distinguish and aggregate various information
distributed in multiple reviews. But they still could not learn enough features
from sparse review data.

6   Conclusion

The heterogeneous preferences from other users help relieve review sparsity in
the review-based recommender system, which is ignored by existing approaches.
Therefore, this paper proposes a model called HRFA to incorporate diverse user
opinions into the model as supplementary reviews. Our model could learn heterogeneous preferences and map them into one semantic space with the help of
a four-way attention network and an adjustment function based on rating differences. The ablation study demonstrates that our model could benefit from heterogeneous preferences carried in the extended supplementary review and avoid
being disturbed by useless information. Additionally, experiments performed on
five sparse datasets illustrate that our model could deal with the problem of
review sparsity better and achieve higher rating prediction accuracy than all
baseline methods.

Acknowledgements. This work was supported by NSFC grants (No. 61972155), the Science and Technology Commission of Shanghai Municipality (20DZ1100300) and the Open Project Fund from Shenzhen Institute of Artificial Intelligence and Robotics for Society, under Grant No. AC01202005020.

References

1. Catherine, R., Cohen, W.W.: Transnets: learning to transform for recommendation. In: RecSys, pp. 288–296 (2017)
2. Chen, C., Zhang, M., Liu, Y., Ma, S.: Neural attentional rating regression with review-level explanations. In: WWW, pp. 1583–1592 (2018)
3. Chin, J.Y., Zhao, K., Joty, S.R., Cong, G.: ANR: aspect-based neural recommender. In: CIKM, pp. 147–156. ACM (2018)
4. Dong, X., Ni, J., Cheng, W., Chen, Z., Zong, B., Song, D., Liu, Y., Chen, H., de Melo, G.: Asymmetrical hierarchical networks with attentive interactions for interpretable review-based recommendation. In: AAAI, pp. 7667–7674 (2020)
5. Guo, L., Shao, J., Tan, K., Yang, Y.: Wheretogo: personalized travel recommendation for individuals and groups. In: IEEE-MDM, pp. 49–58 (2014)
6. Hou, Y., Yang, N., Wu, Y., Yu, P.S.: Explainable recommendation with fusion of aspect information. World Wide Web **22**(1), 221–240 (2018). https://doi.org/10.1007/s11280-018-0558-1
7. Kim, Y.: Convolutional neural networks for sentence classification. In: EMNLP, pp. 1746–1751 (2014)
8. Kingma, D.P., Ba, J.: Adam: a method for stochastic optimization. In: ICLR (2015)
9. Koren, Y., Bell, R.M., Volinsky, C.: Matrix factorization techniques for recommender systems. Computer **42**(8), 30–37 (2009)
10. Li, C., Quan, C., Peng, L., Qi, Y., Deng, Y., Wu, L.: A capsule network for recommendation and explaining what you like and dislike. In: SIGIR, pp. 275–284 (2019)
11. Linden, G., Smith, B., York, J.: Industry report: Amazon.com recommendations: item-to-item collaborative filtering. IEEE Distrib. Syst. **4**(1) (2003)
12. Liu, D., Li, J., Du, B., Chang, J., Gao, R.: DAML: dual attention mutual learning between ratings and reviews for item recommendation. In: SIGKDD, pp. 344–352 (2019)
13. Liu, H., et al.: NRPA: neural recommendation with personalized attention. In: SIGIR, pp. 1233–1236 (2019)
14. Luo, X., Zhou, M., Xia, Y., Zhu, Q.: An efficient non-negative matrix-factorization-based approach to collaborative filtering for recommender systems. IEEE Trans. Ind. Informatics **10**(2), 1273–1284 (2014)
15. McAuley, J.J., Leskovec, J.: Hidden factors and hidden topics: understanding rating dimensions with review text. In: RecSys, pp. 165–172 (2013)
16. Nair, V., Hinton, G.E.: Rectified linear units improve restricted boltzmann machines. In: ICML, pp. 807–814. Omnipress (2010)
17. Rendle, S.: Factorization machines. In: ICDM, pp. 995–1000 (2010)
18. Tang, Y., Guo, K., Zhang, R., Xu, T., Ma, J., Chi, T.: ICFR: an effective incremental collaborative filtering based recommendation architecture for personalized websites. World Wide Web **23**(2), 1319–1340 (2020)
19. Wang, X., Xiao, T., Tang, J., Ouyang, D., Shao, J.: MRMRP: multi-source review-based model for rating prediction. In: DASFAA, pp. 20–35 (2020)

20. Wu, L., Quan, C., Li, C., Ji, D.: PARL: let strangers speak out what you like. In: CIKM, pp. 677–686 (2018)
21. Zheng, L., Noroozi, V., Yu, P.S.: Joint deep modeling of users and items using reviews for recommendation. In: WSDM, pp. 425–434 (2017)

Recommender Systems (2)

MULTIPLE: Multi-level User Preference Learning for List Recommendation

Beibei Li[1,2], Beihong Jin[1,2(✉)], Xinzhou Dong[1,2], and Wei Zhuo[3]

[1] State Key Laboratory of Computer Science, Institute of Software,
Chinese Academy of Science, Beijing, China
Beihong@iscas.ac.cn
[2] University of Chinese Academy of Sciences, Beijing, China
[3] MX Media Co., Ltd, Singapore, Singapore

Abstract. On many streaming platforms, several individual items with certain common correlations are organized into lists, making list recommendation a distinctive and significant task. Multiple data including user-list interactions, user-item interactions and the list-item hierarchical structure are available for list recommendation. However, existing list recommendation models fail to conduct in-depth analyses of user preferences implied in the interactions and also fail to make use of the sequential information among interactions. In this paper, we propose a model named MULTIPLE, which recognizes the structural and compositional complexity of user preferences and explicitly learns user preferences from different perspectives, i.e., list-level, item-level and dual-level. In particular, MULTIPLE designs a gated and attentive sequence learning module to identify the drift of user preferences. By taking the list-item hierarchical structure as a bridge, MULTIPLE designs an attention network to distill essential items from the constituent items of lists and help learn the dual-level user preferences. We conduct extensive experiments on real-world datasets. The performance enhancement on recall@K and NDCG@K verifies the effectiveness of the model.

Keywords: Recommender system · List recommendation · Multi-level user preferences

1 Introduction

In recent years, for attracting users, both the online platforms and mobile Apps bring forth the new in organizing the content, where one of them is to organize individual items with certain correlations into lists [12] or bundles [16] and recommend the lists or bundles to users. For example, Spotify and NetEase Music recommend music lists to listeners [3,7], Amazon and Taobao recommend commodity bundles/lists to customers [1], and MX Player recommends video lists

This work was supported by the National Natural Science Foundation of China (No. 62072450) and the 2019 joint project with MX Media.

Fig. 1. An example: a user interacts with three lists and five items, where the interacted items can be in lists or not. Our task is to predict the next-interact list.

to viewers, where the videos in a list may be either static or dynamic, either personalized or not.

There are two list-related recommendation tasks, one is to generate new lists which conform user preferences, and the other is to recommend previously constructed lists. The former needs to select several items from the candidate item set to maximize the satisfaction of the target user, which is actually a NP-hard problem. Next-basket recommendation [10] and product bundle generation [20] fall into this task. The latter is to predict interaction probabilities between the target user and each candidate list, then recommend the lists with high interaction probabilities to the target user. In this paper, we focus on the latter, which is referred to as list recommendation hereinafter.

Compared to the item recommendation task, the distinctiveness of the list recommendation task is that lists are non-atomic due to the containment relationship between lists and items. The list-item hierarchical structure makes lists and items to be objects of different levels. As a consequence, user preferences on lists are complicated, that is, the user may be interested in the overall intrinsic quality of the list itself or just a few items in it. If we ignore constituent items of lists and treat lists as generalized items, then it is feasible to perform the list recommendation task by existing item recommendation models, including collaborative filtering models [6,17], sequential recommendation models [18,19] and etc. However, such a solution only excavates user preferences on the overall lists while neglecting user preferences on the constituent items of lists.

In the list recommendation scenario, at least three kinds of data, that is, user-list interactions, user-item interactions (which contain the interactions not only between users and items in lists but also between users and items in other contexts, like related items, shown as Fig. 1) as well as list-item hierarchical composition. These data provide an opportunity to learn user preferences from multiple perspectives. However, existing list recommendation methods [4,7] only have simple perception of user preferences or fail to make full use of the sequential property implied in interactions.

Facing all the above challenges, we recognize the user preference learning as a breakthrough and explicitly distinguish the **list-level** and **item-level** user preferences that are learned from user-list and user-item interactions, respectively. Next, we note that the list-level and item-level interactions of the same user in the same time period are entangled with each other, this reminds us that there exists user preferences which integrates both list- and item-level preferences. We call them the **dual-level** user preferences. Further, we borrow the idea of sequential recommendation to model interactions and estimate current user preferences by integrating both general and recent user preferences learned from history interactions. The main contributions of our work are summarized as follows.

First, we propose a list recommendation model MULTIPLE (**MULTI**-level User **P**references for **L**ist **RE**commendation), which learns list-level and item-level user preferences from user-list and user-item interaction sequences respectively, and takes the list-item hierarchical structure as a bridge to discover dual-level user preferences, then dynamically weighs user preferences at each level to produce the final recommendation.

Second, we exploit user-item interactions to distill the essential constituent items of lists for each user, generate personalized list content embeddings, and learn dual-level user preferences. In addition, we construct a gated and attentive sequence learning module that is sensitive to the drift of user preferences, exploring the way of applying the sequential model to list recommendation.

Third, we conduct extensive experiments on real datasets, and the results show that MULTIPLE outperforms other existing item and list recommendation models in recall and NDCG indicators. In addition, we plan to open up the source code and datasets to facilitate other researchers conducting further study on list recommendation.

2 Related Work

Existing recommendation models can be roughly classified into two categories: item recommendation models and list recommendation models.

For item recommendation, the most conventional method is collaborative filtering, which can be exemplified by BprMF [15] and NCF [6]. The former is an implicit feedback collaborative filtering model based on matrix factorization. It uses pairwise loss to learn the embeddings of users and items, then predicts the interaction probabilities by inner dots of user embeddings and item embeddings. The latter proposes a general framework including the generalized matrix factorization and a multi-layer perceptron. Compared with BprMF, NCF explores neural network architectures for collaborative filtering and devises a general framework that learns the nonlinear relationship between users and items. NGCF [17] utilizes the graph neural network to incorporate collaborative signal into the embeddings of model-based CF by leverage high-order connectivity in the use-item integration graph. LightGCN [5] argues the unnecessarily complicated design of GCNs for collaborative filtering, and keeps only two essential components: light graph convolution and layer combination in the model. Item recommendation can also be implemented through sequential models. In [8,9], recurrent neural networks are adopted to model interaction sequences and implement item recommendations. NARM [11] designs an attention mechanism to capture the user's main purpose in the sequence, and STAMP [13] uses a multi-layer perceptron and an attention mechanism to obtain the user's general preferences and recent preferences. Recently, SR-GNN [18] constructs a graph which represents transitions between items from the user-item interaction sequence, and uses graph neural networks to propagate and aggregate information on the graph to learn item embeddings. GC-SAN [19] is a graph contextualized self-attention network based on graph neural network for session-based recommendation.

For list recommendation, despite user-list interaction data, the user-item interaction data and the list-item containment relationship can also be explored. LIRE [14] obtains the user preferences on the list via weighted summation of the user preferences on each item in the list. EFM [2] holds that items in the same list are related and simultaneously excavates user-list interaction data, user-item interaction data, and item-item correlation data. DAM [4] uses the attention mechanism to aggregate the embeddings of constituent items to form bundle embeddings, and then trains the model with user-item interaction data and user-bundle interaction data alternately to obtain the embeddings of users, items and bundles. Besides, AttList [7] is a hierarchical attention mechanism-based model which considers item-list-user hierarchical relationships. It aggregates the embeddings of constituent items to obtain the list embeddings, and further aggregates the embeddings of the interacted lists to obtain user embeddings. BGCN [3] is a graph-based model that re-constructs the two kinds of interactions and an affiliation into the graph, it utilizes the graph neural network's powerful ability to learn the representation of two dependent entities from complex structures. However, the above existing works have not portrayed users from different perspectives, and most of them ignore the orders among interactions.

3   Problem Formulation

For the list recommendation, we denote the list set as $\mathcal{C}$, the item set as $\mathcal{V}$. For each user, we have two kinds of interaction sequences, i.e., the user-list sequence $s = [c_1, c_2, \cdots, c_{|s|}]$ and the user-item sequence $r = [v_1, v_2, \cdots, v_{|r|}]$, where $c_i \in \mathcal{C}, v_i \in \mathcal{V}$. They contain items and lists that user has interacted with, respectively, and are both sorted according to the interaction timestamp in ascending order. Items in list c_i can be expressed as an item list l_i according to their displaying order, i.e., $l_i = [v_1^i, v_2^i, \cdots, v_f^i]$, where $v_*^i \in \mathcal{V}$ and f denotes the number of items in a list. For lists that contain less than f, we pad them with padding items. Based on the above available data, our task is to recommend the lists with top-k highest interaction probabilities to target users. To accomplish the task, we treat the user-list and user-item interactions as sequences and construct a model to learn user preferences from different levels, predicting interaction probabilities between users and each candidate item.

4   The MULTIPLE Model

The Fig. 2(a) shows the framework for multi-level user preference learning in MULTIPLE, where we learn the list-level and item-level user preferences from the user-list and user-item interaction sequences, respectively, and learn dual-level user preferences via interactions and the list-item hierarchical structure. In order to learn user preference embeddings at each level, we propose a gated and attentive module for sequence learning that can perceive user preference drift,

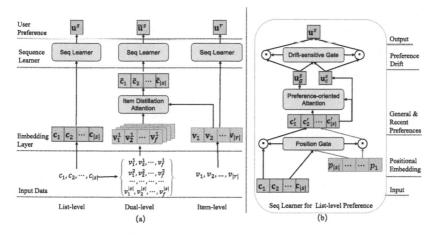

Fig. 2. The illustration of MULTIPLE, where (a) shows how to learn multi-level user preferences from available data and (b) shows the structure of sequence learning module.

which is shown in the right of Fig. 2(b). For each user, we calculate the scores of user preferences at different levels by the learned user preferences embeddings. At last, we obtain the final preference score by weighted summation of the multiple preference scores and apply it for prediction.

4.1   List-Level Preference Score

User preferences for lists are preferences on the intrinsic quality of lists themselves, e.g., the overall topics of lists.

Given any user and the user-list interaction sequence $s = [c_1, c_2, \cdots, c_{|s|}]$, we map each list into an embedding and obtain $\mathbf{S} = [\mathbf{c}_1, \mathbf{c}_2, \cdots, \mathbf{c}_{|s|}]$, where $\mathbf{c}_i \in \mathbb{R}^d$.

Generally, interactions that are farther from current reflect fewer user preferences. In order to differentiate how each interacted list affects the user preference, we fuse the reverse position of lists in the sequence into the list-level preference via the combination of list ID embedding $\mathbf{c}_i$ and reverse position embedding $\mathbf{p}_{|s|-i} \in \mathbb{R}^d$. However, the effect of position differs from list to list. For example, if a user interacts with an unpopular list, then the list may reflect so distinctive user preference that even if it was interacted a long time ago, it still matters much. Therefore, we learn the importance factor of position embedding dynamically by designing a position gate as shown in Eq. (1), which takes both the list ID embedding and the position embedding as input. Then, we obtain the positional list embedding $\mathbf{c}'_i$ by weighted summation of the list ID embedding and the reverse position embedding.

$$
\begin{aligned}
z_i &= \sigma \left(\mathbf{W}_1^T \left(\text{LeakyReLU} \left(\mathbf{W}_2^T \left(\mathbf{c}_i || \mathbf{p}_{|s|-i} \right) \right) \right) \right) \\
\mathbf{c}'_i &= (1 - z_i) \mathbf{c}_i + z_i \mathbf{p}_{|s|-i}
\end{aligned}
\tag{1}
$$

Among them, $||$ means concatenation, σ denotes the sigmoid function, $\mathbf{W}_2 \in \mathbb{R}^{2d \times d}$ and $\mathbf{W}_1 \in \mathbb{R}^{d \times 1}$ are parameters to be learned. After processing the embedding of each list successively, we obtain $\mathbf{S}' = \left[\mathbf{c}_1', \mathbf{c}_2', \cdots, \mathbf{c}_{|s|}' \right]$.

Furthermore, considering that the last interacted list reflects the user's recent intention and preference, we express the user's recent preference as $\mathbf{u}_c^s = \mathbf{c}_{|s|}'$. Each list in the historical sequence has a different degree of relevance to the user's recent intention, and lists completely unrelated to the user's recent intention may bring noise to the user preference modeling. Therefore, we design a preference-oriented attention shown in Eq. (2) to measure how much each interacted list matches user's recent preferences, and obtain user's general preference $\mathbf{u}_g^s$ through weighted aggregation, thereby denoising the unrelated lists and emphasizing strongly related ones.

$$
\begin{aligned}
\theta_i &= \mathbf{q}^\top \sigma \left(\mathbf{W}_3 \mathbf{c}_{|s|}' + \mathbf{W}_4 \mathbf{c}_i' + \mathbf{b} \right) \\
\mathbf{u}_g^s &= \sum_i \theta_i \mathbf{c}_i'
\end{aligned}
\tag{2}
$$

Among them, $\mathbf{W}_* \in \mathbb{R}^{d \times d}, \mathbf{q} \in \mathbb{R}^d$ and $\mathbf{b} \in \mathbb{R}^d$ are parameters, σ denotes the sigmoid function.

As we know, user preferences may drift over time. For example, the user in Fig. 1, who was obsessed with Marvel actions, gets interested in animations later on. It is intuitive that if a user's recent preferences differ a lot from his/her general preferences, it indicates that the user preferences may have changed recently, then the recent preferences are supposed to play a more important role than the general preferences. In order to flexibly learn the importance of the user's recent preferences and general preferences, we design a preference drift gate shown in Eq. (3), where $\mathbf{W}_5 \in \mathbb{R}^{2 \times d}$. The gate takes the difference of user's recent preferences and general preferences that measures user preference drift as input and outputs the weightings $\mathbf{g} = [g_0, g_1]$. Furthermore, the final list-level user preference embedding is calculated by $\mathbf{u}^s = g_0 \times \mathbf{u}_c^s + g_1 \times \mathbf{u}_g^s$.

$$
\mathbf{g} = \operatorname{softmax} \left(\mathbf{W}_5 \left(\mathbf{u}_c^s - \mathbf{u}_g^s \right) \right)
\tag{3}
$$

We refer to the procedure to learn $\mathbf{u}^s$ from $\mathbf{S} = \left[\mathbf{c}_1, \mathbf{c}_2, \cdots, \mathbf{c}_{|s|} \right]$ as $SeqLearner(\cdot)$, which is shown in the Fig. 2(b) and includes three main components, i.e., the reverse position gate, the preference-oriented attention and the preference drift-sensitive gate.

Finally, for each candidate list $c_k \in \mathcal{S}$, we calculate the user preference score by $x_k^s = \mathbf{u}^{sT} \mathbf{c}_k$.

4.2   Item-Level Preference Score

User preferences on lists can also be aggregated by the user preferences on their constituent items. Therefore, we learn item-level user preferences based on user-item interactions, and predict the user preferences on constituent items of each list. Then we obtain the item-level preference scores by aggregating preference scores of constituent items.

Specifically, for a given user-item sequence $r = [v_1, v_2, \ldots, v_{|r|}]$, firstly, we map each interacted item to an embedding and get $\mathbf{R} = [\mathbf{v}_1, \mathbf{v}_2, \ldots, \mathbf{v}_{|\mathbf{v}|}]$, where $\mathbf{v}_* \in \mathbb{R}^d$. We employ $SeqLearner(\cdot)$ with $\mathbf{R}$ as input and obtain $\mathbf{u}^{rT}$. Furthermore, for any candidate list $c_k \in \mathcal{C}$ we get the embeddings of its constituent items denoted by $\left[\mathbf{v}_1^k, \mathbf{v}_2^k, \cdots, \mathbf{v}_f^k \right]$ and calculate the score of user preference on each item $v_j^k, j \in [1, f]$ by $\mathbf{u}^{rT} \mathbf{v}_j^k$.

When the user browses the list, the positions of the items in a list affects its role. For example, items at the front, which are usually easier to be seen and interacted with, are supposed to play a more important role. Therefore, we can calculate the importance of each item according to its position in the list. Specifically, as for the i-th item in a list, we get the position embedding $\mathbf{p}_i \in \mathbb{R}^d$ at first and then calculate the importance factor by Eq. (4), where $\mathbf{W}_6 \in \mathbb{R}^{d \times d}$ and $\mathbf{W}_7 \in \mathbb{R}^{d \times d}$ are parameters.

$$\beta_i^k = \text{softmax}_i \left(\text{LeakyReLU} \left(\mathbf{W}_6^T \text{LeakyReLU} \left(\mathbf{W}_7^T \mathbf{p}_i \right) \right) \right) \tag{4}$$

We calculate the weighted average of the scores of user preferences on each item in a list by Eq. (5) and obtain x_k^r.

$$x_k^r = \sum_{i=1}^{f} \beta_i^k \mathbf{u}^{rT} \mathbf{v}_i^k \tag{5}$$

4.3   Dual-Level Preference Score

Up to now, user preferences have been modeled separately from two different perspectives, i.e., list-level and item-level. Though lists and items are of two different levels, the list-level and item-level preferences of the same user in the same time period are intertwined with each other. Here, we bridge them with the list-item containment relationship and obtain dual-level user preferences that incorporate information at both two levels.

Specifically, it is a common and reasonable way to aggregate the embeddings of constituent items in a list to form the list content embedding due to the hierarchical structure of lists and items [4,7]. Existing work [4,7] uses the non-personalized strategies to aggregate item embeddings, we argue that users may only be interested in just a few of the constituent items, which are supposed to play a more important role during the aggregation. For example, in Fig. 1, the video list named "Everything New On Disney Plus" contains both "The Falcon and the Winter Soldier" based on Marvel Comics characters and "Big Shot" about the basketball. But for the target user, who interacted with lots of Marvel actions, the former is obviously more eye-catching and influential, since it is similar to his interacted items and matches his preference very well. Therefore, the similarities between the interacted items and each constituent item of a list can measure how the users like the constituent item, and we can exploit the user-item interaction to distill essential items in lists.

Specifically, we design a personalized item distillation attention network that incorporates item-level user preferences to obtain embeddings of list content, which includes two modules, i.e., a self-attention module and a personalized attention module.

The self-attention mechanism is used to capture the consistency between items. That is, given any list c_i, after obtaining the embedding lists of its constituent items $[\mathbf{v}_1^i, \mathbf{v}_2^i, \cdots, \mathbf{v}_f^i]$, we fine-tune each item embedding with the self-attention module shown in Eq. (6) and obtain $\left[\mathbf{v}_1^{i\prime}, \mathbf{v}_2^{i\prime}, \ldots, \mathbf{v}_f^{i\prime}\right]$.

$$
\begin{aligned}
\alpha_{pq}^i &= \frac{\mathbf{v}_p^{iT}\mathbf{W}_s\mathbf{v}_q^i}{\sqrt{d}} \\
\mathbf{v}_p^{i\prime} &= \sum_{q=1}^{q=f} \alpha_p^i \mathbf{v}_q^i
\end{aligned}
\tag{6}
$$

We apply the mean embedding of all interacted items as the query vector and construct attention to distill the items that match user preferences from all the constituent items, as shown in Eq. (7), where $\mathbf{W}_9, \mathbf{W}_{10} \in \mathbb{R}^{d\times d}$ are learnable parameters. Since user-item interaction sequences are personalized naturally, the list content embedding $\tilde{c}_i$ is also personalized.

$$
\begin{aligned}
a_p^i &= \mathrm{softmax}_p\left(\sigma\left(\left(\mathbf{W}_9\frac{1}{|r|}(\sum_{j=1}^{|r|}\mathbf{v}_j)\right)^T\left(\mathbf{W}_{10}\mathbf{v}_p^{i\prime}\right)\right)\right) \\
\tilde{\mathbf{c}}_i &= \sum_{p=1}^f a_{jp}^k \mathbf{v}_p^{k\prime}
\end{aligned}
\tag{7}
$$

Content embeddings of each interacted list form $\tilde{\mathbf{S}} = [\tilde{\mathbf{c}}_1, \tilde{\mathbf{c}}_2, \ldots, \tilde{\mathbf{c}}_{|s|}]$. Then, we learn the dual-level user preference $\tilde{\mathbf{u}}^s$ from $\tilde{\mathbf{S}}$ via a $SeqLearner(\cdot)$ which has the same structure but different parameters to those in previous sections. Finally, we calculate the dual-level user preference score by $\tilde{x}_k^s = \tilde{\mathbf{u}}^{sT}\tilde{\mathbf{c}}_k$.

4.4   Prediction and Loss Function

Through the above processes, we obtain preference scores $x_k^s, \tilde{x}_k^s, x_k^r$ from three different levels. The final predicted interaction score is calculated as $\hat{x}_k = w_1 x_k^s + w_2 x_k^r + w_3 \tilde{x}_k^s$. In order to dynamically learn the weightings of each interaction score, we set the weightings w_1, w_2 and w_3 as learnable parameters.

The predicted interaction probability between target user and list $c_k \in \mathcal{C}$ is $\hat{y}_k = \mathrm{softmax}_k(\hat{x}_k)$. The loss function for model training is as Eq. (5).

$$
\mathrm{Loss} = \sum_{k=1}^{|\mathcal{C}|} y_k \log \hat{y}_k + \lambda\Theta
\tag{8}
$$

The main part of the loss function is the cross entropy of the predicted interaction probability $\hat{y}_k$ and the ground truth y_k, where if c_k is the next-interacted list, then $y_k = 1$, otherwise $y_k = 0$. Θ is the sum of L_2 regularization terms of all the parameters in this model.

4.5   Analysis

The most related model to MULTIPLE is AttList [7], which can be seen as a variant of the dual-level preference score calculation part in MULTIPLE. The difference is that AttList adopts a non-personalized attention network to generate list content embeddings without considering item-level user preferences. On the other hand, AttList utilizes the same attention network to aggregate list content embeddings into user preference embeddings, which is not sensitive to the user preference drift. Compared with AttList, MULTIPLE models user preferences from multiple different perspectives and handles user preference modeling more finely.

Table 1. The statistics of two utilized real-world datasets

Datasets	#U	#L	#I	#U-L	#U-I	U-L Density	U-I Density
MTW-20K-1	21443	222	10440	142243	371501	3.00%	0.17%
MTW-20K-2	25298	244	10323	176071	458228	2.85%	0.18%

5 Experiments

In this section, we conduct extensive experiments on real-world datasets to evaluate our model. First, in order to evaluate the performance of our proposed model, we compare it to other existing item and list recommendation models. Then, in order to study the effect of user preferences at different level, we launch ablation experiments for multi-level user preferences. Next, we carry out control experiments and verify the rationality of key modules in MULTIPLE. Finally, through a case, we analyze and understand how user preferences at each level influence the final recommendation.

5.1 Experimental Settings

Datasets. There is no public datasets providing as comprehensive data as MULTIPLE models. We rely on MX Player[1], one of the largest streaming platforms in India that organizes and exposes videos via lists. Some of the lists on MX Player contains personalized items, for example, the personalized list 'Hindi Films For you' contains different Hindi movies when exposed to different users.

We collect the interaction data from the movie tab in MX Player in two periods of time, from Feb. 7. 2021 to Feb. 20. 2021 and from Match. 7. 2021 to Match. 20. 2021. By randomly sampling about 20 thousand users from tens of millions active users, we construct two datasets, MTW-20K-1 and MTW-20K-2, from which we filter out the sequences with a length shorter than 3 and truncate sequences longer than 30 by keeping the latest interacted elements.

The statistics of these two datasets are listed in Table 1. Note that although there are only 200+ candidate lists for each user, there are millions of lists with different content involved in the datasets due to the existence of personalized lists. We preprocess user-list interaction data by sequence splitting, as for

---

[1] https://www.mxplayer.in/.

a list sequence $s = [c_1, c_2, ..., c_{|s|}]$, we generate the sequences and corresponding labels as $([c_1, c_2], c_3)$, $\cdots$, $([c_1, c_2, \ldots, c_{n-2}], c_{n-1})$, $([c_1, c_2, \ldots, c_{n-1}], c_n)$, in which $([c_1, c_2, \ldots, c_{n-1}], c_n)$ is used for testing, $([c_1, c_2, \ldots, c_{n-2}], c_{n-1})$ is used for validating, and the others are used for training. Noted that there is no public dataset for list recommendation that contains the comprehensive data required by MULTIPLE as ours, we plan to open up our datasets.

Metrics. We adopt recall@K and NDCG@K as evaluation metrics. Here, recall@K is equivalent to hit@K.

Implementation. We implement all the models in this paper with PyTorch. For fairness, the dimensions of embeddings in all the models are set to 128, and the batch size is set to 64. All learnable parameters are initialized with the uniform distribution $U(-\frac{1}{\sqrt{d}}, \frac{1}{\sqrt{d}})$. The optimizer used is the Adam optimizer, and the initial learning rates and regularization weightings are grid-searched among $[0.1, 0.01, 0.001, 0.0001]$. We early stop training when Recall@20 on the validation set has not been improved in 4 consecutive epochs, and then make predictions on the test set with the model which has the best performance on the validation set. Besides, for models that require negative sampling, we search the rates of positive and negative samples among $\{1 : 4, 1 : 8, 1 : 12\}$.

5.2 Performance Comparison

In order to measure the performance of MULTIPLE, we choose the following recommendation models as competitors:

1. **BprMF** [15]: BprMF is a classic collaborative filtering model via matrix factorization with pairwise loss.
2. **NCF** [6]: NCF is collaborative filtering model which utilizes multi-layer perceptron to capture a high level of non-linearities.
3. **SR-GNN** [18]: SR-GNN is a graph neural network model for sequential recommendation. It models each interaction sequences as a graph and learns user preference from historical interactions.
4. **GC-SAN** [19]: GC-SAN is a graph contextual self-attention model that utilizes the complementarity between self-attention network and graph neural network to enhance the recommendation performance.
5. **LIRE** [14]: LIRE is a Bayesian-based ranking model that considers to model user's preference in a list by aggregating user preferences over individual items in the list.
6. **DAM** [4]: DAM is a deep attentive multi-task model for both of list recommendation and item recommendation. It proposes a factorized attention network to aggregate the item information of each list.
7. **AttList** [7]: AttList is an attention-based list recommendation model that leverages the hierarchical structure of items, lists and users and reveals additional insights into user preferences.
8. **BGCN** [3]: BGCN is a recent bundle recommendation model based on graph convolutional neural network.

Table 2. The performance comparison results, where the best performance is highlighted in bold and the second best performance is underlined, the last column shows the performance improvement rate between MULTIPLE and the best competitor.

MTW-	Recall	BprMF	NCF	SRGNN	GC-SAN	LIRE	DAM	AttList	BGCN	MULTIPLE	Improv (%)
20K-1	@5	33.42	35.84	37.33	38.37	40.84	_43.48_	42.52	16.62	**56.34**	29.57
	@10	52.38	52.33	55.01	55.82	57.16	59.51	_60.45_	25.22	**75.55**	24.98
	@20	70.69	70.58	75.07	75.24	76.71	78.72	_82.69_	37.44	**90.94**	9.98
20K-2	@5	32.43	35.94	38.68	39.15	41.69	_44.42_	42.62	15.28	**55.87**	25.77
	@10	51.50	52.24	56.97	57.06	57.96	60.00	_62.20_	24.62	**76.07**	22.30
	@20	70.90	71.22	76.83	76.99	77.07	77.20	_82.84_	38.07	**91.03**	9.89
MTW-	**NDCG**	BprMF	NCF	SRGNN	GC-SAN	LIRE	DAM	AttList	BGCN	MULTIPLE	Improv (%)
20K-1	@5	20.50	24.19	25.01	25.93	26.69	28.66	_28.76_	10.53	**38.47**	33.76
	@10	26.61	29.51	30.71	31.48	32.52	33.57	_34.55_	13.31	**44.70**	29.38
	@20	31.26	34.11	35.76	36.31	36.86	38.31	_40.19_	16.36	**48.63**	21.00
20K-2	@5	19.64	23.91	25.74	26.22	26.15	_28.15_	_28.15_	10.00	**37.66**	33.78
	@10	25.82	29.17	31.59	32.00	32.81	32.92	_34.45_	12.98	**44.20**	28.30
	@20	30.73	33.97	36.63	37.10	37.21	37.61	_39.68_	16.38	**48.01**	20.99

Among all the listed competitors, we notice that BprMF, NCF, SR-GNN and GC-SAN are item recommendation models, while LIRE, DAM, AttList and BGCN are list recommendation models. When applying item recommendation models for list recommendation, we treat lists as generalized items. The results are listed in Table 2, from which we can draw the following conclusions.

1. Item recommendation models perform worse than list recommendation models. This verifies that treating the lists as generalized items and applying item recommendation models to solve the list recommendation task are not enough.
2. Compared with DAM, although AttList does not exploit user-item interaction data to train the user embeddings, it still far outperforms DAM. The reason is that AttList learns user preferences via aggregating the dynamic user-list interaction sequences, while DAM learns static user ID embeddings via collaborative filtering. This indicates that learning user preferences by aggregating interaction sequences not only avoids parameters explosion brought by user growth, but also is more accurate and effective than learning static user ID embeddings.
3. The performance of BGCN is very poor and even inferior to BprMF, which is related to the principle of BGCN. BGCN first constructs a heterogeneous graph based on the global interactions and the list-item hierarchical structure, and then performs propagation and aggregation. Since personalized lists are connected to majority of items, the list embeddings are smooth and indistinguishable. Besides, the list embeddings are non-personalized. Therefore, compared with other list recommendation BGCN is not flexible enough to deal with personalized lists.
4. MULTIPLE achieves the best performance. The main reasons are as follows. On the one hand, our model explicitly models user preferences from different levels, predicts the corresponding preference scores, and comprehensively con-

siders each score to produce the final recommendation. On the other hand, we make full use of multi-source data including the orders between interactions, etc.

5.3   Ablation Study for Multi-level User Preferences

In order to verify how user preferences from different levels contributes to the performance, we conduct ablation studies. By assembling different-level user preferences, six model variants are obtained, including the variant List (only utilizes list-level user preferences), Item (only item-level ones), Dual (only dual-level ones), List+Item (both list-level and item-level ones), List+Dual (both list-level and dual-level ones) and Item+Dual (both item-level and dual-level ones). Then, we compare their performance to the original model. The results are as Fig. 3.

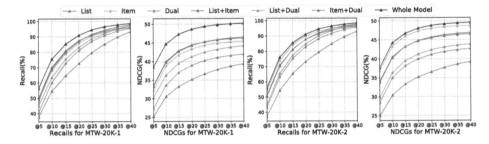

Fig. 3. The ablation study for multi-level user preferences

Table 3. The rationality evaluation for dual-level user preferences learning

Datasets	Models	-SelfAtt		-Variant ListID		MULTIPLE	
MTW-	Metrics (%)	Recall	NDCG	Recall	NDCG	Recall	NDCG
20K-1	@5	54.72	36.89	54.86	37.62	**56.34**	**38.47**
	@10	74.92	43.41	74.73	44.05	**75.55**	**44.70**
20K-2	@5	54.67	37.50	52.61	35.53	**55.87**	**37.66**
	@10	74.22	43.84	71.99	41.81	**76.07**	**44.20**

The results show that the performance of the original model is the best, which demonstrates that user preferences from each level contributes to the recommendation performance. Besides, among all the variants, the variant List performs worst, which further confirms the necessity of modeling user preferences to items in the list.

5.4   Rationality Evaluation

In order to illustrate the design rationality of the main modules in our model, we conduct experiments to verify their contributions to performance. Here, we evaluate the item distillation attention network and the drift-sensitive sequence learning module.

Rationality Evaluation for the Item Distillation Attention. In order to verify the rationality of the item distillation attention network for dual-level user preference learning, we replace the personalized attention with a non-personalized attention that takes list ID embedding as the query (denoted by the variant ListID). In addition, we removed the self-attention module to observe the effect of item consistency learning. The experimental results are as Table 3. The worse performance of variants demonstrate the effectiveness and reasonability of the item distillation attention network.

Rationality Evaluation for the Sequence Learning Module. In this section, we set up multiple experiments to verify the contribution of several main components in the sequence learning module, including the position information (denoted as Pos), the recent preferences (denoted as Recent), the general preferences (denoted as General), drift gate (denoted as Drift), etc. We verify the role of these modules by observing the performance of model variants without them. In particular, in order to verify the necessity of modeling orders between interactions, we construct a model variant that replaces $SeqLearner(\cdot)$ with mean pooling to obtain the user preference embeddings and keeps other settings the same. The experimental results are as Table 4.

Table 4. The rationality evaluation for the sequence learning module

Datasets	Metrics (%)	- Drift	- General	- Recent	- Pos	Mean	OurModel
MTW-20K-1	**Recall@5**	52.61	44.37	55.26	54.40	52.56	**56.34**
	Recall@10	72.20	63.09	74.71	74.03	72.72	**75.55**
	NDCG@5	36.07	29.74	37.65	37.64	35.67	**38.47**
	NDCG@10	42.43	35.79	43.96	44.01	42.19	**44.70**
MTW-20K-2	**Recall@5**	55.39	52.95	55.57	55.10	52.24	**55.87**
	Recall@10	74.29	69.76	75.82	74.53	72.63	**76.07**
	NDCG@5	37.24	37.33	**37.93**	37.23	34.90	37.66
	NDCG@10	43.38	42.78	45.17	43.53	41.50	**44.20**

Table 5. The genre statistics of interacted items

Genres	Count
Drama	9
Thriller	8
Crime	8
Action	7
Romance	4

Table 6. The user-list interaction sequence

Numbers	Interacted Lists
1	Hindi Films For You
2	Critically-Loved Bollywood Hits
3	Top Romantic Films
4	Hindi Films For You
5	Premium Movies
Ground Truth	**Bollywood Hits**

Based on the above experimental results, we have the following findings and conclusions. First, no matter which component is removed, it more or less cause a decrease in performance. Second, removing the general preferences has a greater impact than removing the recent preference. Finally, directly ignoring the orders of interactions makes the performance worse, which confirms our motivation to achieve list recommendation via sequential models.

5.5 Case Study

In order to understand how each level user preferences contribute to the final recommendation, we sample a representative user who interacts with 5 lists and 22 movies and compare the recommendations of the original model and two model variants, i.e., the variant List that only uses the list-level user preferences, the variant List+Item that involves both list-level and item-level user preferences. The genre statistics of the interacted items are listed in Table 5 and the user-list interaction sequence are listed in Table 6. The list "Bollywood Hits" which contains Bollywood produced movies with crime, thriller, drama genres is the actual next-interact list, i.e., the ground truth list. Detailed constituent movies in each list are not listed here due to the limited space.

From results, we have the following findings. First, MULTIPLE and its variants all capture the user preferences on Hindi movies and recommend "Hindi Films For You" to the user. Second, the variant List fails to recall the ground truth list, the variant List-Item recalls the ground truth list successfully but ranks it fourth, while MULTIPLE ranks it first. Third, MULTIPLE outputs totally different recommendations from List and List+Item. Apart from the first list, both of the third and fourth lists recommended by MULTIPLE consist of Bollywood-produced movies in action, thriller, and drama genres. "Top Romantic Films", which has a weaker correlation with the user's item-level preferences, ranks relatively low. Lists containing actions that are not produced by Bollywood are not recommended, which is consistent with the user's item-level preferences on Bollywood and premium films (Table 7).

Table 7. Recommendations of three models/variants, where BOYU is short for 'Based on Your Viewing'

Ranks	List	List+Item	MULTIPLE
1	Hindi Films For You	Hindi Films For You	**Bollywood Hits**
2	Movies BOYU	Movies BOYU	Hindi Films For You
3	MAX HD-30th August	MX Top Picks	Critically-Loved Bollywood Hits
4	Trending Movies on MX	**Bollywood Hits**	Bollywood Chills & Thrills
5	MAX HD-2nd September	Movies BOYU2	Top Romantic Films

6 Conclusion

In this paper, we present a sequential model MULTIPLE for list recommendation. The model can capture user preferences from multiple levels by exploiting user-list and user-item interactions both separately and jointly. MULTIPLE contains a gated and attentive sequence learning module and utilize all the available data including sequence information as fully as possible. The experimental results show that both of user preferences of different levels and sequence information make more or less contributions to the improvement of recommendation performance, which confirms our motivations. Further, our solution for list recommendation provides a reference framework, other complex multi-level preference modeling methods and sequential models can also be explored.

References

1. Bai, J., et al.: Personalized bundle list recommendation. In: The World Wide Web Conference, pp. 60–71. ACM (2019)
2. Cao, D., Nie, L., He, X., Wei, X., Zhu, S., Chua, T.S.: Embedding factorization models for jointly recommending items and user generated lists. In: Proceedings of the 40th International ACM SIGIR Conference on Research and Development in Information Retrieval, pp. 585–594 (2017)
3. Chang, J., Gao, C., He, X., Jin, D., Li, Y.: Bundle recommendation with graph convolutional networks. In: Proceedings of the 43rd International ACM SIGIR Conference on Research and Development in Information Retrieval, pp. 1673–1676 (2020)
4. Chen, L., Liu, Y., He, X., Gao, L., Zheng, Z.: Matching user with item set: collaborative bundle recommendation with deep attention network. In: IJCAI, pp. 2095–2101 (2019)
5. He, X., Deng, K., Wang, X., Li, Y., Zhang, Y., Wang, M.: LightGCN: simplifying and powering graph convolution network for recommendation. In: Proceedings of the 43rd International ACM SIGIR Conference on Research and Development in Information Retrieval, pp. 639–648 (2020)
6. He, X., Liao, L., Zhang, H., Nie, L., Hu, X., Chua, T.S.: Neural collaborative filtering. In: Proceedings of the 26th International Conference on World Wide Web, pp. 173–182 (2017)

7. He, Y., Wang, J., Niu, W., Caverlee, J.: A hierarchical self-attentive model for recommending user-generated item lists. In: Proceedings of the 28th ACM International Conference on Information and Knowledge Management, pp. 1481–1490 (2019)

8. Hidasi, B., Karatzoglou, A., Baltrunas, L., Tikk, D.: Session-based recommendations with recurrent neural networks. arXiv preprint arXiv:1511.06939 (2015)

9. Hidasi, B., Quadrana, M., Karatzoglou, A., Tikk, D.: Parallel recurrent neural network architectures for feature-rich session-based recommendations. In: Proceedings of the 10th ACM Conference on Recommender Systems, pp. 241–248 (2016)

10. Le, D.T., Lauw, H.W., Fang, Y.: Correlation-sensitive next-basket recommendation. In: Proceedings of the Twenty-Eighth International Joint Conference on Artificial Intelligence, pp. 2808–2814. International Joint Conferences on Artificial Intelligence Organization, August 2019

11. Li, J., Ren, P., Chen, Z., Ren, Z., Lian, T., Ma, J.: Neural attentive session-based recommendation. In: Proceedings of the 2017 ACM on Conference on Information and Knowledge Management, pp. 1419–1428 (2017)

12. Liu, G., Fu, Y., Chen, G., Xiong, H., Chen, C.: Modeling buying motives for personalized product bundle recommendation. ACM Trans. Knowl. Discov. Data (TKDD) **11**(3), 28 (2017)

13. Liu, Q., Zeng, Y., Mokhosi, R., Zhang, H.: STAMP: short-term attention/memory priority model for session-based recommendation. In: Proceedings of the 24th ACM SIGKDD International Conference on Knowledge Discovery & Data Mining, pp. 1831–1839 (2018)

14. Liu, Y., Xie, M., Lakshmanan, L.V.: Recommending user generated item lists. In: Proceedings of the 8th ACM Conference on Recommender Systems, pp. 185–192 (2014)

15. Rendle, S., Freudenthaler, C., Gantner, Z., Schmidt-Thieme, L.: Bpr: bayesian personalized ranking from implicit feedback. arXiv preprint arXiv:1205.2618 (2012)

16. Sha, C., Wu, X., Niu, J.: A framework for recommending relevant and diverse items. In: IJCAI, pp. 3868–3874 (2016)

17. Wang, X., He, X., Wang, M., Feng, F., Chua, T.S.: Neural graph collaborative filtering. In: Proceedings of the 42nd International ACM SIGIR Conference on Research and Development in Information Retrieval, pp. 165–174 (2019)

18. Wu, S., Tang, Y., Zhu, Y., Wang, L., Xie, X., Tan, T.: Session-based recommendation with graph neural networks. In: Proceedings of the AAAI Conference on Artificial Intelligence, vol. 33, pp. 346–353 (2019)

19. Xu, C., et al.: Graph contextualized self-attention network for session-based recommendation. In: IJCAI, vol. 19, pp. 3940–3946 (2019)

20. Zhu, T., Harrington, P., Li, J., Tang, L.: Bundle recommendation in ecommerce. In: Proceedings of the 37th International ACM SIGIR Conference on Research & Development in Information Retrieval, pp. 657–666. ACM (2014)

Deep News Recommendation
with Contextual User Profiling and
Multifaceted Article Representation

Dai Hoang Tran[1]([⊠]), Salma Hamad[1], Munazza Zaib[1], Abdulwahab Aljubairy[1],
Quan Z. Sheng[1], Wei Emma Zhang[2], Nguyen H. Tran[3],
and Nguyen Lu Dang Khoa[4]

[1] Computing Department, Macquarie University, Sydney, Australia
[2] Computer and Mathematical Sciences, The University of Adelaide, Adelaide,
Australia
[3] School of Computer Science, The University of Sydney, Sydney, Australia
[4] Data61, CSIRO, Sydney, Australia
{dai-hoang.tran,salma-abdalla-ibrahim-mah.h, munazza-zaib.ghori,
abdulwahab.aljubairy}@hdr.mq.edu.au,michael.sheng@mq.edu.au,
wei.e.zhang@adelaide.edu.au, nguyen.tran@sydney.edu.au,
khoa.nguyen@data61.csiro.au

Abstract. News recommendation is a new challenge in the current age
of information overload. Making personalized recommendations from the
sources of condense textual information is not trivial. It requires the
understanding of both the news article's semantic meaning, and the user
preferences via the user's history records. However, many existing methods are not capable to address the requirement. In this paper, we propose
our novel news recommendation model called CUPMAR, that not only is
able to learn the user-profile's preferences representation in multiple contexts, but also makes use of the multifaceted properties of news articles
to provide personalized news recommendations. The main components
of the CUPMAR model are the News Encoder (NE) and User-Profile
Encoder (UE). The NE uses multiple properties of a news article with
advanced neural network layers to derive news representation. The UE
infers a user's long-term and recent preference contexts via her reading
history to derive a user representation, and finds the most relevant candidate news for her. We evaluate our CUPMAR model with extensive
experiments on the popular MIND dataset and demonstrate the strong
performance of our approach. Our source code is also available online for
the reproducibility purpose.

Keywords: Recommendation systems · News recommendation ·
Neural networks · Attention mechanism · Contextual profile

© Springer Nature Switzerland AG 2021
W. Zhang et al. (Eds.): WISE 2021, LNCS 13081, pp. 237–251, 2021.
https://doi.org/10.1007/978-3-030-91560-5_17

1   Introduction

Reading online news to stay up-to-date with the latest information has been an integral part of modern human culture. However, due to the large amount of news articles as well as the proliferation of news websites, users may get overwhelmed to decide what and where to read. As a result, news recommendation is an effective way to target user reading interests [3,6,15]. News recommendation poses several challenges in comparison to the traditional recommendation problem. First and foremost, unlike movies or shopping items recommendation, news articles are *time sensitive* items. The value and relevance of a news articles deteriorate quickly over a short period of time because fresh news is updated frequently. Due to this time sensitive property, traditional methods like collaborative filtering, which depends on the identity (ID) of users or items, would not work efficiently. Second, the content of a news article contains *dense textual data*, which also encodes the latent preferences of the users. For instance, certain individuals may only read sport news and discard the rest. Thus this is a strong signal indicating the "Sports" category encodes one of their long-term preferences. As another example, some users sporadically click on the latest news that has title or content related to the celebrities, and this behaviour shows a strong short-term interest signal, in which the article's content exhibits certain words or knowledge patterns. This second problem prompts a strong need to measure a user's reading history to infer her latent preferences, either long-term, short-term

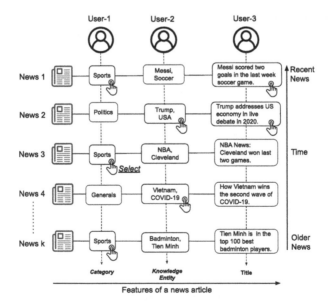

Fig. 1. News recommendation scenario example. The horizontal axis shows multifaceted properties of each news article. The vertical axis shows interactions of each user. Each of them has different choices depending on the properties of the news articles.

or a mixture of the both. Other aspects like *diversity*, also plays a significant role in news recommendation. Users should be able to find the types of news that they have high interest in, but also can explore other news that may pique their curiosity. This diversity can significantly improve users' satisfaction and retain their loyalty to these online news services. Essentially, the core tasks of solving news recommendation problems are capturing a user preferences from their reading history and understanding all of the signals from a news article.

Figure 1 illustrates a scenario where understanding of different contexts between a user and a news article is very important. In this scenario, each news article has several features, such as the category, the knowledge entity inside the news, and the title. Each user can select different news based on the number of news articles that is shown to them. It is very clear that User-1 only selects the articles in the *Sports* category, which defines her long-term preferences. User-2 and User-3 demonstrate different behaviours. User-2 shows attention to the news which contains the knowledge entities about countries, while User-3 only reads the latest news. Therefore, these behaviours display their recent preferences.

Given these challenges, researchers and industry partners resort to deep learning and also spend a significant amount of time to collect the right datasets to facilitate the development of news recommendation. In this paper, we address news recommendation problem by proposing a novel deep learning model named CUPMAR, which is able to learn both the contextual user profile and the news article content representation. The main components of CUPMAR are (i) the News Encoder (NE) and (ii) the User-Profile Encoder (UE). The NE infers representation of a news article based on its important properties such as category, title, and abstract content. Self-attention and attention mechanism are used to learn news content effectively. In addition, due to recent successes of using knowledge entity for news recommendation task [13], we also enrich the learning of news representation by adding knowledge entities taken from WikiData knowledge graph[1] to the feature list of the NE. The UE contains two submodules. The first submodule is the *Long-term Latent Preferences Extraction* (LPE), and the second submodule is the *Recent Latent Preferences Extractor* (RPE). We are strongly motivated by the observation that the reading history of a user always encodes both of her long-term as well as her current interests. Thus by using both LPE and RPE submodules, we can learn the representations of the user's contextual profile. The encoding news representation from the NE's output and the user-profile representation from the UE's output together are used to calculate the *interaction score*. The interaction score helps us to identify highly relevant candidate news articles for each user. Hence, the online news services can recommend a ranked list of suitable news articles to their users, thereby improving their service quality and increasing users' satisfaction.

We perform extensive experiments on the Microsoft News Dataset (MIND) [16], and the results show that our approach improves the efficiency and performance of the news recommendation task.

---

[1] wikidata.org.

2   Related Works

News recommendation is a popular and essential task in the field of natural language processing (NLP) and recommender systems. A number of online businesses rely heavily onto this task to tailor personalized experience for millions of users [6]. The main approach for solving news recommendation problem is to accurately learn the news article and user representation [8]. Henceforth, several popular works rely on different feature engineering strategies to build their own news article and user representation [3,6,14]. Particularly, Latent Dirichlet Allocation (LDA) is used to generate topic distribution features to infer news representation in each session, and user representation is inferred by all the news in her session [2]. Another noteworthy method is the Explicit Localized Semantic Analysis (ELSA) proposed by Son et al. [9] for location-based news recommendation, where location and topic signals are calculated from Wikipedia posts as news representation. Nevertheless, the downside of manual feature engineering is the dependence on expert domain knowledge, which is not always available for many approaches. Additionally, traditional NLP methods do not incorporate word context and word order well enough to derive semantic meaning and learn user and news article representation effectively.

Owning to the popularity of deep learning methods, there have been a lot of efforts during the recent years to address the aforementioned issues pertaining to the task of news recommendations [3,14,15]. For instance, the work of Wang et al. [13] tries to infer news article representation from the news title using a knowledge-entity-aware method with Convolutional Neural Network (CNN) layer, and learn user representation by her browsed history news. Another approach in the work of Okura et al. [6] takes advantages of the denoising autoencoder [12] to learn the news article representation. They use this technique with a Gated Recurrent Unit (GRU) neural network layer to learn the user representation from her history records.

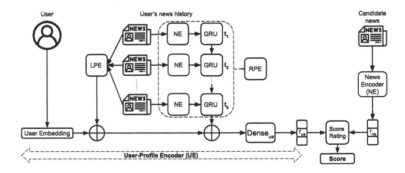

Fig. 2. The CUPMAR model design. CUPMAR has two main components. The first component is NE (News Encoder), which is able to learn a news article representation. The second component is UE (User-Profile Encoder), which is able to derive a contextual user representation thanks to its submodules LPE and RPE. The final interaction score is calculated by the *Score Rating* component.

Our proposed model CUPMAR also takes advantage of deep neural network to solve the news recommendation problem. However, the features that make our model different with the aforementioned models are: (i) the utilization of multi-faceted properties, where each property is first encoded differently and then merged together to derive the final news representations, (ii) the combination of both long-term and short-term interactions to infer the user representations. We rely on an ensemble of multiple advanced neural network mechanisms to automatically capture the similarity between a user and a candidate news article representation.

3   The CUPMAR Approach

In this section, we introduce the CUPMAR (Contextual User-Profile and Multi-faceted Articles Representation) model as showed in Fig. 2. The CUPMAR model comprises of two major components. The first component is NE (News Encoder) that uses multiple neural network mechanisms on its multifaceted properties to learn a news article representation in the form of a news vector. The second component is UE (User-Profile Encoder) that is further sub-divided into two submodules, which are LPE (Long-term Preferences latent Extractor) and RPE (Recent Preferences latent Extractor). Finally, a *Score Rating* component uses inputs of both the candidate news vector and the contextual user profile vector to predict the interaction score between these two entities.

3.1   CUPMAR News Encoder

The task of the CUPMAR News Encoder (NE) is to learn the representation of a news article. A news article contains several pieces of useful information such as the news category, news title, news body content, and the knowledge entities, as depicted in Fig. 3. It is essential to leverage all of these pieces of information to derive a meaningful representation for downstream machine learning tasks. As such, for each news article, we use five main features to encode its representation vector. We denote a news article as $n = \{c, sc, k, t, b\}$, where $c \in C$ is the category feature in the set C of all categories in the dataset, $sc \in C$ is the subcategory feature. We have $k \in K$ as the knowledge entity feature in the set K of all knowledge entities in the dataset. We have t as the news title feature with T words, hence $t = [w_1^t, w_2^t, \ldots, w_T^t]$, where $w^t \in W$ is a word in the title t in the set of all distinct words W in the dataset. Similarly, b is the news body content feature with B words, hence $b = [w_1^b, w_2^b, \ldots, w_B^b]$ where $w^b \in W$ is a word in the body b.

First, we derive the vector $\boldsymbol{r_c}$ from both the category c and subcategory sc of the news article. The category and subcategory features give us clear information about the topic of the news article, and they also serve as strong signals for a user's long-term preferences. The vector $\boldsymbol{r_c}$ is formulated as follows:

$$\boldsymbol{r_c} = \mathrm{ReLU}(\mathbf{W_c} \times [e_c \parallel e_{sc}] + \boldsymbol{b_c}), \tag{1}$$

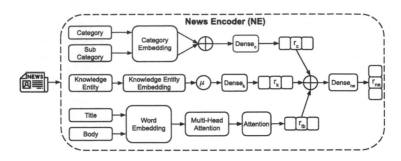

Fig. 3. The News Encoder (NE) component design. In this NE component, multifaceted properties such as the category, knowledge entity, title of the news article are processed in different ways via multiple neural network layers. All of these property vectors are concatenated and go through a Dense layer to derive the final representation vector.

where $\mathbf{W_c}$ and $\boldsymbol{b_c}$ are the weight and bias parameters of the $Dense_c$ (feedforward) layer in Fig. 3, the $[e_c \parallel e_{sc}]$ is the concatenation of the category embedding e_c of category c, and subcategory embedding e_{sc} of subcategory sc, and $ReLU$ is the non-linear activation function [5].

Likewise, we perform a similar procedure to learn the vector $\boldsymbol{r_k}$ of the knowledge entity k of the news article. Since one article can contain multiple knowledge entities, we perform the mean operation on their embedding before feeding them into the $Dense_k$ layer as illustrated in Fig. 3. The formulation is as follows:

$$r_k = \text{ReLU}(\mathbf{W_k} \times \boldsymbol{\mu}(e_{k_1}, e_{k_2}, \ldots, e_{k_n}) + \boldsymbol{b_k}), \tag{2}$$

where $\mathbf{W_k}$ and $\boldsymbol{b_k}$ are the weight and bias parameters of the $Dense_k$ layer, $\boldsymbol{\mu}(e_{k_1}, e_{k_2}, \ldots, e_{k_n})$ is the mean operation of n knowledge entity embeddings e_k in the article.

The most important feature of a news article is the content itself. We want to learn the representation from both the news article's title and body content. Primarily, we want to know how each word interacts with its surrounded nearby words. Therefore, we choose to apply both the attention and multi-head self-attention mechanisms that is popularized by the work of Vaswani et al. [11]. The formulation to learn the representation r_{tb} of the news article content in the title and the body is as follows:

$$r_{\text{tb}} = \mathbf{Att}(\mathbf{Heads}(e_{w_1^t}, e_{w_2^t}, \ldots, e_{w_T^t}, e_{w_1^b}, e_{w_2^b}, \ldots, e_{w_B^b})), \tag{3}$$

where $\mathbf{Heads}(e_{w_1}, \ldots, e_{w_i})$ is a word-level multi-head self-attention layer [11] on each word embedding e_{w_i}. This layer contains k heads, which is a hyperparameter. The head h_k learns the representation of word w_i as follows:

$$h_{i,k}^w = \mathbf{V}_k^w \left(\sum_{j=1}^{T+B} a_{i,j}^k e_j \right), \tag{4}$$

$$a_{i,j}^k = \frac{\exp(e_{w_i}^\mathsf{T} \mathbf{Q}_k^w e_j)}{\sum_{m=1}^{T+B} \exp(e_i^\mathsf{T} \mathbf{Q}_k^w e_m)}, \tag{5}$$

where $\mathbf{Q}_k^w$ and $\mathbf{V}_k^w$ are the weight parameters in the h_k head, $(\cdot)^\mathsf{T}$ is the transpose operation, $T + B$ is the total amount of words in the title and body, and $a_{i,j}^k$ is the interaction weight between i and j words. The final representation for each word w_i is the concatenation of all the self-attention heads, that is $h_i^w = [h_{i,1}^w \parallel h_{i,2}^w \parallel \cdots \parallel h_{i,h}^w]$, hence we have $\mathbf{Heads}(e_{w_1}, \ldots, e_{w_i}) = \{h_1^w, \ldots, h_i^w\}$. Subsequently, the $\mathbf{Att}$(Heads) function of the attention layer then attends to each word after the self-attention representation h_i^w. The formula for deriving the attention weight of each word α_i^w is:

$$b_i^w = q_w^\mathsf{T} tanh(\mathbf{V}_w \times h_i^w + v_w), \tag{6}$$

$$\alpha_i^w = \frac{\exp(b_i^w)}{\sum_{j=1}^{T+B} \exp(b_j^w)}, \tag{7}$$

where $\mathbf{V}_w$ and v_w are the attention weight and bias parameters, q_w is the query vector. After all of these attention weights are calculated, the content vector $r_{\mathbf{tb}}$ of a news article is computed as:

$$r_{\mathbf{tb}} = \sum_{i=1}^{T+B} \alpha_i^w h_i^w. \tag{8}$$

Lastly, we combine all of these multifaceted vectors r_c, r_k and $r_{\mathbf{tb}}$ to learn the *multifaceted news representation* vector $r_{\mathbf{ne}}$ by combining them and let the final $Dense_{ne}$ layer to extract the most prominent patterns of a news article as illustrated in Fig. 3. The formula is as follows:

$$r_{\mathbf{ne}} = \mathrm{ReLU}(\mathbf{W}_{\mathbf{ne}} \times [r_c \parallel r_k \parallel r_{\mathbf{tb}}] + b_{\mathbf{ne}}), \tag{9}$$

where $\mathbf{W}_{\mathbf{ne}}$ and $b_{\mathbf{ne}}$ are the weight and bias parameters of the $Dense_{ne}$ layer, and $[r_c \parallel r_k \parallel r_{\mathbf{tb}}]$ is the concatenation of the multifaceted vectors from a news article.

3.2   CUPMAR User-Profile Encoder

The CUPMAR User-Profile Encoder (UE) is responsible for learning user representation from their reading history. Figure 2 shows the complete architecture of the CUPMAR model, where the left-side portion visualizes the UE and its submodules. A user's reading habit can exhibit both long-term and recent preferences. To extract both of these signals, the UE uses two of its submodules LPE

Algorithm 1: LPE Algorithm

Input: The history news records of k news $N = \{n_1, n_2, \ldots, n_k\}$, threshold L
      for L most frequently chosen categories

Result: The long-term latent preferences vector r_{lpe}

1 **begin**
2     $e_{u_i} \leftarrow$ UserEmbedding(u_i)
3   **for** $i \leftarrow 1$ *to* k **do**
4   count category of news n_i
5   select top L categories (1)
6   **end**
7     $C = \{c_1, c_2, \ldots, c_L\} \leftarrow$ Set C of L categories from (1)
8   count $\leftarrow 0$
9     $r_{\mathsf{lpe}} \leftarrow \{0\}$
10   **for** $i \leftarrow 1$ *to* k **do**
11   **if** n_i *has category in* C **then**
12         $r_{\mathsf{lpe}} \leftarrow r_{\mathsf{lpe}} +$ CateEmbedding(n_i)
13   count $\leftarrow$ count $+ 1$
14   **end**
15   **end**
16    $r_{\mathsf{lpe}} \leftarrow r_{\mathsf{lpe}}/$count
17    $r_{\mathsf{lpe}} \leftarrow [(r_{\mathsf{lpe}} \parallel e_{u_i}]$
18   **return** r_{lpe}
19 **end**

---

and RPE to handle them. By sampling the reading history of the last several days, the LPE can infer the long-term preferences of a user by paying attention to her most frequent reading topics. Likewise, the LPE extracts a user's recent interests by paying attention to the news article title, the body as well as embedded knowledge entities. The following sections dive into the details of each submodule.

LPE Submodule: The sole purpose of LPE is to learn the long-term preferences of a user throughout her history news records. It looks for frequent signals that signify repetitive behaviours of a user. We argue that the news topics from a user's history records serve as a strong indication for a user's general and long-term preferences. Additionally, the unique characteristic of a user also refines her choice. For instance, a basketball fan is more likely to check sports news about NBA rather than checking for badminton news. Therefore, to mimic those long-term preferences scenarios, we decide to assign each user a unique embedding vector based on her Identification (ID), and calculate the accumulation of the most frequent categories of a user's history news records via their categorical embedding. The algorithm for extracting a user's long-term latent preferences vector r_{lpe} is detailed in Algorithm 1.

First, we initialize each user with a unique user embedding vector e_{u_i} using a *UserEmbedding* layer (line 2). Second, we learn the most frequent L categories inside a user history records and store them in the set C (lines 3 to 7). Third, we initially set the long-term latent preferences vector r_{lpe} as a zero-vector, then accumulate r_{lpe} with all of the category embedding vectors that belong in the set C using the category embedding layer *CateEmbedding* (lines 8 to 15). Then, we average the r_{lpe} based on the *count* value, where it counts the total news articles that has the category in set C (line 16). Finally, we concatenate r_{lpe} with the user embedding e_{u_i} (line 17). Using this algorithm, we can extract both user long-term preferences and her unique characteristic into the representative vector r_{lpe}.

RPE Submodule: RPE learns recent preferences of a user via the Gated Recurrent Unit (GRU) neural network layer. We denote K as a variable for the amount of news articles a user has recently read, then for K news articles in chronological order, the set of news records is denoted as $N = \{n_1, n_2, \ldots, n_K\}$. The RPE derives the recent preference latent vector r_{rpe} from a user using the GRU layer as follows:

$$z_t = \sigma(W_z[h_{t-1}, \boldsymbol{NE}(n_t)]), \tag{10}$$

$$r_t = \sigma(W_r[h_{t-1}, \boldsymbol{NE}(n_t)]), \tag{11}$$

$$\widetilde{h_t} = tanh(W_{\widetilde{h}}[r_t \odot h_{t-1}, \boldsymbol{NE}(n_t)]), \tag{12}$$

$$h_t = z_t \odot h_l + (1 - z_t) \odot \widetilde{h_t}, \tag{13}$$

where σ is the sigmoid function, $\odot$ is the item-wise product, W_r, W_z, and $W_{\widetilde{h}}$ are the GRU's network weights, $\boldsymbol{NE}(.)$ is the News Encoder function described in Sect. 3.1. With the initial hidden vector state h_0 initialized as a zero-vector, we repeat the process with the GRU network until we reach the last hidden state vector h_K. Thus, the RPE vector is $r_{rpe} = h_K$.

Contextual User-Profile Representation: Given the two contextual vectors of a user, which are the long-term latent preferences vector r_{lpe} and the recent latent preferences vector r_{rpe}, the final contextual user-profile vector r_{ue}, is calculated as follows:

$$r_{ue} = \text{ReLU}(\mathbf{W_{ue}} \times [r_{lpe} \parallel r_{rpe}] + b_{ue}), \tag{14}$$

where $\mathbf{W_{ue}}$ and b_{ue} are the weight and bias parameters of the *Dense$_{ue}$* layer (Illustrated in Fig. 2), and $[r_{lpe} \parallel r_{rpe}]$ is the concatenation of both contextual vectors that we learn from the aforementioned sections. The usage of both contextual vectors is the key ingredient to help the CUPMAR model achieving better scores as we discuss in the later sections.

3.3   CUPMAR Model Training

To train the CUPMAR, we need to have a scoring function to measure the interactive score between a user and news article representation. One of the effective methods for this requirement is the dot product operation, as applied in the famous work of Okura et al. [6]. Hence, we use the dot product operation to compute the interaction probability inside the final component *Score Rating* of the CUPMAR model, as illustrated in Fig. 2. If we have a user-profile u with its representation vector r_u and a candidate news article n with its representation vector r_n, then we can calculate the interaction score between them as $s(u, n) = r_u^T r_n$.

Additionally, we address our news recommendation problem as a classification task, and use the negative sampling technique during model training [4]. Therefore, when a user is presented with multiple news articles, the articles that are clicked by the user are the positive samples, whereas the other N random sampled articles that are not clicked by the user are the negative samples. Then, the CUPMAR model can learn to infer the interaction probability between the positive and N negative news articles, thus formulating this as a $N+1$ classes prediction for the classification task. The loss function is the negative log-likelihood of the positive samples. As such, the training total loss of all positive samples is calculated as follows:

$$\text{loss} = -\sum_{i=1}^{P} \log \frac{\exp(s(u, n_i^{pos}))}{\exp(s(u, n_i^{pos})) + \sum_{k=1}^{K} \exp(s(u, n_{i,k}^{neg}))} \tag{15}$$

where P is the amount of positive training samples, n_i^{pos} is the ith positive sample in one news session, and $n_{i,k}^{neg}$ is the kth negative sample for this i^{th} positive sample.

4 Evaluation

4.1 The MIND Dataset and Experimental Settings

The recent work of Wu et al. [16] introduced a large-scale MIND dataset, which can serve as a benchmark dataset for news recommendation. We conduct all our experiments on this high-quality dataset. MIND was collected from the user's behaviour logs of Microsoft News website[2]. It has more than 150,000 news articles, and more than 15 million behaviour logs that were generated by one million users. Each news article comes with rich textual attributes such as the category, subcategory, title, body and knowledge entities embedded inside. Additionally, the MIND dataset also comes with a smaller version called MIND-small, which is suitable for quick prototyping and validation. MIND-small accounts for 5% data of the total dataset. Henceforth, the research community has quickly adapted to use MIND as a robust benchmarking dataset for news recommendation, as shown in [3,14,15].

For the evaluation metrics, we use ranking metrics to benchmark the performance of the validation models. The ones that we choose are the Area Under

---

[2] microsoftnews.msn.com.

ROC curve (AUC), Mean Reciprocal Rank (MRR), and the Normalized Discounted Cumulative Gain (nDCG). Each model is evaluated five times and the average scores are reported.

Our CUPMAR model is evaluated against several baseline methods. The first method is Factorization Machines (FM) [7], where we use TF-IDF signals to learn the representations. The second method is Kim et al. CNN model [1], where max pooling layer is used to learn news representation from title and body content. The third method is Deep Knowledge-Aware Network (DKN) model [13], where CNN and knowledge entity attention are used to learn representations. The fourth method is High-Fidelity Archive Network (HiFiArk) [3]. HiFiArk treats user's news history as a compact vector and store them into archives during offline stage. Then during the online stage, these compact vectors are used to infer user interest upon candidate news. The fifth method is Neural News recommendation with Multi-head Self-attention (NRMS) [15], where it uses news encoder with multi-head self-attention to learn words interaction, and attention mechanism on the user encoder to extract user preferences. Additionally, we also create different variants of the CUPMAR model, which are CUPCate, CUPShort, and CUPLong. CUPCate model only considers the category and subcategory features in the news encoder. The user-profile then is encoded by averaging all the history news records representation. CUPShort is identical to CUPMAR, but without the LPE submodule. CUPShort only learns to extract the recent preferences of a user via its GRU layer. By using CUPShort, we can compare the effectiveness of the CUPMAR model when we employ the LPE submodule. Likewise, CUPLong is also identical to CUPMAR, but without the RPE submodule. CUPLong only learns to extract the long-term preferences of a user.

4.2   Performance Evaluation

Table 1 shows the performance results of the CUPMAR model. First, our CUPMAR model achieves the state-of-the-art scores and outperforms all the baseline methods on the MIND dataset. CUPMAR's performance is followed closely behind by NRMS and HiFiArk, which are the two strongly performing models for news recommendation. This result has proven that using multifaceted properties for news encoding and leveraging contextual user profile signals can significantly boost the learning capability of the deep learning model for news recommendation task. Moreover, CUPMAR model also achieves slightly better performance from having the LPE submodule compared to the CUPShort model which only employs the RPE submodule, as can be seen by the small gap in the scores between them.

Second, the deep neural network models clearly show superior performance in comparison to the matrix factorization FM approach. The better performance of neural network models can be explained by their high learning capacity. Due to the high amount of weight parameters, neural network models have the ability to tackle the complicated task of news recommendation. Another evident supporting this statement is the ranking scores of our simple CUPCate model,

Table 1. Performance comparison of CUPMAR model with other methods over the MIND and MIND-small datasets.

Methods	AUC	MRR	NDCG@5	NDCG@10
MIND				
FM	0.5661	0.2416	0.2701	0.3415
CNN	0.5907	0.3071	0.3224	0.3552
DKN	0.5937	0.2614	0.2833	0.3475
HiFiArk	0.6378	0.2749	0.3105	0.3662
NRMS	0.6491	0.2999	0.3298	0.3954
CUPCate	0.5601	0.2312	0.2586	0.3005
CUPShort	0.6428	0.2954	0.3290	0.3913
CUPLong	0.6041	0.2678	0.2965	0.3124
CUPMAR	**0.6582**	**0.3123**	**0.3435**	**0.4084**
MIND-small				
FM	0.5845	0.2785	0.2854	0.3546
CNN	0.6178	0.2987	0.3025	0.3725
DKN	0.6354	0.3142	0.3262	0.3895
HiFiArk	0.6660	0.3287	0.3539	**0.4212**
NRMS	**0.6664**	**0.3321**	**0.3587**	0.4160
CUPCate	0.6239	0.2964	0.3276	0.3867
CUPShort	0.6396	0.2965	0.3272	0.3894
CUPLong	0.6137	0.2844	0.3172	0.3694
CUPMAR	0.6415	0.2961	0.3289	0.3902

which has the lowest scores across all metrics in the MIND dataset. The most likely reason is the low number of parameters it has due to the crude design of only using two categorical features and one dense layer.

Third, we observe an interesting phenomenon. Our CUPMAR model does not perform well when it is trained and evaluated on the MIND-small dataset, which contains about 5% total samples of the MIND dataset. The CUPMAR model ranks in the third place, while the top spots belong to NRMS and HiFiArk models. After careful examination, we believe that due to the usage of multifaceted properties and multiple advanced neural network layers such as self-attention heads, attention layer, GRU layer and several Dense layers in the whole model, the number of the weight parameters in the CUPMAR model increases significantly. We have 40% more weight parameters in comparison to our implemented NRMS model. Although high number of parameters help CUPMAR to make better generalization over large datasets, it is underfit when trained on smaller datasets. This is a little setback we want to improve in the future work.

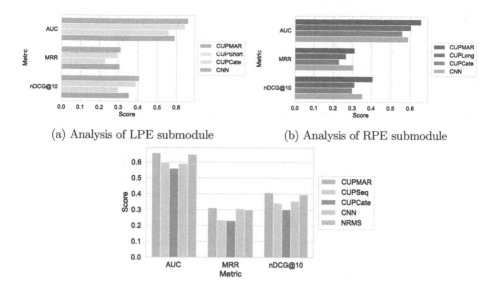

(a) Analysis of LPE submodule   (b) Analysis of RPE submodule

(c) Analysis of using multifaceted properties for news representation

Fig. 4. Evaluation results of different submodules in comparison to other methods.

4.3   Analysis of Using User Contexts

We create two variant models, called CUPShort and CUPLong, respectively. The CUPShort model only uses the RPE submodule inside the UE component to tackle the task, while the CUPLong model only uses the LPE submodule. Then we compare the inference scores of each of them to other models to see the changes in the performance. According to the results shown in Fig. 4a and 4b, we can see that leveraging both the long-term and recent-term contexts can strongly boost the performance of the CUPMAR model. CUPMAR always has higher scores than both CUPShort and CUPLong across all three different metrics. This clearly shows the effectiveness of the contextual information of a user in the news recommendation task.

We also want to answer a further question: which user contextual aspect contributes more to the CUPMAR model. Hence, by looking at the percentage gap in their respective scores to the CUPMAR model, we can confirm that the RPE submodule contributes more to the performance of the CUPMAR model. CUPShort scores (sum of all metrics) lower than CUPMAR by only 3.6%, while the CUPLong's scores witness a gap of 14%. This result shows that the recent-term preferences contribute more to the user representation than the long-term preferences, which does make sense since a user's recent preferences usually also include her long-term preferences as well.

4.4   Analysis of Using Multifaceted Properties

Similar to the analysis of the user contexts, we deploy a model variant called CUPSeq, where we use Seq2Seq [10] architecture on the news title and body to infer the r_{ne} vector, as explained in Sect. 3.1. We then compare the evaluation scores of CUPSeq with other approaches. The experimental results are depicted in Fig. 4c. We can see that using advanced neural network mechanisms outperforms the simple baseline using categorical features, since the CUPMAR, CUPSeq, CNN and NRMS models all score significantly higher than the CUPCate model. Additionally, we observer that the content of a news article contributes more information to the neural models than other signals since both CUPSeq and CNN only employ the news article title and body data. Furthermore, we also understand that more sophisticated architecture such as self-attention layer can learn more effectively than older approaches such as RNN and CNN, because NRMS model achieves better scores than CUPSeq and CNN. Nevertheless, we do see the benefits of using multifaceted properties in the CUPMAR model, as CUPMAR outscores all other models.

5   Conclusion

In this paper, we propose a novel deep neural network called CUPMAR for the challenging task of news recommendation.

At the heart of CUPMAR are the News Encoder and User-Profile Encoder. The News Encoder learns news article representation from various features such as the category, subcategory, knowledge entities inside the article, the article title and body. It uses self-attention, attention and dense layers to effectively combine all the necessary signals to represent a news article. On the other hand, the User-Profile Encoder uses the user's recent historical news data with dense textual information to infer both long-term and recent-term signals for the user representation, thanks to the two submodules the Long-term Preferences latent Extractor and the Recent Preferences latent Extractor with GRU network layer. We perform extensive evaluation of the CUPMAR model on the popular MIND dataset, and CUPMAR shows a better performance against all the baselines.

References

1. Kim, Y.: Convolutional neural networks for sentence classification. In: EMNLP, Doha, Qatar, pp. 1746–1751. ACL (2014)
2. Li, L., Zheng, L., Yang, F., Li, T.: Modeling and broadening temporal user interest in personalized news recommendation. Expert Syst. Appl. **41**, 3168–3177 (2014)
3. Liu, Z., Xing, Y., Wu, F., An, M., Xie, X.: Hi-fi ark: deep user representation via high-fidelity archive network. In: IJCAI, Macao, China, pp. 3059–3065. ijcai.org (2019)
4. Mikolov, T., Sutskever, I., Chen, K., Corrado, G.S., Dean, J.: Distributed representations of words and phrases and their compositionality. In: NIPS, Nevada, USA, pp. 3111–3119 (2013)

5. Nair, V., Hinton, G.E.: Rectified linear units improve restricted boltzmann machines. In: ICML, Haifa, Israel, pp. 807–814. Omnipress (2010)
6. Okura, S., Tagami, Y., Ono, S., Tajima, A.: Embedding-based news recommendation for millions of users. In: SIGKDD, Halifax, NS, Canada, pp. 1933–1942. ACM (2017)
7. Rendle, S.: Factorization machines with libfm. ACM Trans. Intell. Syst. Technol. **3**, 1–22 (2012)
8. Shah, D., Koneru, P., Shah, P., Parimi, R.: News recommendations at scale at bloomberg media: Challenges and approaches. In: Recsys, Boston, MA, USA, p. 369. ACM (2016)
9. Son, J.W., Kim, A., Park, S.: A location-based news article recommendation with explicit localized semantic analysis. In: SIGIR, Dublin, Ireland, pp. 293–302. ACM (2013)
10. Sutskever, I., Vinyals, O., Le, Q.V.: Sequence to sequence learning with neural networks. In: NIPS, Montreal, Quebec, Canada, pp. 3104–3112 (2014)
11. Vaswani, A., et al.: Attention is all you need. In: NIPS, Long Beach, CA, USA, pp. 5998–6008 (2017)
12. Vincent, P., Larochelle, H., Lajoie, I., Bengio, Y., Manzagol, P.: Stacked denoising autoencoders: learning useful representations in a deep network with a local denoising criterion. J. Mach. Learn. Res. **11**, 3371–3408 (2010)
13. Wang, H., Zhang, F., Xie, X., Guo, M.: DKN: deep knowledge-aware network for news recommendation. In: WWW, Lyon, France, pp. 1835–1844. ACM (2018)
14. Wu, C., Wu, F., An, M., Huang, J., Huang, Y., Xie, X.: Neural news recommendation with attentive multi-view learning. In: IJCAI, Macao, China, pp. 3863–3869. ijcai.org (2019)
15. Wu, C., Wu, F., Ge, S., Qi, T., Huang, Y., Xie, X.: Neural news recommendation with multi-head self-attention. In: EMNLP-IJCNLP, Hong Kong, China, pp. 6388–6393. ACL (2019)
16. Wu, F., et al.: MIND: a large-scale dataset for news recommendation. In: ACL (2020)

Intent-Aware Visualization Recommendation for Tabular Data

Atsuki Maruta[1(✉)] and Makoto P. Kato[1,2]

[1] University of Tsukuba, Tsukuba, Japan
s2121653@s.tsukuba.ac.jp, mpkato@acm.org
[2] JST PRESTO, Kawaguchi, Japan

Abstract. This paper proposes a visualization recommender system for tabular data with a visualization intent (e.g., "population trends in Italy" and "smartphone market share"). The proposed method predicts the most suitable *visualization type* (e.g., line, pie, and bar charts) and *visualized columns* (columns used for visualization) based on statistical features extracted from the tabular data, as well as semantic features derived from the visualization intent. To predict an appropriate visualization type, we propose a bi-directional attention (BiDA) model that identifies important table columns by the visualization intent, and important parts of the intent by table headers. To identify visualized columns, we employ a pre-trained neural language model to encode both visualization intents and table columns, and estimate which columns are the most likely to be used for visualization. Since there is no available dataset for this task, we developed a new dataset consisting of over 100K tables and their appropriate visualization. The experiments revealed that our proposed methods accurately estimated suitable visualization types as well as visualized columns. The code is available at https://github.com/kasys-lab/intent-viz .

Keywords: Data visualization · Tabular data · Deep nural networkds

1 Introduction

Visualization is an effective means to gain insights and demonstrate trends of statistical data. As effective visualization requires special skills, knowledge, and deep analysis of data, it is sometimes difficult for end users to produce appropriate data visualization for their purposes. Furthermore, when we have to find necessary data from a large-scale data collection, it would be easy to find the data if the data were properly visualized. Therefore, *visualization recommendation*, which predicts an appropriate *visualization type* (e.g., Line, Pie, and Bar charts) and *visualized columns* (columns used for visualization) for given tabular data, has attracted more attention recently [2,3,6,7,10]. It has been often achieved by a machine learning (ML) approach using features extracted from tabular data, such as means and variances of column values [3].

© Springer Nature Switzerland AG 2021
W. Zhang et al. (Eds.): WISE 2021, LNCS 13081, pp. 252–266, 2021.
https://doi.org/10.1007/978-3-030-91560-5_18

However, there are some limitations in the existing studies on visualization recommender system. First, it has been assumed that only tabular data is given as input for visualization recommendation, while appropriate visualization types are highly dependent on users' *visualization intents*—specific characteristics or content of the data that the visualizer wants to represent. For example, it is not clear which visualization type is the more appropriate for smartphone sales data in 2020: it should be a pie chart if the intent is "to demonstrate the market share", while it should be a line chart if the intent is "to show the growth of the market". Second, in the existing studies, it has been assumed that only columns used for data visualization are input to a visualization recommender system. Given tabular data with a visualization intent, it would be ideal that the system can automatically predict an appropriate visualization type and visualized columns without requiring extra efforts from users.

In this paper, addressing the limitations above, we tackle the problem of identifying an appropriate visualization type and visualized columns for tabular data with a visualization intent. We propose a novel method based on a *bi-directional attention* (BiDA) mechanism for prediction visualization types. BiDA mechanism can automatically identify table columns required for visualization, based on the given visualization intent. Furthermore, as it is bi-directional, our proposed model can attend important terms of the visualization intent based on table headers and important columns in tabular data based on the visualization intent, which are particularly useful for estimating a suitable visualization type. In addition, to identify visualized columns, we propose a method applying a pre-trained neural language model (i.e., BERT [1]) to tabular data. We applied the BERT model as it has shown high performance in a variety of natural language processing tasks. Inputting a pair of a visualization intent and each column header to BERT, we can encode both textual information and use them with statistical features derived from the table for effectively predicting visualized columns.

Since there is no publicly available dataset for the problem setting addressed in this paper, we developed a new dataset consisting of over 100K tables with visualization intents and appropriate visualization types by crawling publicly available visualization of tabular data from Tableau Public, which is a Web service that hosts user-generated data and visualization. To ensure the quality of users' visualization, we manually examined a subset of datasets and found that most of the data visualization was appropriate. We conducted experiments with the developed dataset and found that our BiDA model accurately estimated suitable visualization types, and outperformed baseline methods and those without BiDA. Furthermore, our BERT-based model outperformed baseline methods in the visualized column identification task.

The contributions of this paper are:

1. We tackled a new problem of identifying an appropriate visualization type and visualized columns for tabular data with a visualization intent (Sects. 2 and 3).
2. We developed a new dataset for visualization recommendation (Sect. 4).

3. We demonstrated that the BiDA model and BERT-based model is effective in identifying appropriate visualization types and visualized columns, respectively (Sect. 4).

2 Related Work

This section reviews existing approaches for visualization recommendation, which can be categorized into rule-based and ML-based approaches.

Rule-based approaches [8,9,12,14,15] are mainly based on rules defined by experts. For example, a pie chart is unlikely to be suitable for a table with a large number of rows. Therefore, within the scope of the rules, it is possible to create effective visualizations as if they were created by experts. However, Rule-based approaches is not flexible and cannot be effective when entering data with a large number of columns or rows. And this type of approaches can be accurate when columns used for visualization are also given as input, therefore, it is not clear how to apply those approaches to our problem setting where columns to be visualized are not explicitly given. Moreover, applying this approach requires defining new rules for visualization types that are not used in existing studies such as multi polygon charts, which requires high costs because of consultations with experts. Accordingly, it is difficult to compare rule-based approaches with our approaches.

ML-based approaches are mainly those extracting statistical features from tabular data and training a classifier based on the extracted features [2,3,7,10]. Examples of features include means and variances of column values, and a binary feature indicating whether column values are categorical or numeric. Dibia and Demiralp proposed an encoder-decoder model that translates data specifications into visualization specifications in a declarative language [2]. Luo et al. addressed the visualization recommendation problem by using both rule-based and ML-based (learning-to-rank) approaches [7]. Liu et al. proposed a method for predicting visualization types and visualized columns in the context of a table QA task, in which a table and a *question* are given as input [6]. This method predicts the columns to be used for visualization by identifying the correspondence between the header of the tabular data and question about the tabular data. However, the table QA task requires a specific question for tabular data, which is substantially different from a visualization intent we discuss in this work. Thus, it is not trivial to apply their approach to our problem setting. A study given by Hu et al. [3] is closely related and comparable to ours: they extracted statistical features from tabular data and applied ML models for estimating an appropriate visualization type.

There are several differences between ours and the existing work. Our work predicts an appropriate visualization type and visualized columns based on not only tabular data but also visualization intents. Since those are different types of modals, it is challenging to effectively incorporate both types of input for visualization recommendation. Moreover, our problem setting poses a challenge of selectively using some parts of the tabular data, unlike the existing problem

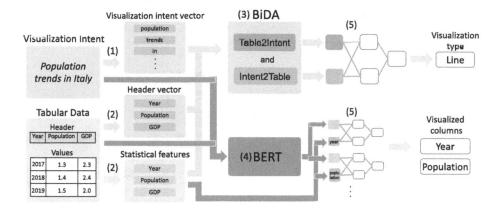

Fig. 1. Our proposed methods with BiDA and BERT.

setting where only a few columns to be used for visualization are given as input for ML approach with the statistical feature in visualization recommendation.

3   Methodology

In this section, we describe the methods for visualization recommendation for tabular data with a visualization intent. Figure 1 illustrates our proposed methods with BiDA and BERT, which comprises five components.

For the visualization type prediction task, **(1) visualization intent embedding** converts each token in the intent into a vector, while **(2) tabular data embedding** encodes table headers based on a word embedding and extracts statistical features from each column. Then, **(3) BiDA** is applied to both embeddings for aggregating information from the visualization intent and tabular data. Finally, **(5) output layer** predicts the most suitable visualization type based on the output of BiDA. Our BiDA model, inspired by a reading comprehension model [13], attends important columns in the tabular data based on a visualization intent, and important terms in the intent based on the tabular data. Therefore, the proposed method enables us to focus only on important columns and tokens in the intent, and be tolerant for redundant columns in a table.

For the other task, visualized column identification, the visualization intent and table headers are directly fed into **(4) BERT** to estimate the relevance of columns to the given visualization intent. This approach was inspired by a Natural-Language-to-SQL model based on BERT [4]. The output of BERT is combined with statistical features derived by **(2) tabular data embedding**, and then used to identify the most relevant columns.

Below, we explain the details of each component in the proposed methods.

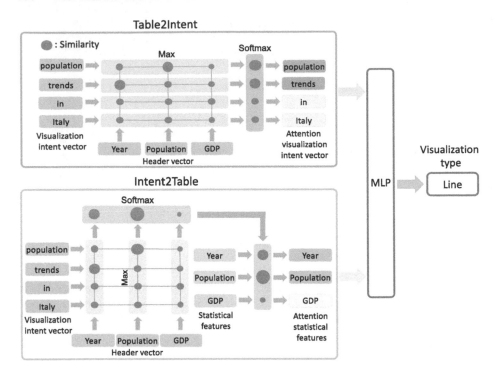

Fig. 2. BiDA computes the weight of each component on one side based on information from the other side, and vice versa.

Visualization Intent Embedding. This component transforms T words in a given visualization intent into word embeddings. The t-th word in the intent is represented by a one-hot representation: $\mathbf{v}_t$ of size $|V|$ where V is an entire set of words or vocabulary. A word embedding matrix $\mathbf{E}_w \in \mathbb{R}^{d_e \times |V|}$ is used to obtain a word embedding $\mathbf{i}_t = \mathbf{E}_w \mathbf{v}_t$ where d_e is the word embedding dimension.

Tabular Data Embedding. Given tabular data consisting of N columns, this component produces a header embedding and extracts statistical features from each column. The header of the j-th column comprises of M_j words and is represented by the mean of their word embeddings, i.e., $\mathbf{h}_j = \frac{1}{M_j} \sum_{t=1}^{M_j} \mathbf{i}_{j,t}$ where $\mathbf{i}_{j,t} \in \mathbb{R}^{d_e}$ is the word embedding of the t-th term in the header of the j-th column. Statistical features are extracted from each column and denoted by $\mathbf{c}_j \in \mathbb{R}^{d_c}$. We extract features following the existing work on visualization recommendation [3] ($d_c = 78$), which include the means and variances of column values and a binary feature indicating whether column values are categorical or numeric. These statistical features were originally used alone to predict appropriate visualization types. In our proposed methods, these statistical features are combined together with textual features for better performances in both visualization type and visualized column prediction tasks.

Bi-directional Attention (BiDA). Figure 2 shows our BiDA model that consists of Table 2Intent and Intent2Table for predicting visualization types. BiDA computes the weight of each component on one side based on information from the other side, and vice versa. In our case, we compute the weight of each column based on the visualization intent (Intent2Table), and that of each term in the visualization intent based on headers of the tabular data (Table 2Intent). Our model then aggregates the word embeddings of the visualization intent, and statistical features of the tabular data with the estimated weights, respectively.

The attentions from both directions are based on the similarity between a column header and a word in the visualization intent, which is defined as follows:

$$s_{tj} = \alpha(\mathbf{i}_t, \mathbf{h}_j) \tag{1}$$

where α is trainable function defined as $\alpha(\mathbf{i}, \mathbf{h}) = \mathbf{w}^\top[\mathbf{i}; \mathbf{h}; \mathbf{i} \circ \mathbf{h}]$. $\mathbf{w}$ is a trainable weight vector, $[;]$ is vector concatenation across row, and $\circ$ is Hadamard product.

Intuitively, columns are considered important if their headers are similar to any of the visualization intent terms. This idea can be implemented as follows:

$$a_j^{(c)} = \text{softmax}(\max_t(s_{tj})) \tag{2}$$

where the maximum similarity between the j-th header and the intent terms is used as the Intent2Table attention $a_j^{(c)}$. Formally, softmax in this equation is defined as:

$$\text{softmax}(\max_t(s_{tj})) = \frac{\exp(\max_t(s_{tj}))}{\sum_{j'=1}^N \exp(\max_t(s_{tj'}))} \tag{3}$$

Statistical features are then aggregated into a single vector with the Intent2Table attentions:

$$\mathbf{c} = \sum_{j=1}^N a_j^{(c)} \mathbf{c}_j \tag{4}$$

The word embeddings of the visualization intent are also aggregated in the same way as $\mathbf{c}$, i.e.,

$$\mathbf{i} = \sum_{t=1}^T a_t^{(i)} \mathbf{i}_t \tag{5}$$

where the Table 2Intent attention $a_t^{(i)}$ is defined as $a_t^{(i)} = \text{softmax}(\max_j(s_{tj}))$. Finally, we obtain the concatenation of the visualization intent and tabular data embeddings: $\mathbf{x} = [\mathbf{i}; \mathbf{c}]$.

BERT. Figure 3 shows the BERT-based model for visualized column identification. While BERT was originally designed for texts, it is applied to the concatenation of the visualization intent and table header in our model. Our proposed model feeds a visualization intent and a column header into the BERT model and combines the output of the BERT with the statistical features of the column for estimating whether the column should be used for visualization.

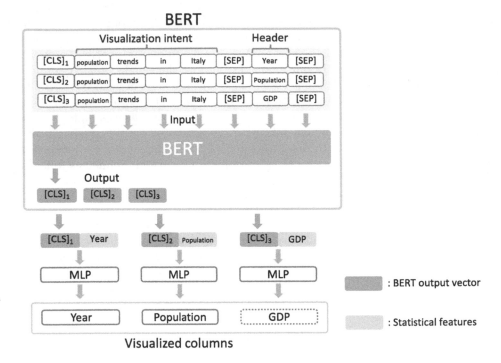

Fig. 3. The BERT model was applied to estimate the visualized columns.

Formally, given a visualization intent consisting of T words, $X = \{x_1, x_2, \cdots, x_T\}$, and the j-th column header containing M_j header words, $Y = \{y_{j1}, y_{j2}, \cdots, y_{jM_j}\}$, the input to the BERT model is as follows:

$$[\text{CLS}]x_1x_2\cdots x_T[\text{SEP}]y_{j1}y_{j2}\cdots y_{jM}[\text{SEP}] \tag{6}$$

where [CLS] and [SEP] are special tokens expected to represent the whole sequence and a separator for BERT, respectively. Letting $\mathbf{y}_{\text{CLS},j} \in \mathbb{R}^{d_{\text{BERT}}}$ be the output of BERT for the [CLS] token ($d_{\text{BERT}} = 768$), we can obtain a vector $\mathbf{y}_j$ as follows:

$$\mathbf{y}_j = [\mathbf{y}_{\text{CLS},j}; \mathbf{c}_j] \tag{7}$$

which is the concatenation of the BERT output and statistical features, and is used to predict whether the j-th column is a visualized column.

Output Layer. The output layers are slightly different for the visualization type prediction and visualized column identification tasks, while they share the same architecture. Given either the output of BiDA $\mathbf{x}$ or that of BERT $\mathbf{y}_j$, we apply a multilayer perceptron with rectified linear units, and use a softmax function for predicting visualization types, and a sigmoid function for estimating whether the j-th column is a visualized column.

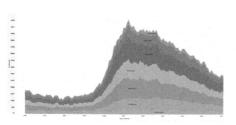

Fig. 4. Example of Area chart.

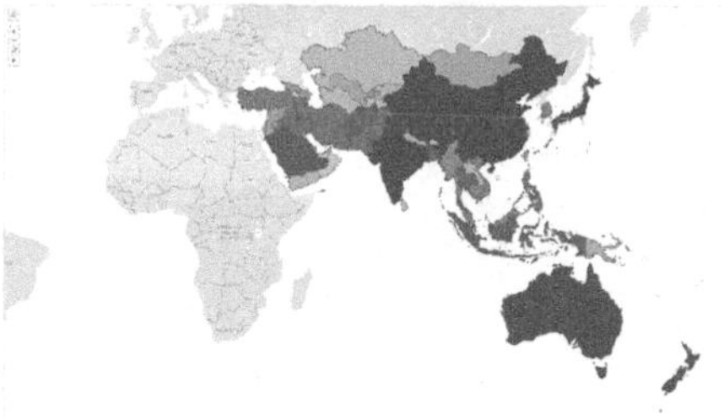

Fig. 5. Example of Multi Polygon chart.

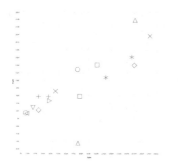

Fig. 6. Example of Shape chart.

Fig. 7. Example of Square chart.

4   Experiments

This section explains the dataset and experimental settings, and discusses experimental results.

4.1   Dataset

Our dataset consists of quartets of tabular data, a visualization intent, an appropriate visualization type, and a set of visualized columns. To develop this dataset, we crawled publicly available visualization of tabular data from Tableau Public[1], which is a Web service that hosts user-generated data and visualization. We used the title of each visualization as a visualization intent, and chart type of each visualization as an appropriate visualization type. Columns used in charts were regarded as visualized columns. Visualization with three or fewer words in the visualization intent was excluded from the dataset, since short titles usually did not make much sense as an intent.

Only eight visualization types were used in this dataset, as the other types are too infrequent. Each visualization type is described as follows:

---

[1] https://public.tableau.com/.

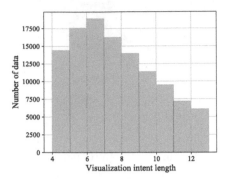
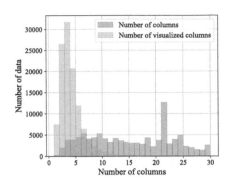

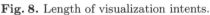

Fig. 8. Length of visualization intents.

Fig. 9. Number of columns and visualized columns.

Area. Figure 4 shows example of Area chart[2]. It usually represents the evolution of multiple numeric variables such as the growth of population.

Bar. It is often used to compare trends and multiple data values. Box-and-whisker plots are included in this category.

Circle. This visualization type is used to represent the distribution of each data. Scatter plots and bubble charts are included in this category.

Line. It is often used to show the evolution of a variable and can be used as an alternative to an area chart for many cases.

Multi Polygon. Figure 5 shows an example of Multi Polygon chart[3]. This chart type can represent a region in a map and is often used to indicate geographical regions.

Pie. It is often used to represent a fraction.

Shape. Figure 6 shows an example of Shape chart[4]. Each instance is independent (it does not represent the evolution of any kind), and denoted by different *shapes* (e.g., a circle, a triangle, or any icon).

Square. Figure 7 shows an example of Square chart[5]. This type represents values by the size of the squares.

Figures 8 and 9 show the distributions of the length of visualization intents and the number of columns, respectively. There are 7.23 words per a visualization intent, and 18.10 columns and 3.76 visualized columns per table on average. After preprocessing the dataset, 115,183 data remained, which were split into training (93,297), validation (10,367) and test (11,519) sets.

---

[2] https://public.tableau.com/app/profile/shahar1185/viz/AREA-CHART/Areachart.

[3] https://public.tableau.com/app/profile/sukumar.roy.chowdhury/viz/Asia-PacificPolygonMap/AsiaPacificPolygonMap.

[4] https://public.tableau.com/app/profile/naveen2129/viz/Shapesscatterchart/Shapesscatterchart.

[5] https://public.tableau.com/app/profile/veita/viz/Recommendation_Square/Chart.

Since the ground truth of the dataset (i.e., visualization types) was based on users' visualization, we investigated the quality of the ground truth by manual assessment. Two annotators, who graduated from universities, were asked to examine a visualization intent and a tabular data and choose appropriate visualization types independently. They were first given an explanation of each visualization type and engaged in a practice session. The annotators were then asked to examine 215 cases. They could skip when they thought there were four or more appropriate visualization types. Otherwise, they could choose up to three appropriate visualization types. We took the union and intersection of their answers for each case, and examined whether the ground truth was in either the union or intersection. As a result, we found the union and intersection contain the ground truth for 63.0% and 51.6% cases, respectively. These results suggest that most of the ground truth in our dataset is considered reasonable.

4.2 Experimental Settings

We used a pre-trained GloVe model [11] for word embeddings, which was trained by the Wikipedia 2014 dump and Gigaword 5 corpus. The size of embeddings is $d_e = 100$.

Visualization Type Prediction. In the BiDA model, the number of layers in the multilayer perceptron was set to six, according to the results with the validation set. The model was trained with a cross-entropy loss function and Adam [5] was used for optimization.

We compared our proposed methods with ML-based approaches used in the related work [3], and could not compare with the other approaches due to the incompatibility of input and output [2,7,10]. Those ML-based approaches are based on 912 statistical features extracted from tabular data, and their hyperparameters were tuned with the validation set. We used Precision, Recall and F1 score as the evaluation metrics in this task.

Visualized Column Identification. we use pre-trained BERT model in hugging face[6] and fine-tuned the three layers on the output side of BERT. The number of layers in the multilayer perceptron was set to two. The BERT model was trained with a binary cross-entropy loss function. We used Adam [5] as an optimization algorithm.

We compared our proposed method with three baselines. "Random" is a weak baseline method that gives each columns a random score. "Word similarity" is a method that calculates the cosine similarity between the mean vector of words in a visualization intent and that of words in a column header. "BM25" uses a BM25 score between a visualization intent and column headers to rank columns. We treated the visualized column identification task as a ranking task rather than a classification task, and used R-precision and nDCG@10 as the evaluation metrics, where visualized columns were regarded as relevant items of relevance grade +1.

---

[6] https://github.com/huggingface/transformers.

4.3   Experimental Results

Visualization Type Prediction. Table 1 shows the results of baselines and our proposed methods in the visualization type prediction task. Some of the baseline models trained with dedicated features performed well, but were not comparable to the proposed models with both visualization intents and table features. "Ours (without BiDA)" are simplified versions of the proposed model in which **i** and **c** were mean vectors of word embeddings i_t and statistical features c_j, respectively. When BiDA was applied to those models, there were significant performance improvements. "Intent" and "Table" indicates the method without statistical features of the table and word embeddings from the visualization intent, respectively. Comparing "Intent" and "Table", "Intent" showed a higher performance than "Table", and "Intent & Table" showed the highest performance, suggesting that both types of features were effective and visualization intents were more effective than tabular features. We conducted a randomized test with Bonferroni correction for the differences between the best model and the others, and found that all the pairs were significant at $\alpha = 0.01$.

Table 1. Performances of the proposed and comparative methods for predicting an appropriate visualization type.

Category	Model	Precision	Recall	F1
Baselines [3]	Naive Bayes	0.222	0.139	0.084
	K-Nearest Neighbor	0.468	0.415	0.434
	Logistic Regression	0.313	0.205	0.201
	Random Forest	0.467	0.420	0.438
	Intent	0.474	0.431	0.444
Ours (without BiDA)	Table	0.463	0.362	0.372
	Intent & Table	0.510	0.481	0.489
	Intent + BiDA	0.491	0.428	0.442
Ours (with BiDA)	Table + BiDA	0.454	0.374	0.395
	Intent & Table + BiDA	**0.536**	**0.500**	**0.512**

Table 2. Performances of Intent & Table + BiDA for each visualization type.

Visualization type	# of examples	Precision	Recall	F1
Area	613	0.515	0.383	0.440
Bar	2,761	0.486	0.596	0.535
Circle	2,301	0.524	0.488	0.506
Line	1,584	0.440	0.506	0.471
Multi polygon	836	0.614	0.663	0.638
Pie	1,365	0.565	0.570	0.568
Shape	1,093	0.588	0.415	0.486
Square	966	0.558	0.381	0.453

Table 3. The most attended terms in the visualization intent for each visualization type. The "Attention" column is the average of the attention values for each word in the test set.

Type	Term	Attention	Type	Term	Attention
	area	0.984		map	1.594
	sheet	0.878		state	0.888
Area	chart	0.779	Multipolygon	sheet	0.878
	time	0.585		region	0.750
	year	0.345		top	0.681
	bar	1.449		pie	1.962
	trend	1.086		donut	1.464
Bar	change	1.069	Pie	sheet	0.874
	area	0.950		gender	0.826
	sheet	0.880		chart	0.819
	bubble	1.530		map	1.649
	map	1.490		sheet	0.880
Circle	box	1.340	Shape	top	0.698
	trend	1.132		time	0.568
	sheet	0.870		year	0.419
	trend	1.040		map	1.735
	trends	1.001		table	1.230
Line	line	0.996	Square	heat	0.946
	sheet	0.883		sheet	0.881
	chart	0.839		top	0.618

We further investigated the performance of the best model with BiDA. Table 2 shows the performances for each visualization type and Table 3 shows the terms that received the most attention in the visualization intent for each visualization type. Formally, the "Attention" column in Table 3 indicates the average of the attention values for each word in the test set. We found that *Multi polygon* chart showed the highest prediction accuracy. The terms receiving the most attention in the *multi Polygon* chart included geographical terms such as *state* and *region*. Since the *Multi polygon* chart was only the geography-related visualization, the prediction could be easier than the other visualization type. The second highest prediction accuracy was achieved by the *Pie* chart. The most attended terms in the *Pie* chart included *pie* and *donut*, which were related to the shape of the visualization. The high prediction accuracy visualization types have unique shapes and features that were not found in other visualizations. On the other hand, area charts and line charts showed lower prediction accuracy, and these charts tend to have a low value of attention on visualization intents.

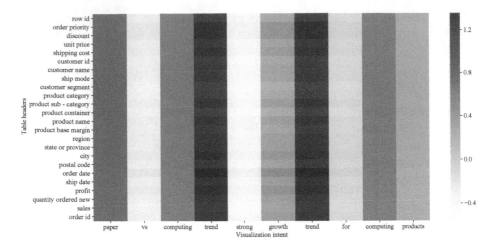

Fig. 10. Visualization of the similarity matrix. The correct visualization type for this example is "line", which was successfully predicted by our model. The user-generated visualization is a line chart comparing the sales of paper and computing products.

Table 4. Performances of the proposed and comparative methods for visualized column identification.

Category	Model	R-Precision	nDCG@10
Baselines	Random	0.335	0.506
	Word similarity	0.356	0.543
	BM25	0.449	0.622
Ours (without BERT)	Table	0.439	0.592
Ours (with BERT)	BERT	0.592	0.719
	BERT & Table	0.603	0.734

This is possibly because there were not particularly effective words for predicting these types of visualization.

Figure 10 visualizes the similarity between a column header and a word in a visualization intent, defined in Eq. 1. The correct visualization type for this example is "line", which was successfully predicted by our model. The user-generated visualization is a line chart comparing the sales of paper and computing products. The figure shows that the depth of each word in the visualization intent changes significantly, and the similarity is not affected by a header, but is greatly affected by a visualization intent. It can be seen that the term *trend* received high attention in the visualization intent. Furthermore, our model also attended visualized columns such as *Order Priority* and *Order Date*.

Visualized Column Identification. Table 4 shows the performance of baselines and our proposed methods in the task of visualized column prediction.

[CLS] most popular car brands [SEP] brand [SEP]

(a) The correct visualized column was successfully predicted by our model. This visualization shows popular car brands.

[CLS] us school shooting total student ##es enrolled by year [SEP] total enrolled [SEP]

(b) This example was successfully predicted by our model. This visualization shows the total number of students enrolled.

[CLS] nba players performance top players and their performance [SEP] mp [SEP]

(c) This example was failed to predict visualized columns. This visualization shows top NBA players.

Fig. 11. Visualization of the BERT attention.

"Table" is a simplified version of our proposed method and used only the statistical features, that is, $\mathbf{y}_j = \mathbf{c}_j$. On the other hand, "BERT" represents the method that did not use statistical features derived from the table. Our proposed model outperformed all the baseline methods, suggesting high effectiveness of using the BERT model for predicting visualized columns. Since the prediction accuracy was slightly higher when used together with statistical features, textual and statistical features were found complementary. A Tukey HSD test found that the differences of all the pairs were statistically significant ($p < 0.01$).

Figure 11 illustrated the strength of attention from [CLS] in the output layer of the BERT model. Figures 11(a) and 11(b) show the case where a visualized column was successfully predicted by our method, and that header words relevant to the visualization intent received high attention. Figure 11(c) shows a failure case where the attention was given to only [SEP], probably because the relationship between the visualization intent and columns is unclear.

5   Conclusions

This paper proposed a visualization recommender system for tabular data with a visualization intent. The proposed method predicts the most suitable visualization type and visualized columns based on statistical features extracted from the tabular data, as well as semantic features derived from the visualization intent. To predict an appropriate visualization type, we proposed a BiDA model that identifies important table columns by the visualization intent, and important parts of the intent by table headers. To identify visualized columns, we employed BERT to encode both visualization intents and table columns, and estimate which columns are the most likely to be used for visualization. Since there is no available dataset for this task, we developed a new dataset consisting of over 100K tables and their appropriate visualization. The experiments revealed that our proposed methods accurately estimated suitable visualization types and visualized columns. Our future work includes prediction of more detailed settings for data visualization such as layouts and styles.

Acknowledgements. This work was supported by JSPS KAKENHI Grant Numbers JP18H03243 and JP18H03244, and JST PRESTO Grant Number JPMJPR1853, Japan.

References

1. Devlin, J., Chang, M.W., Lee, K., Toutanova, K.: Bert: pre-training of deep bidirectional transformers for language understanding. arXiv preprint arXiv:1810.04805 (2018)
2. Dibia, V., Demiralp, Ç.: Data2vis: automatic generation of data visualizations using sequence-to-sequence recurrent neural networks. IEEE Comput. Graphics Appl. **39**(5), 33–46 (2019)
3. Hu, K., Bakker, M.A., Li, S., Kraska, T., Hidalgo, C.: Vizml: a machine learning approach to visualization recommendation. In: Proceedings of the 2019 CHI Conference on Human Factors in Computing Systems, pp. 1–12 (2019)
4. Hwang, W., Yim, J., Park, S., Seo, M.: A comprehensive exploration on wikisql with table-aware word contextualization. arXiv preprint arXiv:1902.01069 (2019)
5. Kingma, D.P., Ba, J.: Adam: a method for stochastic optimization. arXiv preprint arXiv:1412.6980 (2014)
6. Liu, C., Han, Y., Jiang, R., Yuan, X.: Advisor: automatic visualization answer for natural-language question on tabular data. In: 2021 IEEE 14th Pacific Visualization Symposium (PacificVis), pp. 11–20. IEEE (2021)
7. Luo, Y., Qin, X., Tang, N., Li, G.: Deepeye: towards automatic data visualization. In: 2018 IEEE 34th International Conference on Data Engineering (ICDE), pp. 101–112 (2018)
8. Mackinlay, J., Hanrahan, P., Stolte, C.: Show me: automatic presentation for visual analysis. IEEE Trans. Visual Comput. Graphics **13**(6), 1137–1144 (2007)
9. Mackinlay, J.: Automating the design of graphical presentations of relational information. ACM Trans. Graph. (Tog) **5**(2), 110–141 (1986)
10. Moritz, D., et al.: Formalizing visualization design knowledge as constraints: actionable and extensible models in draco. IEEE Trans. Visual Comput. Graphics **25**(1), 438–448 (2018)
11. Pennington, J., Socher, R., Manning, C.D.: Glove: global vectors for word representation. In: Proceedings of the 2014 Conference on Empirical Methods in Natural Language Processing (EMNLP), pp. 1532–1543 (2014)
12. Roth, S.F., Kolojejchick, J., Mattis, J., Goldstein, J.: Interactive graphic design using automatic presentation knowledge. In: Proceedings of the SIGCHI Conference on Human Factors in Computing Systems, pp. 112–117 (1994)
13. Seo, M., Kembhavi, A., Farhadi, A., Hajishirzi, H.: Bidirectional attention flow for machine comprehension. arXiv preprint arXiv:1611.01603 (2016)
14. Stolte, C., Tang, D., Hanrahan, P.: Polaris: a system for query, analysis, and visualization of multidimensional relational databases. IEEE Trans. Visual Comput. Graphics **8**(1), 52–65 (2002)
15. Wongsuphasawat, K., Moritz, D., Anand, A., Mackinlay, J., Howe, B., Heer, J.: Voyager: Exploratory analysis via faceted browsing of visualization recommendations. IEEE Trans. Visual Comput. Graphics **22**(1), 649–658 (2015)

Existence Conditions for Hidden Feedback Loops in Online Recommender Systems

Anton Khritankov[✉] and Anton Pilkevich

Moscow Institute of Physics and Technology,
Dolgoprudny, 141700 Moscow Region, Russia
anton.khritankov@phystech.edu

Abstract. We explore a hidden feedback loops effect in online recommender systems. Feedback loops result in degradation of online multi-armed bandit (MAB) recommendations to a small subset and loss of coverage and novelty. We study how uncertainty and noise in user interests influence the existence of feedback loops. First, we show that an unbiased additive random noise in user interests does not prevent a feedback loop. Second, we demonstrate that a non-zero probability of resetting user interests is sufficient to limit the feedback loop and estimate the size of the effect. Our experiments confirm the theoretical findings in a simulated environment for four bandit algorithms.

Keywords: Hidden feedback loops · Recommender systems · Multi-armed bandits

1 Introduction

As research and applications of machine intelligence progress, more concerns are arising questioning whether the deployed AI systems fair, explainable, and ethical. Recommender systems are ubiquitous in social networks, streaming services, e-commerce, web marketing and advertising, web-search to provide personalized experience. Their algorithms use the new data from the system to improve their future performance. A positive feedback loop results in that only a small subset of available items is presented to the user. This effect is observed when user interests are being reinforced by previous exposure to specific items or item categories, thus producing self-induced concept drift.

The feedback loop effect is studied in real and model systems as an undesirable phenomenon related to reliable and ethical AI. Some of the consequences of the effect are induced shift in users interests [6], loss of novelty and diversity in recommendations [14], presence of "echo chambers" and "filter bubbles" [4,5], induced concept drift in housing prices prediction [8]. Nevertheless, a full description of the feedback loop effect and its existence conditions is still lacking for many cases [1,10].

© Springer Nature Switzerland AG 2021
W. Zhang et al. (Eds.): WISE 2021, LNCS 13081, pp. 267–274, 2021.
https://doi.org/10.1007/978-3-030-91560-5_19

In this paper we extend prior results [6] and explore existence conditions of feedback loops in presence of noise in the user behavior. We specifically consider noise as a random shift in user interests to items and categories the user is exposed to. In Sect. 3 we theoretically derive existence conditions for two noise models and explore our predictions empirically in Sects. 4 and 5 with four bandit algorithms in a simulated environment.

2   Related Work

Multi-armed bandits are commonly used in online recommendation systems and experiment design [3,7]. Earlier results demonstrate how feedback loops influence distribution of recommendations and user preferences. In [12] authors show that posterior distribution in Thompson Sampling algorithms is affected by the feedback loop, which worsens regret performance.

In [6] authors argue that a feedback loop would exist in a stochastic multi-armed bandit recommender system under some mild assumptions unless a set of available items does not grow at least linearly. Whereas noise and sudden shifts in user interest are important phenomena that occur in practice, they are not usually taken into account. An important contribution of this paper is that we show that a feedback loop would not occur even with a bounded set of available items if user interests are affected by specific kinds of random noise.

3   Problem Statement

3.1   A Recommender System Model

Let us consider a recommender system with a single user and available items $\mathbb{M} = \{1, .., M\}$. At time step t the system selects different items $A_t = (a_t^1, .., a_t^l)$ from the set of available items $l < M$ and presents them to the user. User interest to item a_i at step t is described by a function $\mu_t : \mathbb{M} \to \mathbb{R}^M$. Larger values of $\mu_t(a^i)$ correspond to stronger interest. After that the user examines the items and responds with $c_t = (c_t^1, .., c_t^l)$, $c_t^i \in \{0, 1\}$ sampled independently proportionally to the user interest in item $\mu_t(a^i)$. Therefore, we can model the response at step t with a random variable that has Bernoulli distribution: $c_t^i \sim Bern(\sigma(\mu_t(a_t^i))), \sigma(x) = 1/(1+e^{-x})$. In the simple case [6], the evolution of user interests abides to the monotonicity constraint for each element a^i with respect to t and user interest to item a_i is updated according to the rule:

$$\mu_{t+1}^i - \mu_t^i = \begin{cases} \delta_t, & \text{when } a_i \in A_t, c_t = 1, \\ -\delta_t, & \text{when } a_i \in A_t, c_t = 0, \\ 0, & \text{when } a_i \notin A_t, \end{cases} \tag{1}$$

where $\delta_t \sim \text{Uniform}[0, 0.01]$ indicates how much does the user interest change at step t. The initial user interest μ_0^i to item a_i is a random variable with a uniform distribution $\mu_0^i \sim \text{Uniform}[-1, 1]$. A single user assumption is justified when users act independently and recommendations for one user do not affect others.

Following Jiang et al. [6] we define a *positive feedback loop* as a situation in the behavior of a recommender system when a $2-$norm of user interests grows to infinity with step number t:

$$\lim_{t \to \infty} \|\mu_t - \mu_0\|_2 = \infty. \tag{2}$$

3.2   Multi-armed Bandit Problem Statement

Let us sate an online MAB recommendation problem. There is a set of M levers and an agent. Each lever has a probability distribution of reward associated with it. The agent can play $l < M$ levers and get rewards from each of the levers. The goal of the agent is to maximize the reward or, in other terms, minimize the regret, which is difference between the achieved and the maximum total reward. We correspond levers with available items a_i and rewards with user responses c_t^i at step t. User is the environment and the agent executes an item selection policy S. The optimization problem that the agent is trying to solve is $T \cdot l - \sum_{t=1}^{T} \sum_{i=1}^{l} c_t^i \to \min_S$. We study the following bandit algorithms.

Thompson Sampling [13]. We define Bernoulli random variables $\pi_t(\theta_1), .., \pi_t(\theta_M)$ that correspond to the winning $c_t^i = 1$ probability if a lever is played and initialize them at $t = 0$ via prior distribution $\theta_i \sim Beta(1, 1) = \text{Uniform}[0, 1]$. The posterior distribution conditioned on user responses is then given by $Beta(\alpha_t^i, \beta_t^i)$ for each $a^i \in M$. Distribution parameters α_t^i, β_t^i are updated for each $a_i \in A_t$ based on user response as

$$\alpha_{t+1}^i = \alpha_t^i + c_t^i, \beta_{t+1}^i = \beta_t^i + 1 - c_t^i. \tag{3}$$

ϵ-greedy [13]. With probability ϵ the policy decides to explore and selects l levers at random uniformly. With probability $1 - \epsilon$ it returns top l levers with the highest mean rewards d_t^i / n_t^i, d_t^i is total accumulated rewards for item a_i. Policy state is updated from the user feedback $c_t^i \in \{0, 1\}$ as follows:

$$n_{t+1}^i = n_t^i + \mathbb{I}\{a_i \in A_t\}, d_{t+1}^i = d_t^i + c_t^i. \tag{4}$$

Optimal. The policy selects l items with the highest user interest μ_t^i at each step. There is no internal state for the policy that needs to be updated. The policy is optimal in a sense that it knows the actual user interests and selects levers with the highest expected reward.

Random. The policy selects each item from M with the same probability $1/|M|$ at random. The policy does not have any internal state to update.

3.3   Additive Noise Model

Let us drop the assumption made in Sect. 3 that user interests μ_t^i is a known real vector with a stochastic update rule. Instead we model user interest in item a_i as a random function with known mean $\mu_t^i = \bar{\mu}_t^i + \omega_t^i$, where ω_t^i is an unbiased random noise, $\mathbf{E}\,\omega_t^i = 0$ and $\bar{\mu}_t^i = \mathbf{E}\,\mu_t^i$. The interest update rule (1) then becomes:

$$\bar{\mu}_{t+1}^i - \bar{\mu}_t^i = \delta_t c_t^i - \delta_t(1 - c_t^i), \text{ if } i \in A_t \tag{5}$$

$$\bar{\mu}_{t+1}^i - \bar{\mu}_t^i = 0 \text{ otherwise,}$$

where δ_t remains the same as in (1) and c_t^i now depends on the noisy user interest μ_t^i. We consider a case with uniform and bounded noise $\omega_t^i \sim$ Uniform$[-w, w], w > 0$. Then the response $c_t^i \sim Bern\left(\sigma(\bar{\mu}_t^i + \omega_t^i)\right)$. For further analysis we define a *constant best levers* condition when for all $t > t^0$ the set of selected arms remains the same $A_t = A_{t^0}$.

Statement 1 *Let the recommender system with noise (5) satisfy the constant best levers condition. Then*

$$\lim_{t \to \infty} \|\mu_t^i - \mu_0^i\|_2 = \infty, \forall w \geq 0, a_i \in A_{t^0} \tag{6}$$

As follows from this statement any bounded unbiased additive noise in a form of [5] does not prevent a feedback loop from occurring.

3.4 Interest Restarts Model

Users may sometimes lose or forget their interest to items. We call such event an *interest restart*. Restarts may be caused by satisfaction of the interest, change of users agenda outside of the system or introduction of a competing interest, disappearance of the item itself and other reasons. We consider a linear model of the aforementioned effect. When a restart occurs with probability q, then user interest to item a_i is replaced with a new random value that follows the same uniform initial distribution as μ_0^i, or is reduced by $0 \leq s \leq 1$:

$$\mu_{t+1}^i = \begin{cases} \mu_t^i + \Delta_t^i, \text{ with probability } 1 - q, \\ (1 - s)\mu_0 + s(\mu_t^i + \Delta_t^i), \text{ otherwise,} \end{cases} \tag{7}$$

where $\Delta_t^i = \mu_{t+1}^i - \mu_t^i$ as defined in (1). Note that now the update rule is not monotonic.

Statement 2 *Let the recommender system with restarts (7), (1) satisfy the constant best levers condition. If $\mu_0^i > 0$ then the expected user interest is bounded by*

$$\mathbf{E}\,\delta_t \left(\frac{1}{(1 - s)q} - 1\right) > \mathbf{E}\,\mu_t^i \gg 0. \tag{8}$$

4 Experiment

4.1 Experiment Design

We set up an empirical evaluation to test if our assumptions about system behavior are valid and to check whether the theoretical results hold in practical conditions. *RQ 1.* A positive feedback loop occurs even when an unbiased additive noise is added to user interests as demonstrated by Statement 1. *RQ 2.* The system does not need to exhibit exactly the constant best levers regime for results of Statement 1 and Statement 2 to hold. *RQ 3.* Results of Statement 1 and Statement 2 hold for a range of different lever selection polices.

We implement a recommender system [9] described in Sect. 3.1 and compare behavior different lever selection policies. We include Thomson Sampling (TS) and ϵ-greedy trainable selection policies, and Random and Optimal policies as baselines. Both latter policies do not require learning. The following parameters are varied during the experiment: number of items available for recommendation M, how many items are selected l, each of the algorithms parameters. We run the experiment for a maximum of T steps and repeat it with the same parameters in several trials to get an estimate of the confidence interval for results. For TS policy the priors are set to $\alpha_0^i = 1, \beta_0^i = 1, i = \{1, .., M\}$ to reflect the uniform distribution. The policy samples rewards from the prior $r_i \sim Beta(\alpha_t^i, \beta_t^i)$, selects l items with the highest sampled rewards A_t and after that updates the priors based on user feedback (3). The Random and Optimal policies do not require initialization. The Optimal policy returns l items with the highest interests A_t and is updated with new user interests $\bar{\mu}_t^i$ at each simulation step. The ϵ-greedy policy takes ϵ as a constant parameter set before the trial starts, with probability ϵ selects items at random uniformly or returns l items with maximum accumulated rewards. The policy is updated with user responses c_t at each step (4). For Additive noise model we set the amount of noise w parameter. In Interest restarts model we set a probability of restart q and scale s parameters.

The experiment proceeds as follows. A grid of experiment parameters with $l \leq M$ (see the experiment specification) is set up before start and random seed is fixed. Then tuples of parameters are retrieved from the grid and a number of trials is run. At each trial an experiment instance is initialized with parameters taken from the tuple. The selected user interests model and policy are initialized with corresponding parameters. Only one of the models and policies are used in each trial. At each step we store user feedback c_t, interests μ_t^i and the state of the policy: α_t^i, β_t^i for TS policy, d_t^i for ϵ-greedy policy.

4.2   Experiment Results

We show how the total reward changes for the Thompson Sampling (TS) selection policy when the step number $0 \leq t \leq 2000$ at Fig. 1 with different values of additive noise $w \in \{0.0, 0.3, 1.0, 3.0, 5.0, 10.0\}$. Note that the figure uses a logarithmic scale. The maximum value of user interest $\max_i \mu_t^i$ and thus amplitude of the feedback loop $\|\mu_t^i - \mu_0^i\|_2$ grows for all w and this does not contradict Statement 1. Results are averaged and confidence intervals are estimated over 30 runs. Although the growth rate decreases with w for all values of $l < M$ considered in the experiment. Thus for the parameters explored in the experiment we can confirm RQ 1 for TS selection policy.

We found that constant best lever assumption holds for Thompson Sampling (TS) t gets large $t > 1900$ and Optimal selection policies most of the time. While ϵ-greedy ($\epsilon = 0.1$) and Random policies do not exhibit the constant best lever

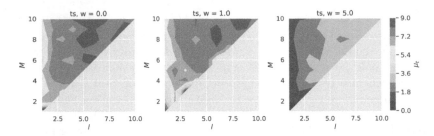

Fig. 1. Maximum user interest (color) for Thompson Sampling (TS) policy with different M and l (y- and x- axes) and strength of additive noise w at step $t = 2000$.

property. At Fig. 2 we compare different selection policies for the additive noise model: Thompson Sampling (TS), Random, ϵ-greedy, Optimal. The figure shows the user interest $\|\mu_t - \mu_0\|_2$ when $0 \leq t \leq 2000$. Results are averaged over and confidence intervals are shown for 30 runs. We can see that all policies exhibit a feedback loop when noise $w = 3.0$. The interest grows much slower for the Random selection policy but the system still exhibits a feedback loop. At Fig. 3 we show the amplitude of the feedback loop $\|\mu_t - \mu_0\|_2$ with respect to restart probability q and scale s. We also plot the expected upper bound predicted by 7 on the same figure. It can be seen that at higher q and s the bound becomes tight. When the restart probability is low, non-optimality of the selection policy and insufficient number of steps $T = 5000$ limits the growth of user interests. That is, the received reward is also bounded by $t \cdot \mathbf{E} \, \delta_t$.

With this we can confirm that Statement 2 even when the constant best levers condition does not strictly hold. We also explore the interest restarts model for different values of available $M \in \{1, .., 10\}$ and selected $l \in \{1, .., 10\}$ levers and obtained similar results. Thus, RQ 2 holds.

Results for different policies with the interest restarts model are also shown at Fig. 3. Results are averaged over and confidence intervals are shown for 30 runs. As we can see, the predicted upper bound 8 holds in all cases, although the Random policy demonstrates smaller growth in user interest than other policies, as expected. Considering the results we can confirm RQ 3 that Statement 1 and Statement 2 do not depend on the selection policy used.

5   Analysis and Limitations

In the experiment we find that additive noise does not prevent a feedback loop from occurring for several selection policies, and that interest restarts does limit the feedback loop. An upper bound is given by Statement 2 and it is tested in the experiments. Therefore, if an interest restart is possible in the system then the feedback loop is limited. The probability of restart and scale parameters should be estimated from the actual user behavior, which is a possible future direction of research. We find the constant best lever assumption useful for theoretical

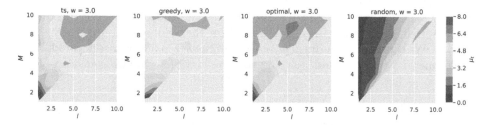

Fig. 2. Maximum user interest (color) for Thompson Sampling (ts), Random, Optimal and ϵ-greedy (greedy) selection policies. Axes: y-available items M, x- selected items l. Total number of steps $T = 2000$, strength of additive noise $w = 3.0$

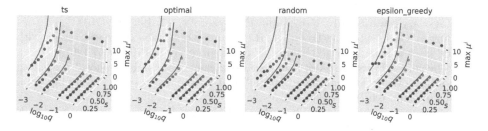

Fig. 3. Maximum user interest averaged over 10 runs (blue dots) with interest restarts model for Thompson Sampling (ts), Random, Optimal and ϵ-greedy (epsilon_greedy) selection policies. Green line—expected maximum user interest (8) for given scale and restart probability. Axes: y- scale parameter s, x- a logarithm of the restart probability $\log_{10} q$. Total number of steps $T = 5000$.

analysis as it greatly simplifies the proof. Results of the analysis are still valid when the assumption is violated in a feedback loop, as experiments show, because the feedback loop just reinforces the best lever already selected by the policy.

Limitations and Validity. We did not consider selection policies that expect data drift in user interests and, therefore, rewards. Non-stationary selection policies, such as Discounted Thompson Sampling (dTS) [2,11], might be specifically suitable in case of sudden changes in user interests, especially for the interest restarts model. Our study is limited to theoretical models, which parameters need to be estimated from the actual user behavior. Nevertheless, our results still hold and useful, because we specify how predictions depend on the parameters even when assumptions are relaxed.

6 Conclusions

We state a problem of existence of feedback loops in presence of noise in user interests. We explore unbiased additive noise model and demonstrate that such type of noise does not affect the existence of the feedback loop both theoretically and experimentally. We also develop an interest restart model that models cases when users partly or completely lose interest recommended items. For this model we show that there exists an upper bound on the feedback loop if a restart is possible. We confirm our findings in the experiment. Further research could focus on parameter estimation and studying the non-stationary multi-armed problem.

References

1. Adam, G.A., Chang, C.H.K., Haibe-Kains, B., Goldenberg, A.: Hidden risks of machine learning applied to healthcare: unintended feedback loops between models and future data causing model degradation. In: Machine Learning for Healthcare Conference, pp. 710–731. PMLR (2020)
2. Besbes, O., Gur, Y., Zeevi, A.: Stochastic multi-armed-bandit problem with non-stationary rewards. Adv. Neural. Inf. Process. Syst. **27**, 199–207 (2014)
3. Burtini, G., Loeppky, J., Lawrence, R.: A survey of online experiment design with the stochastic multi-armed bandit. arXiv preprint arXiv:1510.00757 (2015)
4. Colleoni, E., Rozza, A., Arvidsson, A.: Echo chamber or public sphere? predicting political orientation and measuring political homophily in twitter using big data. J. Commun. **64**(2), 317–332 (2014)
5. DiFranzo, D., Gloria-Garcia, K.: Filter bubbles and fake news. XRDS: Crossroads ACM Mag. Students **23**(3), 32–35 (2017)
6. Jiang, R., Chiappa, S., Lattimore, T., György, A., Kohli, P.: Degenerate feedback loops in recommender systems. In: Proceedings of the 2019 AAAI/ACM Conference on AI, Ethics, and Society, pp. 383–390 (2019)
7. Karimi, M., Jannach, D., Jugovac, M.: News recommender systems-survey and roads ahead. Inf. Process. Manage. **54**(6), 1203–1227 (2018)
8. Khritankov, A.: Hidden feedback loops in machine learning systems: a simulation model and preliminary results. In: Winkler, D., Biffl, S., Mendez, D., Wimmer, M., Bergsmann, J. (eds.) SWQD 2021. LNBIP, vol. 404, pp. 54–65. Springer, Cham (2021). https://doi.org/10.1007/978-3-030-65854-0_5
9. Khritankov, A., Pilkevich, A.: Implementation of online recommender with multi-armed bandits and feedback loops. https://github.com/prog-autom/bandit-loops. Accessed 30 Aug 2021
10. Krueger, D., Maharaj, T., Leike, J.: Hidden incentives for auto-induced distributional shift. arXiv preprint arXiv:2009.09153 (2020)
11. Raj, V., Kalyani, S.: Taming non-stationary bandits: A bayesian approach. arXiv preprint arXiv:1707.09727 (2017)
12. Riquelme, C., Tucker, G., Snoek, J.: Deep bayesian bandits showdown: An empirical comparison of bayesian deep networks for thompson sampling. arXiv preprint arXiv:1802.09127 (2018)
13. Slivkins, A.: Introduction to multi-armed bandits. arXiv preprint arXiv:1904.07272 (2019)
14. Ziarani, R.J., Ravanmehr, R.: Serendipity in recommender systems: a systematic literature review. J. Comput. Sci. Technol. **36**(2), 375–396 (2021)

Retrieval-Based Factorization Machines for CTR Prediction

Xu Wang[1], Yuancai Huang[1], Xiaokai Zhao[2], Weinan Zhao[2], Yu Tang[1(✉)], and Yitao Duan[2]

[1] Beihang University, Beijing, China
{wangxu,huangyc,tangyu}@act.buaa.edu.cn
[2] NetEase Youdao, Beijing, China
{zhaoxiaokai,zhaown,duan}@rd.netease.com

Abstract. Click-through rate (CTR) prediction is a crucial task for personalized services such as online advertising and recommender system. Many methods including Factorization Machines (FM) and complex deep neural models have been proposed to predict CTR and achieve good results. However, they usually optimize the parameters through a global objective function such as minimizing logloss and mean square error for all training samples. Obviously they intend to capture global knowledge of user click behavior, but ignore local information. Therefore, we propose a novel approach of Retrieval-based Factorization Machines (RFM) for CTR prediction, which enhances FM by the neighbor-based local information. During online testing, we also leverage the K-Means clustering technique to partition the large training set to multiple small regions for efficient retrieval of neighbors. We evaluate our RFM model on three public datasets. The experimental results show that RFM performs better than existing models including FM and deep neural models, and is efficient because of the small number of model parameters.

Keywords: Recommender systems · Factorization Machines · Nearest neighbor retrieval · CTR prediction

1 Introduction

It is crucial to predict a user's preference for candidate items for personalized services such as advertising [1] and recommender system [2]. The click-through rate (CTR) prediction can compute the probability that a user will click on an item, and is the core for these services since the probabilities are usually used to rank the items and select the top ones to users.

In the task of CTR prediction, the large-scale historical data often include many discrete and categorical features, such as gender, user/commodity id and location or demographics information, thus can be highly sparse and have complicated feature interactions. Generalized linear models are applied to predict CTR including logistic regression [3] and support vector machines [4], which are

© Springer Nature Switzerland AG 2021
W. Zhang et al. (Eds.): WISE 2021, LNCS 13081, pp. 275–288, 2021.
https://doi.org/10.1007/978-3-030-91560-5_20

difficult to capture high-order feature interactions. Then Factorization Machines (FM) [5] is proposed to model second-order feature interactions by inner product of latent vectors between different features and achieves very promising performance. Based on FM, Field-aware Factorization Machines (FFM) [6] divides features with different fields and extend FM with additional field-aware feature interactions. In addition, High-Order Factorization Machines (HOFM) [7] is presented to model high-order (more than 2) feature interactions.

In order to learn sophisticated feature interactions, deep neural networks are recently proposed to predict CTR [8–10]. Based on the feature embedding, the features are represented by one-hot vectors and embedded to low dimensional dense vector. Wide&Deep [11] feeds these embeddings to a combination of linear model and DNN to achieve both low-order and high-order feature interactions. DeepCross [12] uses a multi-layered residual network to prevent gradient explosion and vanishing problem when the depth of network increases. DeepFM [8] combines FM and DNN through sharing embeddings. NFM [9] obtains second-order feature interaction vectors by FM and feeds them into fully connected layers. AFM [10] applies attention mechanism to second-order interactions vectors, which models the importance of different interactions. HoAFM [13] encodes high-order feature interactions into feature representations in an explicit manner. PIN [14] extends FM with kernel product to tackle the insensitive gradient issue of DNN-based models. These deep learning methods can represent low-order and high-order feature interactions well and obtain good performance for CTR prediction.

However, these methods above [5,8–11] usually train the models and optimize the parameters through a global objective function. Obviously they intend to capture global knowledge of user click behavior, and ignore local information such as the similar samples. The local information has been considered in collaborative filtering based on the memory network [15]. But they only utilize user-item interaction information. In the CTR prediction, there are some contextual information that they miss such as user demographics, time, locations and so on.

In this work, we propose one novel approach of retrieval-based factorization machines (RFM) for CTR prediction, which enhances FM by retrieving similar samples from the training set as the neighbor-based local information during online testing. Specifically, firstly we train a standard FM model to get the sample representation of the second-order feature interaction embedding, and retrieve the similar samples from the training set for one given testing sample. Then we offline preprocess to partition the training set to multiple small regions by the clustering algorithm for efficiency. During online testing, we retrieve the most similar samples as neighbors from the most similar region. Finally, we enhance the FM model by fusing the neighbor-based local information and the original FM output via the weighted sum. We conduct extensive experiments on three public datasets to evaluate our RFM method. The results show that RFM outperforms FM and some existing studies. And it has the same number of trainable parameters with FM, which is much smaller than those of other studies. Therefore, RFM is an efficient and effective approach for CTR prediction.

2   Our Approach

In this section, we provide the approach of Retrieval-based Factorization Machines (RFM) for CTR prediction, which enhances FM with retrieved neighbor-based local information. As shown in Fig. 1, first we train a standard FM to obtain the global knowledge of user click behaviors (Sect. 2.1), and obtain the second-order feature interaction embeddings; second, we offline preprocess to partition the large training set to different small regions by K-Means clustering algorithm based on the second-order feature interaction embeddings (Sect. 2.2); third, during online testing, the small regions can be used to efficiently retrieve similar samples for one given testing sample and get the neighbor-based local information (Sect. 2.3). Finally, we enhance FM for predicting CTR by the weighted-sum fusion of the global and local information (Sect. 2.4).

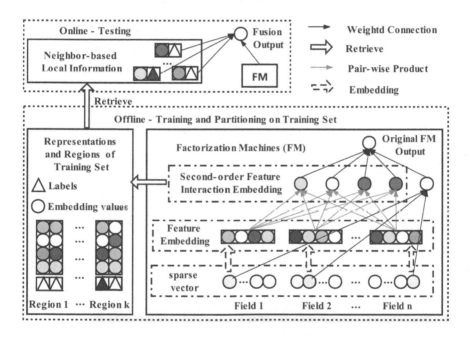

Fig. 1. The architecture of RFM

2.1   Factorization Machines

For the task of CTR prediction, the features of historical click behavior data typically have categorical fields (e.g. gender, commodity categories) and continuous fields after discretization (e.g. cost, age). These fields are usually converted to a set of binary features via one-hot encoding, making the original feature vectors highly sparse.

One common practice is encoding the sparse feature vectors to low dimensional dense vectors by embedding. Given one sample x with n fields, we map each field feature to an embedding vector $v_i \in R^d$, where d is the embedding size. Then we denote the output of the feature embedding as:

$$V(x) = [v_1, v_2, \ldots, v_n] \tag{1}$$

Besides the (first-order) feature embedding described above, the second-order feature interaction embedding is also widely used in FM-based neural models including NFM [9] and AFM [10]. These methods feed feature embedding $V(x)$ into the Bi-Interaction layer [9], and obtain the second-order feature interaction embedding as:

$$S(x) = \sum_{i=1}^{n} \sum_{j=i+1}^{n} x_i v_i \odot x_j v_j \tag{2}$$

where $\odot$ denotes the element-wise product of two vectors, that is $(v_i \odot v_j)_k = v_{ik} v_{jk}$.

Similar to existing FM-based neural models [8–10] for CTR prediction, we first train the feature embedding layer and the second-order feature interaction embedding layer. Instead of feeding the embeddings to upper neural models, we use them to build a standard FM model. Given the sample x as input, the predicted CTR is:

$$\widehat{y}_g(x) = b + \sum_{i=1}^{n} w_i x_i + \sum_{i=1}^{d} S(x)_i \tag{3}$$

where w_i represents the weights of field vectors and b is the bias. The third term can be reformulated [5] by:

$$\sum_{i=1}^{d} S(x)_i = \frac{1}{2} \cdot \sum_{f=1}^{d} \left(\left(\sum_{i=1}^{n} v_{if} x_i \right)^2 - \sum_{i=1}^{n} v_{if}^2 x_i^2 \right) \tag{4}$$

which reduces the computation complexity to $O(nd)$ and can be translated to matrix operation, which can be accelerated by GPU.

Based on Eq. 3, existing work [8,9,13,14] usually trains FM and optimizes model parameters through a global objective function (Eq. 11). Thus FM intends to capture the global knowledge of user click behaviors, and ignores the local information such as the most similar samples in the training set.

2.2   Region Partition

During online testing, for obtaining the local information of one given testing sample, we try to retrieve the similar samples from the training set as its neighbors based on the second-order feature interaction embedding (Eq. 2). However, the training set is often very large. Thus it will incur much overhead to directly compute the similarities among the training set. To solve this problem, we adopt K-Means [16] to offline partition the large training set to multiple small regions as preprocessing. Then we leverage these regions to accelerate the retrieval during online testing. The clustering algorithm runs only once and its result can be used for all testing samples.

Specifically, given all samples X in the training set and the i-th sample X_i, we get the representation of sample X_i based on the second-order feature interaction embedding by:

$$emb_i = BN\left(S\left(X_i\right)\right) \tag{5}$$

where we adopt batch normalization (BN) [17] to normalize the embedding $S\left(X_i\right)$ and keep the distribution of emb_i consistent. Similar to Eq. 4, we reformulate $S(X_i)$ to improve the efficiency as:

$$S\left(X_i\right) = \frac{1}{2} \cdot \left(\left(\sum_{j=1}^{n} v_j X_{ij}\right)^2 - \sum_{j=1}^{n} v_j^2 X_{ij}^2\right) \tag{6}$$

Based on the representation emb_i, we adopt the popular clustering algorithm K-Means [16] to partition all the samples in the training set to multiple regions. In the K-Means algorithm, we compute the Euclidean distance between sample representations and obtain k regions as follows:

$$C = \{c_1, c_2, \ldots, c_k\}, \quad U = \{u_1, u_2, \ldots, u_k\} \quad s.t. \ c_1 \cap c_2 \cap \ldots \cap c_k = \emptyset, \ c_1 \cup c_2 \cup \ldots \cup c_k = X \tag{7}$$

where C is the set of regions, and U is the set of center vectors for different regions. Each sample in region c_i is a tuple described as $(x, \ y)$, where x is the emb_i vector representation and y is the corresponding label. All the regions are disjoint, and their union is the whole training set. The hyper-parameter k is the number of regions.

After clustering, we partition all samples in the training set to k regions, and the center vectors u_i can represent the characteristics of all samples in one same region. We find the most similar region based on the center vectors and then retrieve the similar samples from the region. In this way, we reduce the computation complexity of retrieving.

2.3 Neighbor-Based Local Information

Given one testing sample online, we introduce an efficient approach to retrieve its similar samples as the neighbors based on the disjoint regions of the training set. Instead of computing the most similar samples directly from the large training set, we get the most similar sample region by calculating similarity between the center vectors of regions and the representation of the testing sample. Then we retrieve the most similar samples as neighbors from the region. Finally, we choose top-t ($t \geq 1$) neighbors with the most similarities to capture more local information, and adopt the similarity threshold r to filter out possible noisy neighbors.

Specifically, we first measure the similarity between sample X_i and X_j based on their representations by:

$$sim\left(emb_i, emb_j\right) = \frac{1}{1 + dist_{ed}\left(emb_i, emb_j\right)}, \quad dist_{ed}\left(emb_i, emb_j\right) = \|emb_i - emb_j\|_2 \quad (8)$$

where $dist_{ed}(emb_i, emb_j)$ represents the function computing Euclidean distance between emb_i and emb_j. The smaller the distance between two samples is, the higher the similarity between them is. Then we can get the most similar sample region as follows:

$$g = argmax_{j \in [1,k]} sim\left(u_j, emb_0\right) \quad (9)$$

where emb_0 is the representation of the given testing sample, and g is the index of the most similar sample region in the region set C.

We show how to retrieve top-t neighbors with the similarity threshold r from the region c_g in Algorithm 1.

Algorithm 1: The algorithm of neighbor retrieval

Input: c_g, emb_0, r, t
Output: the list of (y_i, sim_i):N
1 initialize a container $neighbors$;
2 **for** (x, y) *in* c_g **do**
3 $\quad$ $\alpha = sim(x, emb_0)$;
4 $\quad$ **if** $\alpha \geq r$ **then**
5 $\quad\quad$ add (y, α) to $neighbors$;
6 $\quad$ **end**
7 **end**
8 $N = selectTop(neighbors, t)$;
9 $return\ N$;

As shown in Algorithm 1, the threshold r is used to filter out possible noisy neighbors. The output N is a list of tuples (sim, y) containing the similarity and labels of neighbors. The function $selectTop$ selects top t similar samples from $neighbors$. The number of selected neighbors t also influences the performance. The impact of the similarity threshold r and the neighbor number t will be discussed in Sect. 3.

Compared with the global knowledge from all the training set, the retrieved neighbors are only a small subset. But they usually represent the common knowledge of these similar click behavior, which can be treated as the local information of the given testing sample.

2.4 Enhancing FM with Local Information

To improve the FM model, We fuse the retrieved neighbor-based local information N and the global information $\widehat{y}_g(x)$ provided in Sect. 2.1.

Specifically, we add the weighted sum of neighbor information to the original FM output, and normalize the result as follows:

$$\widehat{y}(x) = \frac{\widehat{y}_g(x) + \beta \sum_{i=1}^{t} y_i \cdot sim_i}{1 + \beta \sum_{i=1}^{t} sim_i} \tag{10}$$

where t is the number of retrieved neighbors, y_i and sim_i are the labels and similarities of neighbors, respectively. For balancing the global and the local effect, we add an factor β to control the effect of the local information, which can be manually tuned. However, we set β to 1 in our following experiments to keep the same importance between local and global information.

2.5 Training and Testing

Since the joint training for all the samples in the training set and their corresponding neighbors are very expensive, we only train a standard FM model and fuse neighbor-based local information and the original FM output during testing. In the training phase, we use one global objective function to update trainable parameters for standard FM. Then we offline partition the large training set to multiple small regions based on the second-order feature interaction embeddings. Finally, during online testing, we retrieve the neighbors and enhance FM by fusing the original FM output and the neighbor-based local information. We will provide the details of training and testing in our model, including objective function, dropout [18] and batch normalization [17].

In our task of CTR prediction, we adopt the widely used objective function square loss:

$$L_{reg} = \sum_{x \in X} (\widehat{y}_g(x) - y(x))^2 \tag{11}$$

where X represents the set of instances for training, and $y(x)$ represents the target of instance x.

In order to alleviate overfitting caused by a large d, we introduce dropout [18] for training. In the testing phase, dropout is disabled and all parameters are used for estimating $\widehat{y}_g(x)$.

As described in Sect. 2.2, we normalize the second-order feature interactions embedding vectors through batch normalization (BN) [17] to keep the distribution of emb_i consistent. BN is applied in both the training and testing phases in our RFM model.

3 Evaluation

In this section, we conduct extensive experiments to evaluate our RFM approach on three public datasets and analyze the effectiveness of neighbor-based local information.

3.1 Experimental Setup

We evaluate RFM on three public datasets: Frappe [19], MovieLens[1] [20] and Criteo[2], which are widely used in CTR prediction.

- **Frappe Dataset:** This dataset is often used in context-aware recommendation. It contains 96,203 app usage logs and eight categorical context variables except for user and app ID. We convert each log via one-hot.
- **MovieLens Dataset:** In personalized tag recommendation, it is often used. It contains 668,953 tag applications of users on movies. We also convert each tag application to a feature vector.
- **Criteo Dataset:** This includes 45 million click records, 13 continuous features and 26 categorical features. It has been widely used for the display advertising challenges. We discretize the continuous features by the tool[3].

For Frappe and MovieLens, we treat the log assigned a label of value 1 as "clicked", which means that the user has used the app under the context or applied the tag on the movie. We randomly select the logs representing that the user doesn't use the app or the tag is not applied on the movie as negative samples and assign -1 to their labels. Finally we get 288,609 and 2,006,859 samples, respectively. We randomly split each dataset into three parts: 70% for training, 20% for validation and 10% for testing. For Criteo, we make random sampling and get 550,088 samples. We also split them using the same ratios.

Based on our investigation about parameters in the validation set, we set the default values of parameters in our RFM. We set the embedding size d to 256 and the factor β to 1 by default. The default value of the similarity threshold r in Frappe and MovieLens is 0.8, and it is 0.2 in Criteo. The top-t values are set as 6, 1 and 11 by default in the datasets of Frappe, MovieLens and Criteo, respectively. The default value of the region number is 2 in Frappe and MovieLens, and it is 2^6 in Criteo. We use Adagrad [21] as the model optimizer.

3.2 Evaluation Metrics

Similar to the previous studies [8,11,12], we adopt the widely-used metrics of root mean square error (RMSE) to evaluate the performance. In addition, we use the number of trainable parameters (Param#) to measure the complexity of different models and the training efficiency. If a model has the smaller number of parameters, the training time will cost less.

3.3 Baselines

We compare RFM with the following competitive methods that are designed for sparse data and CTR prediction in recommender systems.

---

[1] https://grouplens.org/datasets/movielens/latest/.
[2] https://labs.criteo.com/2014/02/download-kaggle-display-advertising-challenge-dataset/.
[3] https://github.com/ycjuan/kaggle-2014-criteo.

- FM [5]: FM can effectively capture second-order feature interactions information. It is the infrastructure of many deep neural network models. We use the official C++ implementation[4] for FM.
- HOFM [7]: This is the enhanced version of FM, which can capture high-order feature interaction information. We use the TensorFlow implementation of the high-order factorization machines.
- Wide&Deep [11]: This model consists of wide component and deep component. The wide component is a linear regression model, and the deep component first concatenates embedding vectors and is followed by a MLP [11] to model feature interactions.
- DeepCross [12]: This model concatenates embedding vectors, followed by a multi-layered residual network. With the residual structure, the network can prevent gradient explosion and vanishing problem.
- DeepFM [8]: This model consists of one FM component and one deep component. It combines the power of factorization machines and deep learning to emphasize both low and high order feature interactions. Two components share the embedding parameters.
- HoAFM [13]: It uses a cross interaction layer to update a feature's representation by aggregating other co-occurred features, and performs a bit-wise attention mechanism on the granularity of dimensions.
- PIN [14]: This method extends FM with kernel product methods to learn field-aware feature interactions and adopts a feature extractor to explore feature interactions to tackle the insensitive gradient issue.

3.4 Performance Comparison

We compare the performance of our RFM method with different baselines. Table 1 summarizes the performance and the scale of trainable parameters on the embedding size 256.

Table 1. RMSE and number of parameters for different models. The symbol M means "million".

Model	Frappe		MovieLens		Criteo	
	RMSE	Param#	RMSE	Param#	RMSE	Param#
FM	0.3452	1.38 M	0.4735	23.24 M	0.4679	12.43 M
HOFM	0.3331	2.76 M	0.4636	46.40 M	0.4637	24.86 M
Wide&Deep	0.3661	4.59 M	0.5313	24.71 M	0.4703	23.58 M
DeepCross	0.4071	8.95 M	0.5130	29.01 M	0.4671	35.15 M
DeepFM	0.3305	4.15 M	0.4812	24.58 M	0.4648	27.33 M
HoAFM	0.3466	1.54 M	0.4518	23.37 M	0.4255	13.82 M
PIN	0.3729	2.10 M	0.5207	23.29 M	0.4221	49.55 M
RFM	**0.3225**	**1.38 M**	**0.4509**	**23.24 M**	**0.4113**	**12.43 M**

---

[4] http://www.libfm.org/.

According to Table 1, we have the following observations. RFM has the same scale of trainable parameters as FM. However, RFM performs better than FM by a 6.7% average improvement. This demonstrates the effectiveness of neighbor-based local information. HOFM performs better than FM, since it uses a separated set of embeddings to model high-order feature interactions. In Criteo, HOFM performs better very slightly or even worse. The reason is probably that although the high-order feature interactions can provide useful information, they also introduce noisy information simultaneously. Also, HOFM doubles the scale of parameters and incurs more training overhead. Wide&Deep and DeepCross take the feature embedding (Eq. 1) as the input of deep neural networks, which may miss second-order feature interaction information if the embedding parameters are not initialized by pretrained FM [9]. Thus both of them almost have the worst performance. Furthermore, Wide&Deep and DeepCross have the most parameters introduced by the neural networks. DeepFM combines the knowledge from the feature embedding (Eq. 1) and second-order interactions embedding (Eq. 2) in the FM component, as well as high-order feature interaction from the deep component with sharing embeddings. Thus it has a good performance in three datasets. However, sometimes it performs poorly in the MovieLens dataset, because MovieLens has only three fields and DeepFM may not capture enough useful feature interaction information. Besides, the deep component of DeepFM will lead to more training parameters and decreases the efficiency. HoAFM captures the high-order feature interactions in an explicit manner with the attentive FM, which is worse than RFM. PIN obtains a good performance in Criteo, but performs poorly in Frappe and MovieLens, which shows that sometimes the adaptive embeddings learned by the kernel product may be not effective.

Overall, our proposed RFM achieves the best performance among these models due to the neighbor-based local information. RFM also has the same number of parameters with FM, which can achieve the best training efficiency.

3.5 The Effectiveness of Local Information

We take three examples from the testing sets to qualitatively analyze the effectiveness of neighbor-based local information. Table 2 shows them. In addition, we also count the percentages of testing samples where neighbors correct or worsen the output of FM for three datasets.

Table 2. The effectiveness of neighbor-based local information.

Label	$\widehat{y}_g(x)$	Similarities	Y	Fusion
−1.0	0.0483	[0.8177,0.8064,0.8060,0.8048,0.8008]	[−1.0, −1.0, −1.0, −1.0, −1.0]	−0.7918
1.0	−0.0125	[0.8369,0.8358,0.8347,0.8346,0.8343]	[1.0,1.0,1.0,1.0,1.0,1.0]	0.8044
1.0	−0.0128	[0.8089,0.8074,0.8055,0.8053,0.8049]	[1.0,1.0,1.0,1.0,1.0,1.0]	0.7987

In Table 2, the first column is the ground-truth label of the given samples, the second is the original FM output of $\widehat{y}_g(x)$. The third column represents the similarities between given samples and their corresponding neighbors. The forth column is labels of neighbors. The last is the final output by fusing. Obviously, the original FM outputs $\widehat{y}_g(x)$ deviates from the true labels, which means the global knowledge learned by FM cannot model the click behavior correctly. However, the neighbors whose similarities are more than 0.8 can provide useful local knowledge with correct labels (the fourth column). Then we use such local knowledge to correct the $\widehat{y}_g(x)$ and obtain the final results (the last column). Intuitively, in the real scenario, original FM may predict that a user would not like to click one item because most of users in the training set dislike to click. But several other users who have similar characteristic with the user clicked the item, and the user also intends to click it with a high probability. In this way, the local information can represent the personal and preference knowledge and is effective to enhance the FM model.

Besides, we further record the percentages of three types of results from the testing sets. The type symbols "Better" and "Worse" mean that the fusion result is better or worse than $\widehat{y}_g(x)$, and the "Equal" represents that the fusion result is same as $\widehat{y}_g(x)$. In Frappe, "Better", "Worse" and "Equal" respectively are 63.96%, 34.86% and 1.18%. In MovieLens, they are 76.86%, 18.04% and 5.10%. And in Criteo, 63.17%, 0.96% and 35.87%. We can see that RFM corrects most of the mistakes of $\widehat{y}_g(x)$ in the testing sets, and has a small negative impact at the same time.

3.6 The Impact of Region Number

We explore the impact of the number of regions k with keeping others default. We partition samples in the training set for the efficiency and performance. It can not only accelerate the process of retrieving, but also introduce some better region features. The former increases the efficiency, and the latter improves the effectiveness. We assign 2^n to the region number k, where n is from 0 to 7 with step size 1.

Figure 2 shows the influence of the region number on the performance of RFM for different datasets. We can see that the region partition can influence the performance, and it may improve the performance in some extent with proper numbers of regions. Without region partition (i.e. $n = 0$ and the region number is 1), the model will utilize sample representations to retrieve neighbors by traversing the whole training set, and we find that the performance is always not the best. When partitioning samples to multiple regions, each region has its own center vector for representing the common characteristics of samples in it. Intuitively, the retrieved similar samples may be not the most similar ones from the whole training set when the region number is more than 1 and probably decrease the performance. However, we can see that the region partition can improve the performance as well, that is because the clustering technique can reveal the intrinsic nature and regularity in the training set [22,23], and the most similar samples retrieved from the same region may contain more effective and generalizable information than those from the whole training set.

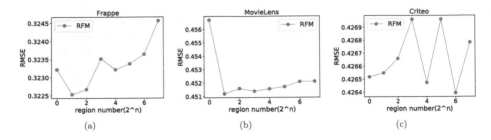

Fig. 2. RMSE for different region numbers

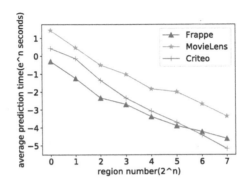

Fig. 3. Average prediction time for different region number

We also measure the efficiency of our RFM for different region numbers. In Fig. 3, we show the average prediction time (APT) for different region number, where APT is the average time to predict one sample in the testing set. For clarity, we take the natural logarithm of APT. As is shown in Fig. 3, the increase of the region number reduces APT roughly in a linear relationship, since the region partition can decrease the target samples for the retrieval by the rate of the region number. The region number is a trade-off between the performance and efficiency.

4   Conclusion

In this work, we propose Retrieval-based Factorization Machine (RFM) which enhances FM with neighbor-based local information. RFM partitions samples in the training set into multiple regions, retrieve neighbor-based local information from similar samples and fuses the local information and the original FM output. We have evaluated RFM on three public datasets, and the experimental results show that our RFM model outperforms existing models with a simple and efficient architecture. Besides, unlike the neural models, RFM is more explicable since the feature interaction of FM and the neighbor-based local information are easy to understand.

Acknowledgements. This work was supported by National Natural Science Foundation of China (No. 62072017, 61702024).

References

1. McMahan, H.B., et al.: Ad click prediction: a view from the trenches. In: SIGKDD, pp. 1222–1230 (2013)
2. Koren, Y., Bell, R., Volinsky, C.: Matrix factorization techniques for recommender systems. Computer **42**(8), 30–37 (2009)
3. He, X., et al.: Practical lessons from predicting clicks on ads at facebook. In: ADKDD@KDD, pp. 5:1–5:9
4. Chang, Y.-W., Hsieh, C.-J., Chang, K.-W., Ringgaard, M., Lin, C.-J.: Training and testing low-degree polynomial data mappings via linear SVM. J. Mach. Learn. Res. **11**, 1471–1490 (2010)
5. Rendle, S.: Factorization machines. In: ICDM, pp. 995–1000. IEEE Computer Society (2010)
6. Juan, Y.-C., Zhuang, Y., Chin, W.-S.: Field-aware factorization machines for CTR prediction. In: RecSys, pp. 43–50
7. Blondel, M., Fujino, A., Ueda, N., Ishihata, M.: Higher-order factorization machines. In: NIPS, pp. 3351–3359 (2016)
8. Guo, H., Tang, R., Ye, Y., Li, Z., He, X.: DeepFM: a factorization-machine based neural network for CTR prediction. In: IJCAI, pp. 1725–1731. ijcai.org (2017)
9. He, X., Chua, T.-S.: Neural factorization machines for sparse predictive analytics. In: SIGIR, pp. 355–364
10. Xiao, J., Ye, H., He, X., Zhang, H., Wu, F., Chua, T.-S.: Attentional factorization machines: learning the weight of feature interactions via attention networks. In: IJCAI, pp. 3119–3125. AAAI Press (2017)
11. Cheng, H.-T., et al.: Wide & deep learning for recommender systems. In: DLRS@RecSys, pp. 7–10. ACM (2016)
12. Shan, Y., Hoens, T.R., Jiao, J., Wang, H., Yu, D., Mao, J.C.: Deep crossing: webscale modeling without manually crafted combinatorial features. In: KDD, pp. 255–262. ACM (2016)
13. Tao, Z., Wang, X., He, X., Huang, X., Chua, T.-S.: Hoafm: A high-order attentive factorization machine for CTR prediction. Inf. Process. Manag. **57**(6), 102076 (2020)
14. Qu, Y., et al.: Product-based neural networks for user response prediction over multi-field categorical data. ACM Trans. Inf. Syst., 5:1–5:35 (2019)
15. Ebesu, T., Shen, B., Fang, Y.: Collaborative memory network for recommendation systems. In: SIGIR, pp. 515–524. ACM (2018)
16. Coates, A., Ng, A.Y.: Learning feature representations with K-Means. In: Montavon, G., Orr, G.B., Müller, K.-R. (eds.) Neural Networks: Tricks of the Trade. LNCS, vol. 7700, pp. 561–580. Springer, Heidelberg (2012). https://doi.org/10.1007/978-3-642-35289-8_30
17. Ioffe, S., Szegedy, C.: Batch normalization: Accelerating deep network training by reducing internal covariate shift. In: ICML. JMLR Workshop and Conference Proceedings, vol. 37, pp. 448–456. JMLR.org (2015)
18. Srivastava, N., Hinton, G.E., Krizhevsky, A., Sutskever, I., Salakhutdinov, R.: Dropout: a simple way to prevent neural networks from overfitting. J. Mach. Learn. Res. **15**(1), 1929–1958 (2014)

19. Baltrunas, L., Church, K., Karatzoglou, A., Oliver, N.: Frappe: understanding the usage and perception of mobile app recommendations in-the-wild. CoRR, abs/1505.03014 (2015)
20. Harper, F.M., Konstan, J.A.: The movielens datasets: history and context. ACM Trans. Interactive Intell. Syst. (tiis), 5(4), 1–19 (2015)
21. Duchi, J.C., Hazan, E., Singer, Y.: Adaptive subgradient methods for online learning and stochastic optimization. J. Mach. Learn. Res. 12, 2121–2159 (2011)
22. Jain, A.K.: Data clustering: 50 years beyond k-means. Pattern Recognit. Lett. 31(8), 651–666 (2010)
23. Gunopulos, D.: Encyclopedia of Database Systems, 2nd edn. Springer, New York (2018). https://doi.org/10.1007/978-1-4614-8265-9

Text Mining (1)

Adversarial Training for a Hybrid Approach to Aspect-Based Sentiment Analysis

Ron Hochstenbach[1], Flavius Frasincar[1], and Maria Mihaela Truşcă[2]($\boxtimes$)

[1] Erasmus University Rotterdam, Burgemeester Oudlaan 50,
3062 Rotterdam, PA, The Netherlands
frasincar@ese.eur.nl

[2] Bucharest University of Economic Studies, 010374 Bucharest, Romania
maria.trusca@csie.ase.ro

Abstract. The increasing popularity of the Web has subsequently increased the abundance of reviews on products and services. Mining these reviews for expressed sentiment is beneficial for both companies and consumers, as quality can be improved based on this information. In this paper, we consider the state-of-the-art HAABSA++ algorithm for aspect-based sentiment analysis tasked with identifying the sentiment expressed towards a given aspect in review sentences. Specifically, we train the neural network part of this algorithm using an adversarial network, a novel machine learning training method where a generator network tries to fool the classifier network by generating highly realistic new samples, as such increasing robustness. This method, as of yet never in its classical form applied to aspect-based sentiment analysis, is found to be able to considerably improve the out-of-sample accuracy of HAABSA++: for the SemEval 2015 dataset, accuracy was increased from 81.7% to 82.5%, and for the SemEval 2016 task, accuracy increased from 84.4% to 87.3%.

1 Introduction

With the ever-increasing popularity of the Web and subsequent increasing abundance of reviews on products or services available, a wealth of interesting information is available to businesses and consumers alike. But due to this increased quantity of reviews available, it becomes more and more infeasible or even impossible to analyse them by hand.

As such, sentiment analysis, concerned with the algorithmic analysis of expressed sentiment, can be of great value. In this research, we are interested in Aspect-Based Sentiment Analysis (ABSA), which entails determining the sentiment with respect to a certain aspect[18]. Specifically, we perform this task at the sentence level. In the sentence "The soup was delicious but we sat in a poorly-lit room" for example, a positive sentiment is expressed with regard to the aspect "Food quality", but the sentiment towards "Ambience" is negative.

[23] approaches this problem using a hybrid model, where first an ontology is used to assign a positive or negative sentiment towards an aspect in a rule-based

© Springer Nature Switzerland AG 2021
W. Zhang et al. (Eds.): WISE 2021, LNCS 13081, pp. 291–305, 2021.
https://doi.org/10.1007/978-3-030-91560-5_21

manner. When this method delivers contradicting or no results, an LCR-Rot neural network as described in [25], extended using representation iterations and called LCR-Rot-hop, is used. [22] further develops this LCR-Rot-hop model by incorporating contextual word embeddings and hierarchical attention. This new model, called LCR-Rot-hop++, delivers state-of-the-art results.

Research into neural networks is rapidly advancing, and novel techniques are developed continuously. One such technique proposed in [5] is Generative Adversarial Network (GAN). Here, two networks are simultaneously trained: a generator tries to generate new input samples, whereas a discriminator tries to discern between real and generated samples. These conflicting objectives converge to a situation where the generator produces samples indiscernible from real data. Furthermore, training a neural network as the discriminator in an adversarial network can increase the robustness of the trained model.

[7] gives an overview of contributions of GANs to the domains of affective computing and sentiment analysis. These mostly fall into the field of image generation, however, which has the benefit that input data is of fixed size. This is not the case in text analysis, as sentences can be of any length. This problem has been worked around in the, so far (to the best of our knowledge), only research where GAN is applied to ABSA in [10]. In this paper, not entire new instances are generated, but rather perturbations are made to real data aimed at increasing loss. These are then used to train a BERT Encoder [24], leading to increased accuracy with respect to the baseline model.

This research contributes to the literature in two ways. First, we will develop a method where adversarial samples are fully generated, rather than obtained from perturbing existing samples as in [10]. Second, HAABSA++ is a more sophisticated methodology with respect to the specifics of the ABSA task than the BERT Encoder used in [10], making it interesting to investigate whether we can achieve a similar increase in accuracy when applying adversarial to HAABSA++.

This research is structured as follows. First, Sect. 2 gives a brief overview of the related works on ABSA and adversarial training. In Sect. 3 the datasets used are described. We discuss the HAABSA++ framework in Sect. 4, as well as how we apply adversarial training to it. Then, Sect. 5 presents the results of the proposed methods. Last, in Sect. 6 we draw conclusions and give suggestions for further research.

2   Related Work

[18] provides a comprehensive survey on Aspect-Based Sentiment Analysis (ABSA) based on three types of methods: knowledge-based, machine learning, or hybrid approaches. In ABSA, one tries to find the sentiment expressed towards a given or extracted aspect in a sentence or full review (in this research we focus on sentences). Whereas ABSA also comprises tasks like target extraction (i.e., the selection of a target word indicative of a certain aspect) and aspect detection (i.e., the detection of aspects towards which a sentiment is expressed in a text), in this research we follow [22] and [23] in focusing on the sentiment classification of sentences in which aspects are explicitly stated in so-called targets.

For example, in the sentence "The food was bland but at least the waiter was allright.", "food" is a target for the aspect "Food Quality", whereas "waiter" serves as a target for the aspect "Service". Towards the former, a negative aspect is expressed, and towards the latter the sentiment is neutral.

In [23], the sentiment classification task of ABSA is addressed by means of a hybrid approach. Here, first a knowledge-based ontology method is used to classify the sentiment towards a target, and a neural network based on [25] is used as back-up when this fails. [22] extends upon this work by adding two features to the neural network part: contextual word embeddings are used rather than non-contextual ones, and a hierarchical attention mechanism is applied. Compared to other methods for ABSA, HAABSA++ is found to provide state-of-the-art results when applied to the SemEval 2015 [14] and SemEval 2016 [15] datasets.

The field of machine learning is in rapid development, and one particular advancement sparking great interest was proposed in [5]. GANs are trained by pitting a generative model against a discriminator. The generator's task is to generate realistic samples when compared to an empirical dataset. At the same time, the discriminator is trained to discern between generated and real samples. These conflicting objectives lead to a minimax game which [5] shows to converge to a situation where generated samples are indistinguishable from real samples.

[7] provides an overview of applications of adversarial training to the field of sentiment analysis. Here, the authors note four benefits of adversarial training. First, it allows for generated emotions to feel more natural than those generated through other techniques. Second, it aids in overcoming the sparse availability of labeled sentiment data. Although plenty of images, speech, and text samples exist, only a small fraction of these are labeled and thus suitable for supervised learning. Third, GANs have the potential to learn in a robust manner, reducing the problems occurring when samples are gathered from different contexts. For example, the word "greasy" related to the domain of restaurant reviews can be understood as positive in a fast-food restaurant, but not in a fine dining establishment. Last, GANs allow for easy quality evaluation of the generated samples. Whereas in prior methods for sample generation, humans needed to assess the quality of the samples, the discriminator built into the GAN framework automates this task. Most of the advancements to GANs listed in [7] focus on generating as realistic as possible samples which are highly relevant for, e.g., image generation, but not as much for the analysis of written reviews.

[13] proposes a semi-supervised GAN, where the discriminator simultaneously acts as a classifier. Namely, whereas a regular discriminator is only given the task to discern between 'Fake' and 'Real' data, in this application the 'Real' class is divided into the different classes of the original task. Such a model is commonly referred to as a Categorical Generative Adversarial Network (CatGAN). When applied to the MNIST dataset (where the discriminator discerns between 'Fake', '0', '1', ... , '8' and '9'), this provided a significant improvement in performance compared to a non-adversarially trained convolutional neural network. Applications of CatGANs are also found in [20] and [17], where again the MNIST dataset is used for testing. In the former, the semi-supervised

CatGAN was found to outperform most competing models, only marginally lagging behind the best performing model. Furthermore, CatGAN outperformed competing GAN-based models by great margins, a fact also revealed in the study of [17].

An application of CatGAN to ABSA is found only recently in [10], but it is not applied in the traditional sense with an explicit generator and discriminator. Rather, real samples are classified by a BERT Encoder (the network intended to train adversarially). The gradient of the loss is then computed with respect to the initial input data, and based on this the original samples are perturbed. These samples are then again fed to the BERT Encoder, and the total loss is computed based on the combined loss of the perturbed and original samples. Although there is no explicit generator or discriminator here, the BERT Encoder is thus adversarially trained in the sense that samples generated with the explicit aim of being difficult to classify, are fed to it. This methodology outperforms the (non-adversarially trained) BERT-PT model [24] when applied to the SemEval 2014 dataset.

3   Data

For our research, we use data posed for the SemEval 2015 and SemEval 2016 contests. Intended to evaluate methods developed for ABSA, these sets contain English review sentences for restaurants. For each of these sentences, one or multiple aspect categories are denoted together with a negative, neutral or positive sentiment.

Table 1. Sentiment frequencies in the used datasets

	SemEval 2015				SemEval 2016			
	Negative	*Neutral*	*Positive*	Total	*Negative*	*Neutral*	*Positive*	Total
Train data	3.2%	24.4%	72.4%	1278	26.0%	3.8%	70.2%	1879
Test data	5.3%	41.0%	53.7%	597	20.8%	4.9%	74.3%	650

Table 1 shows the distribution over sentiment polarities in the considered datasets. Here, we note that for both data sets, 'Positive' polarities are most abundant. Both data sets have a clear minority class. For SemEval 2015, this is the 'Negative' class, whereas for SemEval 2016 this is 'Neutral'. For the SemEval 2015 data set, a larger disparity between the relative frequencies among the polarities can be noted between the train and test data compared to SemEval 2016. Furthermore, we did not notice any remarkable discrepancies between the aspect category frequencies in the train and test sets.

We directly apply the datasets as obtained in [23] after preprosessing. First, the authors deleted all sentences where sentiment is implicit, as the applied methodology requires an explicitly mentioned target. For 2015, this amounted to 22.7% of the train data, and 29.3% of the test data. For 2016, this was 25.0%

and 24.3%, respectively[1]. They then processed the remaining sentences using the NLTK toolkit [2], and, last, tokenized the data and lemmatized the words with the WordNet English lexical database [12].

4   Methodology

In this section, the methodology we employ for sentiment classification of restaurant reviews is discussed. First, we will describe the original HAABSA++ algorithm in Sect. 4.1. Then, Sect. 4.2 puts forward the methodology for the CatGAN and elaborates on how to adapt the standard HAABSA++ algorithm to serve as the discriminator in this methodology. Furthermore, this section describes the training procedure.

4.1   HAABSA++

HAABSA++ is a hybrid approach to aspect-based sentiment analysis where, as proposed in [23], first an ontology is used to determine the sentiment expressed towards a given aspect. If this proves inconclusive, a neural network proposed in [23] and improved upon in [22] is used as back-up.

Ontology. The ontology is based upon [19] and consists of three classes. First, *SentimentMention* contains the expressions of sentiment. It consists of three subclasses, the first of which contains words always expressing the same sentiment value regardless of the aspect, the second containing words always expressing the same sentiment that only adhere to specific aspects, and the third containing words for which the sentiment they express depends on the associated aspect. The *AspectMention* class governs the aspect to which a word adheres, and the *SentimentValue* class labels words as expressing either a positive or negative sentiment. For example, the word 'expensive' could fall into the *Price Negative Sentiment* subclass of the *SentimentMention* class, the *Price Mention* subclass of the *AspectMention* class, and the *Negative* subclass of the *SentimentValue* class. Aspect sentiment is then determined by assessing whether in a given sentence, a word falls into the *AspectMention* class of a given aspect, and if so, in what *SentimentValue* subclass it falls.

The ontology-based method is powerful but might prove inconclusive whenever both negative and positive sentiment are detected for an aspect, or whenever the ontology does not cover the lexicalizations present in a sentence. In these cases, we resort to a back-up neural network approach.

Multi-Hop LCR-Rot Neural Network with Hierarchical Attention and Contextual Word Embeddings. [22] further develops the Multi-Hop LCR-Rot-hop neural network used as back-up in [23] into LCR-Rot-hop++ by adding contextual word embeddings and a hierarchical attention mechanism. Although

---

[1] All values in Table 1 are excluding implicit targets.

they compare multiple ways of implementing these additions, we only consider the BERT contextual word embeddings [3] as these are found to perform best out-of-sample in our datasets. Furthermore, we use the last method out of the four methods for hierarchical attention introduced in [22], as it performs best on average. According to the fourth method, the last layer of attention is applied repetitively and separately for contexts and targets.

First, the words in a sentence are turned into word embeddings using the contextual BERT model. Here, first token vectors that are unique for each word, position embeddings showing the word's location in the sentence, and segment embeddings discerning between multiple sentences present, are averaged. The newly computed embeddings feed into a Transformer Encoder. The entire network is pretrained on the Masked Language Model and Next Sentence Prediction tasks. This method ensures that a word's embedding depends on how it is used in the sentence. For example, the word 'light' could refer to an object's mass or to the visible light coming from the sun or a lamp, depending on the context.

Then, the embedded sentences are split into three parts: Left (with respect to the target), Target, and Right (with respect to the target). Each of these feeds a bi-directional LSTM layer. These outputs are then used to create four attention vectors: a target2context and a context2target for both the left and right contexts. This is done using a two-step rotary attention mechanism, where in each iteration the intermediate context and target vectors are weighted using an attention score computed at the sentence level. This weighting is done separately for the two target vectors and the two context vectors. The four resulting vectors are then concatenated and subjected to another attention layer, which computes attention weights separately for contexts and targets. This process is repeated for a specified number of hops, the final output of which is used as input for a multi-layer perceptron determining the final sentiment classification.

4.2   Classifying Generative Adversarial Network

As described before, a promising advancement in the field of deep learning is the Generative Adversarial Network. First proposed in [5], a GAN generates new data samples by simultaneously training a generative and a discriminative model. While the generative model produces new samples based on a random input vector, the discriminative model tries to discern between the generated samples and real data.

This section first describes how to adapt the neural network in HAABSA++ so as to constitute both a classifier and a discriminator (i.e., a CatGAN). This added objective should increase the robustness of the model, as it is forced to more clearly recognize what input characteristics determine the class to which an instance belongs. This can be made more apparent through an example. Imagine a child tasked with labeling images of vehicles as either 'plane', 'train', or 'car'. It might then be fairly easy for the child to recognize cars, as only they have tires, but such a strategy will lead him to wrongly characterize planes on the ground (with their landing gear out) as cars. If, however, every now and then an image of a bicycle is given to the child, this challenges him to notice that it is not the characteristic "having tires" by which cars can be recognized, but rather "having four tires". This realization will in turn lead him to not label grounded planes as cars anymore, as they have more than four tires.

In the first subsection, we describe the model, and the second subsection gives its implementation details. Then, we present the training procedure and discuss convergence in GAN training. The last subsection describes the procedure for hyperparameter optimization.

Model Formulation. [17] shows how the loss function of the regular GAN model can be adapted so the discriminator simultaneously serves as a classifier, as such constituting a Categorical GAN (CatGAN). Instead of discriminating between real or fake data, the discriminator now discriminates between K+1 classes. In our case, K=3 and the first three classes are 'Negative', 'Neutral' and 'Positive'. The K+1th class is 'Fake'. This yields the loss function:

$$
\begin{aligned}
L_{G,D} = & -\mathbb{E}_{\vec{x},y \sim p_{\text{data}}(\vec{x},y)} \log[p_{\text{model}}(y|\vec{x}, y < K+1)] + \lambda(||\Theta_G||^2 + ||\Theta_D||^2) \\
& -\mathbb{E}_{\vec{z} \sim P(\vec{z})} \log[p_{\text{model}}(y = K+1|\vec{z})],
\end{aligned}
\tag{1}
$$

resulting in the following optimization problem to solve:

$$
\max_G \min_D L_{G,D}.
\tag{2}
$$

In Eq. 1, the first two terms correspond to the loss term as posed in [23][2]. The last term corresponds to the the samples generated by the generator, which are labeled to the new, 'Fake' class ($y = K + 1 = 4$). Furthermore, $y = 1$ means 'Negative', $y = 2$ is 'Neutral' and $y = 3$ is 'Positive'. This loss function can be split into loss functions for the generator and discriminator, respectively, as follows:

$$
L_G = -\mathbb{E}_{\vec{z} \sim P(\vec{z})} \log[p_{\text{model}}(y = K+1|\vec{z})] + \lambda||\Theta_G||^2
\tag{3a}
$$

$$
\begin{aligned}
L_D = & -\mathbb{E}_{\vec{x},y \sim p_{\text{data}}(\vec{x},y)} \log[p_{\text{model}}(y|\vec{x}, y < K+1)] + \lambda||\Theta_D||^2 \\
& -\mathbb{E}_{\vec{z} \sim P(\vec{z})} \log[p_{\text{model}}(y = K+1|\vec{z})].
\end{aligned}
\tag{3b}
$$

Note here that the last term of Eq. 1 is included in both the generator and discriminator loss functions, as fake samples are first processed by the generator and then labeled by the discriminator. Furthermore, one could rewrite $\log[p_{\text{model}}(y = K + 1|\vec{z})]$ as $\log[1 - D(G(\vec{z}))]$ and $\log[p_{\text{model}}(y|\vec{x}, y < K + 1)]$ as $\log[D_y(\vec{x})]$, where $D(\vec{x}) = \sum_{y=1,2,3} D_y(\vec{x})$ represents the probability that the argument $\vec{x}$ stems from the real data. $D_y(\vec{x})$ in turn represents the probability that the argument stems from the real data and has label y.

---

[2] Note that $\lambda||\Theta_G||^2$ and $\lambda||\Theta_D||^2$ correspond to L2-regularization terms and are not included in [17].

Written explicitly as in [23], Eq. 1 becomes:

$$L_{G,D} = -\sum_{j \in \mathbb{J}} \vec{y}_j \times \log(\hat{p}_j) - \sum_{i \in \mathbb{I}} \vec{y}_i \times \log(\hat{p}_i) + \lambda(||\Theta_G||^2 + ||\Theta_D||^2). \quad (4)$$

Here, $j \in \mathbb{J}$ is the batch of real data samples. $\vec{y}_j$ is the real sentiment of the jth sample written in vector form, i.e., when $y_j = 1$, $\vec{y}_j = [1, 0, 0, 0]$, when $y_j = 2$, $\vec{y}_j = [0, 1, 0, 0]$ etc. $\hat{p}_j$ is the predicted sentiment of the jth sample. $i \in \mathbb{I}$ are the generated samples, and $\hat{p}_i$ is the predicted sentiment for the ith generated sample. Note furthermore that $\vec{y}_i$ is always equal to $[0, 0, 0, 1]$. Θ_G and Θ_D contain the parameters for the generator and discriminator, respectively.

A problem occurs, however, when a generator is tasked with forming sentences. Constituted by a neural network, the generator can only output vectors and thus sentences of fixed length. This poses no problem in classical applications of GANs, as for example images generated always have the same resolution and thus output length. Sentences have variable length, however, and the discriminator can therefore possibly distinguish between generated and real samples based on their length. As such, we here only generate the four LCR-Rot-hop++ representation vectors of length $2d$ each, where d is the dimensionality of the embedding. Then, we use the final MultiLayer Perceptron (MLP) of the LCR-Rot-hop++ network simultaneously as classifier and discriminator. The only modifications necessary are then to adapt the loss function to Eq. 4, and let the MLP output a vector of probabilities of length $K + 1$ rather than K.

The generator is constituted by a fully connected, 4-layer MLP. It uses an r-dimensional random vector as input, the first hidden layer will have $2 \times d$ neurons and the second hidden layer will have $6 \times d$ neurons. It then outputs a vector of length $8d$ (according to the LCR-Rot-hop++ model, the four vectors of length 2d are concatenated, resulting a vector of length 8d). This choice of architecture is the result of a trade-off between model complexity and speed of training. Then, the discriminator is constituted by the final MLP layer of the Multi-Hop LCR-Rot++ network in HAABSA++. A schematic of this new algorithm, called HAABSA*, is given in Fig. 1.

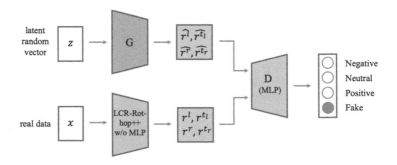

Fig. 1. HAABSA* structure

Implementation Details. Following [23] and [22], we run the optimization algorithm for 200 iterations. For each iteration, a batch of $m = 20$ real samples, as well as 20 fake samples, is selected. We initialize all weights randomly following a U(-0.01,0.01) distribution, and all biases are initialized as zero. To ensure all parameters are included in the regularization terms, Θ_D includes both the parameters in the final MLP layer and the parameters in the LCR-Rot-hop++ network w/o MLP. Θ_G includes the parameters of the generator. The random input for the generator follows a U(0,1) distribution and is of dimension $r = 100$. As in [22], the dimensionality of the word embedding $d = 768$. Last, it is important to note that that, like [23] and [22], we always train the adversarial network on the full training set. This may result in bias, as for the test set, only samples not classified by the ontology are classified by the neural network. However, the increased performance due to the larger training set size outweighs this downside. Our methods are implemented in `Python` using the `TensorFlow` platform. All code can be found at https://github.com/RonHochstenbach/HAABSAStar.

Training Procedure and GAN Convergence. The training procedure for HAABSA*, based on the one put forward in [5], is shown in Algorithm 1. In our implementation, the discriminator is updated every iteration, whereas the generator is updated only every k^{th} iteration [5].

[5] proves the minimax game given in Eq. 2 has a theoretical optimum under two conditions. First of all, the generator and discriminator need enough capacity to, respectively, generate realistic samples and discern them effectively. No formal method to ascertain sufficient capacity is given, however. Second, the discriminator should be allowed to reach its optimum given an instance of the generator, before the generator is updated again. In Algorithm 1, this can be (amongst other later to be discussed options) controlled by the factor k.

However, in [6], it is explained that this only holds true for the convex case. Due to the minimax nature of the game in which a saddle-point is the optimum, convergence is not guaranteed. Namely, the updating procedures for both networks have conflicting aims, and updates with respect to, for example, Θ_G might counteract gains made in prior updates to Θ_D. This means it could also occur that the generator, instead of increasing the loss function, makes the bound in which the discriminator can update looser. This results in the algorithm oscillating instead of converging. [17] describes this problem as well, and further notes that when either the generator or discriminator becomes too good at its task, this prevents the other from learning further. For example, when the discriminator perfectly identifies all fake samples, the generator has no reference for how to change its tactic in order to make samples capable of being misclassified as real by the discriminator.

Hyperparameter Optimization. [17] mentions several intricate possible techniques that could increase the probability of convergence, but careful selection of hyperparameters could also at least prevent divergence. [22] considers multiple hyperparameters, two of which could aid in preventing divergence.

Algorithm 1: HAABSA* training procedure

---

for *number of training iterations* **do**

    if *iteration number divisible by k* **then**

        – Sample minibatch of m noise samples $\{\vec{z}^{(1)}, \ldots, \vec{z}^{(m)}\}$ from noise prior $p_{\text{noise}}(\vec{z})$

        – Sample minibatch of m examples with their associated labels $\{(\vec{x}^{(1)}, y^{(1)}), \ldots, (\vec{x}^{(m)}, y^{(m)})\}$ from dataset $\mathcal{X}$

        – Update the generator by ascending the stochastic gradient of the loss function with respect to Θ_G:

$$\nabla_{\Theta_G} - \frac{1}{m} \sum_{i=1}^{m} [\log(1 - D(G(\vec{z}^i)))].$$

        – Update the discriminator by descending the stochastic gradient of the loss function with respect to Θ_D:

$$\nabla_{\Theta_D} - \frac{1}{m} \sum_{i=1}^{m} [\log(D_y(\vec{x}^i)) + \log(1 - D(G(\vec{z}^i)))].$$

    else

        – Sample minibatch of m noise samples $\{\vec{z}^{(1)}, \ldots, \vec{z}^{(m)}\}$ from noise prior $p_{\text{noise}}(\vec{z})$

        – Sample minibatch of m examples with their associated labels $\{(\vec{x}^{(1)}, y^{(1)}), \ldots, (\vec{x}^{(m)}, y^{(m)})\}$ from dataset $\mathcal{X}$

        – Update the discriminator by descending the stochastic gradient of the loss function with respect to Θ_D

    end

end

First, the learning rate of the employed momentum optimizer governs the pace at which descent takes place in the gradient direction. Whereas too small values might prevent reaching of the optimum in the given number of iterations, smaller values prevent one of the networks from becoming too good too quickly for the other to keep up. Second, the momentum term governs the extent to which gradient values of previous iterations are considered [21]. Lower values for this might, in a similar manner as with the learning rate, prevent divergence.

[22] also considers the L_2-regularization term and the dropout probability (which has been applied to all hidden layers of the network to prevent overfitting) as hyperparameters. Since, however, the optimal value for these was not found to differ much compared to [23], and they do not directly affect the convergence of the minimax game, we leave them at the values found in [22] (i.e. 0.0001 and 0.3 for the L_2-regularization term and keep probability, respectively).

Preliminary trial runs showed that higher values for the learning rate and momentum term led the generator to cause divergence. However, lower values impeded the discriminator from learning. As such, we use different values for the learning rate and momentum term for the generator and discriminator, writing the learning rate and momentum term for the generator as a function of those terms for the discriminator as follows: $lr_{gen} = \mu_{lr} \times lr_{dis}$ and $mom_{gen} = \mu_{mom} \times mom_{dis}$. Here, μ_{lr} and μ_{mom} are multiplier terms for the learning rate and momentum, respectively. We will then consider lr_{dis}, mom_{dis}, μ_{lr} and μ_{mom} as hyperparameters for optimization.

Last, we treat the value of k as a hyperparameter. As this value determines how much more often the discriminator is updated then the generator, it can be of great importance in ensuring one of the networks does not learn too much more quickly than the other.

For the hyperparameter optimization, we use a Tree-structured Parzen Estimator (TPE) approach [1]. This method has a higher convergence speed than other methods like grid search. It optimizes by, for each run, selecting hyperparameters from the options given by maximizing the ratio of fit values of the best set of hyperparameters to the fit values of the previously unconsidered options for hyperparameters.

We run the hyperparameter optimization for the 2015 and 2016 datasets separately. The considered options for the learning rate and momentum term of the discriminator are based upon the found values in [22]. To ensure a manageable runtime, we limit the algorithm to 20 runs. The algorithm is run on datasets obtained by randomly splitting the training set of the given year into 80% train and 20% test data.

5   Results

In this section, we discuss the results obtained from the adversarial training of HAABSA++. First, we provide a brief discussion of the results of hyperparameter optimization in Sect. 5.1. Then, Sect. 5.2 puts forward the obtained accuracies using HAABSA*. Furthermore, we compare these with the accuracies of HAABSA++ and other competing models for ABSA.

5.1   Results Hyperparameter Optimization

To decrease the risk of divergence, we include different values for the learning rate and momentum term of the generator and discriminator. Whereas trial runs with equal learning rates or momentum terms for the generator and discriminator diverged on numerous occasions, for different learning rates and momentum terms the hyperparameter optimization only one out of the total of 40 runs (20 for each dataset) diverged. The results of the hyperparameter optimization are shown in Table 2.

Table 2. Hyperparameter optimization results

Hyperparameter	Considered values	Optimal value	
		2015	2016
lr_{dis}	[0.007, 0.01, 0.02, 0.03, 0.05, 0.09]	0.02	0.03
mom_{dis}	[0.7, 0.8, 0.9]	0.9	0.7
μ_{lr}	[0.1, 0.15, 0.2, 0.4]	0.1	0.15
μ_{mom}	[0.4, 0.6, 0.8, 1.6]	0.4	0.6
k	[3, 4, 5]	3	3

Table 3. Accuracies of HAABSA++ and HAABSA*

		SemEval 2015		SemEval 2016	
		in-sample	*out-of-sample*	*in-sample*	*out-of-sample*
w ontology	HAABSA++	88.8%	81.7%	91.0%	84.4%
	HAABSA*	89.7%	**82.5%**	91.5%	87.3%
w/o ontology	HAABSA++	94.9%	80.7%	95.1%	80.6%
	HAABSA*	**96.6%**	82.2%	**96.2%**	**88.2%**

Note that the accuracies found for HAABSA++ differ slightly from those reported in [22]. This might be due to the stochastic nature of batch selection during training, which was not seeded in the implementation. Best result per dataset is shown in bold

5.2 Performance Analysis

We ran our algorithm using the hyperparameters with the values discussed in the previous section. The results are displayed in Table 3. A few considerations must be made when interpreting these results. HAABSA++ was developed as a hybrid approach, since the ontology accuracy for the cases that it was able to predict was found to be higher than the neural network accuracy out-of-sample (82.8% versus 81.7% and 86.8% versus 84.4% for the SemEval 2015 and 2016, respectively). To get a better understanding of how the neural network parts of HAABSA++ and HAABSA* perform, we also report the accuracies without ontology. Furthermore, training the HAABSA* assumes the classification of data in four classes, whereas HAABSA++ only has three classes to discern between. This hinders direct comparison between the in-sample accuracies of the two models. However, the evaluation of the testing accuracy is more straightforward as we consider only three sentiment options 'Negative', 'Neutral' and 'Positive' for both HAABSA ++ and HAABSA*.

For our comparison between HAABSA++ and HAABSA*, we will thus focus on out-of-sample accuracy. Here, we find HAABSA* to outperform HAABSA++ by a considerable margin. Specifically, when only comparing the neural network parts of both algorithms (i.e., accuracy without ontology), a large increase in accuracy of 7.6% points is found to be achieved by the adversarial training procedure for 2016. For 2015, a smaller but still substantial increase in accuracy of 1.5% points is found.

It is interesting to note that for the SemEval 2016 task, a higher accuracy is obtained when not using the ontology compared to when the ontology is used. For SemEval 2015, using the ontology does yield an increase in accuracy still, but with 0.3% points, the margin is small. After further development of the adversarial training method for the LCR-Rot-hop++ method for ABSA, it might be worthwhile to opt out of using the ontology (and thus, the hybrid approach), instead only using the neural network part.

Table 4. Comparison with competing models

SemEval 2015		SemEval 2016	
HAABSA* (w/o ontology)	**82.2%**	HAABSA* (w/o ontology)	**88.2%**
HAABSA++ [22]	81.7%	XRCE [16]	88.1%
LSTM+SynATT+TarRep [8]	81.7%	HAABSA++ [22]	87.0%
PRET+MULT [9]	81.3%	BBLSTM-SL [4]	85.8%
BBLSTM-SL [4]	81.2%	PRET+MULT [9]	85.6%

Values shown are out-of-sample accuracies as reported in the respective research. Best result per dataset is shown in bold

As can be seen in Table 4, this increase in accuracy leads to HAABSA* outperforming LSTM+SynATT+TarRep (81.7%) for the SemEval 2015 task. Furthermore, when the ontology is not used, HAABSA* outperforms XRCE (88.1%) for the SemEval 2016 task. As such, it performs best among those models considered for the SemEval 2015 and 2016 tasks in [22].

It is also of interest to compare our obtained increase in accuracy from adversarial training with the results of [10]. Although they differ from the original implementation of GANs, instead opting for adversarial perturbations, this is, to the best of our knowledge, the only other research in which a form of adversarial training is applied to ABSA. Whereas direct comparison is not possible since they consider different datasets and a different benchmark model, we note that [10] reports increases in accuracy with a maximum of 1% point for all considered datasets and tasks with respect to the non-adversarially trained benchmark model. We, however, obtain an increase of 7.6% points by training our neural network adversarially on the SemEval 2016 task, while an increase of 1.5% points is found for the 2015 task. This could indicate that the classic implementation of GANs [5] with explicit generator and discriminator delivers better results than the approach with adversarial perturbations.

6   Conclusion

In this work, we applied adversarial training to the neural network part of the method devised for ABSA in [22]. Although convergence might prove troublesome for adversarial networks, we found enabling differing learning rates and momentum terms for the generator and discriminator to reduce the problem of divergence. As such, we achieved an increase overall performance from 81.7% to 82.5% and from 84.4% to 87.3% for the SemEval 2015 and 2016 tasks, respectively. When only using the neural network part, adversarial training increased accuracy from 80.6% to 88.2% for the 2016 task, thus outperforming the hybrid approach with ontology. For the 2015 task, accuracy increased from 80.7% to 82.2%. These increases in accuracy are considerably higher than those found in [10]: the, so far, only other application of adversarial training to ABSA, where the authors opted for adversarial perturbations rather than the classical implementation with explicit generator and discriminator.

In future work, we would like to investigate other methods ensuring stability in GAN training, such as feature matching [11], , minibatch discrimination [17], and historical averaging [17]. Furthermore, we wish to explore whether the generator network can be improved by producing full sentences rather than attention vectors, and experiment with other architectures for the generator. Last, we plan a more direct comparison to adversarial perturbation method of [10] by using the same datasets and benchmark model.

References

1. Bergstra, J., Bardenet, R., Bengio, Y., Kégl, B.: Algorithms for hyper-parameter optimization. In: 25th Annual Conference on Neural Information Processing Systems (NIPS 2011), pp. 2546–2554 (2011)
2. Bird, S., Klein, E., Loper, E.: Natural Language Processing with Python: Analyzing Text with the Natural Language Toolkit. O'Reilly Mcdia, Sebastopol (2009)
3. Dcvlin, J., Chang, M., Lee, K., Toutanova, K.: BERT: pre-training of deep bidirectional transformers for language understanding. In: 2019 Conference of the North American Chapter of ACL: Human Language Technologies (NAACL-HLT 2019), pp. 4171–4186. ACL (2019)
4. Do, B.T.: Aspect-based sentiment analysis using bitmask bidirectional long short term memory networks. In: 31st International Florida Artificial Intelligence Research Society Conference (FLAIRS 2018), pp. 259–264. AAAI Press (2018)
5. Goodfellow, I., et al.: Generative adversarial nets. In: 28th Annual Conference on Neural Information Processing Systems (NIPS 2014), pp. 2672–2680. Curran Associates, Inc. (2014)
6. Goodfellow, I.J.: On distinguishability criteria for estimating generative models. In: 3rd International Conference on Learning Representations (ICLR 2015), Workshop Track (2015)
7. Han, J., Zhang, Z., Schuller, B.: Adversarial training in affective computing and sentiment analysis: recent advances and perspectives. IEEE Comput. Intell. Mag. 14(2), 68–81 (2019)
8. He, R., Lee, W.S., Ng, H.T., Dahlmeier, D.: Effective attention modeling for aspect-level sentiment classification. In: 27th International Conference on Computational Linguistics (COLING 2018), pp. 1121–1131. ACL (2018)
9. He, R., Lee, W.S., Ng, H.T., Dahlmeier, D.: Exploiting document knowledge for aspect-level sentiment classification. In: 56th Annual Meeting of ACL (ACL 2018), pp. 579–585. ACL (2018)

10. Karimi, A., Rossi, L., Prati, A., Full, K.: Adversarial training for aspect-based sentiment analysis with BERT (2020). arXiv preprint arXiv:2001.11316
11. Li, Z., Feng, C., Zheng, J., Wu, M., Yu, H.: Towards adversarial robustness via feature matching. IEEE Access **8**, 88594–88603 (2020)
12. Miller, G.A.: WordNet: a lexical database for english. Commun. ACM **38**(11), 39–41 (1995)
13. Odena, A.: Semi-supervised learning with generative adversarial networks (2016). arXiv preprint arXiv:1606.01583
14. Pontiki, M., Galanis, D., Papageorgiou, H., Manandhar, S., Androutsopoulos, I.: Semeval-2015 task 12: aspect based sentiment analysis. In: 9th International Workshop on Semantic Evaluation (SemEval 2015), pp. 486–495. ACL (2015)
15. Pontiki, M., Galanis, D., Papageorgiou, H., Manandhar, S., Androutsopoulos, I.: Semeval-2016 task 5: aspect based sentiment analysis. In: 10th International Workshop on Semantic Evaluation (SemEval 2016), pp. 19–30. ACL (2016)
16. Pontiki, M., Galanis, D., Pavlopoulos, J., Papageorgiou, H., Androutsopoulos, I., Manandhar, S.: Semeval-2014 task 4: aspect based sentiment analysis. In: Proceedings of the 8th International Workshop on Semantic Evaluation (SemEval 2014), pp. 27–35. ACL and Dublin City University (2014)
17. Salimans, T., Goodfellow, I.J., Zaremba, W., Cheung, V., Radford, A., Chen, X.: Improved techniques for training gans. In: 29th Annual Conference on Neural Information Processing Systems (NIPS 2016), pp. 2226–2234. Curran Associates, Inc. (2016)
18. Schouten, K., Frasincar, F.: Survey on aspect-level sentiment analysis. IEEE Trans. Knowl. Data Eng. **28**(3), 813–830 (2016)
19. Schouten, K., Frasincar, F., et al.: Ontology-driven sentiment analysis of product and service aspects. In: Gangemi, A. (ed.) ESWC 2018. LNCS, vol. 10843, pp. 608–623. Springer, Cham (2018). https://doi.org/10.1007/978-3-319-93417-4_39
20. Springenberg, J.T.: Unsupervised and semi-supervised learning with categorical generative adversarial networks. In: 4th International Conference on Learning Representations (ICLR 2016), Poster Track (2016)
21. Sutskever, I., Martens, J., Dahl, G.E., Hinton, G.E.: On the importance of initialization and momentum in deep learning. In: The 30th International Conference on Machine Learning (ICML 2013). JMLR Workshop and Conference, vol. 28, pp. 1139–1147 (2013)
22. Truşcă, M.M., Wassenberg, D., Frasincar, F., Dekker, R.: A hybrid approach for aspect-based sentiment analysis using deep contextual word embeddings and hierarchical attention. In: Bielikova, M., Mikkonen, T., Pautasso, C. (eds.) ICWE 2020. LNCS, vol. 12128, pp. 365–380. Springer, Cham (2020). https://doi.org/10.1007/978-3-030-50578-3_25
23. Wallaart, O., Frasincar, F., et al.: A hybrid approach for aspect-based sentiment analysis using a lexicalized domain ontology and attentional neural models. In: Hitzler, P. (ed.) ESWC 2019. LNCS, vol. 11503, pp. 363–378. Springer, Cham (2019). https://doi.org/10.1007/978-3-030-21348-0_24
24. Xu, H., Liu, B., Shu, L., Yu, P.S.: BERT post-training for review reading comprehension and aspect-based sentiment analysis. In: 2019 Conference of the North American Chapter of ACL: Human Language Technologies (NAACL-HLT 2019), pp. 2324–2335. ACL (2019)
25. Zheng, S., Xia, R.: Left-center-right separated neural network for aspect-based sentiment analysis with rotatory attention (2018). arXiv preprint arXiv:1802.00892

A Dual Reinforcement Network for Classical and Modern Chinese Text Style Transfer

Minzhang Xu[1(✉)], Min Peng[1(✉)], and Fang Liu[1,2]

[1] School of Computer Science, Wuhan University, Hubei, China
{minzhangxu,pengm,liufangfang}@whu.edu.cn
[2] Department of Information Technology, Wuhan City College, Hubei, China

Abstract. Text style transfer aims to change the stylistic features of a sentence while preserving its content. Although remarkable progress have been achieved in English style transfer, Chinese style transfer, such as classical and modern Chinese style transfer, still relies heavily on manual process. In this paper, we first construct an unsupervised dual reinforcement model to transfer text between classical and modern Chinese styles using a non-disentangled approach, in which the style-transfer-accuracy reward and content-preservation reward are specially designed for model optimization. Meanwhile, we leverage a priori knowledge-based synonym dictionary to build a pseudo-parallel corpus for pre-training to provide a warm start. Experimental evaluations show that our model outperforms state-of-art networks by a large margin.

Keywords: Text style transfer · Reinforcement learning · Priori dictionary

1 Introduction

Text style transfer (TST) is used to convert a particular style of text into another style while keeping the content of the original text intact. Text style refers to content-independent attributes such as sentiment of text [6,18,30], formality [2, 6,18,22], and authorship [2,3,9,29]. The TST task has promising applications in computer-aided writing [12], dialog systems [6], and other fields of natural language generation. Besides, the study of text attributes and semantics can help topic models deal with ambiguous words in short texts [5,8].

Unlike other TST tasks, classical and modern Chinese transfer is more challenging by the fact that style is closely related to content of the sentences, which means that style is hard to extract. Moreover, affected by special Chinese literary phenomena such as Chinese word segmentation and interchangeability of words or characters, the task of converting classical Chinese to modern Chinese will face greater difficulties.

When massive parallel corpora are available, the TST can be easily accomplished by state-of-art seq2seq models [2,9,22]. However, since parallel corpora are very scarce, most recent works have turned to unsupervised approaches.

W. Zhang et al. (Eds.): WISE 2021, LNCS 13081, pp. 306–320, 2021.
https://doi.org/10.1007/978-3-030-91560-5_22

According to the manipulation on style and content, we categorize unsupervised methods in two lines.

The first line of work is to disentangle the style attributes of text from the content: first extract style from content, and then combine the target style vector with content vector to generate text [7,24,30]. Due to the challenge of separating style from content completely, such methods often obtain high style transfer accuracy but poor content preservation.

To preserve content of source text, the second line of methods is proposed to entangle the style and content, including style transformers [4], reinforcement learning(RL) models [6,16,18], etc. Nevertheless, this kind of methods preserves content via cycle reconstruction, which may lose style figures in the process. Besides, existing RL models are hard to converge.

To alleviate problems mentioned above and get a better balance between style transfer and content preservation, we propose a dual reinforcement learning framework for classical and modern Chinese transfer called DRLF. Based on two seq2seq models with self-attention in opposite directions, the framework avoids separating style from content and prevents the problem of low style accuracy caused by the challenge of learning expressive style embeddings. The style transfer reward and content preservation reward are especially designed to balance between style and content. In addition, despite the lack of parallel corpus, certain prior knowledge exists in many TST fields, and thus we construct a priori Chinese thesaurus to generate a pseudo-parallel corpus. That will provide a sufficient warm start for our model.

Our main contributions are as follows:

(1) We propose a dual reinforcement learning text style transfer framework, which guarantees the content preservation while ensuring style transfer accuracy.
(2) We propose a method for constructing a pseudo-parallel corpus, which solves the problem of pre-training for unsupervised reinforcement learning and greatly accelerates the training process.
(3) Validation on classical and modern Chinese datasets shows that our model outperforms other advanced models by a large margin.

The rest of this paper is organized as follows: we discuss related work on TST in Sect. 2. The proposed dual reinforcement learning framework is introduced in Sect. 3. In Sect. 4, we construct the Shiji dataset to validate the effectiveness of our model. Finally, we conclude our paper in Sect. 5.

2   Related Work

Text style transfer is an emergent branch of natural language generation. Although the field of neural style transfer has been established to render images into different styles for a long time [15], unlike image styles that can be discerned intuitively, the styles of text are more obscure and more closely associated with the content, making them difficult to extract [14]. Due to the lack of parallel corpus, some methods have been proposed to construct pseudo-parallel corpora [11,31]. But most of the studies focus on unsupervised methods, which can be divided into two categories based on whether they disentangle the style from content.

The first type is to disentangle content from style in a piece of given text, map them into separate latent spaces, and subsequently combine the target style embedding with the content vector to generate text in target style. Hu et al. [7] adopted a delete-retrieve-generate strategy that first delete style attribute words from the original sentence to obtain content words, and revise the style phrases so that it incorporates the target style thereafter. Finally they combine the two phrases to generate the target style text. In addition, the target style text can also be generated by controlling the separated content and style latent representations, e.g., by cross-aligning the latent distribution of style and content via adversarial learning [23]. A machine translation model [20] is used to translate English sentences into French and German and then translate it back to English, which is supposed to contain only the style-free information. Recently, an advanced framework StyIns [30] learns a more expressive style latent space through a set of style instances to make a better balance between style transfer accuracy and content preservation. However, due to the technological limitation on disentanglement, these methods often fail to achieve satisfactory results in content preservation.

The second category entangles style and content, accomplishing style transfer by modifying the encoder-decoder structure. Inspired by the idea of back translation, bidirectional reinforcement learning frameworks on non-parallel corpora are designed [16,18]. In order to alleviate the contradiction between training and evaluation metrics, Abhilasha et al. [2] and Gong et al. [6] used BLEU and WMD as content preservation rewards to optimize their models. Nevertheless, all these RL models suffer from the hardship to converge. On the other hand, Dai et al. [4] introduced the Transformer into text style transfer, combining original input with style latent vectors to generate utterances of different styles. These methods mentioned above achieve high content preservation but relatively low style transfer accuracy.

Taking all the pros and cons into account, we adopt a non-disentangled reinforcement learning method to balance between style transfer and content preservation. Also, we construct a pseudo-parallel corpus with high quality for pre-training to help our model converge.

3   Methodology

We first define the classical and modern Chinese text style transfer as below: considering two different corpora $D_i = \{x_i\}_{i=1}^M$ and $D_j = \{y_j\}_{j=1}^N$ which are discriminated through two different styles (i.e. classical and modern Chinese), given a sentence $x \in D_i$, the goal of style transfer is to convert this sentence to a new one $y' \in D_j$ via the forward transfer model $f_\theta(x)$. At the same time, we try to keep the content of the original text unchanged.

As mentioned before, a big challenge of text style transfer is that it is difficult for us to train the model via supervised methods due to the lack of parallel corpus. Therefore, in Sect. 3.1 we construct a pseudo-parallel corpus with high quality for pre-training. Two seq2seq models in reverse directions are fine-tuned

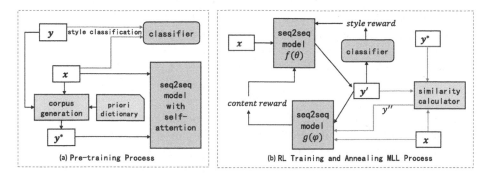

Fig. 1. (a) The pre-training process of our model. y^* represents the generated pseudo-parallel sentence of x. (b)The structure of training process on one direction, where black arrows represent the dual RL training, and blue arrows represent the annealing MLL training. Training on the other direction is the same. (Color figure online)

following the idea of back translation [16,18] to provide supervision after pre-training. Since we do not embed text styles into the hidden space, here lies a problem on how to achieve content preservation while generating text in specific styles when params are discrete between two models. We consequently adopt the reinforcement learning technique, given that one of the advantages of reinforcement learning is to train a model on non-differentiable parameters via methods such as policy gradient [25]. To solve the problem of instability of reinforcement learning, we introduce the annealing maximum likelihood loss (MLL) during the process of reinforcement learning. The model structure is shown in Fig. 1.

3.1 Pretraining with Priori Dictionary

Pseudo Parallel Corpus Construction. Since the warm start for a RL model significantly affects its training efficiency [18], it is important to ensure that the model is effectively and adequately pre-trained. Yet how to build a pseudo-parallel corpus with high quality still remains a challenge. Considering content correlation between different styles of corpora, we leverage Word Mover's Distance(WMD) to construct the pseudo-parallel corpus. Another challenge is that the wording and grammatical construction in classical and modern Chinese are disparate.

Therefore, we build a dictionary of classical and modern Chinese synonyms on Shiji(i.e. *the Grand Scribe's Records*) based on the Dictionary of Common Characters in Classical Chinese [17] to map classical Chinese words with their modern synonyms. Without good solutions for the classical Chinese word segmentation, we cannot update word embeddings with these synonyms. So we substitute the synonyms in target style for words in input sentences, and then use WMD to calculate text similarity. The sentence pair with the highest similarity constitutes the initial pseudo-parallel corpus.

WMD is an algorithm proposed to calculate the distance between two documents [13]. For two sentences s_i and s_j, each word in both sentences is first represented by a word2vec embedding. Given two words w_i and w_j, the word travel cost between them is the Euclidean distance of their word vectors

$$c(i,j) = \|x_i - x_j\|_2 \tag{1}$$

Suppose n is the total number of words in the vocabulary, $t \in \mathbb{R}^{n \times n}$ is a sparse flow matrix, and t_{ij} indicates how many words w_i in s_i are converted to words w_j in s_j. When transferring s_i to s_j, it is guaranteed that the outflow sum of w_i is equal to its word frequency weight d_i in sentence s_i, i.e., $\sum_j T_{ij} = d_i$. Also, the inflow sum of w_j is equal to its word frequency weight d_j in sentence s_j, i.e., $\sum_i T_{ij} = d_j$. The final sentence distance of s_i and s_j is described as:

$$\text{WMD}(s_i, s_j) = \min_{T>0} \sum_{i,j=1}^{n} T_{ij} \cdot c(i,j) \tag{2}$$

The sentences with the smallest distance have the highest similarity.

Pre-training. We use the seq2seq model with self-Attention [27] for pre-training. Two seq2seq models $f_\theta(x)$ and $g_\varphi(y)$ in reverse transfer directions are respectively pre-trained on the pseudo-parallel corpus. The maximum likelihood loss (MLL) is designed as:

$$L = -log(P(y_{1:n} \mid x_{1:m})) = -\sum_{t=1}^{n} log(P(y_t \mid y_{1:t-1}, c_{t-1})) \tag{3}$$

To prevent the model from being trapped into local minima, we use an improved beam search strategy [28] instead of the traditional greedy algorithm to generate text in the decoding process. Unlike the greedy algorithm that finds the word with the highest probability as the current word each step, in beam search, a sequence score function $F(y_t, c_{t-1}, x)$ is defined to calculate a score for every possible word under the historical state $c_{1:t-1}$ and input x. The function $F(y_t, c_{t-1}, x)$ has identical RNN layers as the decoder in experiments, except that the final softmax layer is missing. K words $\hat{y}_t^{(k)}$ that have the K highest possibilities to be the next possible word are then selected and sorted. We define $\hat{y}_{1:t}^{(k)}$ as the sequence with the lowest score among the K-highest-score sequences, i.e. :

$$\left| \{\hat{y}_{1:t}^{(k)} \in \hat{y} \mid F(\hat{y}_t^{(k)}, c_{t-1}, x) > F(\hat{y}_t^{(K)}, c_{t-1}, x)\} \right| = K - 1 \tag{4}$$

Thus the loss function of the F is defined as

$$L_F = \sum_{t=1}^{n} [1 - \text{BLEU}(\hat{y}_{1:n}^{(K)}, y_{1:n})] \cdot [1 - F(y_t, c_{t-1}, x) + F(\hat{y}_t^{(K)}, c_{t-1}, x)] \tag{5}$$

where $\text{BLEU}(\hat{y}_{1:n}^{(K)}, y_{1:n})$ denotes the BLEU score [19] between the least possible sequence $\hat{y}_{1:n}^{(K)}$ among the final K sequences and the pseudo-ground-truth sequence $y_{1:n}$, and the sequence $\hat{y}_{1:n}^{(K)}$ must contain the current subsequence $\hat{y}_{1:t}^{(K)}(t \leq n)$.

If a subsequence $\hat{y}_{1:r}^{(K)}(r<n)$ is followed by a margin sequence [28], i.e., for all the next K words we get $\hat{y}_{r+1}^{(k)} \notin y_{1:n}$, the history sequence before the next search step will be changed to the gold history $y_{1:t}$. At this time, the BLEU$(\hat{y}_{1:n}^{(K)}, y_{1:n})$ in (3) will also be modified to BLEU$(\hat{y}_{r:n}^{(K)}, y_{r:n})$. Finally, the sequence with the highest score among the K sequences is selected as the final result of the decoder.

3.2   Dual Reinforcement Learning

The pre-trained forward model $f_\theta(x)$ converts text x in style s_i into the sentence y' in style s_j, and the backward model $g_\varphi(y)$ converts text y in style s_j into utterance x'' in style s_i. In order that the model can preserve content information while converting text style, two reward functions r_s and r_c are designed for style transfer and content preservation.

After y' is generated, it is feed back into the backward model to reconstruct a sentence x' in style s_i. The two generated sentences are then passed to the evaluator to calculate the style transfer accuracy and content preservation rewards as feedbacks. Subsequently, we optimize parameters of the forward model basing on rewards returned by the evaluator. For the style reward, a pre-trained style binary classifier is used to evaluate the style accuracy of the generated utterances. The content preservation reward, on the other hand, can be divided into two parts: the WMD and back translation reward.

Style Reward. Before training the dual reinforcement learning network, we pre-train a style binary classifier Φ_σ using the Shiji dataset (classical style is marked as s_0 and modern style is denoted as s_1). This classifier is fine-tuned on a Chinese BERT model [26], which is used to evaluate the style accuracy of the generated text.

Given an input sentence $x(x \in D_i)$, the probability that a sentence y' generated by the forward model $f_\theta(x)$ matches style s_j can be obtained by the style classifier Φ_σ. The style reward is then defined as:

$$r_s = P_\Phi(s_i \mid y'; \sigma) \tag{6}$$

where σ is the parameter of the classifier, which is not changed after pre-training.

Content Reward. To balance style transfer and content preservation better, unlike other reinforcement learning models, we design back translation reward and WMD reward to train the model so that it can preserve content information as complete as possible and keep its style transfer accuracy as well.

Since a high-quality pseudo-parallel corpus has been constructed, we first use WMD [13] to compute the textual similarity between the sentence y' generated by $f_\theta(x)$ and the pseudo-ground-truth sentence y. Details about the WMD algorithm is described in Sect. 3.1. While ensuring content preservation, WMD reward can prevent the generated text from being too similar to the original text. It is calculated as:

$$r_{c\_WMD} = norm(-\text{WMD}(y', y)) \tag{7}$$

where the norm function normalizes the WMD reward into $(0, 1)$.

However, given that content mismatches may still exist in pseudo text pairs, the content reward needs to be corrected. Thus, another part of the content preservation reward in our paper draws on the back translation approach.

During the training of forward model $f_\theta(x)$, results of the backward model $g_\varphi(y)$ could be considered reliable after pre-training. If y' preserves the content information completely, it should be easy to reconstruct y' back into x. By converting y' generated by $f_\theta(x)$ back into style s_j, we calculate how much the original content the newly generated utterance x'' contains:

$$r_{c\_back} = norm(log[P_g(x \mid y'; \varphi)]), y' \leftarrow f_\theta(x) \tag{8}$$

Thus the content reward can be written as:

$$r_c = \alpha r_{c\_back} + \beta r_{c\_WMD} \tag{9}$$

where $\alpha, \beta \in (0, 1)$ are hyper-parameters. The total transfer reward is defined as the harmonic mean of the style reward and the content reward:

$$r = (1 + \gamma^2) \cdot \frac{r_s \cdot r_c}{r_s + \gamma^2 r_c} \tag{10}$$

where $\gamma \in (0, 1)$ is the harmonic weight.

Monte-Carlo Policy Gradient with a Baseline. Since we have no access to the distribution of environment state when using a model-free approach, we need to calculate the reward via policy gradient algorithm. To prevent the reward from always being positive, we adopt the Monte Carlo sampling with a baseline to optimize the forward model:

$$\nabla L_f(\theta) = -\nabla \sum_k P_f(y'_k \mid x; \theta) \cdot (r_k - r_{base})$$

$$= -\sum_k P_f(y'_k \mid x; \theta) \cdot (r_k - r_{base}) \cdot \nabla log(P_f(y'_k \mid x; \theta))$$

$$\approx -\frac{1}{K} \sum_{k=1}^{K} (r_k - r_{base}) \cdot \nabla log(P_f(y'_k \mid x; \theta)) \tag{11}$$

where r_{base} is the baseline reward obtained via greedy search strategy and K is the sample number.

During the training process, the two models $f_\theta(x)$ and $g_\varphi(y)$ are trained in rotation periodically in each epoch, so that both models are trained adequately. However, abnormal situations may exist in training, such as the inability to generate fluent text despite the system gets high reward scores. Hence, the annealing maximum likelihood loss (MLL) [21] is also added to training process to avoid such undesirable occurrences.

3.3   Annealing MLL Training

One of the drawbacks of reinforcement learning is the instability. To avoid possible anomalies during model training, the annealing maximum likelihood loss is used to train the model during RL process to ensure the semantic fluency of the generated text.

After one reinforcement learning training step is finished, a pair of temporary pseudo-parallel text (x, y') is available. To prevent the text content or target style from being lost after anomalous reinforcement training step, the pseudo-parallel corpus is updated by matching both y' and the initial pseudo-ground-truth sentence y with input x for WMD similarity as follows:

$$y'' = \begin{cases} argmax_{y^* \in \{y,y'\}} \text{WMD}(x, y^*) & \text{WMD}(x, y) \leq 0.7 \\ y & \text{WMD}(x, y) > 0.7 \end{cases} \quad (12)$$

When similarity between x and y is less than 0.7, it is likely that the pseudo-ground-truth sentence is similar to the input text syntax but their contents are not relative to each other. If so, we will choose the sentence that is more similar to the input between the generated intermediate sentence y' and the pseudo-ground-truth sentence y. Conversely if similarity between x and y is more than 0.7, the initial pseudo-ground-truth sentence is considered to be suitable enough as the parallel sentence to x. The chosen sentence together with input x composes the new pseudo-parallel pair (x, y''). Subsequently, the newly generated sentence pair is used for MLL training based on (3).

The annealing interval is updated by

$$gap = min(maxgap, gap_0 \cdot \varepsilon^{\overline{increase\_gap}}) \quad (13)$$

where $maxgap$ denotes the maximum interval step in annealing training, gap_0 denotes the initial annealing growth rate, $increase\_gap$ denotes that the annealing interval will increase every $increase\_gap$ steps, and i is the current step number. The whole process of training is described in Algorithm 1.

4   Experiments

4.1   Datasets

Since there is no existing corpus on classical and modern Chinese within our knowledge, we leverage *The Grand Scribe's Records* to build a set of non-parallel corpus named Shiji in classical and modern Chinese. Shiji consists of 130 articles with more than 526, 500 words. Compared to other Classical Chinese literary works, it has a distinctive language style, making it very suitable for our task. We select a total of 112 articles from the book as our experimental data. The statistics of each style corpus are shown in Table 1.

4.2   Metrics

We adopt the following metrics to evaluate our network.

Algorithm 1. The dual reinforcement learning algorithm for classical and modern Chinese transfer.

1: Pre-train a binary classifier cl_σ using the original corpora D_i and D_j with style labels
2: Generate pseudo-parallel corpora D_i and D_j with the priori dictionary.
3: Pre-train the two models f_θ, g_φ using corpora D_i and D_j
4: **for** i in range(max_iteration) **do**
5:     # *Train the forward model f_θ*
6:     Sample a sentence x from the src_corpus
7:     Generate a temporary tgt_sentence $y' = f(x; \theta)$
8:     Calculate style reward r_s via (6)
9:     Calculate content reward based on (7),(8),(9)
10:    Calculate total reward r based on (10)
11:    Optimize the model f_θ using r
12:    # *Annealing MLL training for model g_φ*
13:    **if** $i\%p(\varepsilon) = 0$ **then**
14:        Update the temporary pseudo-parallel pair (y'', x) through (12)
15:        Train g_φ) using (y'', x)
16:    Update the annealing gap through (12)
17:    **if** $i\%change\_direction = 0$ **then**
18:        change the direction of models f_θ, g_φ and other parameters conserned
19: **return** Trained models f_θ, g_φ

Table 1. Dataset statistics

Shiji dataset	Classical style	Modern style
Sentences	27,623	28,644
Words	526,786	1,084,284

Style Transfer Accuracy. The style transfer accuracy is usually judged by a pre-trained style classifier [2,4,6,15,16,18,30]. In this paper, we use a Chinese style classifier based on BERT [26] introduced in Sect. 3.2 to judge the style transfer accuracy, which achieves 99.6% accuracy on the classical and modern Chinese test dataset. This classifier-judging approach has been shown to be consistent with human evaluation results [10]. However, it should still be noted that although the style classifier-judging method is widely used for text style transfer, there are still some deviations between the classifier judgments and human evaluation [15].

Content Preservation. Various methods have been proposed to evaluate content preservation, such as BLEU (ref-BLEU or self-BLEU), cosine similarity, and WMD [10,15], but most works use BLEU score [19] for evaluation. Due to the large differences in wording and grammar between classical and modern Chinese, we do not apply the most commonly used self-BLEU score, but obtain the reference text by manual screening and rewriting to calculate the ref-BLEU.

Finally, the harmonic mean (H-Score) and geometric mean (G-Score) of these two metrics are calculated to obtain the comprehensive evaluation results of the experiment.

4.3 Setups

The parameters of our model are tuned on the validation set. The word embedding dimension of the BERT classifier takes its default value of 768, and the word embeddings of the encoder and decoder have 200 dimensions. We adopt the Transformer with 256 hidden size as our seq2seq model. The optimizer is Adam with a pre-training learning rate of 0.001 and a reinforcement learning rate of 0.00001. The content reward weights α and β in (9) are 0.5, and the harmonic weight γ in (10) is 0.6. The two pre-training models are trained for 20 epochs, and the reinforcement learning model is trained for 10 epochs. The batch size is 32 for pre-training and 16 for training process. Part of our code runs on Mindspore [1], which is a new AI computing franmework.

4.4 Baselines and Ablation Study

To verify the effectiveness of our model, we set up the following baseline and ablation experiments. We first compare our model to another reinforcement learning framework DualRL [18], which stands out among other existing models and achieves excellent results especially in tasks where content is hard to extract [15]. Since the delete-retrieve-generate method for pseudo-parallel corpus in DualRL is not applicable to classical and modern Chinese, we generate a pseudo-parallel corpus using a noise strategy (i.e., randomly disordering the sequence of source text words, or randomly adding or deleting words), and the results are almost as good as those which use the delete-retrieve-generate method [18].

In addition, to verify the effectiveness of each module in our network, the pre-training module with the priori dictionary (named as Dic module), the back translation reward module (Back_trans module), and the WMD reward module (WMD module) are removed respectively. The hyperparameters of these experiments are the same. The settings of each part of these models are shown in Table 2. And the experimental results and analysis are presented in Sect. 4.5.

Table 2. Details about model settings

Models	Pretraining	RL content reward
DualRL [18]	Noise	Back_trans
DRLF+Dic+Back_trans	Priori dictionary	Back_trans
DRLF+Dic+WMD	Priori dictionary	WMD
DRLF+Dic+Back_trans+WMD	Priori dictionary	Back_trans+WMD

4.5   Experimental Results

The four models introduced in Sect. 4.4 are evaluate on Shiji dataset, and their results are shown in Table 3. As we can see, the model we proposed achieves the best results both in style transfer accuracy and content preservation. The baseline model cannot have a good warm start due to the low quality of the noisy pseudo-parallel corpus in pre-training. It converges slowly, and the performance is still extremely poor. In the ablation experiments, the style transfer accuracy and content preservation can be greatly improved by adopting the priori dictionary method only. The full model proposed in this paper has a nearly 2% improvement in style accuracy and also gets higher BLEU score compared with the two ablation experiments.

It is worth mentioning that BLEU scores during RL in ablation and full models have a small decrease at the beginning of training, and so as the style transfer accuracy in the ablation experiments. When the model is pre-trained with the priori dictionary through MLL, the modern to classical model achieves better outcomes than the other, which may cause some negative impact on back translation module. After the gap between two models is reduced, the results will improve steadily.

Table 3. Automatic evaluation results on Shiji Dataset. BLEUa and BLEUb measure the content similarity of classical-to-modern model and modern-to-classical model.

	ACC	BLEU	BLEUa	BLEUb	G-Score	H-Score
DualRL [18]	67.650	16.020	14.820	17.220	32.920	25.905
DRLF+Dic+Back_trans	94.600	21.825	20.450	23.200	45.438	35.467
DRLF+Dic+WMD	97.009	25.560	22.550	**28.570**	49.795	40.460
DRLF+Dic+Back_trans+WMD	**98.062**	**26.155**	**24.260**	28.050	**50.644**	**41.296**

We also notice that if we only use back translation reward to preserve content, the generated text will tend to stay unchanged from the original text after several epochs of training. At this time, even though the style accuracy decreases, somehow the model can still get high rewards. However, this is not the result we expect. And if we only use WMD reward, some content information will be lost. But in the full model, the back translation reward helps the model to preserve all the content information, and the WMD reward prevents the generated results from being unchanged, thus guaranteeing a balance between style transfer and content preservation.

At the same time, we also find that the performance of the classical-to-modern model is usually slightly poorer than the modern-to-classical one. The results may lie in that the classical Shiji is extraordinarily concise and has more obscure words, making it more difficult to convert. The generated text shown in Table 4 can give us a more intuitive view on each model.

Table 4. Transferred examples of our experiments. The full model denotes the DRLF+Dic+Back_trans+WMD.

	Model	Classical to Modern
1	Input	又使右贤王入酒泉、张掖，略数千人。
	Reference	又派右贤王入侵酒泉、张掖，掳走了几千人。
	DualRL [18]	又派右贤王入酒泉、张，略数千人。
	DRLF+Dic+Back_trans	又派右贤王进酒泉、张掖，攻取了几千人。
	DRLF+Dic+WMD	又派右贤王进宫廷、张掖，掳走几千人。
	full model	又派右贤王入酒泉、张掖，掳走了几千人。
	meaning of the text	He also sent the right sage king to invade Jiuquan and Zhangye and took several thousand people captive.
2	Input	是岁，贰师将军破大宛，斩其王而还。
	Reference	这年，贰师将军李广利大败大宛，斩杀了大宛王而回来。
	DualRL [18]	这是岁，贰师将军破大宛，斩其王而还。
	DRLF+Dic+Back_trans	这一年，贰师将军打败大宛，斩杀他王而回来。
	DRLF+Dic+WMD	这年，贰师将军率领军队在宛县，斩杀了他们的王而回来。
	full model	这一年，贰师将军在大宛城，斩杀了他的王而回来。
	meaning of the text	In that year, Li Guangli, the general of the second division, defeated Dawan, killed the king of Dawan and returned.
	Model	Modern to Classical
1	Input	呴犁湖单于继位一年后死去。
	Reference	呴犁湖单于立一岁死。
	DualRL [18]	呴犁湖单于位一年后死去。
	DRLF+Dic+Back_trans	呴犁湖单于继一岁后岁死。
	DRLF+Dic+WMD	呴犁湖单于立一岁而死。
	full model	呴犁湖单于立一而死。
	meaning of the text	Monarch Yuliku died one year after succeeding to the throne.
2	Input	燕国人对此觉得很奇怪。
	Reference	燕人怪之。
	DualRL [18]	燕人对此觉得亦奇怪。
	DRLF+Dic+Back_trans	燕人谓此怪其奇。
	DRLF+Dic+WMD	燕人谓之。
	full model	燕人怪之。
	meaning of the text	The Yan people found this very strange.

However, our model still has some shortcomings: first, the priori dictionary is not applicable under all TST tasks. Second, due to the use of several sub-models such as WMD module and the classifier, the number of parameters is greatly increased, which increases the training cost on both time and devices. Besides, the fluency of the generated text cannot be effectively guaranteed because the lack of text fluency detection module.

5   Conclusion and Future Work

In this paper, we propose a dual reinforcement learning model and use it to transfer text between classical and modern Chinese. The experiments demonstrate that our model exceeds other advanced models by a large margin. The high-quality pseudo-parallel corpus constructed through the priori dictionary can provide a warm start for reinforcement learning, which greatly improves the training efficiency of the model. Meanwhile, the WMD text similarity as part of the content preservation reward can avoid the model from over-preserving the original content. That provides a better balance between style transfer accuracy and content preservation.

Since the experiments are conducted only on about 27,000 sentences of Shiji dataset, our model can be evaluated on other corpora to test its adaptability in the future. On the other hand, how to extend the model to multi-style transfer is also a good direction to be explored.

Acknowledgment. This paper was supported by the National key R&D Program of China under Grant No.2018YFC1604003, General Program of Natural Science Foundation of China (NSFC) under Grant NO.61772382 and NO.62072346, and MindSpore.

References

1. Mindspore (2020). https://www.mindspore.cn/
2. Abhilasha, S., Kundan, K., Vasan, B.S., Anandhavelu, N.: Reinforced rewards framework for text style transfer. In: European Conference on Information Retrieval, pp. 545–560 (2020)
3. Carlson, K., Riddell, A., Rockmore, D.: Evaluating prose style transfer with the bible. Roy. Soc. Open Sci. **5**(10), 171920 (2018)
4. Dai, N., Liang, J., Qiu, X., Huang, X.: Style transformer: unpaired text style transfer without disentangled latent representation. In: ACL, pp. 5997–6007 (2019)
5. Gao, W., Peng, M., Wang, H., Zhang, Y., Xie, Q., Tian, G.: Incorporating word embeddings into topic modeling of short text. Knowl. Inf. Syst. **61**(2), 1123–1145 (2018). https://doi.org/10.1007/s10115-018-1314-7
6. Gong, H., Bhat, S., Wu, L., Xiong, J., Hwu, W.M.: Reinforcement learning based text style transfer without parallel training corpus. In: Proceedings of the 2019 Conference of the North American Chapter of the Association for Computational Linguistics: Human Language Technologies, vol. 1 (Long and Short Papers), pp. 3168–3180. Association for Computational Linguistics (2019)
7. Hu, Z., Lee, R.K.W., Aggarwal, C.C., Zhang, A.: Text style transfer: a review and experimental evaluation (2020). arXiv preprint arXiv:2010.12742
8. Huang, J., Peng, M., Wang, H., Cao, J., Gao, W., Zhang, X.: A probabilistic method for emerging topic tracking in microblog stream. World Wide Web **20**(2), 325–350 (2017)
9. Jhamtani, H., Gangal, V., Hovy, E., Nyberg, E.: Shakespearizing modern language using copy-enriched sequence to sequence models. In: Proceedings of the Workshop on Stylistic Variation, pp. 10–19. Association for Computational Linguistics (2017)
10. Jin, D., Jin, Z., Hu, Z., Vechtomova, O., Mihalcea, R.: Deep learning for text attribute transfer: a survey (2021). arXiv preprint arXiv:2011.00416

11. Jin, Z., Jin, D., Mueller, J., Matthews, N., Santus, E.: Imat: unsupervised text attribute transfer via iterative matching and translation. In: Proceedings of the 2019 Conference on Empirical Methods in Natural Language Processing and the 9th International Joint Conference on Natural Language Processing (EMNLP-IJCNLP), pp. 3088–3100 (2019)
12. Klahold, A., Fathi, M.: Computer Aided Writing, pp. 117–130. Springer, Cham (2020). https://doi.org/10.1007/978-3-030-27439-9_9
13. Kusner, J.M., Sun, Y., Kolkin, I.N., Weinberger, Q.K.: From word embeddings to document distances. In: International Conference on Machine Learning (2015)
14. Lample, G., Subramanian, S., Smith, M.E., Denoyer, L., Ranzato, M., Boureau, Y.L.: Multiple-attribute text rewriting. In: ICLR (2019)
15. Li, J., Jia, R., He, H., Liang, P.: Delete, retrieve, generate: a simple approach to sentiment and style transfer. In: Proceedings of the 2018 Conference of the North American Chapter of the Association for Computational Linguistics: Human Language Technologies, vol. 1 (Long Papers), pp. 1865–1874. Association for Computational Linguistics (2018)
16. Li, X., Chen, G., Lin, C., Li, R.: Dgst: a dual-generator network for text style transfer. In: Empirical Methods in Natural Language Processing, pp. 7131–7136 (2020)
17. Wang, L.E.A.: The Dictionary of Common Characters in Classical Chinese. The Commercial Press, Beijing (2016)
18. Luo, F., et al.: A dual reinforcement learning framework for unsupervised text style transfer. In: IJCAI, pp. 5116–5122 (2019)
19. Papineni, K., Roukos, S., Ward, T., Zhu, W.j.: Bleu: a method for automatic evaluation of machine translation. In: ACL '02 Proceedings of the 40th Annual Meeting on Association for Computational Linguistics, pp. 311–318 (2002)
20. Prabhumoye, S., Tsvetkov, Y., Salakhutdinov, R., Black, A.W.: Style transfer through back-translation. In: Proceedings of the 56th Annual Meeting of the Association for Computational Linguistics, vol. 1: Long Papers, pp. 866–876. Association for Computational Linguistics (2018)
21. Ranzato, M., Chopra, S., Auli, M., Zaremba, W.: Sequence level training with recurrent neural networks. In: International Conference on Learning Representations (2015)
22. Rao, S., Tetreault, J.: Dear sir or madam, may i introduce the gyafc dataset: Corpus, benchmarks and metrics for formality style transfer. In: Proceedings of the 2018 Conference of the North American Chapter of the Association for Computational Linguistics: Human Language Technologies, vol. 1 (Long Papers), pp. 129–140. Association for Computational Linguistics (2018). https://doi.org/10.18653/v1/N18-1012
23. Shen, T., Lei, T., Barzilay, R., Jaakkola, T.: Style transfer from non-parallel text by cross-alignment. In: Advances in Neural Information Processing Systems, vol. 30, pp. 6830–6841. Curran Associates, Inc. (2017)
24. Sudhakar, A., Upadhyay, B., Maheswaran, A.: "transforming" delete, retrieve, generate approach for controlled text style transfer. In: Proceedings of the 2019 Conference on Empirical Methods in Natural Language Processing and the 9th International Joint Conference on Natural Language Processing (EMNLP-IJCNLP), pp. 3269–3279 (2019)
25. Sutton, S.R., Barto, G.A.: Reinforcement learning: An introduction. neural information processing systems (1998)

26. Turc, I., Chang, M.W., Lee, K., Toutanova, K.: Well-read students learn better: On the importance of pre-training compact models (2019). arXiv preprint arXiv:1908.08962
27. Vaswani, A., et al.: Attention is all you need. In: Advances in Neural Information Processing Systems 30 (NIPS 2017), pp. 5998–6008 (2017)
28. Wiseman, S., Rush, M.A.: Sequence-to-sequence learning as beam-search optimization. In: Proceedings of the 2016 Conference on Empirical Methods in Natural Language Processing(EMNLP), pp. 1296–1306 (2016)
29. Xu, W., Ritter, A., Dolan, W.B., Grishman, R., Cherry, C.: Paraphrasing for style. In: Proceedings of COLING 2012, pp. 2899–2914 (2012)
30. Yi, X., Liu, Z., Li, W., Sun, M.: Text style transfer via learning style instance supported latent space. In: IJCAI 2020, pp. 3801–3807 (2020). https://doi.org/10.24963/ijcai.2020/526
31. Zhang, Z., et al.: Style transfer as unsupervised machine translation (2018). arXiv: Computation and Language

Representation Learning for Short Text Clustering

Hui Yin[1], Xiangyu Song[1], Shuiqiao Yang[2], Guangyan Huang[1],
and Jianxin Li[1(✉)]

[1] School of Information Technology, Deakin University, Geelong, Australia
{yinhui,xiangyu.song,guangyan.huang,jianxin.li}@deakin.edu.au
[2] Data Science Institute, University of Technology Sydney, Sydney, Australia
shuiqiao.yang@uts.edu.au

Abstract. Effective representation learning is critical for short text clustering due to the sparse, high-dimensional and noise attributes of short text corpus. Existing pre-trained models (e.g., Word2vec and BERT) have greatly improved the expressiveness for short text representations with more condensed, low-dimensional and continuous features compared to the traditional Bag-of-Words (BoW) model. However, these models are trained for general purposes and thus are suboptimal for the short text clustering task. In this paper, we propose two methods to exploit the unsupervised autoencoder (AE) framework to further tune the short text representations based on these pre-trained text models for optimal clustering performance. In our first method Structural Text Network Graph Autoencoder (STN-GAE), we exploit the structural text information among the corpus by constructing a text network, and then adopt graph convolutional network as encoder to fuse the structural features with the pre-trained text features for text representation learning. In our second method Soft Cluster Assignment Autoencoder (SCA-AE), we adopt an extra soft cluster assignment constraint on the latent space of autoencoder to encourage the learned text representations to be more clustering-friendly. We tested two methods on seven popular short text datasets, and the experimental results show that when only using the pre-trained model for short text clustering, BERT performs better than BoW and Word2vec. However, as long as we further tune the pre-trained representations, the proposed method like SCA-AE can greatly increase the clustering performance, and the accuracy improvement compared to use BERT alone could reach as much as 14%.

Keywords: Text clustering · Short text · Representation learning · BERT · Word2vec

1 Introduction

Nowadays, the popularity of social media and online forums (e.g., Twitter, Yahoo, StackOverflow, Weibo) have led to the generation of millions of texts each day. The form of texts such as micro-blogs, snippets, news titles, and

W. Zhang et al. (Eds.): WISE 2021, LNCS 13081, pp. 321–335, 2021.
https://doi.org/10.1007/978-3-030-91560-5_23

question/answer pairs have become essential information carriers [3,4,9,15]. These texts are usually short, sparse, high-dimensional, and semantically diverse [30]. Short text analysis has wide applications, such as grouping similar documents (news, tweets, etc.), question-answering systems, event discovery, document organization, and spam detection [29,32]. Furthermore, a lot of other work also revolves around social media data [10,13,26,27].

Text representation learning is important, and lots of recent studies have demonstrated that effective representations can improve the performance of text clustering [1,11,28]. Traditionally, texts are represented by the Bag-of-Words (BoW) model, where binary weights or term frequency–inverse document frequency (TF-IDF) weights are used for text representation. Later, pre-trained models such as Word2vec [18], GloVe [20] are used to generate word embedding. In addition to obtaining the word vector, the embedding also reflects the spatial position of the word, i.e., words with a common context (semantic similarity) are mapped close to each other in the latent space with similar embeddings. Recently, Bidirectional Encoder Representations from Transformers (BERT) [5] model has caused a sensation in the machine learning and natural language processing communities. It learns the text representation with transformer, an attention mechanism that can learn the contextual relations between words (or sub-words). BERT can understand different embeddings for the same word according to the context. Specifically, if the word represents different meanings in different sentences, the word embedding is also different. For example, "Apple is a kind of fruit." and "I bought the latest Apple." BERT can produce different embedding for "Apple" in different contexts, while word2vec and glove only generate one embedding.

Above advanced pre-trained models can generate effective text representations, and general clustering models, such as K-means, are applied to these text representations to implement text clustering tasks. The more effective the text representation is, the more accurate K-means can classify text into different clusters based on the similarity between text embeddings. Although pre-trained models significantly improve text representation ability compared to the traditional BoW model by incorporating semantic contexts into text representations. However, these models are generally trained on a large scale of text corpus and focused on general-purpose text mining tasks and thus are sub-optimal for the short text clustering purpose. Therefore, we are highly motivated to fine tune the text representations generated by the pre-trained text models like BERT to further improve the performance of short text clustering.

Specifically, we propose to exploit the unsupervised autoencoder framework with different neural network modules to further compress the text representations generated by the pre-trained models into more condensed representations. In our first method, named STN-GAE, we propose to combine the text representations from the pre-trained models and the structural text network to achieve more condensed text representation for short text clustering. To achieve this, we first construct a text network where each node represents a text, and the edge between nodes indicates that the two texts have semantic similarities. The pre-trained text models return the features of the nodes. Then, we exploit graph convolutional networks (GCN) into the autoencoder framework as an encoder to fuse the structure-level and feature-level information for condensed text representation learning. In our second method, named SCA-AE, we exploit a soft cluster assignment constraint [11] into the embedding space of autoencoder for clustering-friendly feature learning. The soft cluster assignment constraint assumes that the latent embeddings are generated from different components. Firstly, a soft cluster assignment distribution P for each data is calculated based on Student's t-distribution [17], where the embeddings close to the component centers are highly confident. Then, an induced distribution Q is further calculated based on P, and a KL-divergence [14] loss between P and Q is adopted to increase the confidence of highly confident embeddings. Based on the two proposed methods, we can further tune the text representations provided by the pre-trained models to improve the short text clustering performance.

In summary, we highlight our contributions as follows:

- We compare the performance of two popular pre-trained text models (Word2vec and BERT) on short text clustering tasks and find that BERT based pre-trained models could provide better representations than Word2vec for short text clustering tasks.
- We propose two methods (STN-GAE and SCA-AE) to fine tune the pre-trained text representations and improve the short text clustering performance. STN-GAE adopts graph convolutional networks to combine text networks and text features for representation learning. SCA-AE exploits a soft cluster assignment constraint into autoencoder to learn clustering-friendly representations.
- Extensive experiments on seven real-world short text datasets have demonstrated that with fine-tuned text representations based on the pre-trained text models, we can achieve much better clustering performance than using the pre-trained model alone. The proposed SCA-AE can improve the clustering accuracy by 14%.

The rest of this paper is organized as follows. Section 2 reviews the related work. Section 3 details the framework and methodology for short text clustering. Section 4 presents the experimental studies. Conclusions are made in Sect. 5.

2   Related Work

In this section, we survey three techniques for short text clustering, e.g., text representation learning, text structure information extracting, and the soft cluster assignment for clustering purposes.

2.1   Text Representation Learning

Traditionally, Bag-of-Words (BoW) is the simplest way to generate a text vector, where the dimension of the vectors is equal to the size of the vocabulary. While for short text, the vector tends to contain many empty values, resulting in sparse and high-dimensional vectors. Also, there is no information about the sentence syntax or the order of words in the text. Word2vec and GloVe are popular unsupervised learning algorithms for learning representations of words, and they are context-independent. These models output just one vector or embedding for each word in the vocabulary and combine all different senses of the word in the corpus into one vector. Word vectors are positioned in the vector space such that words with common contexts in the corpus are located close to each other in the space. Pre-trained Word2vec/GloVe embeddings are publicly available, and word embeddings can also be trained based on special domain corpus. The text embedding is usually represented by the sum or average value of all word embeddings in the text, and the dimensionality of the text vector is equal to the word embedding. For the missing words in the pre-trained word embedding, ignoring unknown words or providing them with random vectors is better.

2.2 Text Structure Information Extracting

Some studies such as PTE [25], Text GCN [31], RTC [16] and SDCN [2] adopt the text structural information to supplement the text representation. The corpus is represented as a network where each node is a text or word and edges are created between nodes according to certain principles, so that each node can obtain additional information from neighboring nodes to enrich its representation. Text GCN [31] and Robust Text Clustering (RTC) [16] use the same strategy to construct a text graph that contains word nodes and document/text/sentence nodes. For the document/text/sentence-word edges, the weight of the edge is the term frequency-inverse document frequency (TF-IDF) of the word in the document, and for the word-word edges, they use the point-wise mutual information (PMI) method to calculate the weight (only the edges with positive PMI value are preserved). Structural Deep Clustering Network (SDCN) [2] constructs a K-Nearest Neighbor (KNN) graph based on the original corpus, and the samples are the nodes in the graph. For all samples, they use heat kernel and dot-product to calculate a similarity matrix, then select the top K similar neighbors for each sample and set edges to connect it with them. After building the network, they employ a GCN model for the network structure, and a DNN model for the text representation, and then combine the loss function of both models for training.

2.3 Soft Cluster Assignment for Text Clustering

The soft clustering assignment algorithm is developed using the K-means algorithm and its variants. The document's assignment is a distribution over all clusters, that is, a document has fractional membership in several clusters. Thus soft cluster assignment techniques are widely used in text clustering to improve performance. The goal is to learn clustering-friendly text representations, where data points are evenly distributed around the cluster centers and the boundaries between clusters are relatively clear. The common method includes the soft cluster assignment loss into the training objectives to optimize the learning models and learn clustering-friendly representations.

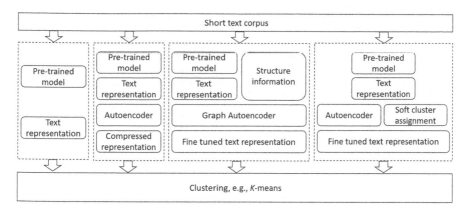

Fig. 1. The general framework of this case study. The input is a short text corpus, and the output is the clustering result.

3 Methodology

In this section, we first introduce the preliminary of the autoencoder framework, then we detail two proposed methods: STN-GAE and SCA-AE (Fig. 3).

3.1 Preliminary of Autoencoder

We use an autoencoder to reduce the dimensionality of the features, and then directly apply K-means to them. Autoencoder compresses the high-dimensional input representation X into a low-dimensional representation Z while retaining most original information. Using these compressed representations can reduce model computational complexity and running time. Figure 2 provides the architecture of the autoencoder. We want to minimize the reconstruction loss to make reconstructed $\hat{X}$ closer to X and adopt Mean Square Error (MSE) as our loss function, as shown in Eq. 1, which measures how close $\hat{X}$ is to X.

$$L(X, \hat{X}) = ||X - \hat{X}||^2 \tag{1}$$

In our experiments, we perform dimensionality reduction preprocessing on BERT based text representation for each dataset and then perform K-means clustering.

3.2 STN-GAE: Representation Learning with Structure Information

We propose a method Structural Text Network-Graph Autoencoder (STN-GAE), which exploits the structural text information among the corpus by constructing a text network and then adopts graph convolutional network into the autoencoder (Graph Autoencoder) to fuse the structural features with the pre-trained text features for text representation learning. This deep neural network

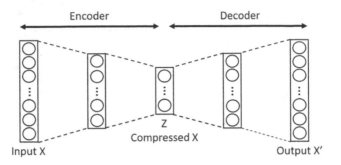

Encoder   Decoder

Input X

Z
Compressed X

Output X'

Fig. 2. Autoencoder architecture. The encoder stage of the autoencoder compresses the high-dimensional input representation X into a low-dimensional representation Z, the decoder stage maps Z to a reconstruction X' with the same shape as X.

(DNN) is able to learn non-linear mapping without manually extracting features to transform the inputs into a latent representation and maintain the original structure. The layer-wise propagation rule of graph autoencoder [12] is defined as follows:

$$H^{(l+1)} = \sigma(\tilde{D}^{-\frac{1}{2}}\tilde{A}\tilde{D}^{-\frac{1}{2}}H^{(l)}W^{(l)}) \tag{2}$$

where $\tilde{A} = A + I_N$ is the adjacency matrix of the undirected graph G with added self-connections, I_N is the identity matrix, D is the degree matrix, σ is a non-linear activation function such as the ReLU, $W^{(l)}$ is the weight matrix for layer l. X is the high-dimensional input representation, $H^{(l)} \in R^{N \times D}$ is the matrix activation in the l_{th} layer, $H^0 = X$.

We first use the same strategy as SDCN [2] to construct a K-Nearest Neighbor (KNN) graph, and each sample is a node in the text network. Then we input the adjacency matrix A and raw features X into Graph Autoencoder (GAE) and get the encoded low-dimensional embedding of the text, and then apply clustering algorithm K-means to them, as shown in Fig. 3(a). We will describe the strategy in detail in the following.

We calculate the cosine score between text embeddings (BERT based embeddings) and construct the semantic similarity matrix S of the whole corpus. After calculating the similarity matrix S, we select the top-K similarity samples of each node as its neighbors to construct an undirected K-nearest neighbor graph. The above strategy ensures that each sample is connected to highly correlated K nodes to ensure the integrity of the graph structure, which is essential for a convolutional neural network. In this way, we can get the adjacency matrix A from the non-graph data. Therefore, the text graph is represented as $G = (V, E, X)$, where $V(|V| = N)$ are the set of nodes and $E(|E| = N * K)$ are the set of edges, respectively, K is the number of neighbors per sample. $X = [x_1, x_2, ..., x_N] \in R^{N \times d}$ is node feature vectors matrix, where x_i represents the feature vector of node i, and d is the dimension of the feature vector, the adjacency matrix is $A \in R^{N \times N}$.

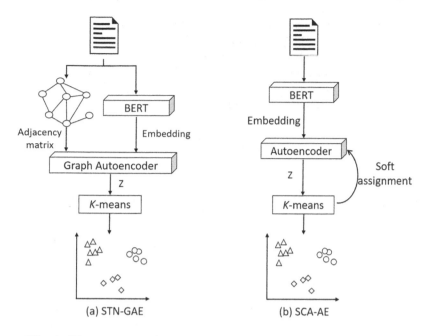

Fig. 3. The structure of proposed methods STN-GAE and SCA-AE.

3.3 SCA-AE: Representation Learning with Soft Cluster Assignment Constraint

A lot of work has been confirmed that using deep neural networks to simultaneously learn feature representation and cluster assignment usually yields better results. In method SCA-AE, we employ cluster assignment hardening loss to optimize the text representation. This method includes three steps: (1) Use BERT model to generate text representation; (2) Use autoencoder to reduce dimensionality to get compressed input embeddings; (3) Use soft cluster assignment as an auxiliary target distribution, and jointly fine-tune the encoder weights and the clustering assignments to improve clustering performance. The third step requires using soft assignment of data points to clusters.

We use the Student's t-distribution Q [17] as a kernel to measure the similarity between embedded nodes z_i and centroids u_j. This distribution q is formulated as follows:

$$q_{ij} = \frac{(1+ \parallel z_i - u_j \parallel^2)^{-1}}{\sum_{j'} (1+ \parallel z_i - u_{j'} \parallel^2)^{-1}} \tag{3}$$

where q_{ij} can be considered as the probability of assigning sample i to cluster j, i.e., a soft assignment. z_i is the i-th row of $Z_{(N)}$, u_j is initialized by K-means on representations learned by DNN model, we treat $Q = [q_{ij}]$ as the distribution of the assignments of all samples.

After obtaining the clustering result distribution Q, we further use an auxiliary target distribution P [28] to improve cluster purity and put more emphasis on samples assigned with high confidence so that the samples are closer to cluster centers. Compared to the similarity score q_{ij}, P is "stricter", which can prevent large clusters from distorting the hidden feature space. The probabilities p_{ij} in the proposed distribution P are calculated as:

$$p_{ij} = \frac{q_{ij}^2 / \sum_{i'} q_{i'j}}{\sum_{j'} q_{ij'}^2 / \sum_{i'} q_{i'j'}} \tag{4}$$

in which each assignment in Q is squared and normalized.

The KL-divergence [14] between the two probability distributions P, and Q is then used as training objective, the training loss L is defined as:

$$L_{soft} = KL(P \parallel Q) = \sum_i \sum_j p_{ij} log \frac{p_{ij}}{q_{ij}} \tag{5}$$

The above process is usually called self-training or self-supervised. By minimizing the KL-divergence loss between P and Q distributions, the target distribution P can help the DNN model learn a clustering-friendly representation, i.e., making the sample closer to the centroid to which it belongs. The structure of method SCA-AE is shown in Fig. 3(b).

4   Experiments

In this section, we report our experimental results regarding the adopted datasets, the compared methods, and the parameter settings.

4.1   Datasets

We choose seven short text datasets with different lengths and categories for experiments. We use raw text as input without additional processing, such as stop word removal, tokenization, and stemming. The statistics of these datasets are presented in Table 1, and the detailed descriptions are the following:

◇ MR[1] [19]: The MR is a movie review dataset for binary sentiment classification, containing 5,331 positive and 5,331 negative reviews.
◇ AGnews[2]: The AG's news topic classification dataset contains around 127,600 English news articles, labeled with four categories. We use the test dataset for experiments, which includes 7,600 news titles.
◇ SearchSnippets [21]: A collection consists of Web search snippets categorized in eight different domains. The 8 domains are Business, Computers, Culture-Arts, Education-Science, Engineering, Health, Politics-Society, and Sports.

---

[1] https://github.com/mnqu/PTE/tree/master/data/mr.
[2] https://www.kaggle.com/amananandrai/ag-news-classification-dataset.

◊ Yahoo[3]: The answers topic classification dataset is constructed using the ten largest main categories from Yahoo. We randomly select 10,000 samples from the dataset.

◊ Tweets[4]: This dataset consists of 30,322 highly relevant tweets to 269 queries in the Text REtrieval Conference (TREC) 2011–2015 microblog track 1. We choose the top ten largest categories for experiments. The total number of samples is 6,988, the sample size of each class is different, the maximum count is 1,628, and the minimum count is 428.

◊ StackOverflow[5]: A collection of posts is taken from the question and answer site StackOverflow and publicly available on Kaggle.com. We use the subset containing the question titles from 20 different categories.

◊ Biomedical[6]: This is a snapshot of one year of PubMed data distributed by BioASQ to evaluate large-scale online biomedical semantic indexing.

Table 1. Statistics of the text datasets.

Dataset	Categories	Samples	Ave. length	Vocabulary
MR	2	10,662	21.02	18,196
AGnews	4	7,600	6.76	16,406
SearchSnippets	8	12,340	17.9	30,610
Yahoo	10	10,000	34.16	23,710
Tweets	10	6,988	7.28	5,773
StackOverflow	20	20,000	8.3	10,762
Biomedical	20	20,000	12.9	18,888

4.2 Compared Methods and Parameter Settings

◊ Word2vec: We use Google's pre-trained word embeddings and then take the average of all word embeddings in the text as the text representation. The words not present in the pre-trained words are initialized randomly, the text vector dimension is 300.

◊ BERT: We employ the sentence-transformer model [22] to generate text embeddings directly. Since these short text datasets come from different sources, and the pre-trained BERT models are trained based on different types of datasets, to avoid the impact of BERT model selection on the performance of text clustering, we run three BERT models on all datasets and choose the best model of the three for each dataset. For datasets Yahoo, SearchSnippets, AGnews, StackOverflow, and Biomedical, the BERT model

---

[3] https://www.kaggle.com/soumikrakshit/yahoo-answers-dataset.
[4] http://trec.nist.gov/data/microblog.html.
[5] https://www.kaggle.com/c/predict-closed-questions-on-stack-overflow/.
[6] http://participants-area.bioasq.org/.

"paraphrase-distilroberta-base-v1" gets the best performance. For dataset MR, BERT model "stsb-roberta-large" gets the best result, and for dataset Tweets, BERT model "distilbert-base-nli-stsb-mean-tokens" is the best one.

◇ Autoencoder: We set the hidden layers to d:500:500:2000:10, which is consistent with other work [2,7,28], where d is the dimension of the input data X and 10 is the dimension of compressed X. We tune other parameters and set the learning rate as 0.001, and the MSE is used to measure reconstruction loss after the decoder decodes the encoded embeddings. We also try other settings of the hidden layer and find that small changes do not affect the performance much. See Sect. 4.4 for details.

◇ STN-GAE: We calculate the similarity matrix S based on the text embeddings generated by the BERT model and use the same strategy in SDCN [2] to construct the K-Nearest Neighbor (KNN) graph, K=10. We set the hidden layer of GAE to d:64:32, and train them with 300 epochs for all datasets, the learning rate is set to 0.002.

◇ SCA-AE: We use the same setting as [8], for instance, the hidden layer size of all datasets to d:500:500:2000:20, 20 is the dimensionality of compressed text embedding. We set the batch size to 64, pre-trained the autoencoder for 15 epochs, and initialize stochastic gradient descent with a learning rate of 0.01 and a momentum value of 0.9.

We adopt the K-means as the clustering algorithm. Two popular metrics are exploited to evaluate the clustering performance, i.e., we employ two metrics: normalized mutual information (NMI) and clustering accuracy (ACC).

4.3   Clustering Performance Analysis

Table 2 indicates the clustering results on seven datasets. For the pre-trained BERT model, we adopt the best one selected from the three pre-trained BERT models, refer to Sect. 4.2 for details. AE, STN-GAE, and SCA-AE use the BERT model to generate text representations as input features. We have the following observations:

◇ In baseline methods (BoW, W2V, and BERT), BERT achieves the best results in all seven datasets for each metric. Compared to Word2vec and BoW, BERT has state-of-the-art semantics representation ability, which can map the text to a more accurate position in the latent space, which has a positive impact on clustering.

◇ AE and BERT achieve similar clustering results. Because the autoencoder retains the most important information in compressing the text embedding, the compressed vector is about the same as the input data.

◇ The method STN-GAE does not perform well as expected, even worse than the AE model and BERT model. The reason is that in the GAE model, the text representation contains structural information, when the structural information in the graph is not clear enough, the performance of the GCN-based method will decrease. This conclusion is consistent with SDCN.

⋄ The use of soft cluster assignment for deep clustering (SCA-AE) is able to achieve the best performance in 5 out of 7 datasets compared to the best baseline method BERT. In the AGnews, Yahoo, and StackOverflow datasets, the accuracy has increased by about 10%, and the NMI has improved slightly.

⋄ For dataset Biomedical, the performance is really poor in all methods, and the possible reason is this dataset belongs to the biology/medicine domain [6,23,24], and the existing pre-trained BERT models are all trained on a common social media dataset. A possible solution is to train a BERT model in this professional field.

Figure 4 displays t-distributed stochastic neighbor embedding (t-SNE) visualizations for clustering on Tweets dataset in (a) Bag of words space, (b) Word2vec space, (c) BERT embedding space, (d) Autoencoder hidden layer space, (e) Graph Autoencoder hidden layer space and (f) Autoencoder hidden layer space with soft cluster assignment. Figures 4a and 4b show that the samples are mixed together in a high-dimensional space without clear boundaries. In Fig. 4c, there are clear boundaries between nodes of different categories, but they are not concentrated enough and somewhat scattered, which proves that the BERT model has excellent representation capabilities. Compared to Fig. 4c BERT embedding space, in Fig. 4f, the sample is more compact from the centroid. This is the feature of soft assignment, which makes the sample closer to the centroid and improves the NMI score.

4.4   Parameters Analysis

The model SCA-AE has achieved the best performance. We further analyze the relevant parameters to verify their impact on the model.

Epochs of SCA-AE. In the soft cluster assignment method (SCA-AE), we vary the number of the epochs in autoencoder while keeping others unchanged. Figure 5 shows the accuracy and NMI values of the seven datasets with different pre-trained epochs. With the increase of epochs in the autoencoder, the accuracy and NMI values of the tweet dataset have improved slightly, while other datasets have not changed much. This demonstrates that the four-hidden-layer autoencoder is able to generate effective low-dimensional text representation in a few epochs.

Autoencoder Layers of SCA-AE. In the SCA-AE model, the hidden layer size of autoencoder is set to d:500:500:2000:20, which is the same as [8]. We change layer numbers and layer sizes of autoencoder hidden layers to verify the impact on the results. d:500:2000:500:20 changes the order of hidden layers, d:500:2000:20 removes a hidden layer, and d:256:512:20 removes a hidden layer while reducing the number of neurons in each layer. Figure 6 shows the accuracy and NMI comparison based on the above hidden layer settings. Although we have reduced the number of hidden layers or reduce the neuron number of the layer,

Table 2. Clustering performance comparison. The results are averaged over five runs.

Dataset	Metric	BoW	W2V	BERT	AE	STN-GAE	SCA-AE
MR	ACC	0.5188	0.526	**0.7848**	0.7492	0.6625	0.7662
	NMI	0.0037	0.002	**0.2511**	0.2013	0.1676	0.2198
AGnews	ACC	0.2809	0.3977	0.5748	0.6748	0.5742	**0.6836**
	NMI	0.0094	0.1193	0.2577	0.3299	0.2914	**0.3414**
SearchSnippets	ACC	0.2459	0.623	0.6675	0.6429	0.4144	**0.6871**
	NMI	0.0893	0.4782	0.4763	0.4619	0.3167	**0.5026**
Yahoo	ACC	0.1507	0.1355	0.4271	0.4803	0.3387	**0.5606**
	NMI	0.0413	0.0264	0.2765	0.3069	0.2529	**0.3472**
Tweets	ACC	0.6161	0.5771	0.8126	0.8424	0.4049	**0.8485**
	NMI	0.7284	0.6084	0.867	0.8875	0.3546	**0.8919**
StackOverflow	ACC	0.4615	0.2289	0.6253	0.6672	0.4049	**0.7655**
	NMI	0.5506	0.1905	0.5962	0.6156	0.4492	**0.6599**
Biomedical	ACC	0.1388	0.2296	**0.4043**	0.3894	0.1550	0.4025
	NMI	0.0887	0.2166	**0.3347**	0.3385	0.1468	0.3329

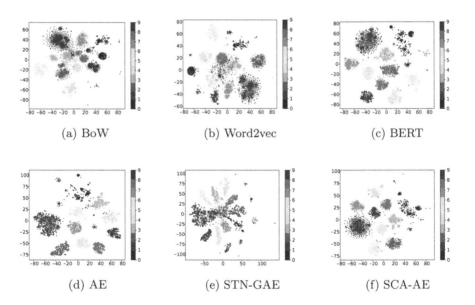

(a) BoW   (b) Word2vec   (c) BERT

(d) AE   (e) STN-GAE   (f) SCA-AE

Fig. 4. t-SNE visualizations for clustering on Tweets dataset in (a) Bag of words space, (b) Word2vec space, (c) BERT embedding space, (d) Autoencoder hidden layer space, (e) Graph Autoencoder hidden layer space and (f) Autoencoder hidden layer space with soft cluster assignment. The true cluster labels are indicated using different colors.

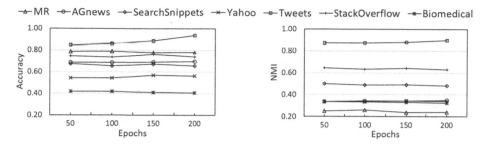

Fig. 5. Epoch size comparison for SCA-AE.

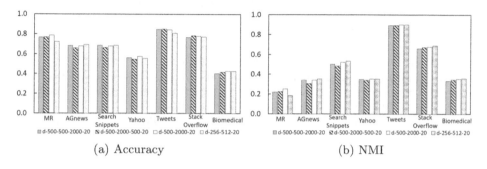

Fig. 6. Hidden layer size comparison for SCA-AE.

the accuracy and NMI do not fluctuate much, so when using autoencoder for text representation dimensionality reduction, simply hidden layers and a small number of the neuron can achieve effective results.

5   Conclusions

In this work, we propose two methods to further enhance the short text representation learning based on the pre-trained text models to obtain the best clustering performance. One is the integration of text structure information into the representation (STN-GAE), and the other is the soft cluster assignment constraint (SCA-AE). We find that the BERT model produces effective text representations compared to traditional models (e.g., Word2vec, bag of words). Experimental results show that the integration of text structure information into text representation (STN-GAE) has no positive impact on short text clustering, suggesting that the nodes in the constructed short text network cannot obtain supplementary information from neighboring nodes. The proposed method, SCA-AE, further fine-tunes the text representation and does improve the clustering performance.

Acknowledgement. This work was mainly supported by the Australian Research Council Linkage Project under Grant No. LP180100750.

References

1. Aljalbout, E., Golkov, V., Siddiqui, Y., Strobel, M., Cremers, D.: Clustering with deep learning: Taxonomy and new methods. arXiv preprint arXiv:1801.07648 (2018)
2. Bo, D., Wang, X., Shi, C., Zhu, M., Lu, E., Cui, P.: Structural deep clustering network. In: Proceedings of The Web Conference 2020, pp. 1400–1410 (2020)
3. Cai, T., Li, J., Mian, A.S., Sellis, T., Yu, J.X., et al.: Target-aware holistic influence maximization in spatial social networks. IEEE Trans. Knowl. Data Eng. (2020). https://doi.org/10.1109/TKDE.2020.3003047
4. Chen, L., Liu, C., Liao, K., Li, J., Zhou, R.: Contextual community search over large social networks. In: 2019 IEEE 35th International Conference on Data Engineering (ICDE), pp. 88–99. IEEE (2019)
5. Devlin, J., Chang, M.W., Lee, K., Toutanova, K.: Bert: Pre-training of deep bidirectional transformers for language understanding. In: Proceedings of NAACL-HLT 2019, pp. 4171–4186, Minneapolis, Minnesota, June 2 - June 7 (2019)
6. Du, J., Michalska, S., Subramani, S., Wang, H., Zhang, Y.: Neural attention with character embeddings for hay fever detection from twitter. Health Inf. Sci. Syst. 7(1), 1–7 (2019). https://doi.org/10.1007/s13755-019-0084-2
7. Guo, X., Gao, L., Liu, X., Yin, J.: Improved deep embedded clustering with local structure preservation. In: IJCAI, pp. 1753–1759 (2017)
8. Hadifar, A., Sterckx, L., Demeester, T., Develder, C.: A self-training approach for short text clustering. In: Proceedings of the 4th Workshop on Representation Learning for NLP (RepL4NLP-2019), pp. 194–199 (2019)
9. Al Hasan Haldar, N., Li, J., Reynolds, M., Sellis, T., Yu, J.X.: Location prediction in large-scale social networks: an in-depth benchmarking study. VLDB J. 28(5), 623–648 (2019). https://doi.org/10.1007/s00778-019-00553-0
10. Jiang, H., Zhou, R., Zhang, L., Wang, H., Zhang, Y.: Sentence level topic models for associated topics extraction. World Wide Web 22(6), 2545–2560 (2018). https://doi.org/10.1007/s11280-018-0639-1
11. Jiang, Z., Zheng, Y., Tan, H., Tang, B., Zhou, H.: Variational deep embedding: an unsupervised and generative approach to clustering. In: Proceedings of the 26th International Joint Conference on Artificial Intelligence (IJCAI 2017), August 2017, pp. 1965–1972 (2016)
12. Kipf, T.N., Welling, M.: Semi-supervised classification with graph convolutional networks. In: ICLR (Poster) (2017)
13. Kong, X., Li, M., Li, J., Tian, K., Hu, X., Xia, F.: Copfun: an urban co-occurrence pattern mining scheme based on regional function discovery. World Wide Web 22(3), 1029–1054 (2019)
14. Kullback, S., Leibler, R.A.: On information and sufficiency. Ann. Math. Stat. 22(1), 79–86 (1951)
15. Li, J., Cai, T., Deng, K., Wang, X., Sellis, T., Xia, F.: Community-diversified influence maximization in social networks. Inf. Syst. 92, 101522 (2020)
16. Liang, Y., Tian, T., Jin, K., Yang, X., Lv, Y., Zhang, X.: Robust text clustering with graph and textual adversarial learning. In: 2020 IEEE Fifth International Conference on Data Science in Cyberspace (DSC), pp. 190–197. IEEE (2020)
17. Van der Maaten, L., Hinton, G.: Visualizing data using t-SNE. J. Mach. Learn. Res. 9, 2579–2605 (2008)
18. Mikolov, T., Chen, K., Corrado, G., Dean, J.: Efficient estimation of word representations in vector space. In: ICLR (Workshop Poster) (2013)

19. Pang, B., Lee, L.: Seeing stars: Exploiting class relationships for sentiment categorization with respect to rating scales. In: Proceedings of the 43rd Annual Meeting of the Association for Computational Linguistics (ACL 2005), pp. 115–124 (2005)
20. Pennington, J., Socher, R., Manning, C.D.: Glove: global vectors for word representation. In: Proceedings of the 2014 Conference on Empirical Methods in Natural Language Processing (EMNLP), pp. 1532–1543 (2014)
21. Phan, X.H., Nguyen, L.M., Horiguchi, S.: Learning to classify short and sparse text & web with hidden topics from large-scale data collections. In: Proceedings of the 17th international conference on World Wide Web, pp. 91–100 (2008)
22. Reimers, N., Gurevych, I.: Sentence-bert: sentence embeddings using siamese bert-networks. In: Proceedings of the 2019 Conference on Empirical Methods in Natural Language Processing and the 9th International Joint Conference on Natural Language Processing, pp. 3982–3992, Hong Kong, China, 3–7 November (2019)
23. Sarki, R., Ahmed, K., Wang, H., Zhang, Y.: Automated detection of mild and multi-class diabetic eye diseases using deep learning. Health Inf. Sci. Syst. 8(1), 1–9 (2020). https://doi.org/10.1007/s13755-020-00125-5
24. Supriya, S., Siuly, S., Wang, H., Zhang, Y.: Automated epilepsy detection techniques from electroencephalogram signals: a review study. Health Inf. Sci. Syst. 8(1), 1–15 (2020). https://doi.org/10.1007/s13755-020-00129-1
25. Tang, J., Qu, M., Mei, Q.: Pte: predictive text embedding through large-scale heterogeneous text networks. In: Proceedings of the 21th ACM SIGKDD International Conference on Knowledge Discovery and Data Mining, pp. 1165–1174 (2015)
26. Tian, Q., et al.: Evidence-driven dubious decision making in online shopping. World Wide Web 22(6), 2883–2899 (2018). https://doi.org/10.1007/s11280-018-0618-6
27. Wang, X., Deng, K., Li, J., Yu, J.X., Jensen, C.S., Yang, X.: Efficient targeted influence minimization in big social networks. World Wide Web 23(4), 2323–2340 (2020). https://doi.org/10.1007/s11280-019-00748-z
28. Xie, J., Girshick, R., Farhadi, A.: Unsupervised deep embedding for clustering analysis. In: International Conference on Machine Learning, pp. 478–487. PMLR (2016)
29. Yang, S., Huang, G., Cai, B.: Discovering topic representative terms for short text clustering. IEEE Access 7, 92037–92047 (2019)
30. Yang, S., Huang, G., Ofoghi, B., Yearwood, J.: Short text similarity measurement using context-aware weighted Biterms. Concurrency Comput. Pract. Experience e5765 (2020). https://doi.org/10.1002/cpe.5765
31. Yao, L., Mao, C., Luo, Y.: Graph convolutional networks for text classification. In: Proceedings of the AAAI Conference on Artificial Intelligence, vol. 33, pp. 7370–7377 (2019)
32. Yin, H., Yang, S., Li, J.: Detecting topic and sentiment dynamics due to Covid-19 pandemic using social media. In: Yang, X., Wang, C.-D., Islam, M.S., Zhang, Z. (eds.) ADMA 2020. LNCS (LNAI), vol. 12447, pp. 610–623. Springer, Cham (2020). https://doi.org/10.1007/978-3-030-65390-3_46

ReAct: a Review Comment Dataset for Actionability (and more)

Gautam Choudhary[1]([✉]), Natwar Modani[1], and Nitish Maurya[2]

[1] Adobe Research, Bangalore, India
{gautamc,nmodani}@adobe.com
[2] Adobe System, Noida, India
nmaurya@adobe.com

Abstract. Review comments play an important role in the evolution of documents. For a large document, the number of review comments may become large, making it difficult for the authors to quickly grasp what the comments are about. It is important to identify the nature of the comments to identify which comments require some action on the part of document authors, along with identifying the types of these comments. In this paper, we introduce an annotated review comment dataset *ReAct*. The review comments are sourced from OpenReview site. We crowd-source annotations for these reviews for *actionability* and *type* of comments. We analyze the properties of the dataset and validate the quality of annotations. We release the dataset to the research community as a major contribution (Full dataset available at https://github.com/gtmdotme/ReAct). We also benchmark our data with standard baselines for classification tasks and analyze their performance.

Keywords: Review dataset · Actionability · Taxonomy · Text-classification

1 Introduction

Review comments play an important role in the evolution of documents. Academic publications routinely go through a peer-review process, where the reviewers provide both their opinion about the suitability of the articles for the publication venue and also feedback to the authors for potentially improving the contributed article. Further, several publication venues are providing the authors a chance to respond to the review comments (for example, ACL, NAACL, EMNLP, etc., in addition to most journals). Therefore, it is important for the authors to be able to quickly digest the review comments so that they can address the concerns of the reviewers and clarify certain points which may not have been communicated adequately by the article itself.

In this work, we focus on two aspects of *understanding* the review comments. First, determining if a review comment requires some action on the part of document authors. This motivates the need for the classification of review comments based on 'actionability'. Second, what type of review comment it is among

© Springer Nature Switzerland AG 2021
W. Zhang et al. (Eds.): WISE 2021, LNCS 13081, pp. 336–343, 2021.
https://doi.org/10.1007/978-3-030-91560-5_24

Agreement, Disagreement, Question, Suggestion, Shortcoming, Statement of Fact, and *Others*, similar to (but not exactly the same as) [11]. We provide the reason for our choice of these specific types and their justification in Sect. 2.2.

Text classification has long been an active area of research, as the classification can help the users efficiently process a large amount of content. Finding actionable comments on social media (tweets) was addressed in [13] using new lexicon features. A *specificity* score was explored in [3] for an employee satisfaction survey and product review settings to understand actionable suggestions and grievances (complaints) for improvements. In yet another work [9], the actionability of review comments for code review is investigated using lexical features. These works address only the actionability aspect of our problem, and the datasets used in these papers are not publicly available except in [9], where the dataset is made available publicly.

Other binary classifications in prior work include Question classification [15], agreement/disagreement classification [1] and suggestions/advice mining [4]. However, such binary classifications only provide information on a single dimension in isolation and fall short in providing a more extensive set of categorization as done in [10], where the authors investigated comments on product reviews in an e-commerce setting. Again, the datasets are not publicly available, and the categories proposed are not comprehensive in nature.

OpenReview is a popular online forum for reviewing research papers and the choice of gathering data from this forum is motivated by a comprehensive study for analyzing the review process [14]. [6] also present *PeerRead* dataset consolidating reviews from a lot of conferences. Our dataset provides finer-grained annotation by providing two labels per review comment sentence and thereby opens up a new research direction.

Our key contributions in this paper are:

- A review comment dataset consisting of $1,250$ labeled comments for identifying actionability and their types. We also have $\sim 52k+$ unlabelled (but otherwise processed) comments in this dataset for future extensions and/or use of semi-supervised approaches.
- A taxonomy for types of review comments.
- Establishing strong baselines for the proposed dataset.

2 Dataset: ReAct

While the prior art focuses on feature engineering and model architecture, we note a lack of publicly available datasets in this problem set. This section describes how we arrive at the proposed annotated dataset, *ReAct*.

In this paper, We use Fleiss' kappa κ [5] as the measure of inter-annotator agreement. It is used to determine the level of agreement between two or more annotators when the response variable is measured on a categorical scale.

2.1 Raw Data Collection and Preprocessing

The proposed dataset is gathered from an online public forum OpenReview where research papers are reviewed and discussed. Multiple anonymous reviewers review the papers and write free-form comments (along with people other than reviewers also writing comments) related to the papers. We extract 911 papers submitted to ICLR (The International Conference on Learning Representations) 2018 from OpenReview and filter out the comments written by people other than the reviewers and get a dataset where each record is a paper associated with its reviews and other metadata (final decision, rating, link to the paper, timestamps, abstract, etc.). Each paper is reviewed by at least 3 reviewers who provide their comments in free-form text. An average review spans about 19 sentences. On manual inspection, we find that giving a specific label to the whole review is not appropriate, since a review often contains some facts about the paper along with some merits/demerits and some questions. A natural choice is to chunk the review into smaller units. We follow a simple approach and choose a sentence as the atomic unit and refer to it as a *comment*. We use a python tool pySBD[1] for splitting and disambiguation of long paragraphs of reviews into sentences.

2.2   Classification Taxonomy

Given the motivating scenario of helping the user quickly be able to respond to review comments, say during rebuttal period, the choice of binary classification (as *actionable* or not) is fairly straightforward and has been used in prior literature as already discussed (although, not in paper review setting). However, the choice of *types* for the finer-grained classification is non-obvious.

To arrive at the appropriate class labels, we randomly selected 50 review comments and three volunteers started categorizing them independently, with an initial *types* seed list of *Suggestion, Agreement, Disagreement,* and *Question,* inspired by [11]. Whenever the volunteers felt that the comment didn't fall in these types, they added a new type. After completing the independent categorization, a pool of all labels for types across the three volunteers was created. Now, a set of labels was consolidated if the volunteers agreed that the individual labels in that set had the same semantics. The final proposed taxonomy of category labels with their initial label assignments in parenthesis were found to be as follows: agreement (appreciation/agreement), disagreement (conflict/disagreement), question (inquiry/question), suggestion (demand/ask/advice/suggestion), shortcoming (problem/issue/shortcoming), fact (opinion/statement of fact), and others (miscellaneous/others).

2.3   Designing the Survey

We selected 125 reviews for the main survey such that they were sufficiently long for having (at least) 10 comments as part of each review, and retained 10 randomly selected sufficiently long comments (having at least 10 words) corresponding to each of these reviews. These 1250 review comments are then annotated by a popular crowdsourcing platform, Amazon Mechanical Turk (AMT)[2].

---

[1] pySBD: https://github.com/nipunsadvilkar/pySBD.

[2] Amazon Mechanical Turk: https://mturk.com/.

Each comment is annotated by 5 different human annotators. The annotators are given a set of instructions for annotating the comments. The task of an annotator is to read the review comments and assign two labels to each comment. The first label $Label_1$ is to be assigned based on the actionability of the comment, i.e., among {*yes, no*} and constitutes $Task_1$. Similarly, $Label_2$ is to be assigned from the proposed taxonomy, one of {*agreement, disagreement, suggestion, question, fact, shortcoming, other*} based on the type of comments and constitutes $Task_2$. We also asked annotators to provide feedback for the survey to mitigate shortcomings, if any. The survey was available to annotators based on certain filters that AMT provides. We restricted the survey to *Mechanical Turk Masters* who had acceptance scores $\geq 95\%$ to get high-quality annotations. The reward of one complete survey (comprising two types of labels for 10 comments) was set to \$ 0.75 based on the feedback received on the pilot surveys floated initially, described next. A time limit of 30 min was set before the survey expired.

2.4   Analyzing Responses

Pilot Survey. Instead of rolling out the survey fully in one go, we followed an iterative approach. A pilot survey was conducted to check if the tasks and instructions were clear and to get an estimate of the quality of responses. We handpicked 5 reviews (different from the ones used in the main survey) having a total of 50 comments using the above survey design. Post completion, we analyzed each of the responses one by one and noted a 'moderate' inter-annotator agreement score (Fleiss kappa), $\kappa \approx 0.48$ among the annotators [7]. A noteworthy thing was the feedback received from annotators which substantially supported our comprehensive, yet simple survey design. The annotators expressed no lack of clarity and also found the survey task to be appropriate and complete. The time to answer was also analyzed and seemed to match with the time taken by the volunteers (close to 10 min per survey). Hence, the feedback received from this survey was sufficiently positive to go ahead with the main survey.

Main Survey. Post successful completion of the pilot survey, we obtain a set of 6, 250 annotations for the above chosen 1250 comments (each labeled by 5 different annotators). Each annotation consists of labels for the comment along with other metadata such as characteristics of the annotator and the survey such as (ID, duration, timestamp, etc.). We found that a total of 33 unique annotators participated in the survey with an average completion time of ~ 10 min.

First, we analyze the Fleiss kappa scores on individual labeling tasks, i.e., for $Label_1$ and $Label_2$. For the $Label_1$ denoting actionability, we observe a 'moderate' inter-annotator score of 0.49 and a slightly higher score of 0.53 for $Label_2$ based on the proposed taxonomy [7]. Next, we analyze at a deeper level by looking at proportions of responses ranging from having a clear agreement to strong ambiguity as shown by proportions of stacked bars in Fig. 1(a) and (b). At an aggregate level, for $Label_1$, almost 50% of annotations have a clear consensus where all 5 annotators vote for the same category label, while 30% of annotations

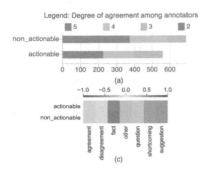
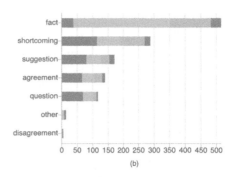

Fig. 1. (a, b) Distribution of review comments based on their count, hued by the fraction of agreement in annotators annotating the same comment for $Label_1$ and $Label_2$ respectively, (c) correlation observed in the two kinds of category labels annotations.

have 4 out of 5 votes and the rest 20% have 3 out of 5 votes as shown in Fig. 1(a) at per category label basis. Similarly, for $Label_2$, more than 70% of annotations have at least 4 out of 5 annotators agree on a specific category label as shown in Fig. 1(b). We observe that *disagreement* is a rare class (with a low agreement between reviewers), suggesting this label may not be essential.

Additionally, the correlation analysis using the Pearson Correlation Coefficient between the two sets of category labels (Fig. 1(c)) strengthens the intuition that suggestions, shortcomings, and questions are more likely to be actionable than the other categories.

2.5   Moderation

We found some noisy responses where annotators labeled shortcomings as non-actionable. Another example of noise found in the data is when annotators annotate an agreement as actionable. The proportion of such noisy responses was very small ($\sim 6.5\%$) in the whole dataset. To improve the quality of data further, we selectively moderated the labeling. In particular, we reviewed $Label_2$ for review comments in the case of maximum disagreement, i.e., when the maximum number of annotators agreeing on that label was 2. In addition, if the maximum number of annotators agreeing on $Label_1$ was 3 for any of *these* comments, then this label was also reviewed. In the cases where the maximum number of annotators agreeing $Label_2$ was 3 (or more), indicating a good level of agreement (3 out of 5), neither of the labels were reviewed (even if for $Label_1$, the maximum number of agreeing annotators were 3, the borderline case).

There were a total of 91 cases where we reviewed $Label_2$, out of which 49 cases where we also reviewed $Label_1$. Ignoring these cases, the inter-annotator agreement for $Label_1$ increased to 0.52, and for $Label_2$, it increased to 0.57. Finally, in review, we ended up changing $Label_1$ for 19 cases. For $Label_2$, there were 34 cases where there was a tie and we picked one of the tied labels. Further, we changed labels for 18 other cases, where we assigned a label as ground truth for *Label*$_2$, which was not voted as (one of) majority label(s) by annotators. Given the small number of cases where we had to change the labels (about 1.5% cases), we believe the annotations' quality is very good.

2.6   Processed Dataset

Based on the above analysis, we assign the ground truth labels based on majority, i.e., out of 5 votes for a given comment, we chose the majority vote as the ground truth label. While a tie is not possible in $Task_1$, it may take place in $Task_2$ (consider the case where all the votes are for different labels or two votes each for two labels). Total number of tied cases for *label*$_2$ was 34. As mentioned before, such ties were resolved through the process of moderation. The final prepared dataset (also available at https://github.com/gtmdotme/ReAct) consists of 1, 250 comments with two sets of labels, *Label*$_1$ and *Label*$_2$. For example, *"It would enhance readability of the paper if the results were more self-contained"* is labeled as actionable and suggestion. A contrasting example of non-actionable and agreement labels is, *"Indeed, the authors have succeed in showing that this is not necessarily the case"*.

3   Benchmarking Experiments

Given a review comment from the proposed dataset, two classification scenarios arise. First, a binary classification ($Task_1$) to identify whether the comment is actionable to the author or not, and secondly, a multiclass classification ($Task_2$) to identify the nature of comment from the proposed taxonomy.

3.1   Feature Extraction and Text Classification Models

Most of the recent works on producing contextual embeddings have been shown to improve results over the human crafted features. Therefore, we experiment with the following state-of-the-art sentence embeddings:

- Universal Sentence Encoder [2] (**USE**) is trained and optimized for text, such as sentences, phrases, or short paragraphs (embedding length = 512).
- **DistilBERT** Embeddings [12] are a distilled version of BERT with faster performance and fewer parameters (embedding length = 768).
- **RoBERTa** Embeddings [8] are built out of tweaking the BERT model hyperparameters to produce robust embeddings that are shown to perform best for STS tasks (embedding length = 1024).

We experiment with the following text classifiers:

- *Baseline-Random*: This model predicts a class uniformly at random (out of 2 for $Task_1$ and out of 7 for $Task_2$).
- *Baseline-Majority*: This model predicts the most frequently occurring class in the training dataset.
- ML-Classifiers: We use standard classification models like Logistic Regression (*LR*), Support Vector Machine (*SVM*), *XGBoost*, and Feedforward Neural Network (*FNN*) with two hidden layers of sizes 128 and 32.

Table 1. Test accuracy and F1 scores of the classification models for each task on our proposed dataset. Here, DistB is DistilBERT and RoB is RoBERTa.

Models	$Task_1$						$Task_2$					
	Accuracy			F1-Score			Accuracy			F1-Score		
	USE	DistB	RoB	USE	DistB	RoB	USE	DistB	RoB	USE	DistB	RoB
Baseline-Random	0.504	0.528	0.500	0.551	0.487	0.480	0.132	0.116	0.136	0.164	0.180	0.159
Baseline-Majority	0.588	0.588	0.588	0.435	0.435	0.435	0.408	0.408	0.408	0.236	0.236	0.236
LR	0.788	0.812	0.812	0.788	0.813	0.813	0.616	0.688	0.688	0.598	0.683	0.683
SVM	**0.796**	0.832	0.832	**0.796**	0.832	0.832	**0.636**	**0.72**	**0.72**	**0.621**	**0.708**	**0.708**
XGBoost	0.784	0.788	0.788	0.785	0.788	0.788	0.604	0.684	0.684	0.591	0.673	0.673
FNN (128, 32)	0.764	**0.848**	**0.832**	0.765	**0.849**	**0.833**	0.6	0.692	0.696	0.594	0.687	0.689

3.2 Experimental Setup

All the experiments involve using the proposed dataset by keeping 80% of the data for training and the rest 20% for evaluating the model. The (random) split is such that the proportions of the $Label_2$ are preserved in the two sets, i.e., stratified split. We use a standard implementation of these models from the scikit-learn python library[3] keeping the default parameters fixed for a fair comparison across variations in models and embeddings.

Evaluation. We evaluate the model performances using the standard metric of accuracy (fraction of correct predictions out of total) on the test set in Table 1. Given the imbalance in our dataset, we also report the f1-scores (weighted average of class-wise F1 scores). We note that the models are able to achieve a fair degree of accuracy (significantly higher than baselines), and also that using larger embeddings (RoBERTa and DistilBERT) results in better accuracy than smaller USE embeddings, although between RoBERTa and DistilBERT, there is no significant difference in performance. We fine-tune the hidden layer parameter for the FNN model and find the combination of using two layers of sizes 128 and 32 as the best configuration. Interestingly, a simple model (SVM) performs at par with a sophisticated model (FNN) as noted from the accuracies in Table 1.

4 Conclusion and Future Work

We present *ReAct*, a novel and carefully annotated dataset for identifying the nature of textual review comments. These comments play an important role in the evolution of many types of documents. We propose a new taxonomy for fine-grained comment classification in a review scenario. We analyze the properties of the dataset along with some baseline systems for the two identified text classification tasks and analyze their performance. Since the two tasks are not completely independent from each other, a multitask learning approach seems desirable. The small fraction of labeled data out of a large pool of unlabelled data also calls for a self-supervised learning algorithm using less data.

---

[3] scikit-learn: https://scikit-learn.org/.

References

1. Ahmadalinezhad, M., Makrehchi, M.: Detecting agreement and disagreement in political debates. In: Thomson, R., Dancy, C., Hyder, A., Bisgin, H. (eds.) SBP-BRiMS 2018. LNCS, vol. 10899, pp. 54–60. Springer, Cham (2018). https://doi.org/10.1007/978-3-319-93372-6_6
2. Cer, D., et al.: Universal sentence encoder. arXiv preprint arXiv:1803.11175 (2018)
3. Deshpande, S., Palshikar, G.K., Athiappan, G.: An unsupervised approach to sentence classification. In: COMAD, p. 88 (2010)
4. Dong, L., Wei, F., Duan, Y., Liu, X., Zhou, M., Xu, K.: The automated acquisition of suggestions from tweets. In: Twenty-Seventh AAAI Conference on Artificial Intelligence (2013)
5. Fleiss, J.L.: Measuring nominal scale agreement among many raters. Psychol. Bull. **76**(5), 378 (1971)
6. Kang, D., et al.: A dataset of peer reviews (peerread): Collection, insights and NLP applications. arXiv preprint arXiv:1804.09635 (2018)
7. Landis, J.R., Koch, G.G.: The measurement of observer agreement for categorical data. Biometrics **33**, 159–174 (1977)
8. Liu, Y., et al.: Roberta: A robustly optimized bert pretraining approach. arXiv preprint arXiv:1907.11692 (2019)
9. Meyers, B.S., et al.: A dataset for identifying actionable feedback in collaborative software development. In: Proceedings of the 56th Annual Meeting of the Association for Computational Linguistics (Volume 2: Short Papers), pp. 126–131 (2018)
10. Mukherjee, A., Liu, B.: Modeling review comments. In: Proceedings of the 50th Annual Meeting of the Association for Computational Linguistics (Volume 1: Long Papers), pp. 320–329 (2012)
11. Sancheti, A., Modani, N., Choudhary, G., Priyadarshini, C., Moparthi, S.S.M.: Understanding blogs through the lens of readers' comments. Computación y Sistemas **23**(3), 1033–1042 (2019)
12. Sanh, V., Debut, L., Chaumond, J., Wolf, T.: Distilbert, a distilled version of bert: smaller, faster, cheaper and lighter. arXiv preprint arXiv:1910.01108 (2019)
13. Spasojevic, N., Rao, A.: Identifying actionable messages on social media. In: 2015 IEEE International Conference on Big Data, pp. 2273–2281. IEEE (2015)
14. Tran, D., et al.: An open review of Openreview: A critical analysis of the machine learning conference review process. arXiv preprint arXiv:2010.05137 (2020)
15. Zhang, D., Lee, W.S.: Question classification using support vector machines. In: Proceedings of the 26th Annual International ACM SIGIR Conference on Research and Development in Informaion Retrieval, pp. 26–32 (2003)

Text Mining (2)

Document-Level Relation Extraction with Entity Enhancement and Context Refinement

Meng Zou[1], Qiang Yang[2], Jianfeng Qu[1(✉)], Zhixu Li[1], An Liu[1], Lei Zhao[1], and Zhigang Chen[3,4]

[1] School of Computer Science and Technology, Soochow University, Suzhou, China
{jfqu,zhixuli,anliu,zhaol}@suda.edu.cn
[2] King Abdullah University of Science and Technology, Thuwal, Saudi Arabia
[3] iFLYTEK Research, Suzhou, China
zgchen@iflytek.com
[4] State Key Laboratory of Cognitive Intelligence, iFLYTEK, Hefei, China

Abstract. Document-level Relation Extraction (DocRE) is the task of extracting relational facts mentioned in the entire document. Despite its popularity, there are still two major difficulties with this task: (i) How to learn more informative embeddings for entity pairs? (ii) How to capture the crucial context describing the relation between an entity pair from the document? To tackle the first challenge, we propose to encode the document with a task-specific pre-trained encoder, where three tasks are involved in pre-training. While one novel task is designed to learn the relation semantic from diverse expressions by utilizing relation-aware pre-training data, the other two tasks, Masked Language Modeling (MLM) and Mention Reference Prediction (MRP), are adopted to enhance the encoder's capacity in text understanding and coreference capturing. For addressing the second challenge, we craft a hierarchical attention mechanism to refine the context for entity pairs, which considers the embeddings from the encoder as well as the sequential distance information of mentions in the given document. Extensive experimental study on the benchmark dataset DocRED verifies that our method achieves better performance than the baselines.

Keywords: Document-level Relation Extraction · Pre-trained model · Hierarchical attention

1 Introduction

Relation Extraction (RE) aims to identify the relation between entities in a given text. It is an important task in Natural Language Processing (NLP) since it can facilitate many downstream tasks, including Knowledge Graph (KG) construction [24] and question answering system [16]. Conventional sentence-level RE only concentrates on relational facts mentioned in **a single sentence**. Although sentence-level RE has achieved great success, the basic assumption of this task limits the relational expression in a single sentence. However, the relation fact

© Springer Nature Switzerland AG 2021
W. Zhang et al. (Eds.): WISE 2021, LNCS 13081, pp. 347–362, 2021.
https://doi.org/10.1007/978-3-030-91560-5_25

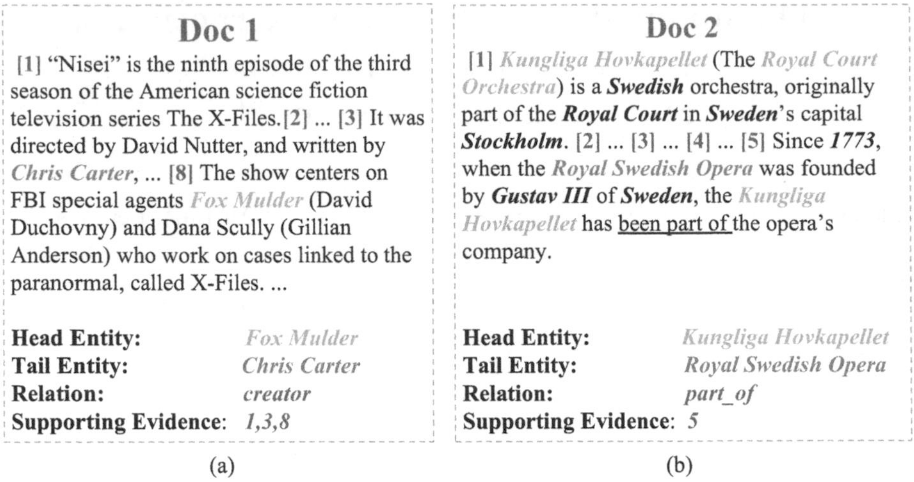

Doc 1	**Doc 2**
[1] "Nisei" is the ninth episode of the third season of the American science fiction television series The X-Files.[2] ... [3] It was directed by David Nutter, and written by *Chris Carter*, ... [8] The show centers on FBI special agents *Fox Mulder* (David Duchovny) and Dana Scully (Gillian Anderson) who work on cases linked to the paranormal, called X-Files. ...	[1] *Kungliga Hovkapellet* (The *Royal Court Orchestra*) is a **Swedish** orchestra, originally part of the **Royal Court** in **Sweden**'s capital **Stockholm**. [2] ... [3] ... [4] ... [5] Since *1773*, when the *Royal Swedish Opera* was founded by **Gustav III** of **Sweden**, the *Kungliga Hovkapellet* has been part of the opera's company.
Head Entity: *Fox Mulder*	**Head Entity:** *Kungliga Hovkapellet*
Tail Entity: *Chris Carter*	**Tail Entity:** *Royal Swedish Opera*
Relation: *creator*	**Relation:** *part_of*
Supporting Evidence: *1,3,8*	**Supporting Evidence**: *5*
(a)	(b)

Fig. 1. Two text examples (head entity in pink & tail entity in green) from DocRED. (Color figure online)

in reality is often expressed across multiple sentences [21,27]. It is reported that 40.7% of the relation instances in DocRED dataset could only be extracted by combining multiple sentences as prediction evidence [27].

To break this limitation, **Doc**ument-level **RE** (DocRE) is proposed to extract all relation triples from a whole document [21,27]. Figure 1(a) shows an example in DocRED, where the relation triple (*Fox Mulder, Chris Carter, creator*) could only be obtained by combining three sentences.

Since the scenario of DocRE is more in line with reality, many efforts have been devoted to this research area, which could be divided into sequence-based models, Graph Neural Network (GNN)-based models, and Pre-trained Language Models (PLMs)-based models. Sequence-based models [11,23] mainly rely on sequential features and syntactic structure extracted from the original text, and then connect different sentences to model the cross-sentence RE. Limited to a few consecutive sentences instead of the document, they cannot capture deep semantic information. Later, GNN-based models [1,3,9,12,20,29] are proposed to construct a document graph, where the nodes represent mentions or entities, the edges indicate various dependencies, and some operations (e.g., graph convolution) are used for information propagation. While benefiting from the advantages of graph structure, the GNN-based models require a sophisticated design for node and edge modeling and require strong computing power.

Recently, PLMs-based models have become a more promising direction in DocRE, as they often achieve higher performance with less complexity compared with GNN-based models. By confirming that the dependency structure and coreference in the text can be well learned by PLMs [18], some DocRE models [17,22] are proposed based on BERT [2]. Beyond this, a recent work [28] proposes CorefBERT, a further pre-trained model for coreference resolution, which achieves remarkable success by combining the *Mention Reference Prediction* (MRP) task and the *Masked Language Modeling* (MLM) task. Another recent work DSDocRE [25] also designs three tasks, which are *Mention-entity Matching* (MM), *Relation Detection* (RD) and *Relational fact Alignment* (RA), for further pre-training. While task RD requires the pre-trained model to distinguish the instance that expresses relation from a pile of instances that do not express any relation, task MM requires the model to capture the coreference information. Task RA forces the embeddings of the co-occurrence sentences of the same pair of entities to be similar. Despite their success, there are two deficiencies remain with the existing models in dealing with the two difficulties mentioned above. (1) The current pre-trained models lack the perceptual ability of relation semantics, such that the encoding of entity pairs cannot provide sufficient information for relation extraction. (2) They tend to ignore the influence of relevant context on the representation of target entity pairs, which is of great significance for relation extraction.

In this paper, we propose a novel PLMs-based model by enhancing the entity embeddings with a task-specific pre-trained encoder, and refining the context for target entity pairs with a hierarchical attention mechanism. Specifically, to tackle the first deficiency with the existing models, we propose to encode the document with a task-specific pre-trained model, where three tasks are involved in the pre-training process. The first task is a novel one, so-called *Relation Semantic Modeling* (RSM), which is designed to capture the relation semantic from diverse expressions by utilizing relation-aware pre-training data. The other two tasks, MLM and MRP, are adopted to enhance the model's capacity in text understanding and coreference capturing. For addressing the second deficiency, we observe that the surrounding text with different mentions of the same entity may have different impacts in describing its relation with the other entities. As the example shown in Fig. 1(b), the surrounding text with the third mention of the head entity provides much more evidence than the first two mentions in indicating its relation with the tail entity. Based on that observation, we need to highlight the influence of more relevant contexts on the representations of target entity pairs. In particular, we craft a hierarchical attention mechanism to refine the context for entity pairs, which considers the embeddings of entities from the pre-trained encoder as well as the sequential distance information of mentions in the given document.

Our contributions are summarized as follows:

- We propose a novel task-specific pre-trained model for DocRE, which not only involves two traditional tasks MLM and MRP, but also considers a novel task specially designed for modeling the relation semantic from diverse expressions by utilizing relation-aware pre-training data.
- We craft a hierarchical attention mechanism to refine the context for entity pairs, such that the model could better capture the relational information implied in the refined context.
- Extensive experiments on the largest benchmark DocRED show that our model has significant advantages over the baselines.

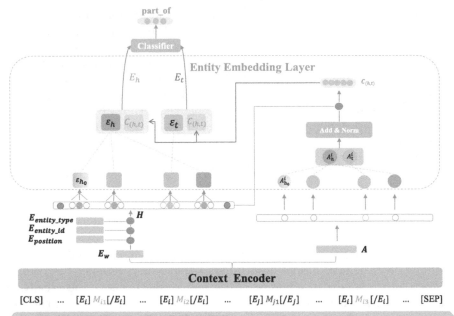

Fig. 2. The architecture of RSCM (Relation Semantic and Context-refined Modeling), its input is a omitted document with the head entity marked in pink while tail entity in green, and their relation is "part of".

Roadmap. We introduce our model in Sect. 2, followed by the experiments in Sect. 3. The related work is covered in Sect. 4. The paper is finally concluded in Sect. 5.

2   Methodology

2.1   Problem Definition

Given a document $\mathcal{D} = \{x_i\}_{i=1}^{l}$ with l words marked with named entities $\mathcal{E}$ and a closed set of relations $\mathcal{R}$. DocRE will extract all relation triples $\{(h, t, r)\}$ composed of head entity $h \in \mathcal{E}$, tail entity $t \in \mathcal{E}$, and relation $r \in \mathcal{R}$. The entity h is expressed by several mentions $\{h_m\}_{m=1}^{J}$ in the document $\mathcal{D}$. The premise of the existence of relation r between h and t is that there exists relation r between at least one pair of mention of h and t, otherwise the relation between h and t is NA.

2.2   Method Overview

Our model is illustrated in Fig. 2. Firstly, we combine three pre-training tasks of RSM, MRP and MLM, and further pre-train the $BERT_{base}$ to obtain a

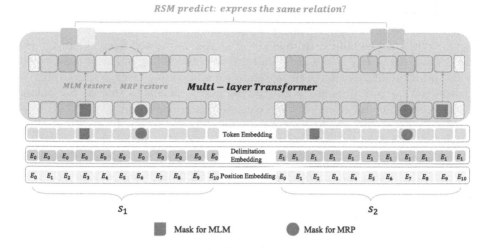

Fig. 3. Pre-training illustration of our context encoder. RSM requires model to judge whether these two instances express the same relation, while MRP and MLM require to select words from context and vocabulary to restore their masked tokens.

context encoder, from which a relation-sensitive document embedding E_w and mention-level attention A is obtained. The pre-training process is introduced in Sect. 2.2(1) and illustrated in Fig. 3. Secondly, the refined context embeddings $C_{(h,t)}$ for target entity pairs are induced with the help of sequential distance clue and attention mechanism, so that some redundant context for this pair can be filtered out. Thirdly, the final representation of an entity pair (h, t) will be interactively enhanced by its refined context embedding $C_{(h,t)}$ in the entity embedding layer. Finally, the classifier can not only get the interactive embedding of the entity pair but also focus on the refined and relevant context.

(1) Context Encoder

Following [13], we introduce $[E]$ and $[/E]$ as the start and end markers of mentions in the text. $[E_i]$ is inserted at the begin of a mentioned entity e_i, while $[/E_i]$ is inserted at the end. We adopt the encoder to encode the input document into a relation-sensitive embedding E_w. At the same time, we could get the token-level attention matrix of each token to other tokens in the input from this encoder, just like the original BERT. Then we draw the attention value of start marker of each mention span to abtain the mention-level attention A.

$$E_w, A = ContextEncoder([x_1, x_2, ..., x_{l'}]) \qquad (1)$$

where l' is the length of the tokenized and marked input document. The enhanced document embedding H will be obtained as follows.

$$x'_i = [E_w(x_i) : E_t(t_i) : E_{id}(id_i) : E_p(p_i))] \qquad (2)$$

$$H = \{x'_i\}_{i=1}^{l'} \qquad (3)$$

where x_i' integrates token embedding $E_w(x_i)$, entity type embedding $E_t(t_i)$, entity id embedding $E_{id}(id_i)$ and position embedding $E_p(p_i)$ of the ith token. $E_w(x_i)$ refers to the hidden state of the ith token extracted from the document embedding. $E_t(\cdot)$, $E_{id}(\cdot)$ and $E_p(\cdot)$ indicate the embedding of entity type, entity id and position information of the token by an embedding layer, respectively. We introduce *None* to represent the entity type and id of tokens that do not belong to any entity. The function $[\cdot : \cdot]$ is a concatenation operation.

Next we first introduce the pre-training data collection process and three pre-training tasks of the encoder, then explain how the output of the encoder is used in our RSCM (Relation Semantic and Context-refined Modeling) model.

Pre-training Data Collection. Following [28], we extract plain text data from Wikipedia[1], and use a ready-made mention recognition and entity linking tools to identify entities. Then the relation between the entity pairs are labeled via distant supervision [8] to construct the pre-training corpus $\tilde{C}$ for pre-training. Instances in pre-training corpus are divided into clusters $\{c^1, c^2, ...\}$ according to the types of relation expressed by them. The instance here represents a short sentence that expresses the fact of the relation between a pair of entity. The formal definition of a instance is $s_i = (Sent_i, r_i, P_h, P_t)$, where $Sent_i$ refers to the sentence sequence, r_i is the relation between h and t in the instance. P_h and P_t indicate the positions of the head entity and tail entity respectively. See Sect. 3 for more details of pre-training.

Relation Semantic Modeling (RSM) is a pre-training task designed by us to strengthen the relation perception ability of the encoder in our RE model. Given embeddings of two instances, the pre-trained model is required to judge whether they express the same relation (i.e., from the same cluster). We mask the mention name under the probability of p to avoid the model learning the name mechanically. For each pair of instances s_i & s_j expressing the same relation, there are K negative instances expressing different types of relation obtained by negative sampling. Based on the idea of contrastive learning [5], this task is trained according to the objective shown in Eq. 4. For two instances expressing the same relation, the task requires the model to embed them with high similarity representation, otherwise to embed them with low similarity representation. In this way, the pre-trained encoder could explicitly model the discriminative relation semantic from diverse expressions.

$$\mathcal{L}_{RSM} = -\sum_{c \subset \tilde{C}} \sum_{s_i, s_j \in c} \log \frac{exp(score(s_i, s_j))}{exp(score(s_i, s_j)) + \sum_{k=1}^{K} exp(score(s_i, s_k))} \tag{4}$$

where $score(\cdot)$ is a scoring function (e.g., Bilinear or cosine function).

Mention Reference Prediction (MRP) has been proved to enhance the pre-trained encoder's coreference reasoning ability for DocRE [28]. We adopt it for pre-training so that the encoder has a robust ability to capture coreferential information. Given the instance sequence s from the corpus $\tilde{C}$, we mask the last

---

[1] https://en.wikipedia.org.

mention m_i of an entity and then restore the masked token by copying other mentions (defined as set C_{m_i}) from the sequence. It is trained by the following objective:

$$\mathcal{L}_{MRP} = -\sum_{s \in \tilde{C}} \log \sum_{m_j \in C_{mi}} Pr(m_j | m_i) \qquad (5)$$

where $Pr(\cdot)$ is the same function as that in CorefBERT, which defines the probability of choosing mention m_j.

Masked Language Modeling (MLM) is used to strengthen the ability of the pre-trained encoder in general language understanding [2]. Following [28], we randomly mask tokens in the sequence and then regard MLM as a cloze task. Formally, given the input sequence s in the corpus $\tilde{C}$ and its masked token set M, this task is trained according to the objective shown in Eq. 6, where $s_{\setminus M}$ means the tokens in s but not in M, and $P(\cdot)$ defines the probability of choosing a token in the vocabulary to restore the masked token m.

$$\mathcal{L}_{MLM} = -\sum_{s \in \tilde{C}} \log \prod_{m \in M} P(m | s_{\setminus M}) \qquad (6)$$

As the pre-training process illustrated in Fig. 2, for the given two input instances from the pre-training corpus $\tilde{C}$, we represent each token by aggregating itself, its delimitation embedding and position embedding. Then we adopt a multi-layer Transformer to encode the representations of input into the final contextual representations, which is used to compute the overall loss for three pre-training tasks according to the objective shown in Eq. 7, where γ_1, γ_2 and γ_3 are parameters to balance the importance of the three tasks.

$$\mathcal{L}_{pretrain} = \gamma_1 \mathcal{L}_{RSM} + \gamma_2 \mathcal{L}_{MRP} + \gamma_3 \mathcal{L}_{MLM} \qquad (7)$$

(2) Entity Embedding Layer
For the embedding H of the input document and the mention-level attention A obtained above, we will next introduce how to utilize them to obtain the final embedding of entity pairs, which is enhanced by their refined context.

For the pair of entities h & t currently considered, we hold that not all mentions of entity h hold relation with another entity t based on the observation mentioned in Sect. 1, and the correlation is inversely proportional to the distance between their respective mentions. So we distinguish the impact of mentions (h_m) on its entity (h) according to their absolute token-distance to the mentions of another entity t. The token-distance here means that when two mentions are separated by t tokens where the distance is t. In this way, the model will pay more attention to the crucial mention pair and their surrounding context. On one hand, the entity-level attention (A_h^i) in the ith attention head is calculated based on its mention-level attention $(A_{h_m}^i)$ as follows:

$$A_h^i = \sum_{h_m \to h} A_{h_m}^i \frac{minDist(h_m, t)}{\sum_{h_m \to h} minDist(h_m, t)} \qquad (8)$$

where $minDist(h_m, t)$ calculates the minimum distance between h_m and mentions of entity t, and $\rightarrow$ indicates the mention h_m belongs to the entity h. $A^i_{h_m}$ is the attention of the start marker of mention h_m in the ith attention head. The attention of tail entity (A^i_t) is calculated in a mirrored way. Then the attention weight of the entity pair h & t to the whole document is calculated and normalized as Eq. 9 & 10. This weight indicates which tokens in the context the current entity pair should pay more attention to.

$$A^i_{(h,t)} = Bilinear(A^i_h, A^i_t) \tag{9}$$

$$\alpha_{(h,t)} = AddandNorm(A^i_{(h,t)})^{\tilde{N}}_{i=1} \tag{10}$$

where $\tilde{N}$ is the number of attention head. Then we can get the refined context embedding related to the pair of entities with the help of entity pair attention $\alpha_{(h,t)}$:

$$C_{(h,t)} = H^\top \alpha_{(h,t)} \tag{11}$$

On the other hand, the interactive embedding ε_h & ε_t of an entity pair is calculated in a similar way as shown in Eq. 12. Then we enhance the final entity embedding with its refined context embedding by a linear layer and an activation layer as shown in Eq. 13 & 14. It is worth mentioning that due to the interaction of distance and context, the same entity is represented differently in different entity pairs.

$$\varepsilon_h = \sum_{h_m \to h} \varepsilon_{h_m} \frac{minDist(h_m, t)}{\sum_{h_m \to h} minDist(h_m, t)} \tag{12}$$

$$E_h = \sigma(W_h \varepsilon_h + W_{ch} C_{(h,t)} + b_h) \tag{13}$$

$$E_t = \sigma(W_t \varepsilon_t + W_{ct} C_{(h,t)} + b_t) \tag{14}$$

where ε_{h_m} is a mention embedding computed by average its token embedding, W_h, W_t, W_{ch}, W_{ct}, b_h, b_t are trainable parameters. The entity embedding layer will enable the model to pay more attention to the mention-pair that more likely to hold the relation facts and the context that is more relevant to the entity pair requiring relation extraction.

(3) Classifier
We formulate the task as multi-label classification task. For the enhanced embeddings of the target entity pair from the previous layer, we send them to a bilinear layer and a fully connected layer to predict their relation:

$$P(r_i|E_h, E_t) = sigmoid(E_h^\mathsf{T} W_{r1} E_t + W_{r2}[E_h : E_t] + b_r) \tag{15}$$

where W_{r1}, W_{r2}, b_r are trainable parameters, $[\cdot : \cdot]$ refers to concatenation operation. In order to ensure the *Recall* metric for DocRE, we consider predicting the relation between each pair of entities in a document. We use binary cross-entropy loss to train our RE model.

$$\mathcal{L} = -\sum_{\mathcal{D} \in S} \sum_{h \neq t} \sum_{r_i \in \mathcal{R}} (r_i) log P(r_i|E_h, E_t) + (1 - r_i) log(1 - P(r_i|E_h, E_t)) \tag{16}$$

Table 1. Statistics on DocRED.

Dataset	Train	Dev	Test	Entities/Doc	Mentions/Doc	Relation
DocRED-Annotated	3053	1000	1000	19.5	26.2	96
DocRED-Distant	101873	–	–	19.3	25.1	96

where $\mathcal{D}$ denotes a document in the corpus S, and r_i is 1 if this relation exists, otherwise 0.

3 Experiments

3.1 Dataset and Evaluation Metrics

The text corpus for pre-training is extracted from Wikipedia[2], then we use an existing entity linking system[3] to link mentions with entities in Wikidata[4]. We further use spaCy[5] to ensure the quality of mention recognition and then match the mention name to Wikidata. If a pair of entities in a sentence have relation r in Wikidata, the sentence will be an instance of the cluster c_r. Each instance is guaranteed to have at least 128 words and express at least one relationship between two entities. We limit the length of the instances within 512 tokens. The pre-training corpus contains 732 relations and 832k instances (approximately 133M words).

We verify the effectiveness of our RSCM model on the largest fine-labeled dataset, DocRED, which is constructed from Wikipedia and Wikidata. This dataset is divided into two parts: human-annotated part and distant-supervised labeled [8] part, both of which provide annotations of mentions, entity types and relation facts (see Table 1 for details). Our segmentation of corpus is follows [27], where train/dev/test set consists of 3053/1000/1000 documents respectively. We use F1 and ign_F1 to evaluate the performance of models in dev and test sets, where ign_F1 refers to not evaluating those triples shared between the train set and the dev/test sets.

3.2 Experiment Settings

For pre-training the encoder, we initialize its parameters with BERT$_{base}$ released by Google[6] to avoid training from scratch. The architecture of the pre-trained encoder is a multi-layer bidirectional Transformer [19]. Its hidden dimension is 768, the size of vocabulary is 28,996, and the parameter scale is slightly larger than that of BERT$_{base}$, about 110M. Additionally, we select some of the same

---

[2] https://en.wikipedia.org.
[3] https://cloud.google.com/natural-language/docs/analyzing-entities.
[4] https://www.wikidata.org.
[5] https://spacy.io/.
[6] https://github.com/google-research/bert.

pre-training hyperparameters as CorefBERT. We adopt Adam optimizer [6] with batch size of 128 and learning rate of 4e-5 after a grid search. The optimization runs 19k steps, where the learning rate is warmed-up over the first 3.9k steps and then linearly decayed. We search the ratio of RSM loss, MRP loss and MLM loss in {1:1:1, 4:4:2, 4:2:4, 2:4:4}, and find the ratio of 4:4:2 achieves the best result.

We train RSCM with contiguous sequences of up to 512 tokens. We use dropout [15] with rate of 0.1 between layers and clip the gradient of model parameters under 1.0. We conduct a grid search on the learning rate in {2e-5, 3e-5, 4e-5} and epoch number in {50, 100, 150}, and finnally adopt AdamW optimizer with batch size of 32, learning rate of 3e-5 and epoch number of 100. All hyper-parameters are tuned on the dev set. Pre-training of encoder and the whole training processes of our RE model are implemented on 4 T V100 GPUs in mixed precision based on the Apex library[7]. We utilize the official evaluation code[8] to evaluate the performance of our RE model.

3.3   Main Result and Analysis

As shown in Table 2, we compare RSCM with the following three types of models:

(1) Sequence-based models. We choose three baselines implemented by [27]. They use CNN or BiLSTM as the document encoder, and then send the embeddings of entity to a simple classifier. The main advantages of RSCM lie in making full use of the general knowledge learned from the large-scale corpus by the pre-trained encoder and more complex mechanisms to capture contextual information.

(2) GNN-based models utilize graph neural network to aggregate and disseminate information [1,9,12,20,29]. GAIN [29] is a state-of-the-art model in this category. From the results, we can see that our model surpasses them in performance even without graph structure. The main reason is that they tend to treat the contribution of mention nodes to their entity node equally when constructing the entity graph, which results in the loss of the relevant context and the negligence of mention pair that actually holds the relation. RSCM can focus on the context that hint the relation facts, which ultimately affect the prediction of the relation between different entity pair.

(3) PLMs-based models. [17,22] predict the relation according to the embeddings of two entities encoded by BERT [2]. Compared with the above two models, CorefBERT achieves the performance gap on DocRE task by further pre-training. This performance gap reveals that coreferential information is of great significant for DocRE. DSDocRE [25] is different from us, as they require the pre-trained model to distinguish the instances that express relation or not, and make the embeddings of the co-occurrence sentences of the same entity pair to be similar. However, their pre-trained model is not sensitive to the type of relation expressed by the instance. In addition, they tend to ignore the influence

---

[7]   https://www.apexlearning.com/resources/library.
[8]   https://github.com/thunlp/DocRED/blob/master/code.

Table 2. Results on DocRED measured by micro ignore F1 (IgnF1) and micro F1. Results are reported in their following citations.

Models	Dev		Test	
	F1	Ign_F1	F1	Ign_F1
Squence-based models				
CNN [27]	43.45	41.58	42.26	40.33
BiLSTM [27]	50.94	48.87	51.06	48.78
Context-Aware [27]	51.09	48.94	50.70	48.40
GNN-based models				
GAT [20]	51.44	45.17	49.51	47.36
GCNN [12]	51.52	46.22	51.62	49.59
EoG [1]	52.15	45.94	51.82	49.48
LSR-GloVe [9]	55.17	48.82	54.18	52.15
LSR-BERT$_{base}$ [9]	59.00	52.43	59.05	56.97
GAIN-GloVe [29]	55.29	53.05	55.08	52.66
GAIN-BERT$_{base}$ [29]	61.22	59.14	61.24	59.00
PLMs-based models				
BERT [22]	54.16	–	53.20	–
BERT-Two-Step [22]	54.42	–	53.92	–
HIN-BERT$_{base}$ [17]	56.31	54.29	55.60	53.70
CorefBERT$_{base}$[28]	57.51	55.32	56.96	54.54
DSDocRE [25]	58.65	57.00	58.43	56.68
RSCM(ours)	**61.77**	**59.58**	**61.60**	**59.55**

of relevant context on the target entity pair. Experiment results show that our model surpasses the previous PLMs-based baselines. There are two main reasons for that: (1) The RSM pre-training task can enhance the ability of the encoder to capture the potential semantic information related to the relation, so that the encoding of entity pairs is enhanced with sufficient information for relation extraction. (2) We adop a hierarchical attention mechanism to make the model focus on the refined context, which is of great significance for relation extraction.

3.4   Ablation Study

We conduct ablation study to investigate the effectiveness of each part of RSCM. The results are shown in Table 3: (i) When we delete RSM/MRP/both of them used for pre-training the encoder, the average performance of the model drops by 2.92, 1.97 and 4.70 respectively. We can see that RSCM is still better than CorefBERT even without RSM or MRP, which shows the efficiency of hierarchical attention mechanism in the entity embedding layer. (ii) After removing entity pair attention and using the original context embedding instead of the

Table 3. Ablation results to illustrate the effectiveness of different parts in RSCM.

Setting	Dev		Test	
	F1	Ign_F1	F1	Ign_F1
RSCM	61.77	59.58	61.60	59.55
w/o RSM	59.02	57.12	58.21	56.49
w/o MRP	60.11	58.27	59.17	57.04
w/o RSM & MRP	57.25	55.28	56.72	54.46
w/o EntityPair Att	58.57	56.58	58.50	56.54

... [6] On 1 December 2001 a joint congress of rival party Unity and *Fatherland - All Russia* decided to <u>merge</u> both parties <u>into</u> a single new political party , *United Russia* . [7] ...

Ground-truth: *founded by, follows*   **CorefBERT:** *founded by*   **RSCM:** *founded by, follows*

[0] *Consort Mei* (died 755) was an <u>imperial consort</u> of the Chinese dynasty Tang Dynasty during the reign of Emperor *Xuanzong* of Tang. [1] She was one of Emperor *Xuanzong* 's favorite <u>concubine</u> ,...

Ground-truth: *spouse*   **CorefBERT:** *child*   **RSCM :** *spouse*

[0] The *Australia – Chile Free Trade Agreement* is a trade agreement between the countries of Chile and Australia. [1] <u>It</u> was signed on July 30 , 2008 and <u>went into effect</u> in *the 1st quarter of 2009* , ...

Ground-truth: *NA*   **CorefBERT:** *NA*   **RSCM :** *inception*

Fig. 4. The results predicted by CorefBERT and RSCM

refined one in the embedding layer, we judge the relation according to the independent embedding of entities extracted from the document embedding, and the metrics drop by 3.08 on average. However, the performance of RSCM is still comparable to that of DSDocRE, which we believe is due to the effectiveness of the pre-training methods.

3.5 Analysis of the RSM Pre-training Task

As shown in Fig. 4, we compare our model with CorefBERT on some cases from DocRED. For the first example, both CorefBERT and RSCM can identify the first relation type *founded by* between *Fatherland-All Russia* and *United Russia*, but CorefBERT cannot discern the second relation type *follows*. It demonstrates the effectiveness of the RSM pre-training task. RSCM can easily perceive the relation type *follows* from the sequence. For the second example, RSCM captures the association between some special words (i.e., *imperial consort* and *concubine*) and the relation type *spouse*, but CorefBERT can not make use of these two special cues used in Chinese, so it predicts wrongly. For the third example, it is obvious that the inception time of the head entity is the tail entity. Our model detects this relation fact, although this is not annotated in DocRED. Since the annotation of data is distant-supervised, and then manually corrected, there is a possibility of wrong labeling. We think that this is because the RSCM can be more sensitive from the expression pattern of *went into effect*, which indicates that we can use the RSM task in the noise reduction of *NA* instances of datasets.

4   Related Work

The current methods of DocRE in general domain can be divided into the following categories:

Sequence-based models use an encoder (such as LSTM) to map all the words in a document into hidden vectors, and then send it to a simple classification layer to predict the relation [14,23]. The length of the input text of them is limited due to the capacity of the encoder. Later, [4,10,11] propose to extract the dependency from the text to connect different sentences to achieve long-distance relation extraction.

GNN-based models generally build a document graph, where the nodes represent mentions or entities, and the edges refer to the dependencies between two nodes. Their classifier judges the relation between entities based on the embedding of the two nodes and edges [1,3,9,12]. Among them, [3] use Attention Guided GCN to construct an edge-weighted graph. [12] introduce dependencies such as adjacent sentences, into the graph for cross-sentence RE. [7] introduce a dynamic span graph framework to disseminate global contextual information. [9] treat the graph as a latent variable and induce relation classification. [29] propose a novel path reasoning mechanism to infer the relation between nodes of entities. In a summary, these models benefit from the inherent advantages of graph structure, but require sophisticated methods to model nodes and edges under the premise of powerful computing capabilities.

PLMs-based models learn general knowledge from large-scale pre-training corpus to strengthen the ability of model in information capturing, which usually achieve better results without complex graph structures. [22] propose to predict whether a relationship exists in an entity pair and then predicts the specific relation types by utilizing BERT. [17] propose a hierarchical model based on BERT. [28] propose MRP to cooperate with MLM, which can effectively enhance the ability of pre-trained model in capturing coreference and achieve remarkable success. [25] require the pre-trained model to judge whether the instance expresses a relational fact, and forcefully map the co-occurring sentences of the same pair of entities into similar embeddings. [26] add entity structure dependencies to the encoder of Transformer [19]. Different from the above work, we design a RSCM pre-training task to enhance the ability of the context encoder to model the relation sematic in diverse expressions. In addition, when predicting the relation between two entities, the model can pay more attention to the relevant context and crucial mention-pair.

A work more related to ours is [13], which proposes a pre-training method for sentence-level RE. The difference between our model and theirs is mainly reflected in: (i) Their RE model can only be used for sentence-level RE, while our model can be applied to DocRE. (ii) In terms of pre-training, they force the instances sharing the same entity pair into similar embedding, regardless of whether these two entities hold relational facts. We adopt distant supervision to align instances that share the same relation, trying to model the general semantic of the same relation in diverse expressions, which is more suitable for relation extraction.

5   Conclusions

In this work, a novel task-specific pre-trained model is proposed for DocRE, which involves a novel task called RSM to model the relation semantic by utilizing relation-aware pre-training data. Attention mechanism is also applied to refine the context for entity pairs, which could enable the model to capture the relational information in the refined context. Extensive experiments on the largest benchmark show that our model has advantages over the baselines.

Acknowledgments. We are grateful to Heng Ye, Jiaan Wang and all reviews for their constructive comments. This work was supported by the National Key R&D Program of China (No. 2018AAA0101900), the Priority Academic Program Development of Jiangsu Higher Education Institutions, National Natural Science Foundation of China (Grant No. 62072323, 61632016, 62102276), Natural Science Foundation of Jiangsu Province (No. BK20191420).

References

1. Christopoulou, F., Miwa, M., Ananiadou, S.: Connecting the dots: Document-level neural relation extraction with edge-oriented graphs. arXiv preprint arXiv:1909.00228 (2019)
2. Devlin, J., Chang, M.W., Lee, K., Toutanova, K.: Bert: Pre-training of deep bidirectional transformers for language understanding. arXiv preprint arXiv:1810.04805 (2018)
3. Guo, Z., Zhang, Y., Lu, W.: Attention guided graph convolutional networks for relation extraction. arXiv preprint arXiv:1906.07510 (2019)
4. Gupta, P., Rajaram, S., Schütze, H., Runkler, T.: Neural relation extraction within and across sentence boundaries. In: Proceedings of the AAAI Conference on Artificial Intelligence, vol. 33, pp. 6513–6520 (2019)
5. Hadsell, R., Chopra, S., LeCun, Y.: Dimensionality reduction by learning an invariant mapping. In: 2006 IEEE Computer Society Conference on Computer Vision and Pattern Recognition (CVPR 2006), vol. 2, pp. 1735–1742. IEEE (2006)
6. Kingma, D.P., Ba, J.: Adam: A method for stochastic optimization. arXiv preprint arXiv:1412.6980 (2014)
7. Luan, Y., Wadden, D., He, L., Shah, A., Ostendorf, M., Hajishirzi, H.: A general framework for information extraction using dynamic span graphs. arXiv preprint arXiv:1904.03296 (2019)

8. Mintz, M., Bills, S., Snow, R., Jurafsky, D.: Distant supervision for relation extraction without labeled data. In: Proceedings of the Joint Conference of the 47th Annual Meeting of the ACL and the 4th International Joint Conference on Natural Language Processing of the AFNLP, pp. 1003–1011 (2009)

9. Nan, G., Guo, Z., Sekulić, I., Lu, W.: Reasoning with latent structure refinement for document-level relation extraction. arXiv preprint arXiv:2005.06312 (2020)

10. Peng, N., Poon, H., Quirk, C., Toutanova, K., Yih, W.T.: Cross-sentence n-ary relation extraction with graph LSTMs. Trans. Assoc. Comput. Linguist. 5, 101–115 (2017)

11. Quirk, C., Poon, H.: Distant supervision for relation extraction beyond the sentence boundary. arXiv preprint arXiv:1609.04873 (2016)

12. Sahu, S.K., Christopoulou, F., Miwa, M., Ananiadou, S.: Inter-sentence relation extraction with document-level graph convolutional neural network. arXiv preprint arXiv:1906.04684 (2019)

13. Soares, L.B., FitzGerald, N., Ling, J., Kwiatkowski, T.: Matching the blanks: Distributional similarity for relation learning. arXiv preprint arXiv:1906.03158 (2019)

14. Song, L., Zhang, Y., Wang, Z., Gildea, D.: N-ary relation extraction using graph state LSTM. arXiv preprint arXiv:1808.09101 (2018)

15. Srivastava, N., Hinton, G., Krizhevsky, A., Sutskever, I., Salakhutdinov, R.: Dropout: a simple way to prevent neural networks from overfitting. J. Mach. Learn. Res. 15(1), 1929–1958 (2014)

16. Sun, H., Ma, H., Yih, W.T., Tsai, C.T., Liu, J., Chang, M.W.: Open domain question answering via semantic enrichment. In: Proceedings of the 24th International Conference on World Wide Web, pp. 1045–1055 (2015)

17. Tang, H., et al.: HIN: hierarchical inference network for document-level relation extraction. In: Lauw, H.W., Wong, R.C.-W., Ntoulas, A., Lim, E.-P., Ng, S.-K., Pan, S.J. (eds.) PAKDD 2020. LNCS (LNAI), vol. 12084, pp. 197–209. Springer, Cham (2020). https://doi.org/10.1007/978-3-030-47426-3_16

18. Tenney, I., Das, D., Pavlick, E.: Bert rediscovers the classical nlp pipeline. arXiv preprint arXiv:1905.05950 (2019)

19. Vaswani, A., et al.: Attention is all you need. arXiv preprint arXiv:1706.03762 (2017)

20. Velikovi, P., Cucurull, G., Casanova, A., Romero, A., Liò, P., Bengio, Y.: Graph attention networks (2017)

21. Verga, P., Strubell, E., McCallum, A.: Simultaneously self-attending to all mentions for full-abstract biological relation extraction. arXiv preprint arXiv:1802.10569 (2018)

22. Wang, H., Focke, C., Sylvester, R., Mishra, N., Wang, W.: Fine-tune bert for docred with two-step process. arXiv preprint arXiv:1909.11898 (2019)

23. Wang, L., Cao, Z., De Melo, G., Liu, Z.: Relation classification via multi-level attention CNNs. In: Proceedings of the 54th Annual Meeting of the Association for Computational Linguistics (Volume 1: Long Papers), pp. 1298–1307 (2016)

24. Weikum, G., Theobald, M.: From information to knowledge: harvesting entities and relationships from web sources. In: Proceedings of the Twenty-Ninth ACM SIGMOD-SIGACT-SIGART Symposium on Principles of Database Systems, pp. 65–76 (2010)

25. Xiao, C., et al.: Denoising relation extraction from document-level distant supervision. arXiv preprint arXiv:2011.03888 (2020)

26. Xu, B., Wang, Q., Lyu, Y., Zhu, Y., Mao, Z.: Entity structure within and throughout: Modeling mention dependencies for document-level relation extraction. arXiv preprint arXiv:2102.10249 (2021)

27. Yao, Y., et al.: Docred: A large-scale document-level relation extraction dataset. arXiv preprint arXiv:1906.06127 (2019)
28. Ye, D., et al.: Coreferential reasoning learning for language representation. arXiv preprint arXiv:2004.06870 (2020)
29. Zeng, S., Xu, R., Chang, B., Li, L.: Double graph based reasoning for document-level relation extraction. In: Proceedings of the 2020 Conference on Empirical Methods in Natural Language Processing (EMNLP) (2020)

TDM-CFC: Towards Document-Level Multi-label Citation Function Classification

Yang Zhang[1,2(✉)], Yufei Wang[2], Quan Z. Sheng[2], Adnan Mahmood[2],
Wei Emma Zhang[3], and Rongying Zhao[1]

[1] School of Information Management, Wuhan University, Wuhan, Hubei, China
yangz10@whu.edu.cn
[2] Department of Computing, Faculty of Science and Engineering,
Macquarie University, Sydney, NSW 2109, Australia
[3] School of Computer Science, The University of Adelaide,
North Terrace, Adelaide, SA 5005, Australia

Abstract. Citation function classification is an indispensable constituent of the citation content analysis, which has numerous applications, ranging from improving informative citation indexers to facilitating resource search. Existing research works primarily simply treat citation function classification as a sentence-level single-label task, ignoring some essential realistic phenomena thereby creating problems like data bias and noise information. For instance, one scientific paper contains many citations, and each citation context may contain rich discussions of the cited paper, which may reflect multiple citation functions. In this paper, we propose a novel task of *Document-level Multi-label Citation Function Classification* in a bid to considerably extend the previous research works from a sentence-level single-label task to a document-level multi-label task. Given the complicated nature of the document-level citation function analysis, we propose a novel two-stage fine-tuning approach of large scale pretrained language model. Specifically, we represent a citation as an independent token and propose a novel two-stage fine-tuning approach to better represent it in the document context. To enable this task, we accordingly introduce a new benchmark, i.e., *TDMCite*, encompassing 9594 citations (annotated for their function) from online scientific papers by leveraging a three-aspect citation function annotation scheme. Experimental results suggest that our approach results in a considerable improvement in contrast to the state-of-the-art BERT classification fine-tuning approaches.

Keywords: Citation function · Masked language model · BERT · Natural language processing

1 Introduction

Citation function is generally considered to be the underlying rationale for why an author utilizes a cited paper as a reference [25]. People usually leverage on

© Springer Nature Switzerland AG 2021
W. Zhang et al. (Eds.): WISE 2021, LNCS 13081, pp. 363–376, 2021.
https://doi.org/10.1007/978-3-030-91560-5_26

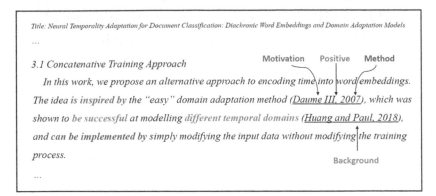

Fig. 1. An example of a citation context extracted from a scientific paper [11]. Words in red color imply that the citation is leveraged as the motivation. Words in color are the linguistic cue for the corresponding citation function labels. Note that in our task, those words are not directly given. (Color figure online)

a cited paper as the main motivation of their own research, for the background information pertinent to several concepts, as the approach undertaken for their baseline model, and sometimes to show their positive or negative reflection about a cited paper. As an essential constituent of the citation content analysis [10, 31], better understanding of a citation function can facilitate us to (a) create better informative citation indexers [25], (b) identify the critical references [26], and (c) search the online scientific paper in a database [3].

Given the enormous volume of online scientific publications proliferating with each passing day, it is impractical to perform the manual labeling of citation functions, and therefore, an automatic citation function classification system becomes essential.

However, in a real scenario, building the citation function classification systems can be a complicated chore. First of all, a single citation sentence within the scientific paper may include multiple citations. Secondly, each citation may encompass a variety of contextual information implying multiple functions. In addition, a citation sentence's linguistic structure can be complex. Figure 1 depicts a concrete example of a complicated situation. It is quite evident that *(Daume III, 2007)*'s idea motivates the citing paper's research and its approach is leveraged as the baseline model of the citing paper. Besides, the citing paper's comment expresses a positive sentiment of the citation. On the contrary, *(Huang and Paul, 2018)* only provide one background information from the cited paper to the citing paper. We also note that the span with blue color implies one function of the citation *(Daume III, 2007)*, whereas, in linguistic structure, it is much closer to the citation *(Huang and Paul, 2018)*. Therefore, it may become difficult for the classification model to tackle the above illustrated scenario.

In addition, we require the function of all citations from the target scientific paper to facilitate further analysis, i.e., to measure the evolution of a scientific field by citation function frames [14]. Previous studies primarily treat the citation function classification task in the form of a sentence-level single-label task [1,3,13,22,23] which, in turn, leads to three major problems: 1) annotating the instance selectively from different paper and not from a document-level perspective which may results in a data bias [3], 2) assigning a single-label to a sentence may cause ambiguity [1,14] and it's not easy to handle the scenario as depicted in Fig. 1, and 3) in the citation context, different context spans may imply different citation functions which cannot be easily covered with a single citation function label.

To address the above stated problems, in this paper, we introduce a new task of *Document-level Multi-label Citation Function Classification (TDM-CFC)* that offers a viable solution with a novel benchmark. To model a realistic citation usage scenario, if a citation context carries diverse contextual information, which implies multiple citation functions, we give the citation multiple function label. To achieve a more nuanced and a precise control of the citation representation, in our approach, we leverage Marked Language Model (MLM) from BERT [5] to create a new masked token to obtain a better representation of citation directly from data. Compared with the standard BERT classification task fine-tuning method which employs special classification token ([CLS]) for representing a sentence, our approach facilitates the model to become more aware of the position of a citation, thereby helping citation build an appropriate relation with the input contextual information. The experimental results depict that our approach outperforms the conventional BERT fine-tuning method for our classification task. Furthermore, we envisage a new annotation scheme which models the three-aspect of the citation function from a document-level perspective by analyzing the purpose of citing the literature. Based on the scheme, we construct a new dataset, *TDMCite* for fine-tuning, via manually annotating around eight thousand resource contexts.

This paper makes three main contributions: 1) we introduce a novel task of modeling document-level multi-label citation function of scientific papers, 2) we develop a two-stage fine-tuning approach for large-scale pre-trained language model (e.g., BERT) to better represent citation in the citation function classification task, and 3) we finally introduce *TDMCite*, i.e., a new dataset encompassing 9594 manually labeled data for citation function classification task from a document-level perspective with ten fine-grained categories[1].

2   Related Work

The study of citation function classification can be traced as early as 1965 when Garfield [6] proposed fifteen categories of citation function, and a similar scheme was also introduced by Weinatoek [29] in 1971. Moravcsik and Murugesa [18] proposed a scheme encompassing four orthogonality facets of citation function: conceptual or operational, evolutionary or juxtapositional, organic or perfunctory, and confirmative or negational in 1975. Those works only provide limited analysis because the manually annotated data scale is small.

---

[1]   https://github.com/young1010/TDM-CFC.

With the development of data mining and NLP techniques in the past decade, automatic citation function classification has become popular [9] and there has been considerable effort in this task [10,12,12,16]. In 2006, Teufel et al. [25] proposed an automatic citation classification scheme with different citation functions and citation sentiment. Using both shallow and linguistically-inspired features, they presented a supervised machine learning framework to classify citation function automatically and acquired good performance. Inspired by Moravcsik and Murugesa's annotation scheme [18], Jochim and Schütze [13] introduced a new feature designed with lexical features and linguistic features and integrated them into the Stanford MaxEnt classifier [17] to improve the classification accuracy. Abu-Jbara et al. [1] proposed an annotation scheme of six categories, mixed the function and sentiment of the citation. They leveraged a sequence labeling model to identify the citation context window. Therefore, they trained a first-order chain-structured CRF (Conditional Random Fields) model with 3,500 manually labeled citation instances. In 2018, Jurgens et al. [14] used the citation frames to measure the evolution of the computer linguistic field by leveraging a six-categories citation function frame. They trained a machine learning classifier with 1,436 manually annotated citation function instances. In 2019, Zhao et al. [33] proposed a new task for modeling the online resource citations' function by leveraging a multitask framework with 3,088 manually annotated data samples. Cohan et al. [3] proposed structural scaffolds, a multitask model to classify citation function by integrating structural information of scientific paper. They conducted a dataset including more than 8000 manually annotated citation instances. However, it only has three categories, and the authors selected the instance from various scientific paper other than acquiring one single paper's all citation instances. Pride and Knoth [20] introduced a new methodology for annotating citations purpoes and finally they constructed a new dataset of 11,233 citations annotated by 883 authors.

These works mainly take the citation function classification task as a sentence-level single-label task. We believe that treating this task as a document-level multi-label task can be a better possible solution.

3   Semi-supervised Multi-label Citation Function Classification

In this section, we first talk about our citation function classification task, and then introduce the annotation scheme of our dataset and the proposed approach.

3.1   Task Definition

We formulate our *Document-level Multi-label Citation Function Classification* task with mathematical notations. Different from the previous task, we propose an approach to building the connection between the citation token and its surrounding contextual information. Therefore, in our method, we take citation as a special token in the input context. The input of our model is a sequence of words $C = [x_1, x_2, \cdots, x_c^1, \cdots, x_c^2, \cdots, x_c^m, \cdots, x_n]$, where x_i, $i \in \{1, 2, \cdots, n\}$

Table 1. The definition and distribution of citation function categories in our dataset.

Aspect	Category	Description	Count
Relationship between papers	Motivation	The cited work serve as the key point/starting point of citing paper	140
	Comparison	All kinds of compare and contrast between different cited paper or with the citing paper	1,084
	Extension	The citing work's research work is a kind of improvement or extension of the the cited work	203
	Application	The citing work adopted the cited work's any kind of knowledge	2,107
Content of cited paper	Background	A citation instance in which the cited work offers background knowledge or factual statements of the citing paper	3,978
	Method	A citation instance in which the cited work offers the method or technique of the cited paper	4,339
	Data	A citation instance in which the citing work mention the data from the cited paper	766
	Result	A citation instance in which the citing work mention the specific results or general findings of the cited paper	555
Sentiment	Positive	The citation in the citing work described affirmative information of the cited paper	655
	Negative	The citation in the citing work was derogatory	440

denotes the word character and x_c^j, $j \in \{1, 2, \cdots, m\}$ denotes for the citation. The expected output is a set of the citation function label $Y = \{y_1, y_2, \cdots, y_w\}$, where y_r, $r \in \{1, 2, \cdots, w\}$ denotes the citation function. Each x_c^j may correspond to single or multiple y_r. The proposed task is to learn the citation function $f : C \rightarrow Y$, i.e., to predict label Y for the citation x_c^j according to the given contextual information C.

This task is nontrivial. In the previous sentence-level citation function classification task [1,3,14], the model encodes the selected citation context sentence sequence and does not take citation as a particular part of the sequence which means the granularity of the method is too coarse. We also note that manually selecting the citation context window out of the citation sentence will significantly affect the performance and cause some noise to the classification model [14]. Some previous works use the sequence labeling method [1] to find the citation context, but they still fail to build the relationship between the citation and its contextual information. In contrast, we note that employing the sequence labelling method to our task maybe to fine-grained and create irreverent information to the citation. Therefore, in our task, we propose a method that focuses on the citation itself to predict its function.

3.2   Annotation Scheme

As depicted in Fig. 1, one citation's function can be complicated, and the function of citation can come from different aspects [16]. Therefore different from the previous annotation scheme, our classification focus on capturing as many broad thematic functions that a citation can serve in the discourse as possible from three different aspects. First, inspired by [14], from the perspective of the relationship between the citing paper and cited paper, we propose four functions: *Motivation, Comparison, Extension,* and *Application.* Second, in the citation content perspective, we extent previous work's scheme [3], and sequentially we get four functions: *Background, Method, Data,* and *Result.* Finally, we note that some of the highly cited previous works leverage the citation sentiment as part of the citation functions. Therefore, we follow them [1,9,15,25,34]. Table 1 shows our entire annotation scheme, along with the description and distribution of each category.

3.3   Partial Masked Language Model

For the text classification task, BERT model is the state-of-the-art and leveraged by many works [4,7,21]. The first token of every input sequence for BERT is always a special classification token ([CLS]). The final hidden state corresponding to this token is used as the aggregate sequence representation for text classification tasks [5]. Masked language model (MLM) pre-training, as epitomized by BERT [5], has proven wildly successful [24]. As the pre-training objective, Masked Language Modeling, different from a traditional language modeling which predicts the next word in a sequence given the history, the masked language modeling predicts a word given its left and right context, therefore, allow each word to indirectly "see itself" [5]. Therefore, in our proposed method, we leverage this character of BERT to help citation aware of its position as well as meaning in the given contextual information. In the training process, MLM randomly mask 15% WordPiece [30] token of the input sequence. If the i-th token is chosen, the model replace the i-th token with (1) the $[MASK]$ token 80% of the time (2) a random token 10% of the time (3) the unchanged i-th token 10% of the time. Then, the output token T_i will be used to predict the original token with cross entropy loss [5].

Recently, MLM has been proved not only helpful during the pre-training, but also can be used as fine-tuning method for downstream tasks [28,32]. Motivated by this, we propose a partially masked language model which is modified from the masked language model [5] to better represent citation for our task. We note that reference mark e.g., [author names],[number] is an important constituent of the sentence when people write their literature. In some sentences, it also can serve as a subject e.g., *"Brack et al.[08] leverage the implementation of..."* or an object e.g., *"We followed [05] to incorporate...".* Figure 2 shows our two-stage classification model architecture. For the first stage, we leverage MLM to fine-tune the BERT model with unlabeled data of citation context. The training objective of stage one is to predict whether there is a citation or not in the given

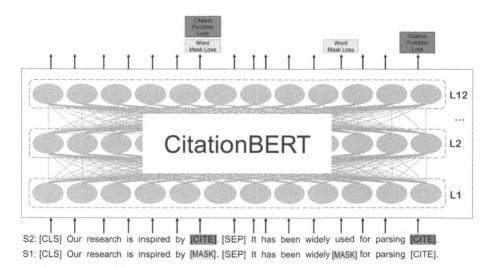

Fig. 2. Our proposed two-stage classification model for identifying citation functions. S1 stands for stage one: we leverage the unlabeled data to fine-tune the model to help the model aware of citation and its position. S2 stands for stage two: we leverage the labeled data to train our model with a multi-label loss function.

input sentences. The purpose is to make the model aware of the citation's position in the input sequence. For the second stage, we continue to fine-tune the model from stage one with labeled training data. To be specific, in stage two, we use *Masked Citation Token* to represent the citation in the sentence and train the model with a multi-label loss function. In the whole process, we leverage the character of MLM, considering citation as an indispensable constituent of a sentence and help the citation token to build the relationship with its surrounding input contextual information. For stage two, following [5], we first encode the input C, then we have embedding sequence $E^c = \{e_1, e_2, \cdots, e_c^1, \cdots, e_c^2, \cdots, e_c^m, \cdots, e_n\}$, where e_i denotes the embedding of word character x_i, e_c^j denotes the embedding of x_c^j. BERT consists of a stack of 12 identical blocks taking the entire sequence as input. Each block contains a multi-head self-attention operation followed by a feed-forward network. As the basic building block of BERT, a self-attentive layer[27] takes a citation sentence sequence of token embeddings, E^c, as inputs and uses k attention heads H^i to model:

$$MH(E^c) = cat([H^1(E^c), \cdots, H^k(E^c)])W' \tag{1}$$

In H, each word embedding E_i^c is projected into Query, Key, and Value vectors. Subsequently, the inter-word attention scores are computed using Query and Key vectors. The value vectors are aggregated accordingly to obtain the attention scores:

$$H^i(E^c) = Softmax(\frac{E^c W_Q^i \cdot (E^c W_K^i)^T}{\sqrt{D_k}}) \cdot E^c W_V^i \tag{2}$$

Finally, the position-wise feed-forward network is computed, followed by a LayerNorm layer [27]:

$$FFN(E^c) = LayerNorm(relu(MH(E^c)W_1 + b_1)W_2 + b_2) \tag{3}$$

We denote the sequence of hidden states at the final layer of BERT as $H_c = \{h_1, h_2, \cdots, h_c^1, \cdots, h_c^2, \cdots, h_c^m, \cdots, h_n\}$. For each citation of the sequence, the output logits of its corresponding citation is defined as:

$$l_c^j = W_s \cdot h_c^j + b_s \tag{4}$$

where l_c^j denotes the output logits. W_s and b_s are parameters.

3.4   Training Loss

In our classification model the training data is given which consist of pairs of citation and its function label, denoted as $= \{(C_1, Y_1), (C_2, Y_2), \cdots, (C_n, Y_n)\}$. Cross entropy loss as the classical loss function is widely used in text classification [5,14,19]. Therefore, we first conduct our experiment by passing the output logits l_c^j of our model to the cross entropy loss:

$$Loss(l_c^j, Y) = -\log\left(\frac{exp(l_c^j[Y])}{\sum_p exp(l_c^j[p])}\right) \tag{5}$$

where p denotes the index. Cross entropy loss only encourages one label for each instance. However, in our task, one instance may have more than one label need to be predicted, thus the recall of cross entropy loss in our task cannot be guaranteed. Therefore, we conduct a multi-label training which may help to improve the recall. We pass the output logits l_c^j of our model to the multi-label soft margin loss [8]:

$$Loss(l_c^j, Y) = -\frac{1}{w}\sum_{q=1}^{w} \cdot Y[q] \cdot \log\frac{1}{1 + exp(-l_c^j[q])}$$
$$+(1 - Y[q]) \cdot \log\frac{exp(-l_c^j[q])}{1 + exp(-l_c^j[q])} \tag{6}$$

where w denotes the number of classes, $Y[q] \in \{0,1\}$, q denotes the index. Multi-label loss function can significantly improve the performance since for each instance it turn multiple results for each label which can help with the recall. We refer the experimental details in next section.

4   Experiments and Results

This section presents the experimental evaluation of our two-stage fine-tuning method in comparison to baseline methods.

4.1   Data Collection Process

Following previous work's strategy [14], our annotation guidelines are created by using a pilot study of 20 papers sampled from the ACL Anthology[2]. Three annotators major in NLP and deep learning completed two rounds of pre-annotation to discuss their process and design guidelines. Our annotation scheme is applied to a randomly selected sample of 230 papers from the proceedings of ACL and EMNLP from 2018–2020. Each paper is processed to extract all context with citation. We first convert PDF files into TXT files and manually fixed all format and grammar mistakes. Then we covert all reference marks e.g., [author names], [number], into the string (citation#). Note that we merge the citation string into one standard string (citation#), since in one citation bracket all reference are present the same function [16]. All citations are then doubly-annotated trained with expertise in NLP and deep learning. The decision is made by a majority rule when a conflict occurs. Finally, 9,594 annotated citation instance are acquired for the fully-labeled 230 papers. Table 1 shows the class distribution in our dataset. For the unlabeled data, we randomly sample 100 papers from ACL Anthology.

4.2   Evaluation Metrics and Implementation

We use Precision, Recall, and F1 score, which are widely used in classification tasks, as our evaluation metrics. In addition, the evaluation metrics for our citation function classification task need to cover all the label sets and need to consider the data imbalance in each category. Accordingly, we leverage the Micro F1 and Marco F1 as our evaluation metrics.

We use pre-trained language model implementation in the *huggingface library*[3]. We begin with the $BERT_{BASE}$ model (uncased, 12-layer, 786-hidden, 110M parameters). The learning rate is set to 5e-5, the batch size is 32, and the epoch is 20. We check out validation after every 200 steps. Our dataset is split into: training set (80%), development set (10%) and test set (10%). For our classification task, we compared our model with the state-of-the-art $BERT_{BASE}$ baseline model by following its standard fine-tuning approach which using [CLS] token to fine-tune [5] the model for the text classification task. To be specific, we first compare the $BERT_{BASE}$ model which only integrates citation sentence as the input ($BERT_{sent.}$). Secondly, we compare the $BERT_{BASE}$ model which takes citation sentence and its context (sentences surrounding the citation) as the input ($BERT_{sent.+con.}$), following previous work [1,3,14]. We compare the results of those approaches with our partial masked language model ($P\text{-}MLM + U$ means with unlabeled data) at a single-label and a multi-label citation function classification task. We note that citations in the same paragraph their functions are more closely related comparing with citations in different paragraphs; thereby we encode the whole paragraph, which contains citation(s) as input.

---

[2]   https://www.aclweb.org/anthology.
[3]   https://huggingface.co/transformers/master/index.html.

4.3   Experimental Results

Our main results for the *TMDcite* dataset are shown in Table 2. We observe that our P-MLM model achieves clear improvements over the state-of-the-art standard BERT fine-tuning approach on our citation function classification task.

Table 2. Experimental results on our dataset (Prec.-Precision; Rec.-Recall).

	Method	Micro			Macro		
		Prec.	Rec.	F1	Prec.	Rec.	F1
Single-label	$BERT_{Sen.}$	76.6	51.0	61.2	62.5	33.2	38.2
	$BERT_{Sen.+con.}$	74.0	49.3	59.2	60.4	27.6	34.6
	$P\text{-}MLM$	76.7	51.2	61.4	64.7	33.9	39.8
	$P\text{-}MLM + U$	**77.7**	**51.8**	**62.1**	**78.3**	**38.9**	**44.8**
Multi-label	$BERT_{Sen.}$	73.1	69.9	71.5	58.8	53.9	55.2
	$BERT_{Sen.+con.}$	72.5	68.1	70.2	62.0	51.6	54.6
	$P\text{-}MLM$	73.0	73.5	73.2	63.8	**58.2**	**60.1**
	$P\text{-}MLM + U$	**75.9**	**73.0**	**74.4**	**66.7**	56.1	58.8

From a document-level perspective, manually selecting citation context may cause ambiguity to the classification model, thus affecting the performance. We first observe that $BERT_{sent.}$ outperforms $BERT_{sent.+con.}$ by 2.0 Micro F1 and 3.6 Macro F1 in single-label classification. It shows that the citation context may cause noise to the classification model in some situations [1,14]. As depicted in Fig. 1, we note that in the real scenario, the citations in one scientific paper can be dense. However, owing to the span of the sentence, the function of the citation may not directly connect to the citation sentence. Therefore we argue that taking the citation function classification as a sentence level task may not be the best option.

Our proposed approach achieves the best performance among all models. We observe that in single-label citation function classification task, our approach receives the best precision, recall, Micro F1 score and Marco F1 score. In multi-label citation function classification task, our $P\text{-}MLM + U$ achieves the best Micro F1 score. In addition, $P\text{-}MLM$ outperforms $BERT_{sent.}$ by 1.4 Micro F1 and 1.2 Marco F1 in single-label classification and by 1.7 Micro F1 and 4.9 Micro F1 in multi-label classification. The rationale is that in our approach, we take citation as a constituent of the sentence by using partial masked language modeling to build the relationship between citation and its contextual information and thus better represent citation.

Multi-label is needed in citation function classification. By comparing the results of all three models between single-label citation function classification and multi-label citation function classification, we note that there is a significant improvement. Our $P\text{-}MLM$ achieves the best performance in multi-label citation function classification by 73.2 Micro F1 score and 60.1 Marco F1 score. We obtain 10.2 Marco F1 and 19.7 Micro F1 improvement in $P\text{-}MLM$ which is the best

Table 3. Detailed category wise classification results (F1 score). For category, we present the abbreviation. (Moti.-Motivation Comp.-Comparison Exte.-Extension Appl.-Application Bac.-Background Meth.-Method Res.-Result Pos.-Positive Neg.-Negative)

	Category	Moti.	Comp.	Exte.	Appl.	Bac.	Meth.	Data.	Res.	Pos.	Neg.
Multi-label	$BERT_{Sen.}$	42.8	71.0	**25**	72.4	75.9	**81.5**	78.2	48.8	48.5	7.8
	$P\text{-}MLM$	62.0	71.1	24.0	**76.3**	75.5	80.2	**86.0**	**51.8**	47.1	**26.8**
	$P\text{-}MLM + U$	**64.0**	**75.8**	10.0	75.0	**77.8**	81.0	85.3	51.2	**50.2**	17.5

performing model. The results prove that taking citation function classification as a document-level multi-label is scientifically right. In this way, we not only obtain the fine-grained citation function but also benefit the classification task.

From Table 2, we further observe that the performance of $P\text{-}MLM$ outperforms $P\text{-}MLM + U$ for multi-label classification. To investigate the possible reasons behind, we report the F1 score for each category in Table 3. Comparing with the baseline model, our model achieves the best results on eight out of ten categories, indicating that our approach is effective to the task. We observe that after leveraging the unlabeled data, the F1 score of Extension (Exte.) and Negative (Neg.) dramatically decrease and those two categories' F1 scores are significantly lower compared with the other categories. Therefore, the Macro F1 score of $P\text{-}MLM + U$ decline. Comparing Table 1's distribution and Table 3's F1 score, we find that the smaller scale of the data of those two categories may directly lead to the unstable performance, because to train the deep leaning model, we need comprehensive data to obtain a stable performance. In our future work, we will put more effort on improving those tail categories' performance.

5   Conclusion

In this paper, we study the problem of citation function classification and propose a novel task named document-level multi-label citation function classification. To handle the complicated scenario of document-level citation function in scientific paper, we further propose a two-stage fine-tuning approach of pre-trained language model by taking citation as an essential constituent of the citation sentence. We introduce a new benchmark consisting of 9594 manually labeled data with ten categories fine-grained annotation scheme.

The experimental results show that our proposed approach is effective. For the future work, we plan to explore pre-trained BERT for scientific domains such as SciBERT [2] using our approach to help more tasks such as the evaluation and prediction. We also plan to integrate more techniques such as multi-task learning [3,33] to model the dependency of different citation function categories that co-occur. Finally, we will investigate how to improve the performance of tail categories.

Acknowledgements. The research work of Yang Zhang is funded under the auspices of the Macquarie University's Cotutelle-International Macquarie Research Excellence Scholarship. The research work of Yufei Wang is funded by a MQ Research Excellence Scholarship and a CSIRO's DATA61 Top-up Scholarship. This research is further funded in part by Australian Research Council (ARC) Discovery Project DP200102298 and National Social Science Fund Major Project of China (No. 18ZDA325).

References

1. Abu-Jbara, A., Ezra, J., Radev, D.: Purpose and polarity of citation: Towards nlp-based bibliometrics. In: Proceedings of the 2013 Conference of the North American Chapter of the Association for Computational Linguistics: Human Language Technologies, pp. 596–606 (2013)
2. Beltagy, I., Lo, K., Cohan, A.: Scibert: a pretrained language model for scientific text. In: Proceedings of the 2019 Conference on Empirical Methods in Natural Language Processing and the 9th International Joint Conference on Natural Language Processing (EMNLP-IJCNLP), pp. 3615–3620 (2019)
3. Cohan, A., Ammar, W., van Zuylen, M., Cady, F.: Structural scaffolds for citation intent classification in scientific publications. In: Proceedings of the 2019 Conference of the North American Chapter of the Association for Computational Linguistics: Human Language Technologies, vol. 1, pp. 3586–3596 (2019)
4. Croce, D., Castellucci, G., Basili, R.: Gan-bert: generative adversarial learning for robust text classification with a bunch of labeled examples. In: Proceedings of the 58th Annual Meeting of the Association for Computational Linguistics, pp. 2114–2119 (2020)
5. Devlin, J., Chang, M.W., Lee, K., Toutanova, K.: Bert: pre-training of deep bidirectional transformers for language understanding. In: Proceedings of the 2019 Conference of the North American Chapter of the Association for Computational Linguistics: Human Language Technologies, vol. 1, pp. 4171–4186 (2019)
6. Garfield, E., et al.: Can citation indexing be automated. In: Statistical Association Methods for Mechanized Documentation, Symposium Proceedings, vol. 269, pp. 189–192. Washington (1965)
7. Garg, S., Ramakrishnan, G.: BAE: BERT-based adversarial examples for text classification. In: Proceedings of the 2020 Conference on Empirical Methods in Natural Language Processing, pp. 6174–6181 (2020)
8. He, J., Li, C., Ye, J., Qiao, Y., Gu, L.: Multi-label ocular disease classification with a dense correlation deep neural network. Biomed. Signal Process. Control **63**, 102167 (2021)
9. Hernández, M., Gómez, J.M.: Survey in sentiment, polarity and function analysis of citation. In: Proceedings of the First Workshop on Argumentation Mining, pp. 102–103 (2014)
10. Hernández-Alvarez, M., Gomez, J.M.: Survey about citation context analysis: tasks, techniques, and resources. Nat. Lang. Eng. **22**(3), 327–349 (2016)
11. Huang, X., Paul, M.J.: Neural temporality adaptation for document classification: diachronic word embeddings and domain adaptation models. In: Proceedings of the 57th Annual Meeting of the Association for Computational Linguistics. Florence, Italy (2019)
12. Jha, R., Abu-Jbara, A., Qazvinian, V., Radev, D.R.: Nlp-driven citation analysis for scientometrics. Nat. Lang. Eng. **23**(1), 93–130 (2017)

13. Jochim, C., Schütze, H.: Towards a generic and flexible citation classifier based on a faceted classification scheme. In: Proceedings of International Conference on Computational Linguistics 2012, pp. 1343–1358 (2012)
14. Jurgens, D., Kumar, S., Hoover, R., McFarland, D., Jurafsky, D.: Measuring the evolution of a scientific field through citation frames. Trans. Assoc. Comput. Linguist. **6**, 391–406 (2018)
15. Li, X., He, Y., Meyers, A., Grishman, R.: Towards fine-grained citation function classification. In: Proceedings of the International Conference Recent Advances in Natural Language Processing RANLP 2013, pp. 402–407 (2013)
16. Lyu, D., Ruan, X., Xie, J., Cheng, Y.: The classification of citing motivations: a meta-synthesis. Scientometrics **126**(4), 3243–3264 (2021). https://doi.org/10.1007/s11192-021-03908-z
17. Manning, C., Klein, D.: Optimization, maxent models, and conditional estimation without magic. In: Proceedings of the 2003 Conference of the North American Chapter of the Association for Computational Linguistics on Human Language Technology: Tutorials, vol. 5, p. 8 (2003)
18. Moravcsik, M.J., Murugesan, P.: Some results on the function and quality of citations. Soc. Stud. Sci. **5**(1), 86–92 (1975)
19. Ohashi, S., Takayama, J., Kajiwara, T., Chu, C., Arase, Y.: Text classification with negative supervision. In: Proceedings of the 58th Annual Meeting of the Association for Computational Linguistics, pp. 351–357 (2020)
20. Pride, D., Knoth, P.: An authoritative approach to citation classification. In: Proceedings of the ACM/IEEE Joint Conference on Digital Libraries in 2020, pp. 337–340 (2020)
21. Qin, Q., Hu, W., Liu, B.: Feature projection for improved text classification. In: Proceedings of the 58th Annual Meeting of the Association for Computational Linguistics, pp. 8161–8171 (2020)
22. Roman, M., Shahid, A., Khan, S., Koubaa, A., Yu, L.: Citation intent classification using word embedding. IEEE Access **9**, 9982–9995 (2021)
23. Roman, M., Shahid, A., Uddin, M.I., Hua, Q., Maqsood, S.: Exploiting contextual word embedding of authorship and title of articles for discovering citation intent classification. Complexity **2021** (2021)
24. Sinha, K., Jia, R., Hupkes, D., Pineau, J., Williams, A., Kiela, D.: Masked language modeling and the distributional hypothesis: Order word matters pre-training for little. arXiv preprint arXiv:2104.06644 (2021)
25. Teufel, S., Siddharthan, A., Tidhar, D.: Automatic classification of citation function. In: Proceedings of the 2006 Conference on Empirical Methods in Natural Language Processing, pp. 103–110 (2006)
26. Valenzuela, M., Ha, V., Etzioni, O.: Identifying meaningful citations. In: Workshops at the Twenty-Ninth AAAI Conference on Artificial Intelligence, vol. 15, p. 13 (2015)
27. Vaswani, A., et al.: Attention is all you need. In: Advances in Neural Information Processing Systems, pp. 5998–6008 (2017)
28. Wang, A., Cho, K.: Bert has a mouth, and it must speak: Bert as a Markov random field language model. In: Proceedings of the Workshop on Methods for Optimizing and Evaluating Neural Language Generation, pp. 30–36 (2019)
29. Weinatoek, M.: Citation indexes. Encycl. Libr. Inf. Sci. **5**, 16–40 (1971)
30. Wu, Y., et al.: Google's neural machine translation system: Bridging the gap between human and machine translation. arXiv preprint arXiv:1609.08144 (2016)

31. Zhang, G., Ding, Y., Milojević, S.: Citation content analysis (CCA): a framework for syntactic and semantic analysis of citation content. J. Am. Soc. Inf. Sci. Technol. **64**(7), 1490–1503 (2013)
32. Zhang, S., Huang, H., Liu, J., Li, H.: Spelling error correction with soft-masked Bert. In: Proceedings of the 58th Annual Meeting of the Association for Computational Linguistics, pp. 882–890 (2020)
33. Zhao, H., Luo, Z., Feng, C., Zheng, A., Liu, X.: A context-based framework for modeling the role and function of on-line resource citations in scientific literature. In: Proceedings of the 2019 Conference on Empirical Methods in Natural Language Processing and the 9th International Joint Conference on Natural Language Processing, pp. 5209–5218 (2019)
34. Zhao, R., Zhang, Y., et al.: Evolution study of sentiment analysis based on bibliometrics of time and space dimensions. Inf. Sci. **36**(10), 171–177 (2018). in Chinese

NOCOL - Nonnegative Orthogonal Constraint Outlier Learning

Thirunavukarasu Balasubramaniam[1]([⊠]), Wathsala Anupama Mohotti[2], Richi Nayak[1], and Chau Yuen[3]

[1] Queensland University of Technology, Brisbane, Australia
thirunavukarasu.balas@qut.edu.au
[2] University of Ruhuna, Matara, Sri Lanka
[3] Singapore University of Technology and Design, Singapore, Singapore

Abstract. Identifying anomalous documents in a text corpus is an important problem that has wide applications. Due to the high dimensional and sparse nature of text data, traditional outlier detection methods fail to identify features that distinguish outliers. Inspired by the capability of Nonnegative Matrix Factorization (NMF) for text clustering, we explore it for text outlier detection. In this paper, a novel NMF-based method called Nonnegative Orthogonal Constraint Outlier Learning (NOCOL) is introduced that learns the outliers effectively during the factorization process. Experimental results show the higher accuracy of NOCOL in identifying text outliers in comparison to the state-of-the-art methods.

Keywords: Nonnegative matrix factorization · Outlier detection · Selective learning · Text outliers · Orthogonal constraint

1 Introduction

Digitalization of news and contextual user interactions have resulted in large digital archives and repositories, in which text outlier detection plays a major role [5]. Outlier detection can be defined as a process of identifying an abnormal behavior that highly deviates from the normal behavior [1]. In the last decade, social media has contributed to the generation of abundant short-text data. It becomes challenging to analyze this sparse data with a lesser number of terms in the posts for identifying semantic dissimilarity between text documents and detecting outliers [18]. As most of the real-world data is unlabelled, unsupervised learning methods like density-based [3], distance-based [11,15], graph-based [6], and distribution-based [7] methods have been commonly used for outlier detection.

Distribution-based methods identify outliers as overfitting to a normal model [7]. This category of methods depends on the assumption of data in the normal model when identifying deviations [8]. They are known not to be scalable

Supplementary Information The online version contains supplementary material available at https://doi.org/10.1007/978-3-030-91560-5_27.

W. Zhang et al. (Eds.): WISE 2021, LNCS 13081, pp. 377–385, 2021.
https://doi.org/10.1007/978-3-030-91560-5_27

to high-dimensional data. Distance-based methods define data points as outliers if they are far from many other points in the dataset considering a minimum distance threshold such as k-NN. However, use of this approach in high-dimensional data where distance differences between points become negligible is challenging [11]. The large computational complexity inherent with these methods makes them less effective for big datasets such as digitized text corpora.

Graph-based outlier detection is another approach used in outlier detection [6]. A graph representation with the local information around each object is used to construct a local information graph and calculates the outlier score by performing a tailored Markov random walks [17]. However, the computational complexity of these graph-based methods is high due to calculating the similarities between each object. Deep learning based techniques such as Autoencoders have been used for outlier detection in supervised or semi-supervised setting [19]. While deep learning based methods are promising for outlier detection in numeric data, it suffers from the complexity issues when dealing with high-dimensional data such as in the text outlier detection.

Lower-dimensional projection were introduced as a remedy to effectively handle high dimension data. The degree of deviation of each observation for its neighbors, after projecting it to the lower-dimensional space, is measured to identify the outliers in [11]. NMF is a commonly used technique for document clustering. However, when the data is sparse, identifying semantic similarity among the documents becomes challenging due to a few sparse features in the lower-rank representation, and NMF fails to capture cluster structure accurately [10].

Leveraging these fine qualities of NMF, in this paper, an NMF-based method, Nonnegative Orthogonal Constraint Outlier Learning (NOCOL) is proposed to learn the text outliers during the factorization process based on a gradient descent-based factorization algorithm. NOCOL learns the lower-rank representation such that the outlier documents are identified in text corpus accurately. The main contributions of this paper are two folded: Firstly, a novel outlier learning factorization algorithm based on gradient descents is proposed to select terms in representing outlier documents and focus on learning the outlier documents rather than non-outlier documents. Secondly, an additional orthogonal constraint on document features/factor matrix is considered in NOCOL to enhance the distinctiveness between outliers and the rest of corpus.

2   Nonnegative Martix Factorization (NMF): Basics

Let $\mathcal{D}$ be the set of documents, $\mathcal{D} = \{d_1, d_2, ..., d_N\}$ and $\mathcal{T}$ be the set of terms, $\mathcal{T} = \{t_1, t_2, ..., t_M\}$. The document-term matrix can be represented as, $\mathbf{X} \in \mathbb{R}^{(N \times M)}$, where N and M represents the dimensions of the matrix (i.e. N documents and M terms). Matrix $\mathbf{X}$ is presented with the weighting scheme of Term Frequency - Inverse Document Frequency (TF-IDF) [16]. The goal of NMF is to factorize $\mathbf{X}$ into two low-rank factor matrices $\mathbf{W} \in \mathbb{R}_+^{(N \times R)}$ and

$\mathbf{H} \in \mathbb{R}_+^{(M \times R)}$ as given in (1) where R represents the rank of the lower-rank matrices.

$$\mathbf{X} \cong \mathbf{W}\mathbf{H}^T \quad s.t., \mathbf{W}_{nr} \geq 0, \ \mathbf{H}_{mr} \geq 0, \ \forall \ n, m, r, \tag{1}$$

where $\mathbf{W}_{nr}$ indicates the $(n, r)^{th}$ element of $\mathbf{W}$ and $\mathbf{H}_{mr}$ indicates the $(m, r)^{th}$ element of $\mathbf{H}$.

3   NOCOL – Nonnegative Orthogonal Constraint Outlier Learning

The overall process of NMF-based outlier detection method NOCOL is shown in Fig. 1. The objective is to learn the values of factor matrices $\mathbf{W}$ and $\mathbf{H}$ based on the document $\times$ term matrix $\mathbf{X}$ as formulated in (1). The process starts with randomly initialized matrices of $\mathbf{W}$ and $\mathbf{H}$. During the factorization process, NOCOL iteratively finds the values of these matrices, such that they produce the best approximation of $\mathbf{X}$.

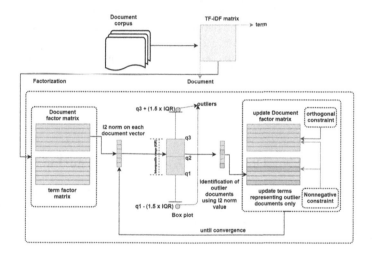

Fig. 1. Overall process of NOCOL.

NOCOL presents an effective method to select terms (relevant vector) representing outlier documents and focus the factorization learning towards outlier documents rather than non-outlier documents. This assists in preserving the distance of outlier documents with the rest of documents while projecting the high-dimensional to a lower-rank representation. An additional orthogonal constraint is considered in NOCOL to enhance the distinctiveness between outliers and non-outliers. NOCOL applies $l2$ norm on each row in $\mathbf{W}$ (i.e. a vector $\mathbf{w}_n$). The vector $\mathbf{w}_n$ presents a document and $l2$ norm on each row returns a positive value, the measure of vector length. Since each component of the vector is squared in measuring length, the outliers receive either higher or lower weighting deviating highly from the inliers. By measuring the vector length of all documents, the dispersion of documents in vector space can be identified. This helps to systematically detect the outlier documents.

For documents that have been recognized as potential outliers, their terms are identified and only those terms are updated in the $\mathbf{H}$ matrix in the next step. During the factor matrices update, nonnegativity is imposed on both factor matrices and orthogonal constraint is imposed on the document factor matrix.

Next, we formulate the objective function for orthogonal constraint NMF and then we provide the algorithmic solution to solve the orthogonal constraint NMF to detect outlier documents.

3.1 Orthogonal Constraint NMF

For NMF to satisfy orthogonality, one of the factor matrices, either $\mathbf{W}$ or $\mathbf{H}$, needs to be orthogonal such that $\mathbf{W}\mathbf{W}^T = \mathbf{I}$ or $\mathbf{H}\mathbf{H}^T = \mathbf{I}$.

Since our objective is to highlight outlier documents during the factorization process, we propose to introduce the orthogonal constraint on $\mathbf{W}$. Hence, the document factor matrix $\mathbf{W}$ will act as the bases to the term factor matrix. This will help to learn the outlier terms distinctly and detect outlier documents based on those terms during the factorization process.

The proposed orthogonal constraint NMF for outlier detection can be formulated using the following objective function as,

$$\min_{\mathbf{W}\in\mathbb{R}_+^{(N X R)},\mathbf{H}\in\mathbb{R}_+^{(M X R)}} f(\mathbf{W},\mathbf{H}) = \left\|\mathbf{X} - \mathbf{W}\mathbf{H}^T\right\|^2 \quad s.t., \mathbf{W}\mathbf{W}^T = \mathbf{I}. \quad (2)$$

3.2 Learning Outlier Documents: Algorithmic Solution

During the optimization process of the objective function formulated in (2), the factor matrices $\mathbf{W}$ and $\mathbf{H}$ are randomly initialized and alternatively updated until convergence. We propose to update the matrices by a selective learning process focusing on learning outliers. It consists of three steps as detailed below.

Step 1: Understanding Data Dispersion and Calculating Outlier Scores. The first task is to understand the data dispersion of documents and calculate the initial outlier score for each document using $l2$ norm and Z-score.

For each iterative update step of $\mathbf{W}$, $l2$ norm of each document vector $\mathbf{w}_n$ in $\mathbf{W}$ is calculated. The $l2$ norm on each document vector $\mathbf{w}_n$ returns a positive value, the measure of vector length. This (positive) score is called an intermediate outlier score and assigned to each document. The higher or lower the value, the further or nearer the document from the origin of vector space, and the higher the chance for it to be an outlier.

The $l2$ norm of each document (i.e. row-wise), say for n^{th} document is calculated as,

$$v_n = \|\mathbf{w}_n\|_2, \quad (3)$$

where v_n is the $l2$ norm value of nth document vector $\mathbf{w}_n$ in $\mathbf{W}$. By calculating $l2$ norm of each document vector as per Eq. (3), the $l2$ norms of all the documents are represented in a vector as,

$$\mathbf{v} = [v_1, v_2, ..., v_n]. \tag{4}$$

The Z-score of each document vector, say for n^{th} document is calculated as,

$$o_n = \frac{v_n - \mu(\mathbf{v})}{\sigma(\mathbf{v})}, \tag{5}$$

where $\mu(\mathbf{v})$ is the mean of $\mathbf{v}$ and $\sigma(\mathbf{v})$ is the standard deviation of $\mathbf{v}$. Similarly, the Z-score of all the documents is calculated and represented in a vector as,

$$\mathbf{o} = [o_1, o_2, ..., o_n]. \tag{6}$$

The higher the Z-score, the higher the chance is of that document being an outlier. Therefore, $\mathbf{o}$ can be used as the potential outlier score of documents.

Step 2: Identifying Outlier Documents and Respective Terms. After calculating outlier scores for all documents, the next task is to detect outliers. The data points of $\mathbf{o}$ are divided into three quartiles $q1$, $q2$ and $q3$ and are calculated as 25th percentile, *median*, and 75th percentile respectively. The interquartile range (IQR) is calculated as,

$$IQR = q3 - q1. \tag{7}$$

Outliers are defined as any points falling below $q1 - (1.5 \times IQR)$ and above $q3 + (1.5 \times IQR)$ following the standard statistical measure [13]. Quartiles of data points can be best visualized using box plot as shown in Fig. 1. Once the potential outlier documents are identified, the terms representing those outlier documents are identified. Suppose document $d1$ is represented by terms $t1$, $t5$, and $t6$. If $d1$ is the document identified as an outlier, then the term vectors for terms $t1$, $t5$, and $t6$ would be identified in the factor matrix $\mathbf{H}$.

Step 3: Updating Factor Matrices. To learn the outliers, we propose to update only the selected term vectors (representing the potential outlier documents) in term factor matrix $\mathbf{H}$. Continuing from the previous example, only three rows ($\mathbf{h}_1$, $\mathbf{h}_5$, and $\mathbf{h}_6$) of $\mathbf{H}$ representing terms $t1$, $t5$, and $t6$ will be updated. After updating selected rows of $\mathbf{H}$, the whole $\mathbf{W}$ will be updated.

Since only the term vectors representing outlier documents are updated, the row-wise Multiplicative Update (MU) [4] rule is defined to update selected rows of $\mathbf{H}$ as follows,

$$\mathbf{h}_s \leftarrow \frac{(\mathbf{H}.(\mathbf{X}^T\mathbf{W}))_s}{(\mathbf{H}\mathbf{W}^T\mathbf{W})_s} \quad \forall s \in \mathbf{s}, \tag{8}$$

where $\mathbf{h}_s$ is s^{th} row of $\mathbf{H}$ and $\mathbf{s}$ is the set of indices of terms representing outlier documents.

The orthogonal constraint update rule to update $\mathbf{W}$ becomes,

$$\mathbf{W} \leftarrow \frac{\mathbf{W}.\mathbf{XH}^T}{\mathbf{WW}^T\mathbf{XH}}. \tag{9}$$

The entire process of steps 1,2 and 3 is iteratively repeated until convergence or the number of iterations limit is reached. This process of NOCOL allows NMF to highlight and learn outliers during the factorization process. Finally, the outlier scores of documents are calculated as per step 1 using the latest updated document factor matrix $\mathbf{W}$.

Table 1. Datasets: sizes and density of document $\times$ term matrix.

	Docs	Unique terms	Total terms	Outliers	Density of $\mathbf{X}$
D1	4909	27882	374642	50	0.0027
D2	5050	13438	200482	50	0.0030
D3	8050	8614	148745	50	0.0021
D4	20100	15851	437914	100	0.0014

4   Experiments and Discussion

Experiments are conducted with real-world datasets to demonstrate the effectiveness of NOCOL in detecting the outliers. Four real-world datasets (Table 1), 20newsgroup (D1), Reuters (D2), SED13 (D3) and SED14 (D4), have been used. All these datasets have labeled information that was used to generate subsets with outliers and clusters of inliers.

4.1   Baseline Methods and Evaluation Criteria

The source code of NOCOL have been made available[1]. We chose a number of relevant methods as benchmarks both traditional and NMF based: 1) k-nearest neighbor based method (KNNO) with outlier percentage in each dataset as the threshold [15], 2) density-based local outlier factor method (LOFO) [3], 3) Pairwise mutual neighbor graph method (MNCO) [6], 4) Generative Adversarial Active Learning Network method (GAAL) [12], 5) NMF [2], and 6) ONMF [9].

We also tested the usefulness of orthogonal constraint employed in NOCOL by experimenting without the constraint (called as NCOL).

Accuracy of outlier detection is analyzed using standard metrics of Receiver Operating Characteristics (ROC) curve and Area Under the ROC curve (AUC). ROC curves in the paper are drawn by taking the optimal threshold point. AUC and ROC focus on correct outlier predictions only and disregard the incorrect inlier predictions. To understand the true accuracy performance, we also need to analyze the inliers' predictions. We report the True Negative Rate (TNR)(closer to 1 is best) and False Negative Rate (FNR)(closer to 0 is best) to measure the inlier predictions' capability of a method.

---

[1] https://github.com/thirubs/NOCOL.

4.2   Accuracy Performance

As shown in Fig. 2, NOCOL outperforms all baseline methods significantly based on *ROC* curves, i.e., *AUC*. NOCOL is the only method that performs well in both the conditions when the document is small-sized (D3 and D4) or medium-sized (D1 and D2). Distance-based and density-based methods like KNNO and LOFO work on D3 and D4 successfully due to the low percentage of outliers in those datasets compared to the size of the total collection. In high dimensionality, these concepts are unable to differentiate points due to distance concentration. MNCO becomes computationally expensive due to the pair-wise comparisons and runs out of memory (o.m) for D4. The performance of MNCO is degraded because of the sparsity of the datasets. With less number of pair-wise comparisons, the similarity calculation becomes challenging in MNCO leading to poor performance. GAAL is an outlier detection method for numeric data and it is not significant for text data. Using the document term matrix as an input for GAAL is a time-consuming task and took nearly seven days to process D1. Therefore, the documents are represented using Glove [14] (standard vector representation technique) to reduce the document dimensions to 200. Due to this dimensionality reduction, GAAL loses information and results in poor performance.

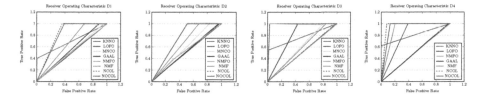

Fig. 2. *ROC* Curve.

Table 2. Accuracy performance; o.m: out of memory.

Method	D1	D2	D3	D4	Avg.	D1		D2		D3		D4	
	AUC measure					FNR	TNR	FNR	TNR	FNR	TNR	FNR	TNR
KNNO	0.70	0.51	0.77	0.81	0.70	0.60	**0.99**	0.98	**0.99**	0.46	**0.99**	0.38	**0.99**
LOFO	0.50	0.52	0.78	0.81	0.65	0.00	0.00	0.00	0.04	0.00	0.57	0.00	0.61
MNCO	0.52	0.58	0.60	o.m	0.57	0.00	0.04	0.00	0.16	0.00	0.20	o.m	o.m
GAAL	0.56	0.52	0.50	0.50	0.52	0.00	0.11	0.00	0.03	0.00	0.01	0.00	0.00
ONMF	0.52	0.54	0.50	0.50	0.52	0.64	0.67	0.62	0.69	0.00	0.00	0.00	0.00
NMF	0.66	0.68	0.58	0.90	0.71	0.00	0.31	0.00	0.32	0.00	0.16	0.00	0.80
NCOL	**0.82**	**0.74**	0.51	**0.96**	0.76	0.00	0.64	0.00	0.49	0.00	0.96	0.00	0.91
NOCOL	0.80	**0.74**	**0.98**	0.94	**0.87**	0.00	0.61	0.00	0.49	0.00	0.97	0.00	0.87

The inherent clustering property of NMF has shown to be an advantage in outlier detection as well. However, the NMF-based method ONMF fails to detect outliers due to the additional parameters that need to be defined. NCOL and NOCOL are also NMF-based methods with sophisticated outlier learning techniques. As shown by the AUC performance in Table 2, NOCOL achieves 16% to 35% improved performance in comparison to baseline methods. Results also show the impact of orthogonal constraint in NOCOL in comparison to NCOL that fails to perform on D3 converging to a local minimum.

5 Conclusion

An effective NMF method for outlier detection is presented in this paper. The proposed NOCOL method first introduces an orthogonal constraint into NMF to learn the factor matrices distinctively and then the factorization process is modified such that the outlier documents are learned effectively. Our experimental results show a significant improvement in detecting outliers in comparison to baseline methods.

References

1. Aggarwal, C.C.: Outlier analysis. In: Data Mining, pp. 237–263. Springer, Cham (2015). https://doi.org/10.1007/978-3-319-14142-8_8
2. Allan, E.G., Horvath, M.R., Kopek, C.V., Lamb, B.T., Whaples, T.S., Berry, M.W.: Anomaly detection using nonnegative matrix factorization. In: Survey of Text Mining II, pp. 203–217. Springer, Heidelberg (2008). https://doi.org/10.1007/978-1-84800-046-9_11
3. Breunig, M.M., Kriegel, H.P., Ng, R.T., Sander, J.: Lof: identifying density-based local outliers. ACM Sigmod Rec. **29**(2), 93–104 (2000)
4. Choi, S.: Algorithms for orthogonal nonnegative matrix factorization. In: 2008 IEEE International Joint Conference on Neural Networks (IEEE World Congress on Computational Intelligence), pp. 1828–1832. IEEE (2008)
5. Dong, X.L., Srivastava, D.: Big data integration. In: ICDE, pp. 1245–1248. IEEE (2013)
6. Ertöz, L., Steinbach, M., Kumar, V.: Finding clusters of different sizes, shapes, and densities in noisy, high dimensional data. In: SDM, pp. 47–58. SIAM (2003)
7. Gokcesu, K., Neyshabouri, M.M., Gokcesu, H., Kozat, S.S.: Sequential outlier detection based on incremental decision trees. IEEE Trans. Signal Process. **67**(4), 993–1005 (2018)
8. Jackson, D.A., Chen, Y.: Robust principal component analysis and outlier detection with ecological data. Environmetrics Off. J. Int. Environmetrics Soc. **15**(2), 129–139 (2004)
9. Kannan, R., Woo, H., Aggarwal, C.C., Park, H.: Outlier detection for text data: an extended version (2017). arXiv preprint arXiv:1701.01325
10. Li, T., Ding, C.c.: Nonnegative matrix factorizations for clustering: a survey. In: Data Clustering, pp. 149–176. Chapman and Hall/CRC (2013)
11. Liu, H., Li, X., Li, J., Zhang, S.: Efficient outlier detection for high-dimensional data. IEEE Trans. Syst. Man Cybern. Syst. **48**, 2451–2461 (2017)

12. Liu, Y., et al.: Generative adversarial active learning for unsupervised outlier detection. IEEE Trans. Knowl. Data Eng. **32**(8), 1517–1528 (2020)
13. McGill, R., Tukey, J.W., Larsen, W.A.: Variations of box plots. Am. Stat. **32**(1), 12–16 (1978)
14. Pennington, J., Socher, R., Manning, C.: Glove: global vectors for word representation. In: Proceedings of the 2014 Conference on Empirical Methods in Natural Language Processing (EMNLP), pp. 1532–1543 (2014)
15. Ramaswamy, S., Rastogi, R., Shim, K.: Efficient algorithms for mining outliers from large data sets. ACM Sigmod Rec. **29**(2), 427–438 (2000)
16. Salton, G., Buckley, C.: Term-weighting approaches in automatic text retrieval. Inf. Process. Manag. **24**(5), 513–523 (1988)
17. Wang, C., Liu, Z., Gao, H., Fu, Y.: Vos: a new outlier detection model using virtual graph. Knowl.-Based Syst. **185**, 104907 (2019)
18. Wang, H., Bah, M.J., Hammad, M.: Progress in outlier detection techniques: a survey. IEEE Access **7**, 107964–108000 (2019)
19. Wang, X., Zheng, Q., Zheng, K., Sui, Y., Cao, S., Shi, Y.: Detecting social media bots with variational autoencoder and k-nearest neighbor. Appl. Sci. **11**(12), 5482 (2021)

Semantic Parsing with Syntax Graph of Logical Forms

Chen Chang[1,2(✉)]

[1] Peking University, Beijing, China
leon_chang@pku.edu.cn
[2] Center for Data Science of Peking University, Beijing, China

Abstract. Semantic parsing aims to convert natural language queries to logical forms, which are strictly structured. Recently neural semantic parsers have paid attention to structure information of target logical forms and set constraints on generating rules. In this work, we propose to use syntax graphs of both query and logical form and to utilize graph neural networks (GNN) in encoder combined with BERT pre-training and decoder with copy mechanism. Besides, we present a predicate review loss function to help GNNs in the decoder capture the syntax graph structure more precisely. Results of experiments on three datasets show that our model outperforms the baseline on MSParS, achieves state-of-art accuracy on ATIS, and has competitive performance on Job.

Keywords: Semantic parsing · GNN · Syntax graph

1 Introduction

Sequence-to-sequence models that are widely used in machine translation have also been proved to be an effective approach. Models based on Long Short-Term Memory network (LSTM) [6] or Transformer [13] achieve great results in semantic parsing. However, models need to face two main challenges that do not exist in machine translation problems.

The first challenge is that target sentence is structured far more strictly than natural language. For example, the lambda calculus for *what do these cost* is `lambda $0 e ( fare $0 )` in ATIS. The word *fare* cannot be substituted by synonyms like *cost*. Nevertheless, making good use of the structure features can be helpful to the generation process.

The second one is OOV (out-of-vocabulary) words in logical forms. Target sentence usually contains entities mentioned in the natural language query. Because entities may vary for each example, no matter how large the training set is, generating entities according to the vocabulary of the training set cannot handle this OOV issue.

In this paper, we introduce a framework of semantic parsing based on Pointer Network [10] and syntax graphs of both natural language queries and target logical forms. Copy mechanism applied in Pointer Network helps to figure out the

© Springer Nature Switzerland AG 2021
W. Zhang et al. (Eds.): WISE 2021, LNCS 13081, pp. 386–393, 2021.
https://doi.org/10.1007/978-3-030-91560-5_28

OOV issue, which is a common approach. Meanwhile, we represent the dependency relation of natural language query sentences and the tree structure of target logical forms as their syntax graphs. Graph neural networks (GNNs) provide a vector representation for each syntax graph node. Masks are applied on the graph while decoding to prevent the model from knowing the complete logical forms before prediction.

Furthermore, we propose a parent review loss function to enhance the ability to extract structure features of logical forms and train the model with both this loss function and generation loss of logical forms.

We evaluate our parser on three datasets: Job, ATIS, and MSParS. Results show that our model performs better than baselines on all these datasets, is competitive with the state-of-art model on Job, and outperforms state-of-art results on ATIS.

2 Syntax Graph

Before discussing model details, we first present syntax graphs of both natural language queries and logical forms, which our model takes as input.

2.1 Graph of Natural Language Queries

The target is to convert a sequence of words $X = \{[BOS], x_1, x_2, \ldots, x_n\}$ to a directed graph $G = (X, E)$. Edge set E is composed of two types of edges: 1) *edges in dependency tree of the sentence* and 2) *edges between adjacent words*. We generate the dependency tree by by HPSG parser [16]. In addition, we set a label for each of the two types of edges.

2.2 Graph of Logical Forms

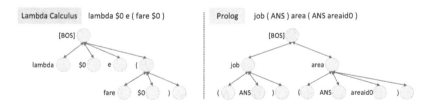

Fig. 1. Syntax graphs of lambda calculus (left) and prolog (right), where edges are all bidirectional.

The three datasets contain logical forms of two types: prolog for Job and lambda calculus for ATIS and MSParS. Figure 1 depicts their syntax graphs.

Prolog. As Fig. 1 shows, the uppercase characters denote variables, and the predicates specify the constraints between variables. In this case, we choose predicates to be the parent nodes of their variables.

Lambda Calculus. Lambda calculus defines an expression that returns true if the entity satisfies the constraints defined in the expressions. We apply a simple rule to convert the expression to a graph, which takes the left parenthesis as a parent node and the corresponding right parenthesis as the end of the subtree.

3   Semantic Parser with Syntax Graphs

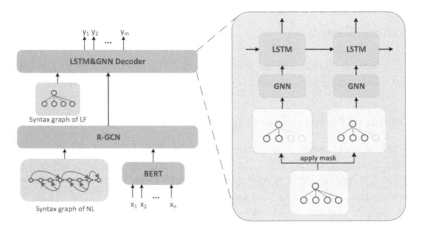

Fig. 2. An overview of our model. We use syntax graphs of both NL and LF. The right part shows the detail structure of the decoder.

Our model follows the encoder-decoder structure. Our target is to convert the query $x = x_1, \ldots, x_n$ into the logical form $y = y_1, \ldots, y_m$.

3.1   Encoder

We obtain the initial representations of words $\{x_i, i = 1, 2, ..., n\}$ by BERT [2], a pre-trained model based on the structure of Transformer, and then the syntax graph is fed into N layers of R-GCN [9]. The output of the last layer of R-GCN $x_i^{(N)}, i = 0, 1..., n$ will be fed into the decoder, where h denotes the hidden size.

3.2   Decoder

The decoder is a combination of M-layer GNN with self-attention [14] and a unidirectional LSTM with attention and copy mechanism.

Temporary Syntax Graph Input. To enable the model to capture the structure feature of logical forms, we feed a temporary syntax graph into the decoder at each step. During training, a mask is applied on the full syntax graph G to ensure the decoder cannot see the nodes after the t-th node at the t-th step. The right part of Fig. 2 depicts this process. The temporary graph is then fed into M layers of Graph Attention Networks (GAT). The node embedding of the last layer is

$$\phi(y_t) = \text{GAT}(y_t) \tag{1}$$

We use an elementwise max pooling to obtain the graph embedding

$$\psi(G_t) = \text{max\_pooling}_i(\phi(y_t)). \tag{2}$$

The usage of the graph embedding will be introduced in Sect. 3.3.

LSTM with Copy Mechanism. At the t-th step, we first calculate the distribution $P_{\text{vocab}}(y_t|y_{<t}, x)$ over vocabulary by LSTM and the attention to the i-th word of the natural language query a_i^t. To solve the OOV issue, we need to enable the decoder to copy the word from the natural language query. We follow the implementation of See et al. [10]. Let $P_{\text{copy}}(y_t = w|y_{<t}, x)$ denote the probability of generating the OOV word w.

$$P_{\text{copy}}(y_t = w|y_{<t}, x) = \sum_{k:x_k=w} a_k^t. \tag{3}$$

A gate variable $g_t \in (0, 1)$ controls whether the decoder generates from vocabulary or copy from the query.

$$z_t = W_z[c_{t-1}; \phi(y_t)] \tag{4}$$

$$g_t = \text{sigmoid}(v_g^T[c_t; s_t; z_t] + b_g) \tag{5}$$

where c_t is the context vector of LSTM and s_t is the cell state of LSTM at t-th step. Thus the final distribution is

$$P(y_t|y_{<t}, x) = g_t P_{\text{vocab}}(y_t|y_{<t}, x)$$
$$+ (1 - g_t) P_{\text{copy}}(y_t|y_{<t}, x). \tag{6}$$

3.3   Predicate Review Loss Function

Like parent feeding in Seq2Tree [3], we propose a parent-aware decoder by a new loss function. As Fig. 3 shows, in addition to predicting the next word of the logical form, the decoder also reviews the nearest predicate, and the distribution over the vocabulary depends on the syntax graph embedding.

$$P(p_t|y_{<t}, x, G_t) = \text{softmax}(W_p[c_t; s_t; z_t; \psi(G_t)] + b_p). \tag{7}$$

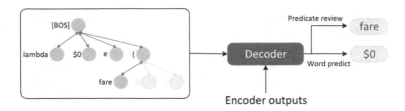

Fig. 3. An overview of predicate review. When the model predicts the next word $0,
it will also try to remember the nearest predicate `fare`.

So the predicate review loss is

$$l_p = \text{cross\_entropy}(P(p_t|y_{<t}, x, G_t), \text{one\_hot}(\hat{p}_t)) \qquad (8)$$

where $\hat{p}_t$ is the gold predicate (the nearest predicate from y_t).

The loss function for training our model is a combination of predicate review
loss and logical form generation loss l_g.

$$loss = \lambda l_p + (1 - \lambda)l_g \qquad (9)$$

where $\lambda \in [0, 1]$ is a hyperparameter.

4   Experiments

4.1   Datasets

We consider three datasets to evaluate our model and replace entity mentions
in datasets with numbered markers.

Job. We use the version provided by [3], where 640 examples are split into 400
training examples and 240 testing examples.

ATIS. ATIS is the abbreviation of Air Travel Information System consisting of
queries about air travel. We use the version by [4], which contains 4434 training
examples and 448 testing examples.

MSParS. MSParS (v1.0) [5] is a large-scale dataset for semantic parsing. We
use the single-turn queries consisting of over 60,000 pairs of natural language
queries and logical forms. We keep the original train/dev split in which there
are 54690 training examples and 7703 development examples.

4.2   Experimental Setup

Model Configuration. We use 1280-dimensional word vectors for decoder,
1280 hidden units for encoder and 640 units for decoder. The embedding vectors
for the query side are computed by $\text{BERT}_{\text{BASE}}$. Number of layers of the GNNs
is set to 3 for encoder and 2 for decoder. We balance the two loss functions and
set λ to 0.5.

Training. We use AdamW optimizer [7] with starting learning rate 0.003 and the same warmup and decay strategy as Vaswani et al. [13]. The number of warmup steps is set to 2000 and the batch size is selected from {16, 32} for each task.

Evaluation. We evaluate accuracy by exact match accuracy with gold logical forms for all datasets used in the experiment, directly comparing with output words.

4.3   Results and Analysis

Main Results. Accuracies on Job, ATIS, and MSParS are illustrated separately in Table 1, 2.

Table 1. Test accuracies on Job and ATIS

Model	ATIS	Job
Seq2Tree [3]	84.6	90.0
Pointer network [10]	85.7	87.1
Rabinovich et al.(2017) [8]	85.9	**91.4**
Graph2Seq [15]	85.9	91.2
Cao et al.(2019) [1]	89.1	–
Shaw et al.(2019) [11]	89.7	–
Ours	**90.4**	**91.4**

Table 2. Accuracies on MSParS

Model	Acc
Seq2Seq [12]	73.1
Baseline(BERT + Pointer)	76.8
Ours	**78.4**

Job and ATIS. Our model achieves competitive performance on Job and state-of-art results on ATIS. On both two datasets, our model significantly outperforms Seq2Tree, with 5.8% and 1.4% improvement of accuracy respectively. One possible reason is that our model takes the temporary graph that contains all nodes generated before into consideration. However, Seq2Tree only considers nodes that are on the same path and generation is independent between nodes on different paths.

In comparison with Graph2Seq, which uses rich syntactic information of natural language queries, the additional syntactic graph of logical forms enables our model to improve the accuracy by more than 4% in ATIS. Besides, Shaw et al. also applies GNN layers in the decoder and construct the input graph by the sequential order of generation. Instead of this limited structure feature, the syntax graph we build covers more structure features, fully utilizing the GNN layers.

MSParS. On MSParS, we compare our model with Seq2Seq model and baseline model to demonstrate that syntactic information may improve performance. Baseline model is based on Pointer Network, replacing the embedding layer of encoder with $BERT_{BASE}$. As Table 2 shows, our model outperforms these two models.

Ablation Study. In this section, we explore the contributions of various components in our system.

Table 3. Ablation study on ATIS

Model	Acc
Ours (full model)	**90.4**
- w/o Predicate Review	88.8
- w/o Predicate Review & Syntax Graph	87.3

Table 4. Acc. on ATIS with λ.

λ	Test accuracy
0	88.8
0.3	90.2
0.5	90.4
0.8	89.5

Predicate Review. We train our model only by logical form generation loss to investigate how much our new loss function affects performance on ATIS.

The test accuracy is improved by 1.6% with predicate review loss function. We also adjust λ and list the result in Table 4. Accuracy drops when we set λ too large, and it is evident that if λ is set to 1, the model will collapse and fail to learn from training.

Syntax Graph of Logical Forms. We use the single word as input of the decoder instead, replacing the GNN layers with an embedding layer. Since the predicate review needs syntax graph embedding as input, only logical form generation loss is used for training. As Table 3 shows, test accuracy drops by 1.5% and 3.1% lower than the full model. One explanation for this drop is that the decoder is more challenging to learn syntax information from a sequence of words than from a graph.

5   Conclusion

In this paper, we present a semantic parsing framework based on syntax graphs of both natural language queries and logical forms, enabling the parser to learn from syntactic information through GNNs. We also propose a novel loss function to improve the ability of our parser to capture structure features so that the parser can generate reasonable logical forms. Experimental results show that syntax graphs can enhance performance across datasets.

References

1. Cao, R., Zhu, S., Liu, C., Li, J., Yu, K.: Semantic parsing with dual learning. In: Proceedings of the 57th Annual Meeting of the Association for Computational Linguistics, pp. 51–64 (2019)
2. Devlin, J., Chang, M.W., Lee, K., Toutanova, K.: Bert: Pre-training of deep bidirectional transformers for language understanding. arXiv preprint arXiv:1810.04805 (2018)
3. Dong, L., Lapata, M.: Language to logical form with neural attention. In: Proceedings of the 54th Annual Meeting of the Association for Computational Linguistics (Volume 1: Long Papers), pp. 33–43 (2016)
4. Dong, L., Lapata, M.: Coarse-to-fine decoding for neural semantic parsing. In: Proceedings of the 56th Annual Meeting of the Association for Computational Linguistics (Volume 1: Long Papers), pp. 731–742 (2018)
5. Duan, N.: Overview of the NLPCC 2019 shared task: open domain semantic parsing. In: Tang, J., Kan, M.-Y., Zhao, D., Li, S., Zan, H. (eds.) NLPCC 2019. LNCS (LNAI), vol. 11839, pp. 811–817. Springer, Cham (2019). https://doi.org/10.1007/978-3-030-32236-6_74
6. Hochreiter, S., Schmidhuber, J.: Long short-term memory. Neural Comput. 9(8), 1735–1780 (1997)
7. Loshchilov, I., Hutter, F.: Decoupled weight decay regularization. In: International Conference on Learning Representations (2018)
8. Rabinovich, M., Stern, M., Klein, D.: Abstract syntax networks for code generation and semantic parsing. In: Proceedings of the 55th Annual Meeting of the Association for Computational Linguistics (Volume 1: Long Papers), pp. 1139–1149 (2017)
9. Schlichtkrull, M., Kipf, T.N., Bloem, P., van den Berg, R., Titov, I., Welling, M., et al.: Modeling relational data with graph convolutional networks. In: Gangemi, A. (ed.) ESWC 2018. LNCS, vol. 10843, pp. 593–607. Springer, Cham (2018). https://doi.org/10.1007/978-3-319-93417-4_38
10. See, A., Liu, P.J., Manning, C.D.: Get to the point: Summarization with pointer-generator networks. In: Proceedings of the 55th Annual Meeting of the Association for Computational Linguistics (Volume 1: Long Papers), pp. 1073–1083 (2017)
11. Shaw, P., Massey, P., Chen, A., Piccinno, F., Altun, Y.: Generating logical forms from graph representations of text and entities. In: Proceedings of the 57th Annual Meeting of the Association for Computational Linguistics, pp. 95–106 (2019)
12. Sutskever, I., Vinyals, O., Le, Q.V.: Sequence to sequence learning with neural networks. In: Advances in Neural Information Processing Systems, pp. 3104–3112 (2014)
13. Vaswani, A., et al.: Attention is all you need. In: Advances in Neural Information Processing Systems, pp. 5998–6008 (2017)
14. Veličković, P., Cucurull, G., Casanova, A., Romero, A., Lio, P., Bengio, Y.: Graph attention networks. arXiv preprint arXiv:1710.10903 (2017)
15. Xu, K., Wu, L., Wang, Z., Yu, M., Chen, L., Sheinin, V.: Exploiting rich syntactic information for semantic parsing with graph-to-sequence model. In: Proceedings of the 2018 Conference on Empirical Methods in Natural Language Processing, pp. 918–924 (2018)
16. Zhou, J., Zhao, H.: Head-driven phrase structure grammar parsing on Penn treebank. In: Proceedings of the 57th Annual Meeting of the Association for Computational Linguistics, pp. 2396–2408 (2019)

Case Study of Few-Shot Learning in Text Recognition Models

Jianzong Wang, Shijing Si[(✉)], Zhenhou Hong, Xiaoyang Qu, Xinghua Zhu, and Jing Xiao

Ping An Technology (Shenzhen) Co., Ltd.,, Shenzhen, China

Abstract. Optical text recognition models are widely applied in document processing systems. However, a high-quality text recognition model usually requires large number of samples, extensive amount of time and computation resources. In this paper, we propose a few-shot learning framework for unsegmented text recognition, which comprises of a conventional encoder-decoder recognition module, as well as a generative module for convolutional feature generation. In the meta-training stage, a base model for general text recognition and feature vector generation is trained with large synthesized text image dataset. In the meta-testing stage, the base model is adjusted with a small number of authentic samples. With the complementation of synthesized feature vectors, the base model is adapted to the target dataset distribution. The proposed framework only requires a few authentic samples. It is both data- and time-efficient in adapting existing models to new target datasets. Experimental results on authentic datasets used in industrial applications show that the proposed meta-testing approach outperforms conventional transfer learning by up to 5.84%.

1 Introduction

In modern document processing systems, optical text recognition plays an important role in digitizing text documents [2]. To achieve satisfactory accuracy in text recognition, practitioners often make attempts in two directions. First, increasingly complicated models have been proposed for variable-length text recognition. For example, deep neural networks (DNNs) are widely adopted as the core components in text recognition models [6,9]. Secondly, researchers collect efficient data to train the latest models [10]. Common practice for training a text recognition model is to optimize an entire neural network from scratch, over tens of thousands of text images. Various datasets, such as CTW, SVHN, and COCO-Text, have been published to facilitate research on text recognition. For the complicated models and large datasets, the training often consumes days or months. Additionally, the trained networks usually do not generalize well to other datasets [7]. For a new dataset with different lexicon, fonts or layout configurations, the recognition network often needs to be re-trained with significant amount of labeled data and time.

W. Zhang et al. (Eds.): WISE 2021, LNCS 13081, pp. 394–401, 2021.
https://doi.org/10.1007/978-3-030-91560-5_29

In practice, the requirements for large number of authentic samples and extensive amount of training time hinder many industrial applications [1]. It is often the case when we are required to prove the proficiency of the recognition models within a day, given only a few conceptual samples from the target dataset. In this paper, we investigate the application of synthetic data and generative few-shot learning approaches for Chinese text recognition to improve data and time efficiency. A few-shot learning framework, based on recurrent encoder-decoder recognition model, is proposed. We train the base model with extensive synthetic Chinese text images, and adjust the base model to adapt to target datasets with under 50 authentic samples. Contributions of this paper include:

- Analysis of factors affecting text recognition accuracy in different training and testing sets.
- We propose a few-shot learning framework based on recurrent encoder-decoder recognition model.
- Demonstration of few-shot text recognition model training on authentic datasets used in industrial applications.

2 Proposed Methods

2.1 Model Architecture

In this section, we elaborate the overall architecture of the proposed few-shot learning framework, which is a generative few-shot learning model. A generator is trained to synthesize the feature vectors of text images, to compensate the lack of real samples during meta-testing. Our text recognition model is depicted in Fig. 1. It consists of a regular attention encoder-decoder text recognition module and a feature vector generator module.

Recognition Module. The design of recognition module is based on the sequence recognition network (SRN) proposed in [8]. It is a classic encoder-decoder framework for unsegmented text recognition. Preprocessed images are fed to a backbone CNN that serves as a low-level feature extractor. The 2-layer bi-directional LSTM encoder takes the row-vectors of the feature maps as time sequences. Encoded sequence features are then decoded by the attention gated recurrent unit (GRU) decoder. The structure of the recognition module is illustrated in the upper part of Fig. 1.

Generator Module. The feature generator module synthesizes feature vectors from given text labels and source images. In the generation module, $G1$ extracts attribute vectors from input images, while $G2$ converts the embedding of random text labels into latent vectors representing the underlying labels. The attribute vector and latent vector are then fused to form a synthesized feature vector in the same form as extracted from the CNN. Feature vectors from CNN and the generator are both translated into text labels by the encoder-decoder text recognition module.

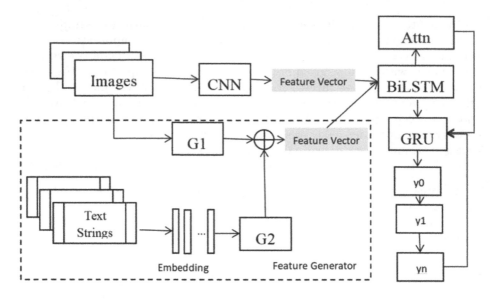

Fig. 1. Overall architecture of the proposed base text recognition model.

2.2 Regularization of Feature Extractor

As shown in [3], regularizing the feature extractor effectively improves classification accuracy for novel classes. In this paper, we propose a novel regularization technique for the text image feature extractor.

The proposed regularization method is illustrated in Fig. 2. Given preprocessed text $\mathcal{I}$ of fixed size $W \times H$, the CNN converts $\mathcal{I}$ to a feature vector ξ of size $W/s \times 1$, where s is the total stride of the CNN. Let $\{\mathbf{b}_i = \{x_i, y_i, w_i, h_i\}, i = 1, ..., T\}$ be the bounding boxes the characters in $\mathcal{I}$, their corresponding locations in the feature vector ξ is

$$\hat{\mathbf{b}}_i = \{\frac{x_i}{s}, 0, \frac{w_i}{s}, 1\}, \ i = 1, ..., T. \quad (1)$$

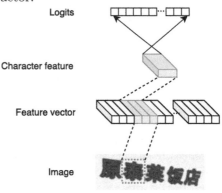

Fig. 2. The architecture of character feature regularization.

Then, the sub-regions $\xi(\hat{\mathbf{b}}_i)$ are cropped and resized to a fixed length character feature vector, $\{\rho_i, i = 1, ..., T\}$. We use a fully connected layer FC to transform ρ_i's into one-hot character labels. The character feature regularization loss function is then given by

$$\mathcal{L}_{\text{char}} = \sum_{\mathbf{x}, \mathbf{y} \in \mathcal{D}} \sum_{i=1}^{|\mathbf{y}|} \log y_i^T \cdot c(\rho_i), \quad (2)$$

where $c(\rho_i) = \texttt{softmax}(FC(\rho_i))$.

3   Training Procedure

In this section, we elaborate the implementation details of the proposed text recognition few-shot learning framework.

In our experiments, synthetic text images are generated following the procedure in [4]. Specifically, texts are generated in 9 different approaches: text generation, font rendering, boarder/shadow rendering, base coloring, projective distortion, padding, natural data blending, resizing and noise. Examples of generated text images are shown in Fig. 3.

All neural networks are implemented under the GPU accelerated Tensorflow framework, released by Google. Experiments are performed on a workstation with 2 NVIDIA Tesla V100 GPUs (16 GB VRAM).

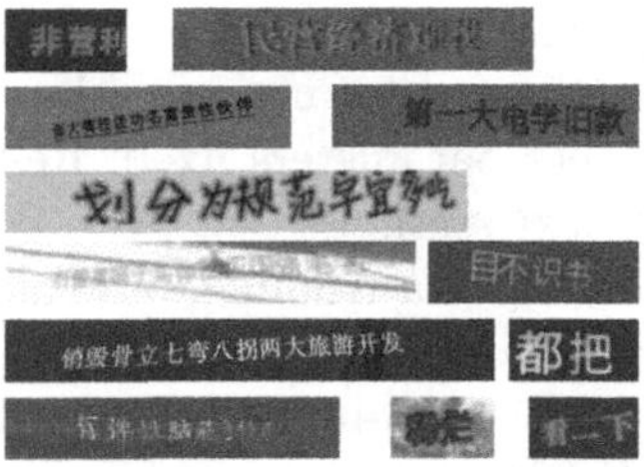

Network parameters are optimized to minimize the total losses

$$\mathcal{L}_{\text{total}} = \mathcal{L}_{\text{rec}} + \lambda \mathcal{L}_{\text{char}}, \qquad (3)$$

Fig. 3. Examples of synthesized text images.

where λ is a predefined hyper-parameter. In experiments without character feature regularization, $\lambda = 0$. When character feature regularization is enabled, we set $\lambda = 0.001$ in our experiments.

Text images must be resized and padded to the same size to enable batch processing in Tensorflow. For the recognition module input, the images are first proportionally resized to have maximum width of 192 pixels, and maximum height of 32 pixels. The resized images are then padded with zeros to fixed size 192×32 pixels. For the generator module input, on the other hand, images are directly resized to 192×32 pixels without preserving aspect ratios. Pixel values are centered and normalized to $[-1, 1]$. Similar with meta learning methods, the application of the proposed model is divided into three stages: *meta-training*, *meta-testing* and *inference*.

Meta-Training: Base Model. In the meta-training stage, the full network structure is trained over a large number of synthetic text image samples. The text recognition module as well as the generator module are trained with respect to the total loss (Eq. 3), including

$$\mathcal{L}_{\text{rec}} = \sum_{\mathbf{x}, \mathbf{y} \in \mathcal{D}} \log p \left(T_\Theta \circ f_{\Theta_F}(\mathbf{x}) = \mathbf{y} \right) + \sum_{\mathbf{x}, \eta \in \mathcal{D}} \log p \left(T_\Theta \circ g_{\Theta_G}(\mathbf{x}, \eta) = \eta \right), \quad (4)$$

$$\mathcal{L}_{\text{char}} = \sum_{\mathbf{x}, \mathbf{y} \in \mathcal{D}} \sum_{i=1}^{|\mathbf{y}|} \log y_i^T \cdot c_{\Theta_c} \left(\rho_i (f_{\Theta_F}(\mathbf{x})) \right). \quad (5)$$

In the above loss functions, f is the feature extraction function defined by the CNN structure, g is the feature generator function and T the encoder-decoder translator. Θ, Θ_F, and Θ_G are parameters of T, f, and g, respectively. Θ_c are parameters of the fully connected layer in character regularizer. $\mathbf{x}$ and $\mathbf{y}$ are image samples and their correspondent text label from the training set $\mathcal{D}$. η are random text labels sampled from the same distribution as $\mathbf{y}$ in $\mathcal{D}$.

In our experiments, the size of the corpus is 5,577 characters, including 10 digits, 26 alphabets, 6 punctuation/symbols, and 5,535 frequently used Chinese characters. Note that the proposed framework can be trained end-to-end, where Θ, Θ_F, Θ_G, and Θ_c are jointly optimized by minimizing $\mathcal{L}_{\text{total}}$. However, as shown in our experiments, a better training strategy is to train the parameters Θ, Θ_F, and Θ_c first, without the generator module. The parameters in Θ_G are then optimized to minimize the recognition loss of the generated samples only, with other parameters fixed. Results of both training strategies are elaborated in the next section.

During meta-training, base models are optimized with the ADADELTA optimizer [5], with initial learning rate 1.0. The batch size is set to 64. The same amount of text labels are fed to the generator module for fake feature synthesis. For 100,000 synthetic text images, the end-to-end model converges at 50 epochs. When training the recognizer and generator modules separately, both modules are trained for 50 epochs, respectively.

Meta-Testing: Domain-Specific Model Adaptation. In the meta-testing stage, the base text recognition model is adjusted to fit the distribution of low-sample dataset. Strictly speaking, a distribution cannot be estimated from only a few samples. As discussed in the previous section, the underlying distribution of a text dataset $\mathcal{D}$ involves two random variables, the text label η and imaging attribute $\mathbf{z}$. Usually, the distribution of text labels η can be obtained by counting the frequency of characters or N-grams in certain type of documents. Since such statistical information based on large samples are not privacy related, it is often cheap or free to obtain. On the other hand, the distribution of imaging attributes $\mathbf{z}$ is not analytically describable. There is no reliable way of representing their distribution, let alone taking samples from it. In our proposed framework, the imaging attributes are extracted from sample images $\mathbf{x}$ in the generator module $G1$. We simulate the distribution of the target dataset by fusing imaging attributes of the few real samples in the target dataset with text labels drawn from N-grams of the target document type.

During meta-testing, parameters of feature extractor f and feature generator g are fixed, while parameters Θ of T are fine-tuned to minimize $\mathcal{L}_{\text{rec}}$ over the target dataset. Stochastic gradient descent optimizer with learning rate 0.01 is used to fine-tune Θ. In our experiments, the batch size of recognizer module is set to 1 during meta-testing. For each batch, 100 fake feature vectors are generated with randomly chosen text labels. Meta-test runs for 1,000 epochs for each target dataset.

4   Experiment Results and Analyses

In this section, we present experimental results of the proposed few-shot learning framework on authentic data from industrial applications.

Table 1. Ablation study of the proposed framework over three datasets.

Model	Vehicle owner	ID name	Total CRA
SRN	68.18%	81.48%	75.78%
SRN + Character regularization	69.96%	81.19%	76.38%
SRN + Character regularization + generator (e2e)	70.80%	**81.71%**	77.03%
SRN + Character regularization + generator (separate)	**72.46%**	81.18%	**77.44%**

Datasets used in our experiments are:

- *Synth100K* - we synthesized 100,000 text images following the procedure described in the previous section. This dataset is used to train the base model of the proposed framework.
- *Vehicle owner* - authentic data cropped from 4,000 scanned copies of vehicle registration cards. The images typically contain 10–20 Chinese characters representing group or company names.
- *ID name* - authentic data cropped from 5,000 scanned copies of Chinese citizen ID cards. The images contain 2–4 Chinese characters representing individual citizen names.

During meta-testing, samples are randomly chosen from the training sets of Vehicle owner and ID name datasets.

Table 2. Meta-test results on vehicle owner and ID name datasets.

Dataset (Base model)	n = 10	n = 20	n = 30	n = 40	n = 50
Vehicle owner - fine tune	76.01%	80.10%	81.46%	79.34%	83.06%
Vehicle owner (e2e)	74.07%	77.23%	79.64%	80.68%	81.25%
Vehicle owner (separate)	**77.83%**	**80.31%**	**81.74%**	**81.13%**	**83.38%**
ID name - fine tune	77.95%	77.86%	78.92%	80.00%	79.77%
ID name (e2e)	79.69%	80.74%	81.13%	81.42%	81.50%
ID name (separate)	**82.02%**	**82.41%**	**83.02%**	**83.33%**	**83.34%**

In Table 1 and Table 2, "e2e" stands for the base model trained with the end-to-end strategy, while "separate" stands for the base model where recognition and generation modules are trained sequentially. The fine-tune results are based on the separately trained model (Fig. 4).

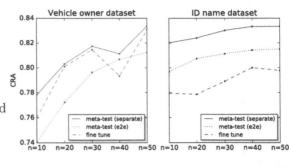

Fig. 4. Meta-test results on vehicle owner and ID name datasets.

Ablation study of the proposed framework is shown in Table 1. The total character recognition accuracy (CRA) increases when the character regularization term is added. By including the generator module, the CRA further increases on both datasets. The separately-trained model outperforms the end-to-end version by 0.44%.

Table 2 compares meta-test and fine-tune results on vehicle owner and ID name datasets with varying number of real samples. Overall, the proposed meta-test strategy significantly increases the target-dataset CRA on both tested datasets. The meta-test results based on "separate" base model outperforms the ones based on "e2e" base model in all cases, suggesting a better suited generation module using the separate training strategy.

Compared to the conventional fine-tune methods, the proposed few-shot learning framework shows significant advantage over small-scale authentic data sets. When n = 10, the "separate" meta-test results outperform the fine-tune ones by 1.82% and 4.07% on vehicle owner and ID name datasets, respectively. As the number of real samples increases, the gap between the fine-tune and meta-test results decreases. But the "separate" meta-test results still outperforms the fine-tune results in all cases.

5   Conclusion

Training a text recognition model on large corpora such as Chinese characters often require large amount of data and training resources. In addition, the trained text recognition models cannot be effectively adapted to new datasets without similar amount of data and effort as the first model. In this paper, we discuss the hindrance of model transferability in text recognition models, and propose a comprehensive framework for few-shot learning in unsegmented text recognition. Experiments show that our proposed few-shot learning methods can effectively increase the character recognition accuracy of a base model in the target datasets, with much smaller samples and training time compared to conventional methods. Due to time and resource limits, we only tested the performance of the proposed methods on small-scale datasets.

Acknowledgment. This work is supported by National Key Research and Development Program of China under grant No.2018YFB0204403. Corresponding author is Shijing Si from Ping An Technology (Shenzhen) Co., Ltd.

References

1. Chen, X., Jin, L., Zhu, Y., Luo, C., Wang, T.: Text recognition in the wild: a survey. ACM Comput. Surv. (CSUR) **54**(2), 1–35 (2021)
2. de Sousa Neto, A.F., Bezerra, B.L.D., Toselli, A.H., Lima, E.B.: Htr-flor++ a handwritten text recognition system based on a pipeline of optical and language models. In: Proceedings of the ACM Symposium on Document Engineering 2020, pp. 1–4 (2020)
3. Hariharan, B., Girshick, R.: Low-shot visual recognition by shrinking and hallucinating features. In: Proceedings of the IEEE International Conference on Computer Vision, pp. 3018–3027 (2017)
4. Jaderberg, M., Simonyan, K., Vedaldi, A., Zisserman, A.: Synthetic data and artificial neural networks for natural scene text recognition. In: Proceedings of the Conference on Neural Information Processing Systems (2014)
5. Kingma, D.P., Ba, J.: Adam: a method for stochastic optimization. In: Proceedings of the International Conference on Learning Representations (2015)
6. Luo, C., Zhu, Y., Jin, L., Wang, Y.: Learn to augment: Joint data augmentation and network optimization for text recognition. In: Proceedings of the IEEE/CVF Conference on Computer Vision and Pattern Recognition, pp. 13746–13755 (2020)
7. Rey-Area, M., Guirado, E., Tabik, S., Ruiz-Hidalgo, J.: Fucitnet: improving the generalization of deep learning networks by the fusion of learned class-inherent transformations. Inf. Fusion **63**, 188–195 (2020)
8. Shi, B., Bai, X., Yao, C.: An end-to-end trainable neural network for image-based sequence recognition and its application to scene text recognition. IEEE Trans. Pattern Anal. Mach. Intell. **39**(11), 2298–2304 (2017)
9. Yousef, M., Hussain, K.F., Mohammed, U.S.: Accurate, data-efficient, unconstrained text recognition with convolutional neural networks. Pattern Recogn. **108**, 107482 (2020)
10. Zharikov, I., Nikitin, P., Vasiliev, I., Dokholyan, V.: Ddi-100: Dataset for text detection and recognition. In: Proceedings of the 2020 4th International Symposium on Computer Science and Intelligent Control, pp. 1–5 (2020)

Service Computing and Cloud Computing (1)

Detecting Document Versions and Their Ordering in a Collection

Natwar Modani[1]([⊠]), Anurag Maurya[2], Gaurav Verma[3], Inderjeet Nair[1],
Vaidehi Patil[2], and Anirudh Kanfade[4]

[1] Adobe Research, Bangalore, India
{nmodani,inair}@adobe.com
[2] IIT Bombay, Mumbai, India
[3] Georgia Tech, Atlanta, USA
gverma@gatech.edu
[4] NITK, Mangalore, India

Abstract. Given the iterative and collaborative nature of authoring and the need to adapt the documents for different audience, people end up with a large number of versions of their documents. These additional versions of documents increase the required cognitive effort for various tasks for humans (such as finding the latest version of a document, or organizing documents), and may degrade the performance of machine tasks such as clustering or recommendation of documents. To the best of our knowledge, the task of identifying and ordering the versions of documents from a collection of documents has not been addressed in prior literature. We propose a three-stage approach for the task of identifying versions and ordering them correctly in this paper. We also create a novel dataset for this purpose from Wikipedia, which we are releasing to the research community (https://github.com/natwar-modani/versions). We show that our proposed approach significantly outperforms state-of-the-art approach adapted for this task from the closest previously known task of Near Duplicate Detection, which justifies defining this problem as a novel challenge.

Keywords: Version detection · Near Duplicate Detection · FCN · Wikipedia based dataset

1 Introduction

Creation of documents is an iterative process. For instance, legal contracts need iterative editing to incorporate comments from stakeholders (this process is typically called red-lining), research papers go through several revisions based on feedback from reviewers and collaborators, etc. This leads to many versions of documents getting created. When one needs to find the latest version of a document or organize their documents, these additional versions of documents present added cognitive load for the user. In absence of any tool, they may have to open and skim through the documents one-by-one, which leads to inefficiency and

© Springer Nature Switzerland AG 2021
W. Zhang et al. (Eds.): WISE 2021, LNCS 13081, pp. 405–419, 2021.
https://doi.org/10.1007/978-3-030-91560-5_30

fatigue. Using the names and timestamps may not always be a reliable mechanism, as names are given by people and most people do not strictly adhere to any naming convention. Also, the timestamps get altered by even minor and unintentional edits and saves (and sometimes due to moving/copying the files), and hence, they also do not render themselves as viable solution to the problem. Besides the human-centric motivation, several automated methods that operate on a collection of documents, like systems that cluster similar documents [11,16] and recommend related ones [13,14,18–20] may perform sub-optimally in situations where the collection comprises of unnecessary documents, say, by recommending redundant and/or outdated documents.

While there have been works on somewhat related problems, such as near-duplicate document detection [4], document similarity [6], clustering [16], and plagiarism detection [8], etc., we believe that version identification is a different problem for two reasons. First, the notion of similarity of documents need to be defined differently for identifying versions as we will discuss later. Second, we also need to determine the adjacency of versions and their directionality (from older to newer version). To the best of our knowledge, our work is the first attempt to define this problem formally, provide a dataset and an approach for identifying and ordering versions of documents from a collection of documents.

Our problem statement is as follows. Given a *collection* of documents as input, identify the sets that are *versions* of each other and provides an *order of documents* within each set. Please note that the collection may comprise more than one set of versions along with some documents that have only one version.

We propose a three-stage approach for solving this problem. The first stage of our approach divides the document collection into sets of documents that can potentially be versions of each other for efficient computation. We use a two-step process, where we first find very similar paragraphs across documents using MinHash-LSH [2]. We then use the number of such (very similar) paragraphs common to a pair of documents (relative to the number of paragraphs in the shorter of the two documents) to determine if the pair of documents can be potentially versions of each other. The second stage of our approach classifies the candidate pairs of documents as versions-or-not using a fully convolutional neural network (FCN) [12] based binary classifier. We use pairwise lexical and entity-based similarities to create heatmap images, which are used by our proposed FCN-based classifier that identifies structures that indicate local similarity patterns. Once we determine which documents are versions of each other, in the third stage we find the order of creation among the versions. We create a graph with nodes as document instances and edges between a pair of document instances if the previous stage determined these document instances as potentially versions of each other. The edges have a weight based on the extent of alignment (based on minimum transformations required to convert one document instance into the other) between two document instances. We look for a directed chain which has minimum weight to find the order of versions in this graph.

We create a suitable dataset from Wikipedia articles which we are releasing to enable further research in this direction. We show experimentally that our proposed approach significantly outperforms the appropriate baselines adapted from closest task of Near Duplicate Detection on this task. This indicates that the problem of version detection is sufficiently different from the closest problem addressed in prior work.

In summary, our contributions in this paper are:

- We propose the problem of identifying and ordering versions of documents from a given collection of documents.
- We propose a three-stage approach for efficiently solving the problem.
- We create a novel real world data set from Wikipedia suitable for this task, which we are releasing to the research community.
- We show experimentally that our proposed approach significantly outperforms the appropriate baselines adapted from closest task of Near Duplicate Detection on this task.

The paper is organized as follows. We briefly review the relevant prior work in Sect. 2. We then describe our proposed methodology in Sect. 3. We describe the dataset and its construction in Sect. 4 and report the experimental results in Sect. 5. Finally, we conclude in Sect. 6.

2   Related Work

Although the specific problem of identifying versions in a collection of documents has not been addressed in previous literature, there has been prior work in related areas of near-duplicate document detection, paraphrase detection and plagiarism detection. While none of these related tasks require the ordering part, the detection task can be thought of as finding document 'similarity', where 'similarity' definition depends on the specific task. For Near Duplicate Detection (NDD), the similarity is defined at lexical level, i.e., documents that are near duplicates of each other would have nearly the same set of words. MinHash-LSH [2] is document fingerprinting method from the family of Locality Sensitive Hashing (LSH) functions for estimating the lexical Jaccard similarity between documents. [9] introduced sectional MinHash, a modification that divides the document in sections and uses the matching for both the characteristics (words occurring in the document) and in which section they occur. This enhancement to MinHashLSH with sectional information achieves state-of-the-art performance on the task of near-duplicate document detection. Most of the literature on NDD [1,4,7,9] attempts to find a good hash function hypothesizing that a better hash function will improve the detection accuracy.

Paraphrase detection is a similar task to NDD where similarity is defined at semantic level instead of lexical level. Several methods that use deep learning have been proposed to compute a representation in the embedding space that can capture semantic similarity more effectively irrespective of the lexical differences [3,17]. However, even though consecutive versions go through some paraphrasing of already existing content, the common content across versions is largely preserved lexically. However, different versions of same document are not always near-duplicates or paraphrase of each other and may involve considerable addition/deletion of content, which NDD/paraphrasing methods fail to consider.

For plagiarism detection, similarity can be thought of as having overlapping segments between the documents. Accordingly, [5,8,15] focus on finding overlapping segments across documents along with the extent of overlap. [15] used sentence-level tf-idf to quantify similarity at sentence level and proposed an algorithm to find maximal length overlapping segments. [5] used a Vector Space Model (VSM) with Parts-Of-Speech (POS) tagging, Named Entity Recognition (NER), etc., to generate potential candidates for plagiarism detection. [8] used a similar approach but only with POS tags and n-grams. While these approaches can be useful in detecting versions in a pair-wise manner, they do not solve the task of finding appropriate *sets* of documents and do not attempt to identify the ordering. Also, these methods rely largely on tf-idf frequencies and do not capture the patterns inside matching versions.

Given the motivation, the need for being able to address the problem of identifying versions from a collection is validated, and therefore, both an approach and dataset for this task are missing links we attempt to address in this paper.

3   Proposed Methodology

The key intuition we use to detect the versions of a document is that typically, across versions, the local lexical structure is preserved to a degree, even if at global level, it changes significantly. Typical edits to evolve the document from a version to the next version are insertion of new content, deletion of some of content, and replacing small parts of content. Sometimes, parts of content are also moved around with small degree of edits. In most cases, the edits do not replace very large parts of documents across versions. Hence, versions of a documents are not only semantically similar, even the words, sentences and paragraphs are largely preserved across versions. Further, even in case of paraphrasing, typically the entities are preserved even if the other words are changed.

Leveraging this insight, we design a three-stage approach. In the first stage, we want to *efficiently* find candidate set of documents that can *possibly* have a version relation among them, while retaining a high recall. Following that, we want to do more precise classification on a pairwise basis to determine if two documents are versions of each other. Finally, once we have pairwise version relation determined, we order all the versions correctly together. The reason why this task is required is due to the fact that not all predictions on the document pair level are going to be correct. For example, some additional pairs may be deemed to be versions of each other, which makes it difficult to know what is the correct order of these versions of the document.

3.1   Candidate Selection

Given that for a collection with n documents, the number of pairs of document are $O(n^2)$, which can be computationally expensive if we naively compare every pair of documents. So we use MinHash-LSH [2] based approach to find these candidate sets efficiently. However, instead of a direct application of MinHash-LSH on document level, we propose a two-step approach for better accuracy. We take the paragraphs as unit and compute the MinHash for each of the paragraphs. We use Locality Sensitive Hashing (LSH) to put the paragraphs in the same bucket if they are lexically very similar (defined as having a Jaccard similarity higher than a threshold when considering the paragraphs as sets of words). LSH can be tuned to put similar items in the same bucket and putting dissimilar items in different buckets with certain level of probability [2].

We then generate the candidate sets in the following manner. We scan the buckets one at a time. For each pair of paragraphs in one bucket, we increment count of votes for the pair of documents from which these paragraphs come from. Now, after scanning all the buckets, we check the number of votes for each pair of documents and compare that with the length of the documents in terms of paragraphs. We deem the document pairs as a candidate if the number of votes for this pair are larger than a certain fraction of the number of paragraphs in the smaller of the two documents.

The idea of dividing the document into smaller units (sections) was used in [9] also. However, their approach was to divide the document into a fixed (and typically, small) number of *sections* and look for match not only for words, but also for in which section of document are they found. By using this additional information (about location of the words), they improve the error metric compared to a vanilla MinHash-LSH based Near Duplicate Detection system. However, for our setting, there can be significant changes between versions of the document, we want to not only look for existence of similar paragraphs but want to allow for addition/deletion of content (in addition to small changes to content within preserved paragraphs).

3.2   FCN-Based Binary Classification

The next step is more precise binary classification for version relation of candidate pairs of documents. While our eventual task is to identify document instances that are versions of a document, and arrange them in an order, we believe that trying to solve a simpler version of this problem and then using post-processing to recover the order (as discussed in next subsection) may perform better.

Hence, we take three different definitions of a pair of documents being related by version relation. Consider a set of v documents that are versions of a document, say $D_1, D_2, ..., D_v$. Here, D_1 is the first version of a document, D_2 is second version, and so on, till D_v which is the latest version. The three definitions are as follows:

Definition 1. *Undirected (and not necessarily adjacent): Any two of these documents D_i and D_j, $1 <= i, j <= v, i \neq j$ are considered as related.*

Definition 2. *Directed (but not necessarily adjacent): Two of these documents D_i and D_j are considered as related if $1 <= i < j <= v$.*

Definition 3. *Adjacent (and directed)*: *Two of these documents D_i and D_j are considered as related if $1 <= i < j <= v$ and $i + 1 = j$.*

In addition to the local lexical structure preservation, we believe that typically, the entities present in the document tend to be preserved across versions. Based on these two hypotheses, we design our feature set for the classifiers for a candidate pair of documents. We construct a matrix of dimensions $m \times n$, where m and n are the lengths of the first and second document respectively in the pair in terms of number of sentences. Please note that while for candidate generation, we operated at paragraph level, for the binary classification task, we are operating at a sentence level.

We compute similarity between every sentence i of first document with every sentence j of the second document based on lexical structure, denoted by S_{ij}^L and based on entities involved, denoted by S_{ij}^E. Lexical similarity S_{ij}^L is computed by representing the sentences in terms of sentence level TF-IDF (used earlier in [15] for plagiarism detection, essentially treating each sentence as a document and computing TF-IDF based on this treatment) vector representation and taking the cosine similarity. Entity based similarity S_{ij}^E is taken as the Jaccard similarity between the entities present in the two sentences.

While it is not obvious if two documents are versions of each other by analyzing the individual values within these matrices, it is fairly straightforward to observe emergent patterns in the 2D heatmaps constructed using these raw values. We can clearly see presence of local diagonals like patterns in the heatmap in Fig. 1a. On the other hand, Fig. 1b shows example of document pair which may have overlapping words, but the clear local diagonal patterns are missing. We train FCN-based binary classifier on these heatmaps.

Fully Convolutional Networks [12] have proved to be capable of modeling complex patterns in visual data by extracting meaningful features from images given supervised data. This motivated us to leverage FCNs for our usage for identifying the version relatedness between two documents. We specifically choose to

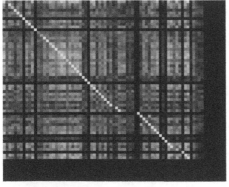

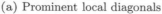

(a) Prominent local diagonals   (b) Local diagonals are absent

Fig. 1. Examples of pair of documents that are (a) versions of each other, and (b) that are not

go with FCN (as opposed to other convolutional networks) as they are invariant to the size of the heatmap and they can provide equally rich representation as the standard CNN architecture. They have a clear advantage while working with images which have a skewed aspect ratio. Our heatmap is clearly not square in shape, hence, if we use interpolation technique over it to fit to certain size (as required by conventional CNNs), the image has a high chance of getting distorted, i.e., the patterns of local diagonal in the heatmap can get diluted due to bi-linear interpolation. However, with FCN the exact size does not matter if it is less than the max_size. Even if the image dimensions do exceed a certain maximum, we can easily pool it back to the required size. In the case of max-pool it only changes the granularity of sentence level heatmap, i.e., in that case we can think of the heatmap as sentence similarity matrix with a window and stride of 2 sentences. We train the FCN directly on two channel heatmap (one channel corresponding to lexical similarity and another for entity-based similarity) to predict if the two documents are related by the version relation. This allows the model to automatically extract the useful patterns.

As originally proposed [12], the FCN is composed of both encoders and decoder parts. On the other hand, for our formulation we use just the encoder for compressing the similarity heatmap to a feature vector which is then used for binary classification. This is a little different to what FCNs are used for in general, that is semantic segmentation. But we do borrow and preserve an interesting property of convolution networks which is the fact that they respond to specific local patterns in an image. The heatmap constructed by us may have patterns like "local diagonals" which may be useful for classification. Another motivation for using FCN instead of a normal CNN is that FCNs are robust to the dimensions of the image. Since the documents have variable sizes, this means that the heatmaps themselves are of different aspect ratios (often becoming really extreme). FCN helps address the above problem. In practice the encoder also needs a fixed size image (200×200 in our case), so we pad the rest of the image by zeros. If the input has a larger length than 200 in any dimension, we max pool it to bring it down inside the 200×200 box (We can also apply max pool across 2D spatial-dimension after applying FCN for images of larger dimensions). Note that we also need to pad in case of CNNs but it's ability to detect patterns can be hampered by various aspect ratios which is not the case for FCNs. The only difference in our case when compared to a standard CNN is that we are using 1×1 convolution in place of a fully connected layer.

3.3   Finding the Order Among Versions

If we think of each instance of every document as a node and adjacent directed version relation as a directed edge, then the versions of a document will form a directed chain, if there is only one evolution path of the document. If the document is repurposed, we may see multiple outgoing edge from some nodes, and hence may get a tree. Our objective is to partition the set of all nodes as the trees/directed chains of versions of documents. The pairwise labelling may not provide sufficient information to construct this tree/chain if we use *Undirected* (Definition 1) or *Directed* (Definition 2) definitions of version related pairs. *Adjacent* (Definition 3) definition provides sufficient information to construct the tree/directed chains in theory. However, in practice, the accuracy of models for this definition is lower (as the task is harder). Hence, we have many more false positives (which may create a graph more complex than trees/chains, including cycles) and false negatives (which may break the group of versions of a document into multiple chains/subgraphs) as we see in our experimental results.

Hence, we introduce a third part of the processing, where we take the binary classifier output and construct the graph. We then find all connected components within this graph and create fully connected subgraph (clique) for each connected component by adding all the missing edges. This step helps us in further **increasing our recall** by providing additional candidates for our final objective. Then the edges are assigned weight by computing a 'version score' between the pair. Finally, we compute the maximum weighted spanning tree on this graph.

We propose two ways of calculating the version scores. First is directly taking the FCN output as a number in range $[0, 1]$ instead of the binary output, which we call as FCN-Predict.

Another way to define the version score for the edge weight is as follows (called as alignment version score):

$$V = S - \lambda_1 \times I - \lambda_2 \times D \tag{1}$$

where, S is the alignment score (explained later), I denotes the number of sentence insertions, and D denotes the number of sentence deletions, and λ_1 and λ_2 are hyperparameters. Version Score can be understood to be a measure of alignment (or inverse of 'distance', in terms of changes required to convert one document to the other, between them) between two documents. Once we compute the version scores, for each connected component, we make the smallest document as the root of the tree and then do a topological ordering of the above tree to use that as a chain. This chain provides us with the final ordering among documents that are versions of each other.

To compute S, I, and D, we construct a heatmap M of sentence-level similarity by taking a linear combination of heatmaps that capture lexical (S^L) and entity-based (S^E) overlap. Alignment Score S is computed using Dynamic Programming algorithm over the matrix M with elements $S_{ij} = (S_{ij}^L + \lambda_3 S_{ij}^E)$ (representing the overall similarity between sentence i of document 1 with sentence j of document 2, with λ_3 being a hyperparameter) to find maximum possible reward to go from the index $(0,0)$ to index (m,n) where m and n are number of sentences in documents i and j respectively.

At each element of the matrix M, we have 3 options, move diagonally with reward $R_d = S_{ij}$ (corresponding to the sentence being modified from sentence i of document 1 to sentence j of document 2), move horizontally (corresponding to deleting a sentence from document 1, counted in D) or move vertically (corresponding to inserting a sentence into document 2, counted in I). The goal of this approach is to quantify the extent of alignment between two documents based on minimum transformations required to convert document 1 to document 2 by finding the highest reward traversed path from the index $(0, 0)$ to (m, n). If the documents were exactly same, the path would lie along the diagonal with all S_{ij} values as 1. However, if a new sentence j is added to document 2, the traversal would involve a horizontal movement. Similarly, if a sentence has been deleted from document i, it would involve a vertical movement. Consequently, we quantify the number of sentence insertions I and sentence deletions D as the number of horizontal moves and vertical moves in the traversed path, respectively, and subtract it from the alignment score S to get a version score V.

The intuition behind these two definitions of version score is as follows. The FCN based score is going to capture the level of preservation of local lexical structures and entities. However, since it is trained for a binary classification task, it is not clear if the score would be appropriate for the task we are using it for. On the other hand, the direct alignment based score is specifically designed for this task. However, it would not capture the similarity present between documents if some segments of the document are moved within it. Hence, we use both of these and determine the appropriate method empirically.

4   Wikipedia Versions Dataset

Due to lack of an existing large-scale dataset suitable for our problem, we curated a new dataset using revision history of Wikipedia articles. We took Wikipedia revision dumps that contain all revisions of Wikipedia pages on a certain topic. Each revision is accompanied by some metadata (like timestamp, parent_id, etc.). To ensure that we consider versions that are sufficiently different from each other, we use timestamps in an automated filtering process to create the dataset.

Each edit made by the user to a Wikipedia page can be considered a separate version of that Wikipedia page (these are referred to as revisions in the Wikipedia terminology). However, often there are some artifacts in the revisions – for example, troll edits are insincere edits made by users. We observe that these edits do not last, and the page is reset to the last relevant version by the moderators within a short duration. Based on this observation, we remove all edits that do not last more than $AVE/10$ seconds, where AVE is defined as the average time between successive edits for a given Wikipedia page. Instead of setting a global threshold, this threshold varies for every page because some pages are edited more frequently than others. We use timestamps obtained along with metadata for the purpose of filtering. Next, we filter out pages that are rarely edited. This is done by setting a hard threshold – i.e., pages should consist of at least 10 versions. After this we stochastically sample $\lfloor U(min, width) \rfloor$ revisions from all versions, U is the uniform distribution, min is the minimum number of samples, $min + width + 1$ is the maximum number of samples, and $\lfloor . \rfloor$ is the floor function (greatest integer no larger than the number). We take min to be 7 and $width$ as 5. However, since it is still possible that the set of revisions obtained via uniform sampling don't have considerable differences between them, we stochastically filter out more revisions to encourage differences between them. This is done by having higher probability of sampling versions which have more difference in terms of length of the documents with the previously selected versions. This increases the probability of having the set of versions that differ more in content.

Once we obtain the list pages and the versions being considered for each of them, we scrape the Wikipedia article corresponding to each version. We remove all additional information in sections like References, Category, etc. from the scraped articles. This is done to ensure that the model doesn't utilize this information to learn undesirable patterns, and to ensure that the identification of relations is solely based on the content.

Our final dataset comprises $1,755$ unique Wikipedia articles with a total of $10,267$ versions; each article having 5.85 versions on average. We split the dataset into train, validation, and test sets in the ratio of $0.7 : 0.1 : 0.2$, respectively. As mentioned in the introduction, we are releasing a dataset generated in this manner to the research community.

5   Evaluation and Results

In this section, we present our experimental results. First, we discuss the baseline we design to compare its performance against that of our approach.

Baselines: We use the Sectional MinHashLSH that achieves state-of-the-art results on near-duplicate detection task [9] as the baseline for binary classification. For ordering, we use the content length as baseline, i.e., a longer document is deemed as more recent version compared to a shorter ones.

Parameter Choices: All the hyperparameters were assigned via a grid search.

Candidate Generation Parameters: The threshold in MinHash-LSH for deeming two paragraphs as near duplicate was 0.5, and number of permutation functions used was 1024. The threshold on fraction of paragraphs (of the shorter document) that need be near-duplicate for including the pair of documents are candidate for being a version related pair was taken as 0.3.

FCN Architecture: There are 7 (2D-)Convolution layers followed by a Sigmoid layer in our implementation of FCN. The stride for all the convolution layers is 1×1 and activation function is *ReLU*. All convolution layers are followed by (2D-)batch normalization [10] (except last convolution layer) and (2D-)max pooling (with kernel size 2×2) except last two convolution layers. The kernel size is 5×5 for first, second and fifth convolution layers, 7×7 for third, 6×6 for fourth, 3×3 for sixth and 1×1 for the seventh convolution layer (and therefore, it is like a fully connected layer). A padding of 2×2 is used in first two layers and 3×3 for third layer. Other layers do not use padding.

Version ordering related parameters: λ_1 and λ_2 (weight of insertions and deletions in Eq. 1, respectively) were taken as -0.05 and 0.15, respectively, and λ_3 (weight of entity-based similarity with respect to lexical similarity) was taken as 1.4.

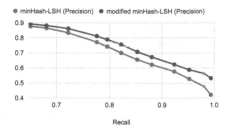

Fig. 2. Comparison of Sectional MinHash-LSH and modified MinHash-LSH

Candidate Selection: We first evaluate the effectiveness of MinHash-LSH for candidate selection. Given the size of our dataset ($2,068$ versions of 351 documents in the test set), an all-pair comparison would lead to more than 4.2 million (for **Adjacent**; for **Undirected** and **Directed**, it is half as many) pairwise checks. The ground truth number of version pairs are $10,426$ for definition 1 (**Undirected**), $5,213$ for definition 2 (**Directed**) and $2,053$ for definition 3 (**Adjacent**). Figure 2 shows the precision and recall trade-off curves for the Sectional MinHash-LSH based approach and our proposed approach for finding the candidate pairs. Based on the operating point we choose (shown by $\star$ in Figure), we end up with $19,307$ pairs for **Undirected**, $9,667$ for **Directed** and $3,876$ for **Adjacent**, which are significantly lower than the naïve quadratic number of pairs. Also, Fig. 2 shows that our proposed two step method achieves up to 10% higher precision for a given recall value compared to Sectional MinHash-LSH method. Also, the gap in performance of Sectional MinHash-LSH and our proposed method increases as we go towards the higher recall side, which are the preferred operating point, as we do not want to discard too many actual version pairs in this stage.

Binary Classification: Now we compare the performance of our proposed approaches for binary classification task. We also train a Logistic Regression (LR) classifier using hand-crafted features obtained from the similarity heatmaps for comparison purpose. The features were obtained by quantifying the correlation ($corr(.)$) between the two indices of the lexical (S^L) and entity-based (S^E) similarity matrices discussed above. We use $corr(S^L)$, $corr(S^E)$, $corr(S^E \circ S^L)$, $corr((S^L)^{16})$, $corr((S^E)^{16})$, $corr((S^L + S^E)^4)$, and $corr((S^L - S^E)^4)$; $\circ$ denoting element-wise multiplication. We also experiment with several other higher powers in the range of $[2, 16]$ and we find that the mentioned set of features perform the best. While training the logistic regression model with all the higher powers, we note that the learned β-coefficients for powers which are not noted above were insignificant and could be dropped without a noticeable change in the final precision or recall scores.

To evaluate the performance of models, we consider the three settings defined earlier: *(a)* undirected and not necessarily adjacent (*Undirected*), *(b)* directed but not necessarily adjacent (*Directed*), and *(c)* directed and adjacent (*Adjacent*). As it is evident from the discussion earlier, these three settings progressively become more challenging, but also get closer to the actual task at hand. For instance,

Table 1. Performance after the binary classification stage (average of 5 runs). Columns in the left indicate the performance of various models on the end-to-end task, whereas the ones in the right indicate the performance of binary classifiers alone on the tasks they were trained for, respectively

Type	Model	Candidate Gen + Classifier wrt GT			Classifier only for respective tasks		
		F_1	Precision	Recall	F_1	Precision	Recall
Undirected	Baseline	0.3361	0.3210	0.3527	0.4163	0.4367	0.3977
	LR-LS	0.3718	0.4089	0.3410	0.7216	0.7342	0.7094
	LR-Full	0.3817	0.4169	0.3520	0.7813	0.7914	0.7714
	FCN-LS	0.6012	0.6373	0.5690	0.7862	0.8006	0.7723
	FCN-Full	**0.6339**	**0.7087**	**0.5734**	**0.8743**	**0.9031**	**0.8472**
Directed	Baseline	0.3361	0.3210	0.3527	0.3642	0.3743	0.3546
	LR-LS	0.3519	0.4063	0.3101	0.4179	0.4332	0.4036
	LR-Full	0.3853	0.4371	0.3461	0.4354	0.4679	0.4071
	FCN-LS	0.6191	0.6371	0.6020	0.6672	0.6817	0.6533
	FCN-Full	**0.6547**	**0.7216**	**0.5991**	**0.7246**	**0.8312**	**0.6422**
Adjacent	Baseline	0.3361	0.3210	0.3527	0.3361	0.3210	0.3527
	LR-LS	0.3819	0.4197	0.3504	0.3976	0.4420	0.3613
	LR-Full	0.4011	0.4562	0.3579	0.4139	0.4837	0.3617
	FCN-LS	0.6376	0.6551	0.6210	0.6306	0.6306	0.6317
	FCN-Full	**0.6987**	**0.7832**	**0.6306**	**0.7036**	**0.7583**	**0.6562**

detecting adjacent relations is harder than detecting directed relations, as it involves not only identifying that document A comes before B, but also whether it comes *immediately* before B. For each of the three settings, we construct negative and positive examples according to the above formulation and quantify the performance of our models. Since the number of negative samples are much more in number, we sample three times the number of positive samples from the set of all negative samples to get a reasonably balanced training data set.

We run the Logistic Regression and FCN networks on only the lexical similarity heat map (LR-LS and FCN-LS, respectively), as well as, on both the lexical and entity-based similarity (LR-Full and FCN-Full, respectively), in addition to the baseline. Table 1 shows the performance comparison. The left-hand side columns give result for the eventual task of identifying consecutive versions in correct order. Please note these results compare the performance with respect to the ground truth, so the loss of recall due to candidate selection is also reflected in the loss of recall in these columns. Of course, the baseline operates directly on the full set and does not have corresponding loss of recall due to any pre-processing. In all the three settings FCN outperforms LR, and they both outperform the baseline significantly. Further, one can also see that by including the entity-based similarity (in FCN-Full and LR-Full), the performance for the classifier improves significantly.

The right-hand side columns of Table 1 give results for the tasks the classifiers were trained on. Here, the performance is with respect to the candidate pairs (and not ground truth). As one can see, our proposed FCN-Full classifier performs considerably better than other models, consistently on all metrics and

Table 2. Performance of models trained on different task on the end-to-end task of detecting adjacent versions (average of 5 runs).

Type	Model	F_1	Precision	Recall
Baseline		0.4231	0.3997	0.4494
Undirected	FCN-Predict	0.6643	0.7012	0.6310
	FCN-Full	**0.6761**	**0.7128**	**0.6430**
Directed	FCN-Predict	**0.7013**	**0.7281**	**0.6764**
	FCN-Full	0.6942	0.7273	0.6640
Adjacent	FCN-Predict	0.7116	**0.7560**	0.6721
	FCN-Full	**0.7142**	0.7333	**0.6860**

Table 3. Performance of models trained on different task on task of detecting undirected/directed/adjacent versions, as applicable (average of 5 runs)

Type	Model	F_1	Precision	Recall
Undirected	Baseline	0.7731	0.7518	0.7956
	FCN-Predict	0.8933	0.9110	0.8763
	FCN-Full	**0.8961**	**0.9122**	**0.8805**
Directed	Baseline	0.7374	0.7140	0.7623
	FCN-Predict	**0.8291**	**0.8648**	0.7962
	FCN-Full	0.8261	0.8431	**0.8012**
Adjacent	Baseline	0.4231	0.3997	0.4494
	FCN-Predict	0.7116	**0.7560**	0.6721
	FCN-Full	**0.7142**	0.7333	**0.6860**

tasks. Since the performance of LR depends on features that are symmetric with respect to document order, it performs a lot worse in identifying directed and adjacent version relations. We also note again that including entity-based similarity also helps in detecting version, as the performance of FCN-Full (LR-Full) is better than FCN-LS (LR-LS), which only uses lexical similarity heatmaps. We also note that baseline performs significantly below the proposed approaches, even though, these tasks (particularly, Undirected setting) are closer to what the baseline was designed for.

Version Ordering: We now discuss the result of arranging the versions in full ordered sequence. Since for this dataset, the versions form a chain instead of general directed acyclic graph, we restrict our algorithm to find directed chains with maximum weights. Comparing the results given in Table 1 (for stage 1 and 2) with Table 2 (for stage 1, 2 and 3) shows that our proposed step helps improve the results significantly for Undirected and Directed settings ($\approx$6%), whereas for Adjacent, the improvement is modest ($\approx$2%), as expected. While using Adjacent setting still leads to best performance on the overall task, the gap in performance is reduced significantly. Also, please note that the two methods that we use to compute the version scores - using the prediction scores of FCN (FCN-Predict) and using the edge weights based on alignment based version score as described in Sect. 3.3 Eq. 1 (FCN-Full), yield similar results consistently across different tasks and metrics.

While our objective was to find directed ordered chains of versions of documents, we also present the result for a task based on *Undirected, Directed* and *Adjacent*. In Table 3, we see that the performance of our proposed approach is consistently better than the baseline approach across all the three settings and the three metrics. This shows that even though the baseline methods were closest to *Undirected*, our method still outperforms it significantly. Of course, for the objective we set out to achieve, the ratio of performance is even more impressive in favour of our proposed approach. As expected, we also observe a consistent drop across the three metrics as we move from the *Undirected* setting to the *Directed* setting, and then to the *Adjacent* setting. We attribute this drop to the progressive increase in task's difficult, as discussed earlier.

6   Conclusion

In this work we introduce a novel problem, namely, detecting and ordering versions of documents (in which they were created) from a document repository. We created a dataset, which we share with the research community, and also provided an end-to-end approach for the task. The dataset will help further the research in this domain. Our proposed approach involves three parts where we start by finding candidate documents using MinHash-LSH and then classify the selected candidates using an FCN-based classifier. The last step of our approach involves identifying the order of versions by finding the maximum spanning tree in the graph where documents are considered to be nodes and the edge-weights denote the version score between them. Our empirical results show that the individual parts of our proposed approach perform better than standard approaches and the end-to-end approach outperforms the state-of-the-art approach for near-duplicate document detection (NDD). This also indicates that the problem is sufficiently different than NDD, and therefore worthy of further research.

References

1. Alsulami, B.S., Abulkhair, M.F., Eassa, F.E.: Near duplicate document detection survey. Int. J. Comput. Sci. Commun. Netw. **2**(2), 147–151 (2012)
2. Broder, A.Z.: Identifying and filtering near-duplicate documents. In: Giancarlo, R., Sankoff, D. (eds.) CPM 2000. LNCS, vol. 1848, pp. 1–10. Springer, Heidelberg (2000). https://doi.org/10.1007/3-540-45123-4_1
3. Cho, K., et al.: Learning phrase representations using RNN encoder-decoder for statistical machine translation. In: Proceedings of the 2014 Conference on Empirical Methods in Natural Language Processing (EMNLP), pp. 1724–1734 (2014)
4. Chowdhury, A., Frieder, O., Grossman, D., McCabe, M.C.: Collection statistics for fast duplicate document detection. ACM Trans. Inf. Syst. (TOIS) **20**(2), 171–191 (2002)

5. Ekbal, A., Saha, S., Choudhary, G.: Plagiarism detection in text using vector space model. In: 2012 12th International Conference on Hybrid Intelligent Systems (HIS), pp. 366–371. IEEE (2012)
6. Elsayed, T., Lin, J., Oard, D.W.: Pairwise document similarity in large collections with mapreduce. In: Proceedings of ACL-08: HLT, Short Papers, pp. 265–268 (2008)
7. Ertl, O.: SuperMinHash - A New Minwise Hashing Algorithm for Jaccard Similarity Estimation. arXiv e-prints arXiv:1706.05698, June 2017
8. Gupta, D., Vani, K., Leema, L.: Plagiarism detection in text documents using sentence bounded stop word n-grams. J. Eng. Sci. Technol. **11**(10), 1403–1420 (2016)
9. Hassanian-esfahani, R., Kargar, M.J.: Sectional minhash for near-duplicate detection. Expert Syst. Appl. **99**, 203–212 (2018)
10. Ioffe, S., Szegedy, C.: Batch normalization: accelerating deep network training by reducing internal covariate shift. In: International Conference on Machine Learning, pp. 448–456. PMLR (2015)
11. Liu, X., Gong, Y., Xu, W., Zhu, S.: Document clustering with cluster refinement and model selection capabilities. In: Proceedings of the 25th Annual International ACM SIGIR Conference on Research and Development in Information Retrieval, pp. 191–198 (2002)
12. Long, J., Shelhamer, E., Darrell, T.: Fully convolutional networks for semantic segmentation. In: Proceedings of the IEEE Conference on Computer Vision and Pattern Recognition, pp. 3431–3440 (2015)
13. Lv, Y., Moon, T., Kolari, P., Zheng, Z., Wang, X., Chang, Y.: Learning to model relatedness for news recommendation. In: Proceedings of the 20th International Conference on World Wide Web, pp. 57–66 (2011)
14. Park, K., Lee, J., Choi, J.: Deep neural networks for news recommendations. In: Proceedings of the 2017 ACM on Conference on Information and Knowledge Management, pp. 2255–2258 (2017)
15. Sanchez-Perez, M.A., Sidorov, G., Gelbukh, A.F.: A winning approach to text alignment for text reuse detection at pan 2014. In: CLEF (Working Notes), pp. 1004–1011 (2014)
16. Sherkat, E., Nourashrafeddin, S., Milios, E.E., Minghim, R.: Interactive document clustering revisited: a visual analytics approach. In: 23rd International Conference on Intelligent User Interfaces, pp. 281–292 (2018)
17. Socher, R., Huang, E.H., Pennin, J., Manning, C.D., Ng, A.Y.: Dynamic pooling and unfolding recursive autoencoders for paraphrase detection. In: Advances in Neural Information Processing Systems, pp. 801–809 (2011)
18. Weng, S.S., Chang, H.L.: Using ontology network analysis for research document recommendation. Expert Syst. Appl. **34**(3), 1857–1869 (2008)
19. Xu, X., et al.: Understanding user behavior for document recommendation. In: Proceedings of The Web Conference 2020. p. 3012–3018. WWW 2020. Association for Computing Machinery, New York, NY, USA (2020). https://doi.org/10.1145/3366423.3380071
20. Yang, X., Lo, D., Xia, X., Bao, L., Sun, J.: Combining word embedding with information retrieval to recommend similar bug reports. In: 2016 IEEE 27th International Symposium on Software Reliability Engineering (ISSRE), pp. 127–137. IEEE (2016)

Focus on Misinformation: Improving Medical Experts' Efficiency of Misinformation Detection

Aleksandra Nabożny[1] [ID], Bartłomiej Balcerzak[2] [ID], Mikołaj Morzy[3(✉)] [ID], and Adam Wierzbicki[2] [ID]

[1] Gdańsk University of Technology, Gdańsk, Poland
[2] Polish-Japanese Institute of Information Technology, Warsaw, Poland
[3] Poznań University of Technology, Poznań, Poland
Mikolaj.Morzy@put.poznan.pl

Abstract. Fighting medical disinformation in the era of the global pandemic is an increasingly important problem. As of today, automatic systems for assessing the credibility of medical information do not offer sufficient precision to be used without human supervision, and the involvement of medical expert annotators is required. Thus, our work aims to optimize the utilization of medical experts' time. We use the dataset of sentences taken from online lay medical articles. We propose a general framework for filtering medical statements that do not need to be manually verified by medical experts. The results show the gain in fact-checking performance of expert annotators on capturing misinformation by the factor of 2.2 on average. In other words, our framework allows medical experts to fact-check and identify over two times more non-credible medical statements in a given time interval without applying any changes to the annotation flow.

Keywords: e-health · Misinformation · Text-mining · Human-in-the-loop · Credibility assessment · Natural language processing · Machine learning

1 Introduction

The spread of medical misinformation on the World Wide Web has become a significant social problem. We face a global "infodemic" of dubious medical claims, distrust in medical science, conspiracy theories, and outright medical falsehoods circulating in social media. The recent SARS-CoV-2 pandemic has exacerbated the existing problem of low confidence in medical institutions, pharmaceutical companies, and governmental agencies responsible for public health [13,18]. At the same time, we observe a growing trend of relying on online health information for self-treatment [5]. Given the possible consequences of using online health advice ungrounded in medical science, the task of assessing the credibility of online health information becomes pressing.

W. Zhang et al. (Eds.): WISE 2021, LNCS 13081, pp. 420–434, 2021.
https://doi.org/10.1007/978-3-030-91560-5_31

Distinguishing between reliable and unreliable online health information poses a substantial challenge for lay Internet users [1]. Labeling source websites as either credible or non-credible is not sufficient as false claims can be a part of an article originating from a credible source and vice versa. Often, credible medical statements can serve as the camouflage for disinformation woven into otherwise factually correct statements. Even subtle changes to the overtone, wording, or strength of a medical statement can change its meaning, for instance, by exaggerating the side effects of a drug or by conflating relative and absolute risks of a medical procedure. As an example, consider the following phrase: *"Aspirin should not be consumed during pregnancy"*. This phrase is generally true but does not apply when an early pregnancy is at the risk of miscarriage when consuming small doses of aspirin can significantly lower the risk.

Even experienced medical professionals find it challenging to assess the truthfulness of online medical information. What is considered to be "true" in the domain of medicine is often subject to a very complex context. This context is provided by external medical knowledge and clinical practice. Medical professionals often focus on the possible impact of health information on the choices made by patients rather than evaluate the factual correctness of a statement. In other words, a factually correct statement may still inflict health damage on patients when presented mischievously or in isolation. The phrase *"For starters, statin drugs deplete your body of coenzyme Q10 (CoQ10), which is beneficial to heart health and muscle function"*, despite factual correctness, would raise objections from medical professionals as it may discourage a patient from taking statins. In this example, the expert uses external knowledge from their clinical practice that for patients requiring statin therapy, its benefits far outweigh the potential risks associated with coenzyme Q10 deficiency. This additional context of online health information evaluation makes it extremely difficult to frame the task in terms of machine learning.

An additional problem that arises when evaluating online health information stems from the involvement of human judges. Whether these are annotators who curate training data for statistical models or subject matter experts (SMEs) who provide final scores, it is paramount that trained medical practitioners perform these tasks. Unfortunately, data labeling for online health information assessment, to the large extent, cannot be crowd-sourced due to the unique competencies required to provide the ground truth labels. Over-worked medical practitioners struggle to secure the time required for debunking online medical falsehoods and cannot keep up with the flood of online medical misinformation.

Scarce human resources are the bottleneck stifling the development of automatic online health information assessment methods. To address this issue, we propose to frame the problem of online health information evaluation as a machine learning problem, with the business objective being the optimization of the utilization of medical experts' time. Firstly, though, we have to change the definition of the machine learning task. As we have stated above, assessing the truthfulness of medical statements is subjective, context-dependent, and challenging. Instead, we propose to develop machine learning models that assess the *credibility* of medical statements. We define a medical statement to be credible if the statement is in accord with current medical knowledge and does not entice a patient to make harmful health-related decisions or to inspire actions contrary to the current medical guidelines. We do not try to discover the intention of an author of online health information. Thus we use the general term "misinformation" to represent both malicious and unintentional deception.

The business objective of optimizing the utilization of medical experts' time has yet to be framed in terms of an objective function driving the training of statistical models. We treat the time budget allocated by a medical expert to debunking online medical information as a fixed value. Similarly, we treat the average time required by a medical expert to evaluate a single medical statement as a fixed value. The only intervention that can influence the utilization of medical experts' time is the re-ranking of medical statements for annotation. We propose to focus medical experts' attention on statements that are possibly non-credible and contain medical misinformation. This, in turn, requires the development of methods for the automatic discovery of credible statements. The objective function is to maximize the recall of credible medical statements at a fixed high precision threshold. In this way, we can extract a large set of medical statements which are guaranteed to contain credible medical information due to fixed precision, and remove these statements from the queue of statements for human annotation, allowing medical experts to focus their limited time on the discovery of non-credible statements. Our experiments show that this approach increases the utilization of medical experts' time by the factor of 2.

Our main contributions presented in this paper include:

- introduction of the general approach for the optimization of the utilization of human annotators' time using machine learning,
- evaluation of the approach using the task of annotating online medical information credibility,
- developing a set of statistical classifiers for assessing the credibility of medical statements with the precision ranging from 83.5% to 98.6% for credible statements across ten different medical topics,
- developing a new method of data and label augmentation for improving the accuracy of the credibility classifiers,
- experimental evaluation of the augmentation method proving its efficacy.

2   Related Work

There are multiple strategies for improving the credibility of online health information. They include information corrections, both automatically-generated and user-generated [4], and the manipulation of the visual appeal and presentation of medical information [8]. A recent meta-analysis [23] shows, however, that the average effect of correction of online health information on social media is of weak to moderate magnitude. The authors point out that interventions are more effective in cases when misinformation distributed by news organizations is debunked by medical experts. When misinformation is circulated on social media by peers, or when non-experts provide corrections, interventions have low impact.

The approaches to automatic classification of online medical misinformation differ depending on the media and content type. Most studies employ content analysis, social network analysis, or experiments, drawing from disciplinary paradigms [24]. Online medical misinformation can be effectively classified by using so-called peripheral-level features [29] which include linguistic features (length of a post, presence of a picture, inclusion of an URL, content similarity with the main discussion thread), sentiment features (both corpus-based and language model-based), and behavioral features (discussion initiation, interaction engagement, influential scope). Peripheral-level features proved to be useful for detecting the spread of false medical information during the Zika virus epidemic [7,22]. Stylistic features can be used to identify hoaxes presented as genuine news articles and promoted on social media [19]. Along with identifying hoaxes, it is possible to identify social media users who are prone to disseminating these hoaxes among peers [9]. An applied machine learning-based approach, called *MedFact*, is proposed in [21], where the authors present an algorithm for trusted medical information recommendation. The *MedFact* algorithm relies on keyword extraction techniques to assess the factual accuracy of statements posted in online health-related forums.

More advanced methods of online medical information evaluation include video analysis (extracting medical knowledge from YouTube videos [15]), detecting misinformation based on multi-modal features (both text and graphics [25]), and website topic classification. The latter approach was successfully applied by [2,14] using topic analysis (either Latent Dirichlet Annotation or Term-Frequency). In addition, Afsana *et al.* use linguistic features, such as word counts, named entities, semantic coherence of articles, the Linguistic Inquiry Word Count (LIWC), and external metrics such as citation counts and Web ranking of a document.

We consider the full article's credibility prediction as burdened with source bias, as well as not precise enough to perform the targeted decision explanations. That is why, instead of the articles, we chose to classify smaller chunks of text (triplets of sentences, in particular). In previous approaches, the classifiers rated entire documents. For example, in the study evaluating entire articles [2], they were assessed against 10 criteria, none of which directly determines whether the content is credible or not. Our method differs from the approaches presented in the literature earlier in two important aspects: we leverage the context of medical expert's annotation by data and label augmentation, and we modify the objective function to optimize for the recall of the positive class given the fixed precision threshold.

3   Methods

3.1   Dataset

Our dataset consists of over 10 000 sentences extracted from 247 online medical articles. The articles have been manually collected from health-related websites. The choice of major categories (cardiology, gynecology, psychiatry, and pediatrics) has been dictated by the availability of medical experts participating in

Table 1. Number of sentences from each class by the topic.

Category	Topic	CRED	NEU	NONCRED
Cardiology	Antioxidants	375	175	144
Cardiology	Heart supplements	221	124	78
Cardiology	Cholesterol and statins	1058	565	406
Gynecology	Cesarean section vs. natural birth	275	53	31
Pediatry	Children & Antibiotics	298	52	82
Pediatry	Diet and Autism	236	71	124
Pediatry	Steroids for kids	560	101	40
Pediatry	Vaccination	730	223	309
Pediatry	Allergy testing	790	398	214
Psychiatry	Psychiatry	1194	676	402

the experiment. After consulting with medical experts, we have selected certain topics known to produce controversy in online social networks. For each topic, we have collected a diversified sample of articles presenting contradicting views (either supportive or contrarian) and we have extracted statements for manual evaluation by medical experts. The dataset is open-sourced and publicly available[1]. For the detailed description of the dataset, we refer the reader to [16].

Nine medical experts took part in the experiment, including 2 cardiologists, 1 gynecologist, 3 psychiatrists, and 3 pediatricians. All experts have completed 6-years medical studies and then a 5-year residency program. The experts were paid for a full day of work (approximately 8 h each). Each medical expert had at least 10 years of clinical experience, except for the gynecologist who was a resident doctor. We have accepted his participation in the experiment due to his status as a Ph.D. candidate in the field of medicine. One of the psychiatrists held a Ph.D. in medical sciences. Given the high qualifications of participants, we consider their judgments as the ground truth for medical statement evaluation. The experts were allowed to browse certified medical information databases throughout the experiment. Each expert evaluated the credibility of medical statements only within their specialization.

Collected online articles were automatically divided into sentences and presented to the medical experts in random order. Sentence segmentation has been done using the dependency parser from the spaCy text processing library. Since input text follows closely the general-purpose news style, the default spaCy processing pipeline produces very robust sentence segmentation. Along with each sentence we have displayed a limited number of automatically extracted keywords. If the medical expert decided that a sentence could not have been assessed due to insufficient context, he or she could have expanded the annotation view by showing preceding and succeeding sentences. Each medical expert was asked to

---

[1] https://github.com/alenabozny/medical_credibility_corpus.

annotate approximately 1000 sentences. Medical experts evaluated the credibility of sentences with the following set of labels and the corresponding instructions:

- CRED (credible)—a sentence is reliable, does not raise major objections, contains verifiable information from the medical domain.
- NONCRED (non-credible)—a sentence contains false or unverifiable information, contains persuasion contrary to current medical recommendations, contains outdated information.
- NEU (neutral)—a sentence does not contain factual information (e.g., is a question) or is not related to medicine.

Table 1 presents the number of sentences in each class summarized by category and topic. Within the four larger topical categories (cardiology, gynecology, psychiatry, or pediatrics), our dataset is divided into smaller subsets (topics). Considering these topics separately dramatically improves the performance of the classifiers. However, some topics included in the dataset were too small for training a classifier. Thus, we do not consider them further in this article.

3.2 Data Augmentation

The annotation of the dataset by medical experts has revealed the importance of context for providing a label (see Table 2). Over 25% of non-credible sentences required the surrounding context of one sentence, with 20% of credible sentences and 12% neutral sentences requiring similar context. To provide this context for statistical models, we have decided to transform single sentences into sequences of consecutive non-overlapping triplets of sentences. Since individual sentences have already been labeled by medical experts, we have transferred ground truth sentence labels to triplet labels in the following way:

- negative: a triplet is negative if any of the sentences constituting the triplet has the label NONCRED,
- positive: a triplet is positive if all of the sentences constituting the triplet are either CRED or NEU.

Figure 1 depicts the idea of label transfer applied to the dataset.

Table 2. Number m of surrounding sentences needed to understand the context and evaluate the credibility of a sentence for credible, non-credible, neutral, and all sentences.

m	Credible [%]	Non-credible [%]	Neutral [%]	**All [%]**
0	80.07	71.27	88.30	**80.43**
1	18.83	26.60	11.03	**18.39**
>1	0.18	0.37	0.04	**0.18**

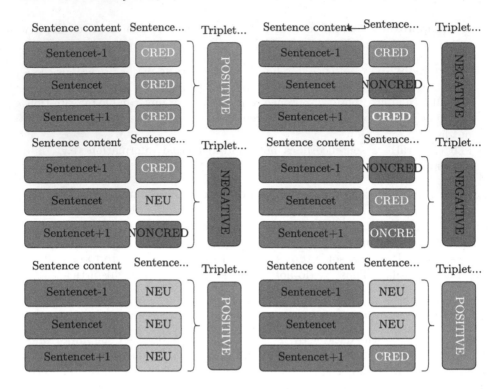

Fig. 1. Label transfer for augmented data

Example of a positive triplet (from "Statins & cholesterol"):

"Not smoking could add nearly 10 years and quitting increases life expectancy by reducing the chances of emphysema, many cancers, and heart disease. Although my doctor checks my cholesterol every year, it remains low and taking a statin will have a very small, if any, effect on my life expectancy. What's worse, my doctor has never asked if I smoke cigarettes, exercise regularly, or eat a healthy diet."

Example of a negative triplet (from "Statins & cholesterol"):

"OK, maybe the benefits of taking a statin are small, but many smart doctors say a reduction of five-tenths or six-tenths of 1% is worthwhile. Yet the few published observations on people over the age of 70 do not show any statistically significant statin-related reductions in deaths from any cause. Of course, not everyone is like me."

3.3   Feature Set

Features that have been selected for credibility classification purposes are based on the qualitative analysis of the dataset concerning the findings reported in

Sect. 2. The ultimate number of features varies between categories. The feature set has been created manually and feature selection methods have been used to remove non-informative features. The choice of traditional NLP features has been deliberate as we want to maintain the explainability of credibility classification models. Also, somewhat contrary to popular belief, initial experiments have shown that these traditional features are superior to sentence embeddings computed using BERT [6]. It remains to be seen if using a language model fine-tuned to the medical domain [3] would make the embeddings a better source of features for the credibility classifier.

Uncased TF-IDF (Number of Features: Varying from 920 to 4103). Bag of words, n-gram, term frequency (TF), term frequency inverted document frequency (TF-IDF) are the most commonly used textual features in natural language processing [28]. In this work, we chose TF-IDF values to account for the importance of each word. We use the Python package spaCy to perform sentence tokenization.

Dependency Tree-Labels Count (Number of Feaures: Up to 45). Overly complex sentences have a higher probability of containing the hedging part than simple sentences (the base of a sentence may contain a factually false statement, but the other part would soften its overtone so that it seems credible). Thus, we count the base elements of dependency trees to model the potential existence of such phenomena.

Named Entities Counter (Number of Features: Up to 18). There are some indicators of conspiratorial and/or science-skeptical language (hence the popularity of using agent-action-target triples in the study of conspiratorial narratives [20]) Those narratives may be captured by counting named entities of specified categories, such as false authority (PERSON), Big Pharma blaming (ORGANIZATION, PRODUCT), distrust to renowned institutions (ORGANIZATION), facts and statistics (NUMBER). In the experiment we have used the NER labeling scheme available in the English language model offered by the spaCy library.

Polarity and Subjectivity (Number of Features: 2). Sentiment analysis is a broadly-used feature set for misinformation detection classifiers. It has been used, for example, for detecting anti- and pro-vaccine news headlines [27]. Highly polarized and/or emotional language can indicate misinformation.

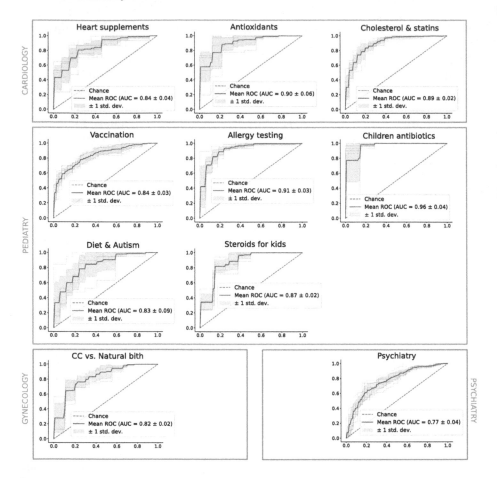

Fig. 2. ROC curves of cross-validated classification results for each medical topic.

LIWC (Number of Features: 93). Aggressive, overly optimistic, advertising language (e.g. for a drug or novel therapy) or other patterns can affect the credibility of textual information [12]. The LIWC offers a corpus-based sentiment analysis approach by counting words in different emotion categories. Empirical results using LIWC demonstrate its ability to detect meaning in emotionality. In addition, it has been employed to extract the sentiment features for the detection of misinformation in online medical videos [11]. LIWC provides features regarding emotional dimensions, the formality of the language, spatial and temporal features, as well as structural information (e.g. word per sentence count).

3.4 Feature Selection and Model Training

The workflow for training statistical models is identical for each topic and includes two steps: feature selection and model selection. Feature selection is performed using Logistic Regression and Recursive Feature Elimination (RFE) [10]. RFE conducts a backward selection of features, starting from a predictive model using all available features. For each feature, the importance score is computed, and the least important feature is removed. The model is retrained with remaining features and the procedure is repeated until the desired number of features remains. We use Logistic Regression as the baseline model for RFE, limiting the number of features to 30% of the number of samples in a given topic. Due to the lack of space we cannot provide a detailed description of selected feature sets for each medical category, but we will include this information in the extended version of the paper. In this paper we also assume that the list of topics is known in advance and that each sentence is already assigned to a topic. This, of course, raises the question of practical applicability of our method when the topic of an article is unknown. Recent advances in automatic medical subdomain classification [26] suggest that the topic of the article can be successfully extracted from the text.

For training the model we use the TPOT library [17]. TPOT uses a genetic algorithm to optimize the workflow consisting of feature pre-processing, model selection, and parameter optimization, by evolving a population of workflows and implementing mutation and cross-over operators for workflows. To constrain the space of considered models we use Logistic Regression, XGBoost, and the Multi-layer Perceptron as the initial pool of available models. The optimization is driven by the F_1 measure.

4 Results

The main objective of our method is the maximization of the utilization of medical experts' time when annotating online medical statements. We optimize statistical models to find credible statements, thus increasing the number of non-credible statements that can be presented to medical experts. The results below analyze the efficiency of trained statistical models in finding credible statements. Recall from Sect. 3.2 that statistical models are trained on a binary dataset consisting of positive (credible and neutral) and negative (non-credible) triplets of sentences.

Figure 2 presents ROC curves for cross-validation. The number of folds depends on the number of samples in a given topic. Based on the ROC curves we have empirically adjusted the cutoff threshold for each classifier's prediction of the positive class. Our goal was to maximize the recall of the positive class while preserving fixed high precision for the positive class. In other words, samples that fall above the cutoff threshold are assumed to contain only credible or neutral sentences, and will not be presented to medical experts for manual evaluation. We have selected the cutoff threshold for each topic using the following criteria:

- the difference between the proportion of true negative samples and the proportion of negative samples in the entire test set should be maximized, with minimum variance,

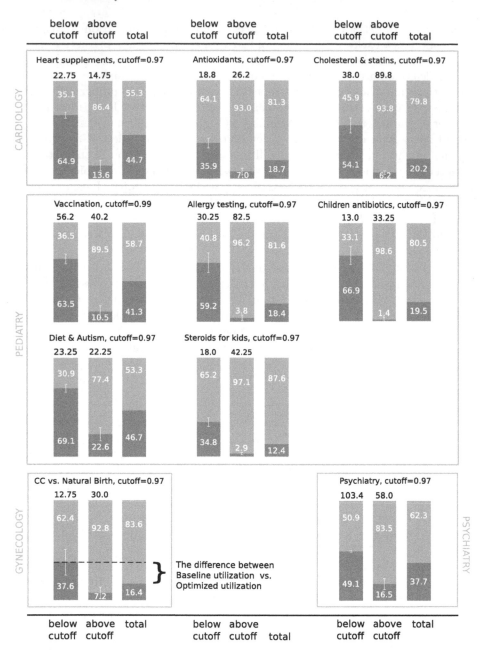

Fig. 3. Cross-validated proportions of positive and negative samples a) below the cutoff b) above the cutoff c) in the entire test set. Black labels indicate the mean number of samples in each group

- the precision for the true positive class should be maximized,
- the number of samples above the cutoff should be maximized.

The results of the cutoff filtering are presented in Fig. 3. For each topic, we show the distribution of positive and negative samples in the entire topic (the *total* column) and in the subsets defined by the cutoff. For instance, there are 44.7% of negative samples and 55.3% of positive samples in the *Heart supplements* topic. The subset of samples defined by the cutoff point of 0.97 contains only 13.6% of negative samples, and the remaining subset contains 64.9% of negative samples. In other words, by removing the samples above the cutoff threshold from manual experts' evaluation we are increasing the number of negative samples that the experts may annotate from 44.7% to 64.9%. We refer to the proportion of negative samples in the topic as the *baseline utilization*, and the proportion of negative samples after the intervention (i.e., below the cutoff threshold) as the *optimized utilization*.

Table 3 presents the main results of our experiment. We report baseline and optimized utilization, the difference in percentage points, and the factor of improvement of medical experts' time utilization.

Table 3. Comparison of baseline and optimized utilization of medical experts' time.

Category	Baseline utilization [%]	Optimized utilization [%]	pp. diff	Factor
Heart supplements	44.7	64.9	↑ 20.2	1.5
Antioxidants	18.7	35.9	↑ 17.2	1.9
Cholesterol & Statins	20.2	54.1	↑ 33.9	2.7
Vaccination	41.3	63.5	↑ 22.2	1.5
Allergy testing	18.4	59.2	↑ 40.8	3.2
Children antibiotics	19.5	66.9	↑ 47.4	3.4
Diet & Autism	46.7	69.1	↑ 22.4	1.5
Steroids for kids	12.4	34.8	↑ 22.4	2.8
CC vs. Natural birth	16.4	37.6	↑ 21.2	2.3
Psychiatry	37.7	49.1	↑ 11.4	1.3
Mean	–	–	↑ **25.9**	**2.2**

5   Discussion

Evaluation of the credibility of online medical information is a very challenging task due to the subjective assessment of credibility, and the specialized medical knowledge required to perform the evaluation [16]. Fully automatic classification of online medical information as credible or non-credible is not a viable solution due to the complex externalities involved in such classification. For the foreseeable future, keeping a human judge in the annotation loop is a necessity. At the same time, qualified human judges are the scarcest resource and their time must be utilized efficiently. Previous approaches to automatically assessing the credibility of medical texts did not take into account the need to weave a human judge into the real-time verification process.

In our work, we present a framework for the optimization of the utilization of medical experts' time when evaluating the credibility of online medical information. To prioritize the evaluation of non-credible information by medical experts, we train classifiers that can filter out credible and neutral medical claims with very high precision exceeding 90% for most medical topics considered in our study (vaccination, allergy testing, children antibiotics, steroids for kids, antioxidants, cholesterol & statins, and C-section vs. natural birth).

Table 3 depicts the key benefit for the potential human-in-the-loop fact-checking system that our solution provides—an increase in the probability that a medical expert will encounter a non-credible medical statement in the annotation batch. As we can see, for all topics the improvement in the utilization of medical experts' time is substantial. The average improvement over all topics is 25.9% points, which means that within the same amount of time and at the same average time needed to annotate a single sentence, medical experts using our method annotate over two times as many non-credible medical statements on average. It is a "pure win" since this improvement does not require any changes to either the annotation protocol or the annotation interface, we simply make much better use of the valuable experts' time allocated to data annotation.

6   Conclusions and Future Work

One limitation of our method is a small number of statements that contain misinformation, but would not be seen by experts. However, we need to keep in mind that medical experts may not be able to see all statements anyway, as their time and attention are limited and may not be enough to process all suspicious information.

In a realistic use-case, medical experts would continually evaluate a stream of statements derived from the ever-growing set of online articles on medical and health topics, as well as information from social media. Our method allows increasing the efficiency of misinformation detection by debunking medical experts, who will discover more than twice as much misinformation without increasing the time spent on evaluation (or the number of evaluating experts), and without introducing any changes to the annotation workflow. Our method can be regarded as a universal filter for medical web content.

In our future work, we will focus on gathering more data by introducing the demo expert crowd-sourcing system in a few medical universities. We will put special emphasis on the iterative process of adjusting proper annotation protocol and professional training for medical students to gain the annotation accuracy as close as possible to the experts (medical practitioners with at least a few years of experience), thus further reducing costs of expert medical credibility annotation.

References

1. Abramczuk, K., Kakol, M., Wierzbicki, A.: How to Support the Lay Users Evaluations of Medical Information on the Web? (2016).https://doi.org/10.1007/978-3-319-40349-6_1
2. Afsana, F., Kabir, M.A., Hassan, N., Paul, M.: Automatically assessing quality of online health articles. IEEE J. Biomed. Health Inform. **25**(2) (2021). https://doi.org/10.1109/JBHI.2020.3032479
3. Alsentzer, E., et al.: Publicly available clinical BERT embeddings. arXiv preprint arXiv:1904.03323 (2019)
4. Bode, L., Vraga, E.K.: See something, say something: correction of global health misinformation on social media. Health Commun. **33**(9), 1131–1140 (2018). https://doi.org/10.1080/10410236.2017.1331312
5. Chen, Y.Y., Li, C.M., Liang, J.C., Tsai, C.C.: Health information obtained from the internet and changes in medical decision making: questionnaire development and cross-sectional survey. J. Med. Internet Res. **20**(2), e47 (2018)
6. Devlin, J., Chang, M.W., Lee, K., Toutanova, K.: BERT: pre-training of deep bidirectional transformers for language understanding. arXiv preprint arXiv:1810.04805 (2018)
7. Dito, F.M., Alqadhi, H.A., Alasaadi, A.: Detecting medical rumors on twitter using machine learning. In: 2020 International Conference on Innovation and Intelligence for Informatics, Computing and Technologies, 3ICT 2020. Institute of Electrical and Electronics Engineers Inc. December 2020. https://doi.org/10.1109/3ICT51146.2020.9311957
8. Ebnali, M., Kian, C.: Nudge Users to Healthier Decisions: A Design Approach to Encounter Misinformation in Health Forums (2020). https://doi.org/10.1007/978-3-030-20500-3_1
9. Ghenai, A., Mejova, Y.: Fake cures. In: Proceedings of the ACM on Human-Computer Interaction 2(CSCW), November 2018. https://doi.org/10.1145/3274327
10. Guyon, I., Weston, J., Barnhill, S.: Gene selection for cancer classification using support vector machines. Technical report (2002)
11. Hou, R., Perez-Rosas, V., Loeb, S., Mihalcea, R.: towards automatic detection of misinformation in online medical videos. In: 2019 International Conference on Multimodal Interaction. ACM, New York, NY, USA, October 2019. https://doi.org/10.1145/3340555.3353763
12. Jensen, M.L., Averbeck, J.M., Zhang, Z., Wright, K.B.: Credibility of anonymous online product reviews: a language expectancy perspective. J. Manage. Inf. Syst. **30**(1) (2013). https://doi.org/10.2753/MIS0742-1222300109
13. Latkin, C.A., Dayton, L., Yi, G., Konstantopoulos, A., Boodram, B.: Trust in a COVID-19 vaccine in the U.S.: a social-ecological perspective. Soc. Sci. Med. **270** (2021). https://doi.org/10.1016/j.socscimed.2021.113684
14. Li, J.: Detecting False Information in Medical and Healthcare Domains: A Text Mining Approach (2019). https://doi.org/10.1007/978-3-030-34482-5_21
15. Liu, X., Zhang, B., Susarla, A., Padman, R.: YouTube for Patient Education: A Deep Learning Approach for Understanding Medical Knowledge from User-Generated Videos. ArXiv Computer Science, July 2018
16. Nabożny, A., Balcerzak, B., Wierzbicki, A., Morzy, M.: Digging for the truth: the case for active annotation in evaluating the credibility of online medical information. JMIR Preprints, November 2020

17. Olson, R.S., Urbanowicz, R.J., Andrews, P.C., Lavender, N.A., Kidd, L.C., Moore, J.H.: Automating biomedical data science through tree-based pipeline. Optimization (2016)

18. Pollard, M.S., Lois, M.: Davis: decline in trust in the centers for disease control and prevention during the COVID-19 pandemic. Technical report (2021). https://doi.org/10.7249/RRA308-12

19. Purnomo, M.H., Sumpeno, S., Setiawan, E.I., Purwitasari, D.: Biomedical engineering research in the social network analysis era: stance classification for analysis of hoax medical news in social media. Procedia Comput. Sci. **116** (2017). https://doi.org/10.1016/j.procs.2017.10.049

20. Samory, M., Mitra, T.: The government spies using our webcams: the language of conspiracy theories in online discussions. In: Proceedings of the ACM on Human-Computer Interaction 2(CSCW), November 2018. https://doi.org/10.1145/3274421

21. Samuel, H., Zaïane, O.: MedFact: Towards Improving Veracity of Medical Information in Social Media Using Applied Machine Learning (2018). https://doi.org/10.1007/978-3-319-89656-4_9

22. Sicilia, R., Lo Giudice, S., Pei, Y., Pechenizkiy, M., Soda, P.: Twitter rumour detection in the health domain. Expert Syst. Appl. **110** (2018). https://doi.org/10.1016/j.eswa.2018.05.019

23. Walter, N., Brooks, J.J., Saucier, C.J., Suresh, S.: Evaluating the Impact of Attempts to Correct Health Misinformation on Social Media: A Meta-Analysis. Health Communication, August 2020. https://doi.org/10.1080/10410236.2020.1794553

24. Wang, Y., McKee, M., Torbica, A., Stuckler, D.: Systematic literature review on the spread of health-related misinformation on social media. Soc. Sci. Med. **240** (2019). https://doi.org/10.1016/j.socscimed.2019.112552

25. Wang, Z., Yin, Z., Argyris, Y.A.: Detecting medical misinformation on social media using multimodal deep learning, December 2020

26. Weng, W.H., Wagholikar, K.B., McCray, A.T., Szolovits, P., Chueh, H.C.: Medical subdomain classification of clinical notes using a machine learning-based natural language processing approach. BMC Med. Inform. Decis. Making **17**(1), 1–13 (2017)

27. Xu, Z., Guo, H.: Using text mining to compare online pro- and anti-vaccine headlines: word usage, sentiments, and online popularity. Commun. Stud. **69**(1), 103–122 (2018). https://doi.org/10.1080/10510974.2017.1414068

28. Zhang, X., Ghorbani, A.A.: An overview of online fake news: characterization, detection, and discussion. Inf. Process. Manage. **57**(2) (2020). https://doi.org/10.1016/j.ipm.2019.03.004

29. Zhao, Y., Da, J., Yan, J.: Detecting health misinformation in online health communities: incorporating behavioral features into machine learning based approaches. Inf. Process. Manage. **58**(1) (2021). https://doi.org/10.1016/j.ipm.2020.102390

Sensory Monitoring of Physiological Functions Using IoT Based on a Model in Petri Nets

Kristián Fodor[✉] and Zoltán Balogh

Department of Informatics, Faculty of Natural Sciences, Constantine the Philosopher University in Nitra, Trieda Andreja Hlinku 1, 949 74 Nitra, Slovakia
{kristian.fodor,zbalogh}@ukf.sk

Abstract. Emotion recognition relies heavily on physiological reactions and facial expressions. There may be a few research that deal with the collection of physiological data from users using various Internet of Things (IoT) devices, but even fewer studies exist that deal with the classification and prediction of emotional states based on collected physiological data. In the majority of studies, research use invasive devices during monitoring methods such as electroencephalography (EEG) and electrocardiography (ECG). This paper investigates the collection of physiological data by creating a complex sensory network. The network uses some non-invasive IoT devices such as smart bands and different modules that are connected to the Raspberry Pi microcomputer and Arduino microcontroller. The goal is to make the sensory network as less invasive as possible. The Petri net model simulating how certain emotional states affect physiological data has already been created by us, however it is necessary to customize that model to include the physiological data, that the created sensory network can collect. The aim is to be able to collect enough physiological data from users so a dataset can be created, which will later serve as an input to a classification and prediction model. The physiological data will be sent to a server to process and store the data in a database. For the heart rate data, a mobile application will be created to partially automatize the collection and storage of data.

Keywords: Sensory monitoring · Petri Nets · Physiological states · Emotional states

1 Introduction

In this paper we are investigating the creation of a prototype sensory network that will be able to gather physiological functions that later can be used to identify the emotional states of the users. A Petri net model will be created that will allow the simulation of how physiological changes can also mean changes in emotional states. The input places of the model will be the same physiological functions that we are going to collect with our prototype. The main idea is to find the hardware and software solutions which together can form a complex solution of sensory monitoring. Heart rate, body temperature and galvanic skin response will be monitored using different IoT devices, such as Raspberry

© Springer Nature Switzerland AG 2021
W. Zhang et al. (Eds.): WISE 2021, LNCS 13081, pp. 435–443, 2021.
https://doi.org/10.1007/978-3-030-91560-5_32

Pi, Arduino and smart bands. The measured values will be sent to a server for processing and storing the data in databases. After verifying the functionality of the created prototype, an experiment can be performed on subjects in a real environment.

2 Related Work

Physiological and emotional monitoring is a very popular topic among researchers, mainly in the fields of stress and anxiety monitoring [1]. During some of the experiments, in order to provoke emotional reactions, different methods were used. Such as using movies during experiments with the assumption that specific genres of movies evoke specific emotions in the subjects [2]. For example, we can assume that horror movies can evoke the emotion of fear or disgust in the user, while comedies can make the user feel happy. This experiment can also be done in the virtual reality environment, which can trigger stronger emotional responses [3].

Fuzzy Petri nets allow us to simulate the ambiguity that exists in both the real world and emotional states [4]. The authors [5] also addressed fuzzy logic in IoT. They used fuzzy logic and heuristic techniques to reduce costs, energy consumption and to control gas and interruptible appliances.

The authors [6] used visual and auditory signals to induce emotions and collected four types of physiological signals, namely body temperature, galvanic skin response, blood volume fluctuations, and electrocardiogram. The resulting average classification accuracy is 61.8%. Chanel et al. [7] performed 100 inductions of high and low arousal emotions in four subjects and recorded the EEG response, blood pressure, and skin conductivity in the subjects. Leu and his colleagues conducted research on the recognition of emotions based on physiological signals [8]. Previous research also suggests that higher decreases in electrodermal activity correlate with more negatively balanced situations [9, 10].

Other methods included the usage of music or music videos [11]. All these methods can be used to gain new knowledge about emotion detection based on physiological functions and in the future, the research can be used to make machines emotionally intelligent [12].

3 Materials and Methods

In this paper, a conceptual model in Petri Nets is created alongside a prototype of a sensory network. With this model we were able to generate random decreases and increases in physiological functions and by using IF-THEN rules, we were able to tell what basic emotions the user might be feeling. We were identifying six basic emotions, also called as the "Big Six" by Ekman's research [13] and also a neutral state if we were not able to detect any of the six basic emotions. Earlier this year, we have created a conceptual model in Petri Nets [14], however one of the input physiological functions was Electroencephalography (EEG), which could not be measured in a non-invasive method.

The input places of the current model are Galvanic Skin Response (GSR); Heart Rate (HR) and Body Temperature (BT).

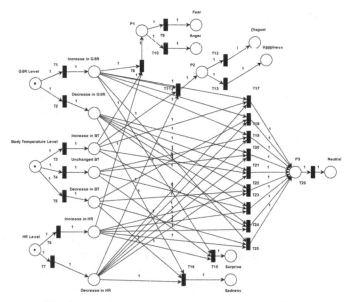

Fig. 1. The Petri net model with changed input places

Body temperature can also be used for emotion classification as slight changes can be observed during different emotional states [15–17]. The new version of the model, where the EEG input signals are replaced with body temperature measurements can be seen in the figure (Fig. 1).

A neutral state is added to ensure that no deadlocks are present in the inside of the model (Fig. 2).

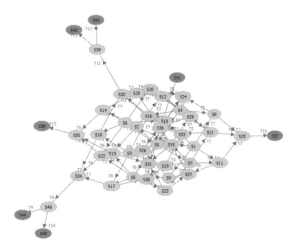

Fig. 2. Graph of reachability for the latest version of the model

4 Research Methodology

The first module that we added to our Raspberry Pi 4 microcomputer is a camera module, namely NoIR Camera V2. This camera module is enriched by the lack of an infrared filter and all captured images will look quite unusual in daylight. However, we gain the ability to capture clearer video at night, making it possible to detect emotions not only in adequately illuminated places, but at night time and in darker places as well.

We used A. Balaji's [18] repository to be able to detect and recognize the emotions using our camera. The model also uses Ekman's classification to detect basic emotions. The list included happiness, sadness, fear, surprise, anger, and disgust, which are still the most commonly accepted candidates for basic emotions among emotion theorists [19]. A seventh emotional state – a neutral state is added to the classification in case the model is not able to detect any of the six basic emotional states.

The model has a few dependencies that needs to be installed on the Raspberry Pi in order to use the model: 1. Python 3 (we used Python 3.6); 2. OpenCV; 3. Tensorflow (as of the time of writing, the repository only supports tensorflow-2.0).

The machine learning model was trained using 35887 pictures of faces from the Facial Expression Recognition 2013 (FER-2013) dataset [1]. This dataset contains 48 × 48 grayscale images labelled with the emotion depicted in the image. A random sample can be seen in the figure (Fig. 3).

Fig. 3. Sample of images from the FER-2013 dataset [20]

After training the model, a 63.2% test accuracy can be reached in 50 epochs when using a simple 4-layer convolutional neural network (CNN).

The model uses the haar cascade method to detect faces in each frame of the camera feed. Object detection using a cascade classifier based on Haar features is an effective object detection method proposed by Paul Viola and Michael Jones in their 2001 paper "Rapid Object Detection using a Boosted Cascade of Simple Features" [21]. This method is a machine learning-based approach where a cascade function is trained from multiple images (positive and negative). It is then used to detect objects in other images.

The detected region of the image containing the face is resized to 48 × 48 pixels to match the size of the images in the dataset and is passed as input to the 4-layer convolutional neural network supported by Keras, which acts as an interface for the TensorFlow library.

The network then outputs a list of softmax scores for the seven classes of emotions. Softmax converts the numerical output of the final linear layer of a multi-class classification neural network (logits) to probabilities by calculating the exponents of each output and then normalizing each number by the sum of those exponents such that the total output vector adds up to one. The sum of all probability should be exactly one. For such a multi-class classification task, the loss function is often cross entropy loss. Softmax is commonly added to the last layer of an image classification network. The emotion with the highest softmax score is then overlayed on the camera feed as seen in the figure (Fig. 4).

Fig. 4. Emotion detection output from the model

The first physiological function we are collecting is heart rate data, which is a measure of cardiovascular activity along with interbeat interval (IBI), blood pressure (BP, heart rate variability (HRV) and blood volume pulse (BVP) [22]. Heart rate reflects emotional activity and can be used to distinguish positive and negative emotions [23].

Recently we have conducted a research at our university to test the accuracy of the heart rate sensor inside different smart bands on the market. The heart rate measurement was performed by considering a certified pressure holter (A&D BOSO TM -2430), as a reference device, which is commonly used in the medical environment. We chose holter because the accuracy of the heart rate measurement is ±5%. Simultaneously with the pressure holter, we also measured the heart rate using smart bands at precisely set time intervals. Subsequently, we evaluated the measured data from the holster and individual smart bands and compared their accuracy based on comparative statistics.

When we summarized all the achieved results and compared the measured values, we could confirm that the most accurate of all tested devices was the Mi Band 4 smart band by Xiaomi. The current newest version from this manufacturer is the Mi Band 6, but we assume that the accuracy of the sensors in Mi Band 6 is the same as in Mi Band 4, which was used in the research.

All measured values are recorded after connecting the band in the Mi Fit application created by Huami technologies. Pairing with the application is secured using bluetooth technology. The low-energy chipset of bluetooth 5.0, enables fast and reliable connection and at the same time lower battery consumption. This technology also ensures long battery life and reliable data transmission. Twenty days of durability and longevity are made possible by a high-density lithium-polymer battery.

In the Mi Fit application, it is possible to export heart rate data from a specific interval, however it is a long process. The user has to log in and choose which data he wants to export. The data is then sent to the user's email address as a CSV file.

A faster method for accessing heart rate data is to use a third-party application called Tools & Mi Band, which also allows the user to export the heart rate data in a CSV file, but without the need to log in and wait for the CSV file to arrive in the email. The application also allows the user to share this CSV file via messaging applications or to save it on their personal cloud storage like Google Drive.

To store the measured heart rate more efficiently, we created a simple application that is able to open a CSV file (e.g., from Google Drive), iterate through each row and create JSON objects. These objects are then used to create a JSON array that is sent to a server. The PHP script on the server then decodes the JSON array and uses SQL to create new records in our database table. This way we can partially automate the process of storing the data in a database and we can also store data from multiple sensors from different sources in the same database. This is especially efficient if we want to synchronize the data from different sources for the same user. Later, if we will have more data from different sources (e.g., from Arduino or Raspberry Pi), we can use this application to read the values in the database from all of the sources with their respective sensors.

The next physiological function we measured was galvanic skin response. GSR refers to changes in sweat gland activity that reflect the intensity of emotional states, also known as emotional arousal. The emotional arousal level will vary with the environment. If something is scary, threatening, joyful or something else significantly related to emotions, then the subsequent changes in the emotional response will also increase the eccrine activity of sweat glands. Research has shown how this is linked to emotional arousal [24–26] To measure the GSR of the user, we used the Arduino Uno microcontroller with the GSR sensor by Grove. The GSR sensor applied a constant voltage to two electrodes that are in contact with the skin of the user (usually placed against the inside parts of two fingers). Compared with the skin resistance in series with the voltage supply and the electrodes, the circuit also contains a very small resistance. The purpose of the circuit is to measure the conductance of the skin and its changes by applying Ohm's law (voltage = intensity × resistance = intensity/conductance). Since the voltage remains constant, the skin conductance can be calculated by measuring the current flow through

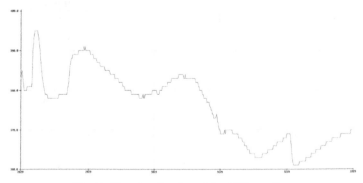

Fig. 5. The visualization of the GSR readings

the electrode. We measured the GSR of the user every 5 ms then averaged every 10 readings (to avoid glitches in the data) and printed out the average in the console. Thanks to Arduino's Serial Plotter in the integrated development environment (IDE), we were able to visualise the readings aswell (Fig. 5).

Then, we have altered the code to print out possible emotion changes. In order to do that, we have calculated a threshold value at the beginning by calculating the sum of 500 readings and divided it by 500 to get an average threshold value. Then in order to be able to send these sensor values to our database, we have added a Wifi Shield by Seeed Studio to our Arduino Uno and added the WiFly library to our IDE in order to be able to connect to a nearby Wifi access point. Using the HTTPClient library we were able to connect to the PHP script on our server. The PHP script processed the POST data (sensor values) and added the records to the database table. The GSR values are stored in the same database, but a new table has been created for these values.

The third physiological function we are collecting is body temperature. To collect the data, we are using the MLX90640 by Pimoroni, which is a fully calibrated 32 × 24 pixels thermal IR array in an industry standard 4-lead TO39 package with digital interface. According to the official site [27], this thermal camera can detect temperatures from −40 to 300 °C with approximately 1 °C accuracy and up to 64FPS.

The thermal camera was connected to the Raspberry Pi 4 microcomputer and after installing all of its dependencies, the output from the camera can be seen in the figure (Fig. 6).

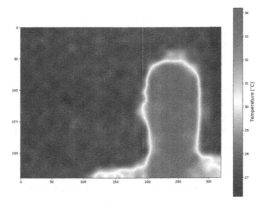

Fig. 6. Thermal camera output

The used thermal camera is an affordable solution, however the 1 °C accuracy is questionable in our context as different emotion changes can mean changes in the user's body temperature in the size of decimal numbers, therefore for the experiment we will need a camera with the highest possible accuracy. At the time of purchase, all the cameras on the market for Raspberry Pi and Arduino had a 1 °C accuracy.

Instead of a database table, the figures created by the thermal camera and the emotion detection software can be stored on the server by using File Transfer Protocol (FTP). Then a prediction model can use Optical Character Recognition (OCR) to read the data from the figures.

5   Conclusion and Future Work

A complex sensory network was created, which is able to use a camera module to detect faces and recognize basic emotions. A thermal camera has also been added to measure the body temperature of the users. This sensory network also measures the GSR of the user to detect emotion changes.

Next, we developed an application that partially automatizes the process of storing heart rate data from a smart band inside a database table on a server. The application can be then used to read all data stored in the database, which can be from multiple sources and sensors.

These common IoT devices can be used to do an experiment, where we will acquire physiological functions to create a dataset that can later be prepared (preprocessed and transformed) for quantitative and qualitative analysis. The measured values in the dataset can be then compared with limit values and after adding machine learning to the project, the dataset will later serve as an input to train a predictive model. The predictive model will be able to classify and predict emotional states according to the input physiological functions. The created model and the sensory network can later be applied in industrial production, in the automotive industry, in smart homes, or in education.

Acknowledgments. This research has been supported by the projects KEGA 036UKF-4/2019, Adaptation of the learning process using sensor networks and the Internet of Things.

References

1. Goodfellow, I.J., et al.: Challenges in representation learning: a report on three machine learning contests. Neural Netw. **64**, 59–63 (2015). https://doi.org/10.1016/j.neunet.2014.09.005
2. Li, L., Chen, J.-H.: Emotion recognition using physiological signals. In: Pan, Z., Cheok, A., Haller, M., Lau, R.W.H., Saito, H., Liang, R. (eds.) ICAT 2006. LNCS, vol. 4282, pp. 437–446. Springer, Heidelberg (2006). https://doi.org/10.1007/11941354_44
3. Magdin, M., Balogh, Z., Reichel, J., Francisti, J., Koprda, S., Molnár, G.: Automatic detection and classification of emotional states in virtual reality and standard environments (LCD): comparing valence and arousal of induced emotions. Virtual Reality (2021). https://doi.org/10.1007/s10055-021-00506-5
4. Balogh, Z., Turčáni, M.: Possibilities of Modelling Web-Based Education Using IF-THEN Rules and Fuzzy Petri Nets in LMS. In: Abd Manaf, A., Zeki, A., Zamani, M., Chuprat, S., El-Qawasmeh, E. (eds.) ICIEIS 2011. CCIS, vol. 251, pp. 93–106. Springer, Heidelberg (2011). https://doi.org/10.1007/978-3-642-25327-0_9
5. Khalid, R., Javaid, N., Rahim, M.H., Aslam, S., Sher, A.: Fuzzy energy management controller and scheduler for smart homes. Sustain. Comput. Inform. Syst. **21**, 103–118 (2019). https://doi.org/10.1016/j.suscom.2018.11.010
6. Brady, K., et al.: Multi-modal audio, video and physiological sensor learning for continuous emotion prediction. In: Proceedings of the 6th International Workshop on Audio/Visual Emotion Challenge, pp. 97–104 (2016). https://doi.org/10.1145/2988257.2988264
7. Chanel, G., Kronegg, J., Grandjean, D., Pun, T.: Emotion assessment: arousal evaluation using EEG's and peripheral physiological signals. In: Gunsel, B., Jain, A.K., Tekalp, A.M., Sankur, B. (eds.) MRCS 2006. LNCS, vol. 4105, pp. 530–537. Springer, Heidelberg (2006). https://doi.org/10.1007/11848035_70

8. Leu, F., Ko, C., You, I., Choo, K.-K. R., Ho, C.-L.: A smartphone-based wearable sensors for monitoring real-time physiological data. Comput. Electr. Eng. **65**(C), 376–392 (2018). https://doi.org/10.1016/j.compeleceng.2017.06.031

9. Herbon, A., Peter, C., Markert, L., Meer, E., Voskamp, J.: Emotion studies in HCI - a new approach. In: Proceedings of the 2005 HCI International Conference, vol. 1. Las Vegas (2005)

10. Ward, R.D., Marsden, P.H.: Physiological responses to different WEB page designs. Int. J. Hum. Comput. Stud. **59**(1), 199–212 (2003). https://doi.org/10.1016/S1071-5819(03)000 19-3

11. Soleymani, M., Koelstra, S., Patras, I., Pun, T.: Continuous emotion detection in response to music videos. In: 2011 IEEE International Conference on Automatic Face and Gesture Recognition (FG), pp. 803–808 (2011). https://doi.org/10.1109/FG.2011.5771352

12. Picard, R.W., Vyzas, E., Healey, J.: Toward machine emotional intelligence: analysis of affective physiological state. IEEE Trans. Pattern Anal. Mach. Intell. **23**(10), 1175–1191 (2001). https://doi.org/10.1109/34.954607

13. Ekman, P.: Basic emotions. In: Handbook of Cognition and Emotion, Wiley, New York, NY, USA, pp. 45–60 (1999)

14. Fodor, K., Balogh, Z.: Process modelling and creating predictive models of sensory networks using fuzzy petri nets. Procedia Comput. Sci. **185**, 9 (2021)

15. Leonard, J.: What does anxiety feel like and how does it affect the body? (2018). https://www.medicalnewstoday.com/articles/322510. Accessed 16 June 2021

16. Anger - how it affects people. BetterHealth website. https://www.betterhealth.vic.gov.au/health/healthyliving/anger-how-it-affects-people#bhc-content. Accessed 16 June 2021

17. Stevenson, R., et al.: Disgust elevates core body temperature and up-regulates certain oral immune markers. Brain. Behav. Immun. **26**, 1160–1168 (2012). https://doi.org/10.1016/j.bbi.2012.07.010

18. Balaji, A.: Emotion detection. GitHub repository, 2018. https://github.com/atulapra/Emotion-detection. Accessed 15 June 2021

19. Kowalska, M., Wróbel, M.: Basic Emotions. In: Zeigler-Hill, V., Shackelford, T. (eds.) Encyclopedia of Personality and Individual Differences, pp. 1–6. Springer, Cham (2017). https://doi.org/10.1007/978-3-319-28099-8_495-1

20. Carrier, P.-L., Courville, A.: The facial expression recognition 2013 (FER-2013) dataset (2013). https://datarepository.wolframcloud.com/resources/FER-2013. Accessed 15 June 2021

21. Viola, P., Jones, M.: Rapid object detection using a boosted cascade of simple features. In: Proceedings of the 2001 IEEE Computer Society Conference on Computer Vision and Pattern Recognition. CVPR 2001, vol. 1, p. I (2001). https://doi.org/10.1109/CVPR.2001.990517

22. Mandryk, R.L., Atkins, M.S.: A fuzzy physiological approach for continuously modeling emotion during interaction with play technologies. Int. J. Hum. Comput. Stud. **65**(4), 329–347 (2007). https://doi.org/10.1016/j.ijhcs.2006.11.011

23. Winton, W.M., Putnam, L.E., Krauss, R.M.: Facial and autonomic manifestations of the dimensional structure of emotion. J. Exp. Soc. Psychol. **20**(3), 195–216 (1984). https://doi.org/10.1016/0022-1031(84)90047-7

24. Boucsein, W.: Electrodermal Activity, 2nd edn. Springer, New York (2012). https://doi.org/10.1007/978-1-4614-1126-0

25. Salimpoor, V.N., Benovoy, M., Longo, G., Cooperstock, J.R., Zatorre, R.J.: The rewarding aspects of music listening are related to degree of emotional arousal. PLoS ONE **4**(10), e7487 (2009). https://doi.org/10.1371/journal.pone.0007487

26. Critchley, H.D.: Review: electrodermal responses: what happens in the brain. Neuroscience **8**(2), 132–142 (2002). https://doi.org/10.1177/107385840200800209

27. Pimoroni: MLX90640 Thermal camera breakout – standard (55°). https://shop.pimoroni.com/products/mlx90640-thermal-camera-breakout?variant=12536948654163. Accessed 15 June 2021

Service Computing and Cloud Computing (2)

A Multi-perspective Model of Smart Products for Designing Web-Based Services on the Production Chain

Ada Bagozi, Devis Bianchini, and Anisa Rula$^{(\boxtimes)}$

University of Brescia, Brescia, Italy
{ada.bagozi,devis.bianchini,anisa.rula}@unibs.it

Abstract. In this paper, we propose a multi-perspective model for smart products as a basis to implement advanced web-based services in the smart factory production chain. The model relies on three perspectives that are: the *product*, the *production process* and the *work centers* involved in the production chain. For each perspective, the physical world is connected with the cyber world, where collected sensor data is properly organised to enable data analysis and exploration according to the three perspectives in an interleaved way. A portfolio of web-based services at production chain level is also described according to the model. Among them, a data access dashboard has been designed in order to enable controlled exploration of production chain data for different actors involved in the production activities, ranging from the suppliers to the producer of the final product. The approach is devoted to the production of costly and complex products, where the conceptualisation of a Smart Product as the integration of product, process and infrastructure sensor data aims at ensuring high product quality levels, long lasting operativity, less frequent and efficient maintenance activities and constant performances over time.

Keywords: Smart product · Multi-perspective model · Web-based services · Production chain · Cyber-physical production systems · Industry 4.0

1 Introduction

The ever-growing application of digital technologies in modern smart factories has enabled the integration of product design, manufacturing processes and general collaboration across factories over the supply chain with the exploitation of service-oriented architectures and web-based systems [1]. Information and communication technology promoted the design and widespread integration of Cyber Physical Systems (CPS) with production lines. Cyber Physical Systems are hybrid networked cyber and engineered physical elements that record data (e.g., using sensors), analyse them using connected services (e.g., over cloud computing infrastructures), influence physical processes and interact with human

© Springer Nature Switzerland AG 2021
W. Zhang et al. (Eds.): WISE 2021, LNCS 13081, pp. 447–462, 2021.
https://doi.org/10.1007/978-3-030-91560-5_33

actors using multi-channel interfaces. In industrial production environments CPS interact with each other and with operators who supervise the machines, according to the Human-In-the-Loop paradigm [2], in the so-called Cyber Physical Production Systems (CPPS).

In this scenario, sensor data may refer to multiple aspects of the production chain, ranging from measures taken on the product (e.g., during quality control procedure), information regarding the phases of the production process (e.g., unexpected delays), measures gathered from single machines and work centers involved in the production process in order to detect anomalies or monitor machine performances. In recent research literature models, methods and architectures have been thoroughly proposed to connect simple work centers to the cloud in order to implement predictive maintenance, energy savings, self-configuration and other capabilities of a manufacturing production chain [3–5]. Sensor data collection and analysis may also have a positive impact on the Product Lifecycle Management (PLM), that is, the business activity of effectively and efficiently managing products across their entire life cycle, from the product design all the way through product manufacturing until it is retired and disposed of [6,7]. This vision is also coherent with the RAMI 4.0 Reference Architectural Model for Industrie 4.0 [8]. The perspective of life cycle value stream (IEC 62890), spanning over the development, production and maintenance phases, is interleaved with the IEC 62264/IEC 61512 hierarchy levels (from product to the connected world level) and with the smart factory layers (starting from industrial assets up to the business layer). Web-based systems may serve as a meeting point for all the actors across the production chain, to be able to collect and visualise data collected under products, production processes and machines perspectives and dispatch information to actors who have the permissions to access it through multiple channels and exploiting the modularity of service-oriented architectures.

Beyond the design of cloud-based services, focused on single data perspectives in isolation, a common facility that is able to aggregate and keep track of all the required information in a multi-perspective manner is fundamental. We identify this facility as a *Smart Product*, that is associated with all the information at the shop floor level (both process and involved work centers) as long as the product itself is moved forward throughout the production line. The approach proposed in this work, is devoted to the production of costly and complex products, where the conceptualisation of a Smart Product as the integration of product, process and infrastructure sensor data aims at ensuring high product quality levels, long lasting operativity, less frequent and efficient maintenance activities and scalable performances over time. To test the effectiveness of the dashboard in a real case study, the production of valves to be used in deep and ultra-deep water applications has been considered. Valves are placed in prohibitive environments and, once installed, are difficult to remove and maintain over time and require high quality levels. Our idea is that the quality of such products can be ensured through a careful provisioning of services along the entire production chain modelled as a Cyber Physical Production System, constantly checking the advancement of production process and the conditions of work centers operating in the production chain through web-based tools and systems.

The contributions of this work are threefold: (i) the definition of a multi-perspective Smart Product model in the production chain, identifying the conceptual elements that can be considered to describe a Smart Product; (ii) the design of an ecosystem of web-based services at the production chain level, based on the multiple perspectives in an interleaved way; among designed services, a web-based dashboard has been conceived to collect data from the production chain, display data to authorised actors only and dispatch data over a distributed architecture to implement the other services in the ecosystem; (iii) evaluation of the capability of web-based services to enable data sharing across the actors of the production chain, driven by the proposed Smart Product model.

The paper is organised as follows: Sect. 2 discusses related work; in Sect. 3 the multi-perspective Smart Product model is described; Sect. 4 presents the ecosystem of web-based services, based on the model, designed at the production chain level; in Sect. 5, the focus is on the dashboard as the meeting point to collect data, display it to authorised actors and dispatch it to the other services; usability experiments to test the effectiveness of the dashboard in the deep and ultra-deep water application scenario are discussed in Sect. 6; finally, Sect. 7 closes the paper and sketches future research efforts.

2   Related Work

A CPS is a complex system with different elements and specific system structures engaged in different applications [9,10]. According to different authors the architecture of a CPS may vary from 2 to 6 layers [11,12]. We adopt the abstract model which considers a three layer architecture since the aim of our work is to connect several CPS and overcome the problem of the composition and interoperation between them. In a production chain there are many actors involved. However, each of them presents a different unit-level CPS [13] for the lifecycle of its own product. The author in [14] reviews the lifecycle models that can be relevant for the CPS. Only a few approaches have been proposed for improving the communication of different CPS in the network [15–17]. For this reason, a CPPS can be employed as a composition of several unit-level CPS. A CPPS aggregates products, resources and equipment to enhance the efficiency of a manufacturing production chain. In recent years, there has been a particular attention on CPPS [3–5]. However, most of the approaches focus on the aggregation at the process level; while others focus on the composition of services. Differently from these approaches, we use the services to coordinate applications and connections. Authors in [3] propose a generic architecture based on digital twin that relies on four layers (physical, network, virtual and application layer). Moreover, authors propose a CPPS information modelling that is represented by the AutomationML format. With respect to [3], our approach is focused on a multi-perspective data model that relies on three perspectives (product, production process, working centers) in order to implement advanced web-services for the entire production chain. In [5] a middleware-based approach is proposed in order to enable operators to model a CPPS according to their need. Through the CPPS modelling, operators are able to extend CPPS functionalities and add new smart services. Differently from our approach, authors in [5] consider mainly the physical components of the CPPS extended with smart services. However, the product and the production line are not taken into account. Complementary to our work, the authors in [4] propose a CPPS and an information model that includes data from: product, process, plan, plant and resource. Differently from [4], we propose a data access dashboard designed in order to enable controlled exploration of production chain data for different actors involved in the production activities, ranging from the suppliers to the producer of the final product.

3 A Model for Smart Products

As a prerequisite for creating smart manufacturing products we propose to design and implement the manufacturing of submarine valves as a Cyber-Physical Production System (CPPS). A CPPS is conceived as an ecosystem of single Cyber-Physical Systems (CPS), providing real-time sensing, information feedback, dynamic control and other services through the physical, the computing and the communication components also known as the 3-C. The CPPS provided here requires different manufacturing activities and as such we model it at the *system-level* constituted by multiple *unit-level* CPS. Each unit-level corresponds to the smallest unit participating in the manufacturing activities. We consider three unit-level CPS: the work center, the valve product and the production process. The work centers have an important role, where each work center may contain one or more machines to process the components in order to achieve the final product. In the work centers, sensor data on machine behaviour (such as electrical consumption, vibrations, accelerations, rotating or linear speed of machine components, temperatures and so forth) can be collected to implement specific recovery actions on the work centers (e.g., energy efficiency optimisation, predictive maintenance applications). In addition, the sensor data on work centers correlated with the final product quality and production process delays can be used to detect problems/anomalies, thus providing a mean to control processes and product quality through work centers condition monitoring. The product necessary to build the valve are mainly tools and consumables (raw materials, the semi-worked or fully worked products). Data collected from product parts (e.g., during quality check) or tools (e.g., tool wear) can be correlated with work centers anomalies and production process delays as previously explained. Finally, during the production process all components are assembled together, wrought and built. The information about process steps (e.g., delays) can be gathered and fruitfully integrated with other data, conceiving the whole CPPS as a completely interleaved ecosystem of unit-level CPS.

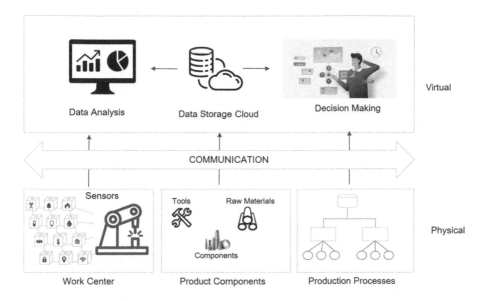

Fig. 1. Architecture of a CPPS.

3.1   System-Level Perspective of CPPS

Figure 1 shows the integration of the 3-C. The **physical space** consists of different manufacturing resources, that is, the machines, the tools, the materials of the product, and the production processes. Considering the heterogeneity of this application scenario, different sensors are applied in the working center to gather data such as the machine speed and the energy consumption. Wired or wireless sensors can be deployed on tools to achieve the tool wear and, for equipment with high degree of automation, the data of the embedded modules. RFID can be used for lifecycle tracking of materials and environmental sensors to detect the real-time changes of the environment. To keep it simple, we do not consider the humans in the physical space of this process, but the architecture depicted in Fig. 1 is general and flexible enough to include also on-field operators working nearby the machines if it is worth collecting some information also about them, such as their position, the machine they are supervising and the working hours. This may help to correlate problems on the production chain regarding the presence/absence of operators, suggesting for example to move more expert operators nearby machines that perform more critical steps in the production process. Real-time data are gathered in the physical space and may be sent to the virtual space. In return, controls are sent back to the physical space to perform (manually or automatically) some configurations adjustments or specific setup.

Figure 1 shows also the **virtual space** which consists of computing resources and tools such as data storage in Cloud, data analysis and decision making. The interactions in CPPS are managed as one-to-many relationships between the

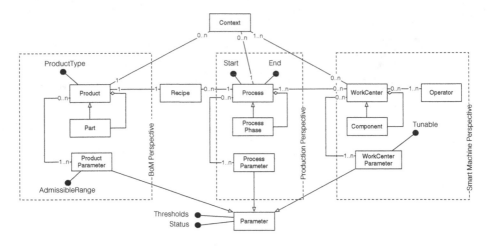

Fig. 2. Multi-perspective data model of the production chain CPPS.

physical and the virtual space. Meaning that a computing resource may influence one or more physical components at a time. In the following section we focus on presenting the multi-perspective model as part of the data exchange in the production chain.

3.2   Multi-perspective Data Model

As part of the CPPS virtual space, we provide a multi-perspective data model to organise the data produced by the unit-level CPS, which compose the CPPS, into three analysis perspectives: the *product*, the *production process* and the *work centers* perspectives. The data model is shown in Fig. 2. In the following we describe each perspective separately and how they relate to each other.

Product. Each product is composed of a set of parts which are identified by a part code and can be composed of other sub-parts. This relationship between product parts is represented through a hierarchy making the navigation structure of a product flexible. The hierarchy represents what is commonly referred to as *Bill of Material* (BoM). In fact, each product part can be: i) the final product, i.e. a product ready for marketing; ii) semi-worked product, i.e., a product part that is the result of assembly steps, but it requires further manufacturing processes to be included within the final product (this usually holds for complex products such as the valves considered in the application scenario of this paper); iii) raw material, i.e., a part that is in its primitive state, without being processed. In Fig. 2, a part is also as a subclass of product, due to the fact that we are addressing a production chain, where different involved actors are playing different roles, namely the final product producer and the suppliers. A specific part can be conceived as semi-worked or as a raw material for the producer, but as a final product for the supplier of that part. We will return on this issue when presenting the web-based dashboard and we will discuss future research issues.

Production Process. In the product perspective, we are focused on the BoM, but not on the process of how different parts are assembled, according to the so-called *recipe* in manufacturing industry. The production process represents the relationships between the various processing phases to obtain the final product: each processing phase includes different sub-phases. A hierarchical structure is used to model this relationship as for the BoM. In this way it is possible to model a recursive composition of phases and sub-phases, reaching a level of detail ranging from macro to micro industrial processes. A production phase is conceivable as an instance of an operation to create a product or a product part as specified in the recipe. Specific parameters are collected on such instances (*process parameters*). For example, among the process parameters we mention the cycle time, on which possible delays could be detected.

Work Centers. A work center comprises one or more machines. The hierarchical relationship in this case is between the work center and the component machines, but each machine in a recursive way may be composed of other parts (for example, an oil pump, electrical engines, spindles, and so forth). There is a connection between the production process and the work center through the execution phase. In other words, a work center will perform many production phases during its lifecycle. A production process can be executed and distributed on several work centers. The *operator* represents the human who participates in the production process, supervising the work center.

Parameters. Different kinds of parameters are used to monitor the behaviour of a CPPS according to the three perspectives. On each item (final product, semi-worked part, raw material) in the BoM some *product parameters* are measured, concerning the quality properties of the item. Values of these parameters must stay within acceptable ranges. On each process phase, proper *process parameters* are measured as well, concerning the phase duration, that must be compliant with the end timestamp of each phase, as established by the production schedule. *Work center parameters* are gathered to monitor the working conditions of each work center or of the production environment. Parameters are monitored through proper thresholds, established by domain experts who possess the knowledge about the production process. Parameters bounds are used to establish if a critical condition has occurred on the monitored CPPS. In particular, we distinguish between two kinds of critical conditions: i) *warning conditions*, that may lead to breakdown or damage of the monitored system or production process failure in the near future; ii) *error conditions*, in which the system can not operate.

4   An Ecosystem of Web-Based Services

4.1   Multi-perspective Smart Product Access Services

To effectively implement the exploration of data collected throughout the production chain, a data visualisation *dashboard* is described in this section, based on the multi-perspective data model. The user of the dashboard will be supported during the exploration, to easily find the data corresponding to anomalous situations, exploring them and moving between different critical situations through the exploitation of the three Smart Product perspectives. Different functionalities are offered by the dashboard according to the user skills, responsibilities and needs. The users can be divided into two main categories: *analysts*, i.e., expert users of the domain who use the dashboard for detailed exploration of the production (e.g., to identify the causes from which abnormal situations occurred); *observers*, i.e., users with limited skills and responsibilities, who use the dashboard for consultative purposes, with the aim to quickly find specific information.

The advantage of the multi-perspective data model representation is to analyse the data according to different perspectives. The different views are created being inspired by multi-dimensional OnLine Analytical Processing (OLAP) databases. Operating through slice and dice actions (selection and projection of data), it is possible to obtain different ways of displaying information, by accessing:

- *data relating to one or more valves*; for each valve it is in fact possible to analyse in detail the precise data and indicators relating to all the processing phases, as well as data relating to the work centers in which the processing phases have been executed;
- *data relating to one or more work centers*; it is possible to view precise and aggregated data of all the phases that are carried out in each work center, together with the sensor data on the work center conditions and the product data relating to the valves that have been assembled during the processing phases;
- *data relating to one or more processing phases*; for each phase it is possible to access the product data of the valves that have been assembled in the processing phase; in addition, data about the work centers where the processing phases have been executed can also be accessed.

Pivoting operations allow users to further analyse the data by focusing only to specific dimensions. By excluding one of the three perspectives and focusing on the other two, it is possible to obtain different views:

- *specific data about processing phases and/or work centers*, excluding the product perspective;
- *specific data for product and work center*; through this perspective, it is possible to view the temporal trends of the data and thus highlight any found anomaly;
- *indicators* calculated on the production phases of a single valve; this data can provide a general view about the progress of the production process.

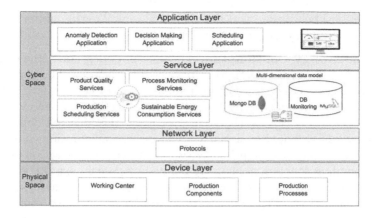

Fig. 3. Architecture of the production chain services.

4.2   Production Chain Services Based on the Smart Product Model

Based on the multi-perspective data model of the CPPS, some important web-based services can be designed to support important aspects of the production chain such as the production scheduling, the process monitoring, the sustainable energy consumption and product quality check. All services interact through the dashboard previously described, according to three distinct stages named *Collect*, *Dispatch* and *Display*. Figure 3 provides an overview of the ecosystem of web-based services, sketched in the following.

Production Scheduling Services. To implement these services, the dashboard interacts with the schedulers of each actor involved in the production process. During the *Collect* stage, the dashboard gathers the data necessary to identify the product, the resources and the steps necessary for its realisation. The dashboard then receives the planned production process generated by the scheduler of the final product producer, i.e., the delivery date of the order, the start and end time of each production phase and the deadlines of the suppliers for the delivery of raw materials or semi-worked parts. Therefore, the dashboard sends the expected delivery dates and materials to the suppliers (*Dispatch*). Once the suppliers have received the required delivery dates from the dashboard, they use their own schedulers for planning internal production processes required to supply the raw materials or semi-worked products. Their schedulers may confirm or propose alternative dates, which are forwarded to the other actors involved in the overall production chain of the final product, for a possible rescheduling, if it is necessary. As the various stages of production are completed, the progress of production is updated and monitored on the dashboard (*Display*). Any delay is transmitted to the schedulers, always with the aim of putting in place a rescheduling that plans the new delivery time, which represent the new deadline. We remark here that the interaction between the dashboard and the schedulers can be recursive, that is, a supplier might have in turn her own suppliers to be notified. In this vision, the first-level supplier might be equipped with its own dashboard that works in the same way. We will discuss about this in the concluding section about future research issues.

Sustainable Energy Consumption. During the _Collect_ stage, the dashboard gathers data necessary to identify the resources to monitor. This data may be collected as sensor data directly from monitored work centers/machines (e.g., electrical energy consumption, temperature and so forth). The energy consumption optimisation service is in charge of processing the collected data and possibly sending back to monitored machines proper reconfiguration commands to reduce or limit energy consumption (_Dispatch_). Energy optimisation algorithms and corresponding countermeasures are defined by domain experts who possess the knowledge about monitored work centers or machines. The description of these algorithms is out of the scope of this paper. Upon request from an authorised user, the energy consumption optimisation service transmits the data of the energy performance report to the dashboard (_Display_).

Process Monitoring. During the _Collect_ stage, the dashboard gathers the data necessary to identify the product, the resources and the phases necessary for its realisation. The process monitoring service is in charge of processing the data needed to monitor the production process. Upon the occurrence of certain events or deviations from expected process execution times, the process monitoring service communicates to the dashboard the information about the detected delays. The dashboard is in charge of notifying the delays to the schedulers of all actors involved in the production process, thus triggering the production scheduling services (_Dispatch_). Upon request of an authorised user, the process monitoring service transmits the data of the process performance report to the dashboard (_Display_).

Product Quality Services. During the _Collect_ stage, the dashboard gathers the data that corresponds to some anomalies detected by product quality check. The product quality service is in charge of processing the collected data and possibly requiring some rescheduling and alternatives of the production process that might (totally or partially) balance the anomalies that have been detected on products (_Dispatch_). These reconfiguration strategies are defined by domain experts and designers who know the involved product and the corresponding production process. These strategies are out of the scope of this paper. Upon request of an authorised user, the dashboard may display results of product quality check (_Display_).

5   Dashboard Implementation

The dashboard is part of an application context that has a strong heterogeneity, given the need to integrate data from different information systems into a single platform and interact with the above presented web-based services. The implementation of the dashboard is designed to allow both vertical and horizontal extensions between different production chains. Hence, a supplier can also be part of other production chains where it can become the main actor, as already previously remarked.

The dashboard is intended for very different end users, with different needs and privileges. The need to manage different access permissions leads to *customised interfaces*. Therefore, it is possible through the dashboard to configure what to show and what to hide for each category of users. In this way, the data visualisation interface is capable of *dynamically adapting* to the user who is using the platform, without having to foresee the different possible configurations during the development phase. A data domain expert is required to define *access rules*, but the implemented dashboard also includes a *user-friendly configuration panel* to support the domain expert to define these permissions and no additional technical expertise is required. The dashboard implementation includes:

- *frontend*, for the representation of the visible part of the dashboard to the user, including the interface graphics, and for the representation of information and the acquisition of the data entered by the user;
- *backend*, which allow the application to work, by hosting all the modules necessary to communicate with the services described above and in charge of interacting with the frontend.

The user interface was developed based on the React JS library[1], while for the backend Node.js[2] was adopted. React JS uses API calls for the interaction with the backend. As previously described, the dashboard is fed with two types of data: parameters and aggregated values. In addition, personal data are also added to give context to the information collected. The data are therefore designed to be stored in two different databases, each with a very specific purpose: i) a MySQL relational database for product, production process, working production cycles and anything else that can be imported from the various information systems; ii) a NoSQL database in MongoDB for data collected from the various working centers present in the production chain.

From the homepage of the frontend, it is possible to start the data exploration by choosing one of three tiles, that represent the three perspectives in the Smart Product model (Fig. 4): i) *the valve synoptic* tile allows an exploration from the product perspective, ii) *the production process phases* tile allows an exploration from the process perspective, iii) *the working centers* tile allows exploration from the work centers perspective. There are different functionalities implemented to interact with the three perspectives such as filtering, browsing and analysing data e.g., through graphics.

6   Usability Experiments

In this section we present a preliminary evaluation of the dashboard usability to test the effectiveness of the proposed web-based system to enable sharing Smart

---

[1] https://reactjs.org/.
[2] https://nodejs.org/en/.

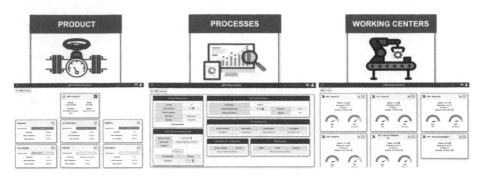

Fig. 4. Screenshots of the dashboard, based on the three smart product perspectives.

Product data across the actors of a typical production chain, which may possess different technological skills, and to facilitate Smart Product data exploration under the different perspectives that have been proposed. Data exploration is very limited in a platform covering one perspective at a time and it is too time consuming, since it requires access and query over several platforms. Combining information across multiple platforms it is difficult, as a result of the lack of knowledge about how these platforms are interconnected in order to analyse different occurring problems. Therefore, our usability study will be focused mainly on the multi-perspective platform. Therefore, the main goal of this evaluation is to prove that the dashboard supports users in data exploration better than each other platform explored separately.

6.1   Design and Methodology

We propose a user interaction experiment based on the *within-subjects design* approach, where all the participants take part to the experiment in every condition. The users will perform an activity in an environment similar to the one in which they actually operate. We provide tasks performed by the users that include multiple perspectives e.g., *Which is the work center responsible for this particular operation in the processing phase?*, *When is the expected delivery date for the valve having serial number 532232161?*. The experiment is: *significant*, meaning that the experimental setup, the users involved in the experiment and the experimental context must be representative; *reliable*, meaning that, in this particular context, poor reliability is linked to users' variability. The experiment comprises four steps: (i) *test preparation*, verify and check that everything is working correctly before submitting to users; (ii) *introduction*, provide a welcome message to the users and explaining the purpose of the test; the users will be required to complete a questionnaire at the end of the test; the questionnaire contains an introduction section of and a short presentation of the test; (iii) *execution*, remotely fulfil the test, in which the examiner operates, for not influencing the user during the experiment; (iv) *debriefing*, perform through questionnaires the collection of opinions following the test.

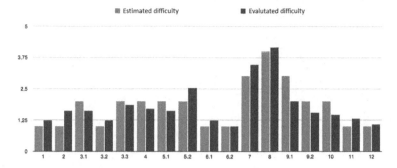

Fig. 5. Comparison between the degree of estimated and evaluated difficulty of tasks required to users participating to the experiments.

6.2 Participants

The profile of involved participants ranges from less experienced to high qualified users. In particular, users belong to three categories based on their level of expertise, or the familiarity with data visualisation tools, such as synoptics and dashboard, and in a more general sense with management systems for knowledge issues relating to industrial processes:

- *beginner user* category comprises users who are not familiar with the topics covered in the project;
- *intermediate user* category comprises users who are fairly familiar with the issues of the management systems and related software, but without having a great experience;
- *expert user* category comprises users who work daily with the management systems and are familiar with the use of the software.

The participants involved in the experiments are 13 persons, taken from industrial production contexts, divided in 2 beginner users, 3 intermediate users and 8 expert users.

6.3 Results

The results obtained from the experiments have been analysed considering three features: i) the degree of the difficulty of tasks required to the users from the examiner's viewpoint (*estimated difficulty*); ii) the correctness of answers; iii) the degree of the difficulty of requests as evaluated by the users (*evaluated difficulty*). In the first experiment, a comparison has been made between the estimated and the evaluated difficulty. Figure 5 shows that the distance between the estimated and the evaluated difficulty degree of the questions is low, meaning that the estimated values reflect the reality.

Figure 6 shows the correctness of the answers having a high rate of 89% of correct answers. Among the wrong answers we may notice the answer number 8

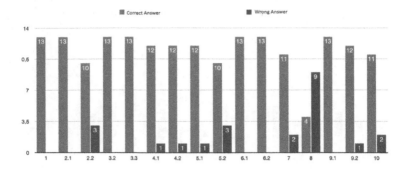

Fig. 6. Comparison between correct and wrong answers.

to be mistaken by most of the users. This question consists in searching problems occurred during a processing phase and corresponds to one of the most complicated tasks. The difficulty to perform this task is because the users must exploit a multi-perspective view, since this task requires to consult the data of the work center where the considered production phase is carried out. Consequently, the need to further highlight the problems, making them more visible and easily identifiable by operators, emerges from these experiments. The occurrence of an anomaly must also be objective, and not subjective depending on the user's interpretation.

Overall, the users performed the experimental tasks as required. This demonstrates that the multiple perspectives provided through the dashboard supports accurately the operator in carrying out the tasks and consequently improves the exploration of available information. The comparison with the single platforms used previously to the dashboard was not possible. For this reason, the tasks assigned to the user regard the exploration of the data in case a multi-perspective is needed.

7   Concluding Remarks

In this paper, we proposed of a multi-perspective Smart Product model in the production chain, identifying the conceptual elements that can be considered to describe a Smart Product. An ecosystem of web-based services at the production chain level has been designed, based on the multiple perspectives in an interleaved way. Among designed services, a web-based dashboard has been conceived to collect data from the production chain, display data to authorised actors only and dispatch data over a distributed architecture to implement the other services in the ecosystem. The research described here is part of an ongoing project to move the implementation of Industry 4.0 services at the production chain level in the application domain of deep and ultra-deep water valves. Future efforts are required to test the proposed solution in an intensive way when the massive collection of data will start in the project, further extending the ecosystem of services with tools for facing big data volume, velocity and variety issues. Usability experiments will be extended as well, comparing the usage of the proposed solution with the usage of isolated, not yet integrated systems.

References

1. Lee, J., Kao, H.-A., Yang, S.: Service innovation and smart analytics for industry 4.0 and big data environment. Procedia CIRP **16**, 3–8 (2014). product Services Systems and Value Creation. Proceedings of the 6th CIRP Conference on Industrial Product-Service Systems
2. Nunes, D., Silva, J.S., Boavida, F.: A Practical Introduction to Human-in-the-Loop Cyber-Physical Systems, Wiley, Hoboken (2018)
3. Zhang, H., Yan, Q., Tong Wen, Z.: Information modeling for cyber-physical production system based on digital twin and AutomationML. Int. J. Adv. Manuf. Technol. **107**, 1927–1945 (2020)
4. Park, K.T., Lee, J., Kim, H.-J., Noh, S.: Digital twin-based cyber physical production system architectural framework for personalized production. Int. J. Adv. Manufact. Technol. **106**, 1787–1810 (2020)
5. Stock, D., Schel, D., Bauernhansl, T.: Middleware-based cyber-physical production system modeling for operators. Procedia Manufact. **42**, 111–118 (2020)
6. Stark, J.: Product Lifecycle management. In: Product Lifecycle Management (Volume 2). DE, pp. 1–35. Springer, Cham (2016). https://doi.org/10.1007/978-3-319-24436-5_1
7. Gerhard, D.: Product lifecycle management challenges of CPPS. In: Biffl, S., Lüder, A., Gerhard, D. (eds.) Multi-Disciplinary Engineering for Cyber-Physical Production Systems, pp. 89–110. Springer, Cham (2017). https://doi.org/10.1007/978-3-319-56345-9_4
8. Hankel, M., Rexroth, B.: The reference architectural model Industrie 4.0 (rami 4.0), ZVEI **2**(2), 4–9 (2015)
9. Yao, X., Zhou, J., Lin, Y., Li, Y., Yu, H., Liu, Y.: Smart manufacturing based on cyber-physical systems and beyond. J. Intell. Manuf. **30**(8), 2805–2817 (2019)
10. Tao, F., Qi, Q., Wang, L., Nee, A.: Digital twins and cyber-physical systems toward smart manufacturing and industry 4.0: correlation and comparison. Engineering **5**(4), 653–661 (2019)
11. Dong, J., Xiao, T., Zhang, L.: A prototype architecture for assembly-oriented cyber-physical systems. In: Xiao, T., Zhang, L., Fei, M. (eds.) AsiaSim 2012. CCIS, pp. 199–204. Springer, Heidelberg (2012). https://doi.org/10.1007/978-3-642-34384-1_24
12. W. Xiao-Le, H. Hong-Bin, D. Su, C. Li-Na, A service-oriented architecture framework for cyber-physical systems. In: In: Qian, Z., Cao, L., Su, W., Wang, T., Yang, H. (eds.) Recent Advances in Computer Science and Information Engineering, vol. 126, pp. 671–676. Springer, Heidelberg (2012). https://doi.org/10.1007/978-3-642-25766-7_89
13. Tao, F., Zhang, M.: Digital twin shop-floor: a new shop-floor paradigm towards smart manufacturing. IEEE Access **5**, 20418–20427 (2017)
14. Thoben, K.-D., Pöppelbuß, J., Wellsandt, S., Teucke, M., Werthmann, D.: Considerations on a lifecycle model for cyber-physical system platforms. In: Grabot, B., Vallespir, B., Gomes, S., Bouras, A., Kiritsis, D. (eds.) APMS 2014. IAICT, vol. 438, pp. 85–92. Springer, Heidelberg (2014). https://doi.org/10.1007/978-3-662-44739-0_11

15. Pigosso, D.C., Zanette, E.T., Filho, A.G., Ometto, A.R., Rozenfeld, H.: Ecodesign methods focused on remanufacturing. J. Cleaner Prod. **18**(1), 21–31 (2010)
16. Ranasinghe, D.C., Harrison, M., Främling, K., McFarlane, D.: Enabling through life product-instance management: solutions and challenges. J. Netw. Comput. Appl. **34**(3), 1015–1031 (2011)
17. Kiritsis, D.: Closed-loop PLM for intelligent products in the era of the internet of things. Comput. Aided Des. **43**(5), 479–501 (2011)

A Multi-view Learning Approach for the Autonomic Management of Big Services

Fedia Ghedass$^{(\boxtimes)}$ and Faouzi Ben Charrada

LIMTIC, Faculty of Sciences, Tunis Elmanar University, Tunis, Tunisia

Abstract. Big services have recently emerged as a solution to process, encapsulate and offer huge volumes of data as a service. However, its management operations are beyond the ability of human administrators, due to several challenges including big services' large-scale nature and complexity, the heterogeneity of its components, the dynamicity and uncertainty of its hosting cloud environments. To cope with these challenges, we endow big services with self-* capabilities and we propose an autonomic computing architecture for big services. We also take advantage of two recent technologies called knowledge graphs and multi-view learning, to represent the managed big service's information (service descriptions, services' and data sources' quality levels, management policies) as a heterogeneous information network. Finally, a decision mechanism to select and trigger the appropriate management policies is defined and validated through a set of experiments.

Keywords: Big service management · Autonomic computing · Knowledge graph · Multi-view learning · Graph neural network

1 Introduction

Recently, the combination of big data processing capabilities and cloud computing services has led to the emergence of a new big data-centric service model, called *big service* (BS) [26]. Introduced for the first time in 2015, big services are also seen as a collection of interdependent services from virtual (e.g., cloud services) and physical (e.g., transportation services) domain-specific resources for the analysis and processing of big data [26]. Recent attempts from leading providers to implement big data services include Alibaba *MaxCompute*, *Oracle Big Data Cloud Service*, and IBM *Big Data Service on Silo* system.

The complexity and large-scale nature of big services raise a serious question regarding the ability of human administrators or cloud/service providers to efficiently manage the eventual execution deviations of big services. In fact, these latter's management operations inherit, not only traditional Web service management challenges (e.g., failures, privacy and security risks, black Quality of services (QoS) degradation [18], but also data management issues [6]. Maintaining a stable execution of big services through triggered management and adaptation actions has not yet been seriously addressed by researchers [16]. Indeed,

© Springer Nature Switzerland AG 2021
W. Zhang et al. (Eds.): WISE 2021, LNCS 13081, pp. 463–479, 2021.
https://doi.org/10.1007/978-3-030-91560-5_34

existing approaches focus mainly on specific lifecycle phases, such as big service modelling and representation [15,22,27], big service selection and composition [9,10,19,28], big service provisioning [5]. Also, processing and analyzing the big monitored data that result from big services execution is beyond the capacity of human administrator and traditional monitoring frameworks. Nowadays, the analytical technology is increasingly offered in cloud environments with powerful and distributed computing capacities [13]. Hence, learning from big services' data should be an essential capability of big service management (BSM) solutions, in order to trigger appropriate big service adaptation and evolution actions.

One way to deal with the above problems, is to combine *autonomic computing* (AC) [11] and *multi-view learning* (MVL) [17] paradigms. The first will be used as a powerful means to hide the complexity of the big service management process and to carry out management operations autonomously. As for multi-view learning techniques, the idea is to infer new knowledge regarding big services (e.g., conflict between management policies, new management rules, etc.) and to efficiently process their huge consumed/monitored data.

- We model the managed big service as an autonomic computing system [11].
- We model the autonomic managers' knowledge, i.e. big service's related data and its management policies, as a multi-view knowledge graph [3].
- We define a decision mechanism that allows parsing the big service's knowledge graph (BSKG), to select the appropriate management operations.

In Sect. 2, we first summarize the related works on big service management. Section 3 presents an autonomic computing architecture for big services. In Sects. 4 and 5, we propose a multi-view learning approach to model big services' information and management policies using knowledge graphs. In Sect. 6, we define a mechanism for the selection of big service management plans. Section 7 presents the experimental results. Section 8 concludes the paper.

2   Related Work

The need for big service management operations is not limited to its heterogeneous component services or complex workflows, but it encompasses the huge amount of data processed by those services. Existing approaches either they proposed frameworks for service management and adaptation, like in [4,24], or they design frameworks for managing and processing big data [20]. Most existing approaches have addressed specific lifecycle phases, like the selection and composition in the works of [9,10,19]. In all these works there was a lack of understanding regarding the big services' capabilities and behavior. In fact, big services have been treated as data-centric Web services, and hence were specified using traditional QoS models, like in the work of [10]. Add to that, there is no consensus on the big service model, as some authors consider big services as a complex workflow of Web services, while other researchers treat the managed big service as a combination of Web and cloud services [16]. In most of these works, big services' issues have been treated as a combinatorial optimization problem, which is explained by the use of evolutionary techniques like in [1,10], in order to cope with the complex and dynamic nature of big services environments.

To deal with evolution issues in service ecosystems, Liu et al. [15] adopted knowledge graphs, joint learning, and natural language processing (NLP) to construct a multi-layer network-based service ecosystem model (MSEM). To trigger the service evolution in a sequential or causal way, service-related events which are collected from public news are incorporated into the MSEM. In [5], Ding et al. have designed a quality assurance framework for big data services. In their online big data service, also called Cell Morphology Assay (CMA), machine learning is combined with metamorphic testing technique for the evaluation of scientific software in CMA. Wang et al. [25] have dealt with big service provisioning in the context of enhanced living environments (ELE), also called cyber-physical social systems (CPSSs). To efficiently integrate, process, and analyze ELE big data, they adopted high-order singular value decomposition. Other researchers have employed Hadoop and Spark frameworks to realize some management operations, mainly service selection and composition [1,10,19]. However, the adoption of big data frameworks in big service management is still in its infancy.

Existing researches propose frameworks that mainly deal with service deviations, without concentrating on the consumed data's management facilities. Focusing only on the service level while ignoring the data level makes the proposed solutions similar to traditional Web/cloud service processing tools. Despite that the quality of big data-centric services is affected by the quality of the consumed data [19], ignoring data quality is a common limitation of current big service approaches. Finally, none of the existing approaches has incorporated autonomic management policies into the description of big services. Besides the above issues, four major differences between big services and existing service delivery models justify the need for new management solutions for big services. These include the heterogeneity of big services' composition units, their cross-domain nature, their data massiveness, and the large-scale environment [16].

3   Autonomic Big Service Fabric

In this section, we introduce the concept of *autonomic big service*. By inheriting the autonomic computing principles [11], a big service' components (heterogeneous services, data sources, etc.) can be seen as managed resources, whereas autonomic managers endow those components with self-* capabilities, including self-configuration, self-healing, self-optimization, and self-protection.

A number of autonomic managers is instantiated according to the complex workflow structure of the managed big service (see Fig. 1). This latter is composed by a set of heterogeneous and autonomous entities capable of modifying their behavior according to the changes in their environment. Each autonomic manager has a set of managed resources, including the executing service itself as well as its consumed data. Knowing the huge number of value-added services that form several workflows domain-specific workflows, the autonomic managers tasks are coordinated by other autonomic managers at the workflow level.

The cornerstone to the autonomic big service management framework is a fabric of autonomic managers. Some of these latter are instantiated to monitor and manage the involved services from different domains, whereas others called autonomic cloud brokers (ACB) are associated to the cloud zones. Each cloud is enhanced to an autonomic element and is designed as an ACB. Each ACB is responsible for monitoring, allocating and sharing the available cloud resources, to ensure SLA (Service Level Agreement) compliance. Multiple ACBs form a self-managed cross-domain multi-cloud overly network that enables composing, managing and providing big services in a distributed approach. Figure 1 also shows that services from different domains are encapsulated as autonomic elements to collectively form a cross-domain big service ecosystem. In the workflow and service layer, autonomic managers interact with ACBs through their manageability interfaces to realize big service management operations.

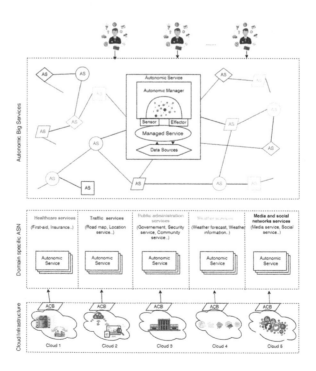

Fig. 1. Autonomic environment of managed big services

Given the huge number of big service components that must be continuously monitored by autonomic managers, and Knowing that each of these latter has its own knowledge expressed in terms of service capabilities, QoS and QoD information, management policies, etc., we propose in the next section a multi-relational and a multi-view modelling of autonomic big service knowledge based on the concept of knowledge graphs [3].

4   Big Service Knowledge Graph

Proposed for the first time by Google in 2012, knowledge graphs have been exploited by leading companies (e.g., LinkedIn, Wikimedia, Facebook, etc.) to boost their services and improve the users' quality of experience. Such promising concept is becoming a key element for artificial intelligence and knowledge-intensive applications, and has been exploited in different services, including contextual information retrieval, question answering, location-based services [3].

We define the big service knowledge graph (see Definition 1) as a large and multi-relational heterogeneous information network. The connections between its different entities (services, data sources, produced events, management policies, QoS and QoD data, etc.) are typed relations. Other important knowledge included in the BSKG concerns the monitoring policies and the auxiliary information, such as conflicts between management plans.

Definition 1. *(Big Service Knowledge Graph) A BSKG is an information network denoted as $\mathcal{G} = (\mathcal{V}, \mathcal{E}, \mathcal{F})$, where $\mathcal{V} = <\mathcal{V}_s \cup \mathcal{V}_d \cup \mathcal{V}_e \cup \mathcal{V}_a \cup \mathcal{V}_p>$ is a set of vertices (services, data sources, events, actions, management policies), $\mathcal{F}$ is a set of features, and $\mathcal{E} \subseteq \mathcal{V} \times \mathcal{V}$ is a set of edges. $\mathcal{V} = <\mathcal{S}, \mathcal{D}, \mathcal{P}>$ denotes respectively the subsets of services, data sources, and management policies associated to the big service. $\mathcal{E}$ denotes the set of typed relations between vertices.*

Definition 2. *(Fact) a fact in the BSKG is a triple $f = (h, r, t)$, where $h, t \in \mathcal{V}$ are service, data, or policy entities, $r \in \mathcal{E}$ is a typed relation between h and t. The set of facts in the BSKG is denoted by $\mathcal{D}^+ = \{(h, r, t)\}$.*

Definition 3. *(Relation) a relation $r \in \mathcal{E}$ is a typed connection between two entities in the BSKG. $\mathcal{E}_k : \mathcal{V}_i \rightarrow \mathcal{V}_j$, where $\mathcal{V}_i, \mathcal{V}_j \in \mathcal{V}$. Depending on the entities types, the relation function $f(\mathcal{E}_k)$ is defined as follows:*

$$f(\mathcal{E}_k) = \begin{cases} SIMILARTO & if\ e_i, e_j \in \mathcal{V} \\ CONSUME & if\ e_i \in \mathcal{V}_s \wedge e_j \in \mathcal{V}_d \\ INVOKE & if\ e_i, e_j \in \mathcal{V}_s \\ MANAGEDBY & if\ e_i \in \mathcal{V}_s \wedge e_j \in \mathcal{V}_p \\ PRODUCEDBY & if\ e_i \in \mathcal{V}_e \wedge e_j \in \mathcal{V}_s \\ TRIGGER & if\ e_i \in \mathcal{V}_e \wedge e_j \in \mathcal{V}_a \\ PRODUCE & if\ e_i, e_j \in \mathcal{V}_e \\ ISCONFLECTING & if\ e_i, e_j \in \mathcal{V}_a \end{cases} \tag{1}$$

Several elements are involved in the BSKG construction. These include the component services together with their functional and QoS descriptions, the consumed data sources and their QoD levels, the management policies formed of the eventual events and their triggered management plans. Relations are also the core elements of the BSKG. Examples of connections between services, data sources, and management policies include *service-data source*, *service-service*, *service-policy*, *event-action*, etc. The BSKG is constructed according to the big service workflow structure, which expresses the dependencies between its component services, as well as the consumed data sources. Given the deviations that may affect the big service execution and the changes in its environment, the BSKG structure may undergo several updates by autonomic managers, like adding/removing nodes, updating information on edges (e.g., QoS/QoD values, events causality, conflicts between policies, etc.).

5   Autonomic Management via Multi-view Learning

In this section, we take advantage of graph embedding, as a widely used operation to infer additional knowledge from knowledge graphs [23]. In our work such operation helps predicting the latent connections between BSKG entities. These include the similarities between entities, the conflicts between management policies, etc. By this way, the big service's autonomic managers will have more alternatives to decide about the suitable management plans, such as excluding a conflicting policy or a highly sensitive data source.

Motivated by the similarities between knowledge graphs and multi-view data, we combine Graph Neural Networks (GNN) [29] and multi-view representation learning [21] to infer the missing knowledge from the BSKG. This latter is split into a set of interdependent views (see Definition 4), from which we learn the partial embeddings to, finally, aggregate them and provide a unified and representation of BSKG entities, as well as the predicted connections between them.

Definition 4. *(View) A view $\phi_i = (\mathcal{V}_i, \mathcal{E}_i, \mathcal{F}_i)$ is a sub-network of the BSKG, where $\mathcal{V}_i \in \mathcal{V}$, $\mathcal{E}_i \in \mathcal{E}$, and $\mathcal{F}_i \in \mathcal{F}$ denote, respectively, the subsets of entities, relations and features specific to ϕ_i. The relations $r^t \in \mathcal{E}_i$ can be of the same type $t \in T$. The aggregation $\phi = \{\phi_1 \cup \phi_2 \cup ... \cup \phi_k\}$ of all views corresponds to the whole BSKG.*

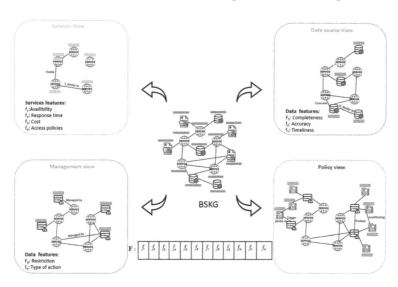

Fig. 2. Multi-view representation of the big service knowledge graph

Figure 2 shows that the *service view* ϕ_s is extracted by considering the *INVOKE* and *S-SIMILAR* relations. These relations refer to the service invocation paths within a managed big service, as well as the similarities between this latter's components and other substituting services. The relations *CONSUME* and *D-SIMILAR* are exploited in the extraction of the facts that constitute the *data source view* ϕ_d. As for the *policy view* ϕ_p, its related facts are extracted based on the relations *TRIGGER* and *CONFLICTING*.

To predict the latent relations within the multi-view BSKG, we define a three-stage multi-view learning process, as shown in Fig. 3: (1) encoding of the BSKG content into a low-dimensional vector embeddings space using a GNN approach, (2) fusion (weighted aggregation) of the view-specific embeddings by

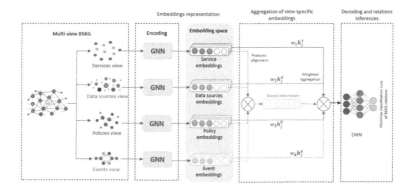

Fig. 3. Multi-view learning process

considering the dependencies between the BSKG views, (3) augmenting the initial BSKG structure with the inferred links between its nodes, based on the proximity (closeness) degrees between nodes.

The whole multi view learning process is detailed in Algorithm 1.

Algorithm 1 : Multi-view learning for the Big Services Management

---

1: **Input:** $\phi = \{\phi_1, ..., \phi_k\}$ Set of views denoting the BSKG.
2: **Output:** $\{\mathcal{E}_t^p\}_{t=1}^T$ Predicted relations in the BSKG.
3: **Begin**
4: $\{\mathcal{E}_t^p\}_{t=1}^T \leftarrow \emptyset$
5: **for each** view $\phi_i \in \phi$ **do**   ▷ Compute view-specific embeddings.
6:     **for each** node $v_j^i \in V_i$ **do**
7:         $\mathcal{N}(v_j^i) \leftarrow$ Sample Neighborhood $(k, v_j^i) : (v_j^i, v) \in V_i$ and $v \in \mathcal{N}(v_j^i)$ ▷ $\mathcal{N}(.)$
   is a fixed size k-depth neighborhood for the node v_j^i.
8:         $\mathbf{h}_j^i \leftarrow ENC(v_j^i)$ ▷ Compute embedding $\mathbf{h}_j^i = [\mathbf{s}_j^i, \mathbf{x}_j^i]$ of the jth node in ϕ_i
9:     **end for**
10: **end for**
11: **for each** edge type $t \in T$ **do**
12:     Learn the view-specific representations for nodes $\mathbf{h}_j$
13:     $\mathbf{h}_j \leftarrow \frac{1}{|\phi|} \cdot \sum_{i=1}^{|\phi|} w_i \times \mathbf{h}_j^i$ ▷ Weighted aggregation of the jth node's embeddings
14:     **for each** relation $r_i \in \mathcal{E}$ **do**
15:         $\mathcal{L} \leftarrow$ Minimize relation's inference loss for r_i using Eq. (8)
16:         $\mathcal{E}_t^p \leftarrow \mathcal{E}_t^p \cup r_i$
17:     **end for**
18: **end for**
19: **Return** $\{\mathcal{E}_t^p\}_{t=1}^T$

Encoding of BSKG Embeddings. The goal of this first step is to learn the partial representation, i.e. view-specific embedding, of each node in the view ϕ_k. To do so, each view is fed into a distinct two-layer GNN module, as depicted in Fig. 3. The learned representation $\mathbf{h}_j^i$ of each node $v_j^i \in \phi_i$ (e.g., the jth policy in the management view ϕ_p), is defined by two parts: (1) the node embedding $\mathbf{s}_j^i$ with its structural context (neighbor nodes), and (2) the node features $\mathbf{x}_j^i$. By combining $\mathbf{s}_j^i$ and $\mathbf{x}_j^i$, the view-specific representation of $v_j^i \in \phi_i$ is learned as the concatenation $\mathbf{h}_j^i = [\mathbf{s}_j^i, \mathbf{x}_j^i]$ of v_j^i's context and features.

To select the neighborhood $\mathcal{N}(v_j^i) : (v_j^i, v_n^i \in \phi_i)$ of the embedded node, a sampling step is performed while considering the data distribution in each BSKG view. Then, a forward propagation algorithm is applied to generate the view-specific embedding of the node v_j^i. This step is achieved by aggregating the features of the k-depth neighborhood (nodes in $\mathcal{N}(v_j^i)$) in an iterative manner. $\mathbf{s}_j^i$ is computed using the mean aggregator [7] defined in Eq. (2).

$$s_j^i(l) = \delta(AGGREGATE_l \cup \{s_j^i(l-1), \forall v_j^i \in \mathcal{N}(v_j^i)\})) \qquad (2)$$

where $s_j^i(l)$ is the representation of v_j^i at the $l(1 \leq l \leq k)$ iteration, $\delta(.)$, is the sigmoid function, $AGGREGATE_l = \widetilde{\mathbf{W}}.MEAN(.)$ corresponds to the mean aggregator functions, with $\widetilde{\mathbf{W}}$ is the weight parameter matrices of the aggregator functions responsible for propagating information between the layers of the GNN model, $\mathcal{N}(v_j^i)$ is the sampled neighborhood for the node v_j^i.

Weighted Aggregation of View-Specific Embeddings. The partial low-dimensional representations that resulted from the encoding step are specific to their containing views. For instance, the embedding of a service node in the view ϕ_s differs from that specific to the view ϕ_s, as the former depends on the service view features (e.g., QoS attributes), while the latter depends on the data source view features (e.g., QoD dimensions). Since the overall representation of each node must take into account the complementarity between the partial embeddings and the dependencies between views, we propose to fuse the partial view-specific node representations by applying a weighted aggregation strategy. This latter ensures that each view will have a distinct proportion in the final representation, as it allows the BSKG views to reinforce each other.

The aggregation of view-specific embeddings is defined in Eq. (3).

$$\mathbf{E}_j = \left\{ \sum_{i=1}^k w_i \mathbf{h}_j^i, w_i \geq 0, \sum_{i=1}^k w_i = 1 \right\} \tag{3}$$

Here w_i represents the ith view's weight, and $\mathbf{h}_j^i$ is the partial embedding of the jth node in the view ϕ_i. Depending on the value of w_i, the portion of $\mathbf{h}_j^i$ will contribute or not in the final node representation E_j.

To correctly fuse these partial embeddings, the shared information must be first approximated and aligned from k views ($iif \ w_i \neq 0$). For this purpose, Eq. (4) is used as a multi-view loss function:

$$\mathcal{L}_{MV} = \frac{1}{k} \sum_{i=1}^k \alpha_i \left\| \mathbf{h}_j^i \mathbf{M}^i - \mathbf{H}^i \right\|_2^2 \tag{4}$$

Here $\mathbf{H}^i \in \mathbb{R}^{m_t \times d_i} \ (d_i = d^r/d^t)$ denotes the ith node $(1 \leq i \leq k-1)$/edge $(i = k)$ feature view, and m_t means the total edge number of $\mathcal{E}^t$. $\mathbf{P}^i \in \mathbb{R}^{d_i \times d}$ is the projection matrix. $\mathbf{H}^i$ Is the shared knowledge across multiple node embeddings.

To achieve the aggregation task, we determine the contribution (weight w_i) of each view in the BSKG using the self-attention mechanism [14] (see Eq. 5).

$$w_i = softmax(tanh(\mathbf{C}^i \mathbf{W}^i)) \tag{5}$$

Here, $\mathbf{W}^i$ denotes the patterned weight parameter matrix w.r.t. the view ϕ_i. w_i is learned during the model training. Aggregated embeddings are defined as:

$$\mathbf{h} = \frac{1}{k} \sum_{i=1}^k w_i C^i. \tag{6}$$

At this stage, the aggregated embeddings are exploited in the prediction of latent relations (e.g., *INVOKE, MANAGEDBY, CONSUME, CONFLICT*), based on the common Deep Neural Network (DNN) method [12]. This step is achieved using Eq. 7.

$$\widetilde{\mathbf{A}}_{ij}^t = f(\mathbf{a}^T \mathbf{c}_{ij}^t) \tag{7}$$

where $\mathbf{a}$ denotes the weighted vector of the predicted relation r_{ij}^t, $\mathbf{c}_{ij}^t$ is the aggregated vector w.r.t. r_{ij}^t, and $f(.)$ is a softmax function.

Unified Representation and BSKG Completion. In this final step, the goal is to augment the BSKG's initial structure with the inferred links between its nodes. To do so, the proximity (closeness) degrees between nodes are used to determine the predicted relations. The overall objective function of the proposed multi-view learning model is defined in Eq. 8:

$$\mathcal{L} = \mathcal{L}_{MV} + \mathcal{L}_{CE} \tag{8}$$

Here, $\mathcal{L}_{MV}$ refers to the loss that resulted from aggregating view-specific node/edge embeddings, whereas $\mathcal{L}_{CE}$ is the cross-entropy loss, which consists in minimizing the negative log-likelihood, as defined in Eq. (9). $\mathcal{L}_{CE}$ is obtained by learning the parameters of the aggregator functions, and by determining the classification loss of the DNN model.

$$\mathcal{L}_{CE} = \sum_{i=1}^{n} \sum_{j=1}^{n} - \left[\gamma \mathbf{A}_{ij} log \widetilde{\mathbf{A}}_{ij} + (1 - \mathbf{A}_{ij}) \, log \left(1 - \widetilde{\mathbf{A}}_{ij} \right) \right] \tag{9}$$

where $\widetilde{\mathbf{A}}_{ij}$ and $\mathbf{A}_{ij}$ denote, respectively, the predicted likelihoods and the ground truth for the ith relation w.r.t, the jth entity in the BSKG.

The model parameters are computed using the common gradient descent method [8] for optimization until the convergence of the overall model loss.

6   Selection of Big Service Management Policies

During execution, big services are often subject to behavioral changes. These latter are expressed as events in the BSKG, and may trigger various management operations, such as service reinvocation, data source substitution (see Fig. 4).

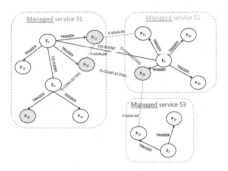

Fig. 4. Example of events sub-graph

The monitored and detected events are first located in the BSKG, so that to determine their triggered management policies (see Fig. 4). Since some policies may lead to incompatibility and SLA violation issues, the concerned entities and relations in each autonomic manager's knowledge subgraph (detected events, management policies, similarity and conflict relations) will be integrated by the coordinating autonomic manager, so that to resolve conflicts and decide on the suitable management policies at a higher level. By this way, the conflicting policies (e.g., P_{11} and P_{32}) will not be part of the final big service management plan. The decision process is detailed in Algorithm 2.

Algorithm 2 starts by locating each occurred event $e \in E$ and its context, i.e. co-events, associated management policies and the relationships between them (lines 6–7). The purpose of this first step is to construct the events subgraph $\mathcal{G}^E$ (see Fig. 4), by combining each event's context (line 13). After that, each management action a_i that may be triggered by e is used to get the *CONFLICT* facts, which contain in their tail part the conflicting actions w.r.t. a_i. To do so, the algorithm gets the neighbor plans based on the *CONFLICT* relation (line 9). If no conflicting action is related to the current plan, this latter's score is computed, and the plan is saved in the set P_i denoting the candidate plans to solve the event e_i. Algorithm 2 uses the equation in line 10, to evaluate the candidate action in terms of QoS, history of successful executions, and the number of conflict relations. In line 10, $|F_i|$ is the number of *CONFLICT* triples in the event subgraph (see line 9), n_{ij} denotes the number of times when the ith management action a_i is triggered for the jth service, n_{ij}^f is the number of failed executions of the management action a_i, QoS_{ij} is the quality level of the jth service after being repaired by the management operation a_i. The first part of the equation serves as a penalty value for the score of the current management action. The lower are n_{ij}^f and $|F_i|$, the higher is the score of a_i, hence its ability to correctly repair the failed service. The evaluation and selection routine is repeated until checking all the triples having the relation *CONFLICT* (lines 21–23). Finally, the management actions with the highest score will be returned to achieve the big service management task (line 27).

Algorithm 2 : Selection of big service management actions

1: **Input:** Set of detected events E, BSKG $\mathcal{G}$.
2: **Output:** Management plan P.
3: **Begin**
4: $P \longleftarrow \emptyset$
5: $\mathcal{G}^E \longleftarrow \emptyset$
6: **for each** event $e \in E$ **do**   ▷ Event subgraph construction
7:     Locate e in $\mathcal{G}$
8:     **for each** action $a_i \in Context(e)$ **do**   ▷ Get management actions for event e
9:         $F_i \longleftarrow getTriples(a_i, CONFLICT)$
10:        $s_i \longleftarrow \frac{1-n^f_{ij}}{n_{ij}.|F_i|} \cdot \sum_{k=1}^{n_{ij}} QoS_{ij}$ ▷ Computer a_i's score w.r.t. the conflict relations
11:        Label a_i with s_i and update $Context(e)$
12:    **end for**
13:    $\mathcal{G}^E \longleftarrow \mathcal{G}^E \cup Context(e)$ ▷ add event e's triples to $\mathcal{G}^E$
14: **end for**
15: Sort actions in $\mathcal{G}^E$ ▷ Sort management actions according to their scores
16: **for each** event $e_x \in E$ **do**
17:    **if** $< a_i, CONFLICT, a_j > \not\subset \mathcal{G}^{\mathcal{E}}$ **then**   ▷ Check conflict between actions a_i
    and a_j, where $< e_x, TRIGGER, a_i >$ and $\exists e_y \in \mathcal{G}^E, < e_y, TRIGGER, a_j >$
18:        Add a_i to P ▷ Select a_i as a triggered action by e_x
19:        Remove e_x from E
20:    **else**
21:        **if** $< a_j, CONFLICT, a_k > \not\subset \mathcal{G}^{\mathcal{E}}$ **then**   ▷ Check conflict between actions
    a_j and a_k, where $a_k \neq a_i$ and $\exists e_z \in \mathcal{G}^E, < e_z, TRIGGER, a_k >$ and $e_z \neq e_x$
22:            $i \leftarrow i + 1$
23:            GOTO 17
24:        **end if**
25:    **end if**
26: **end for**
27: **Return** P

---

7   Implementation and Experiments

To study the complexity of BSKG processing and the impact of multi-view learning on the selection of management policies, we implemented our approach using Python programming language on Google Colab[1]. Due to the absence of datasets on big data-centric services and to cover the BSKG entities mentioned in Sect. 4, we constructed the BSKG by collecting data from several sources of information. These include (i) WS-DREAM[2] dataset, which we translated its content into *service*-related entities, (ii) a taxonomy of faults across service-based application layers [2], which we used to instantiate and associate the event/action entities as management policies for each service in WS-DREAM. The remaining entities and their features (e.g., data sources, data quality dimensions) were randomly generated due to the unavailability of such type of information.

---

[1] http://shorturl.at/wOZ48.
[2] https://wsdream.github.io/.

Evaluation of the Multi-view Learning Method. To study the utility of learning the BSKG embeddings from multiple views, and its impact on the management policies' selection process, we generated four views: service view ϕ_s, data view ϕ_d, management view ϕ_m, and policies view ϕ_p, and we fed them into four GNNs. As for the traditional embedding method, we fed the whole BSKG content ($\mathcal{G}$) into one GNN. Accuracy measures, in terms of F1-score and PR-AUC are recorded in Fig. 5 under the parameters: $k = 2$, $\lambda = 0.001$, $d = 128$.

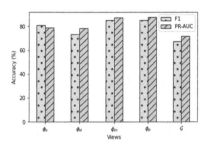

Fig. 5. Accuracy measures for single-view and multi-view learning

Figure 5 shows that the MVL approach has generated more useful embeddings for the BSKG completion task. PR-AUC and F1 scores exceed those produced by the single-GNN embedding method by $[6.9, 16.1]\%$ and $[5.7, 17.8]\%$, respectively. In fact, the neighborhood samples generated for the MVL method are characterized by a lower heterogeneity, compared to those produced for the single-view GNN, in which several typed relations can belong to a node context, i.e. neighborhood. This uniform sampling caused a high variance during the training and inference, which has led to sub-optimum accuracy. Hence, the low accuracy scores in the case of single GNN embedding are explained by the incapacity of the model to cope with the heterogeneity of BSKG content.

Impact of MVL on the Management Policies' Selection. In this test series, we varied the number of entities that represent the injected events/faults. The event types are randomly selected from the events subset $\mathcal{V}_e$. Figure 6a depicts the total number of candidate policies, as well as the number of *CONFLICT* relations discovered with and without using the MVL step.

For all the test cases, the MVL approach allowed detecting the highest number of conflict relations (in $[1, 8]$). Without MVL step, only 5 relations have been identified in the best case. In fact, correctly learning the representations of BSKG entities had a positive impact on the evaluation of proximity degrees between management policies and the inference of latent connections between them. That can be seen from Fig. 6a, where the number of candidate policies with MVL is always lower than that without the learning step. In fact, Algorithm 2 excludes the triples of the form $< a_i, CONFLICT, a_j >$ and limits the selection of big service management actions to the non conflicting ones. Therefore, the higher is

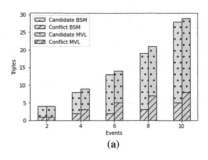 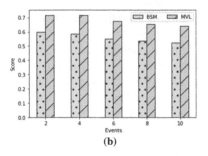

Fig. 6. (a) Numbers of conflict triples, and (b) scores of management actions

the number of excluded *CONFLICT* triples, the better the management plan's score is (in $[0.639, 0.717]$). In contrast, the lowest scores (in $[0.522, 0.597]$) were produced by the traditional approach (see Fig. 6b), which consists in populating the BSKG without inferring conflict or similarity facts.

Model Complexity and Computation Time. In this test series, we studied the computational complexity of the BSKG processing and parsing, by varying the number of injected events ($[2 : 10 : 2]$).

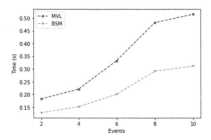

Fig. 7. Computation time with different numbers of injected events

In Fig. 7, it is obvious that the time devoted to run through the BSKG structure is proportional to the number of injected events. The computation time is, not only dependent on the complexity of locating an event, but also on the amount on relations between the candidate management policies. In fact, selecting the suitable actions requires checking the existence of conflicts between them, which may cause an extra time due to the additional verified facts (case of $|E| \geq 8$). But in the end, the distribution of the big service knowledge over the instantiated autonomic managers will considerably reduce the number of processed triples. Regardless of the time taken to select management policies, the time complexity of the MVL-based BSKG completion task mainly depends on two steps: the learning of view-specific embeddings and the fusion of those partial embeddings for the purpose of relations prediction. As the first step considers the entities/relations types (R and T) and the sample ($N_j^i(k)$) fixed-size, it takes $O(\prod_{k=1}^{K} |\mathcal{N}(v_j^i)|)$. The second step takes $O(tnd^2)$, as it depends on the learning of parameter matrices ($\mathbf{P}^i \in \mathbb{R}^{d_i \times d}$ and $\mathbf{C}^i$). Therefore, the MVL complexity is in the order of $O(\prod_{k=1}^{K} |\mathcal{N}(v_j^i)| + tnd^2)$.

8 Conclusion and Future Work

In this paper, we addressed the management issues of a new emerging service model called *big service*. We exploited autonomic computing principles [11], to design the managed big service as an autonomic system. Given the complex nature of big services, we modeled its related data as a knowledge graph, which is continuously updated by autonomic managers according to the big service's current state. To infer additional knowledge (e.g., conflicts between management policies) for autonomic managers, we proposed a multi-view learning method that allowed predicting latent relations, by embedding the BSKG into a low-dimensional vector space. Finally, a decision mechanism was defined to select and trigger the appropriate management policies in response to a detected problem.

Additional experiments are currently underway, to confirm the promising results. Also, given the dynamic and evolving nature of the BSKG, incremental learning could be an effective solution to cope with the complexity and cost of re-embedding the whole BSKG. Finally, big data frameworks could be good candidates to deal with the massiveness of big services [26] and to manage them in a parallel fashion, hence increasing the autonomic managers' performance. For this purpose, we intend to implement the proposed algorithms and the big services operations using Apache Spark.

References

1. Bhaskar, B., Jatoth, C., Gangadharan, G., Fiore, U.. A mapreduce-based modified grey wolf optimizer for qos-aware big service composition. Concurr. Comput. Pract. Exp. **32**(8), e5351 (2020)
2. Bruning, S., Weissleder, S., Malek, M.: A fault taxonomy for service-oriented architecture. In: 10th IEEE High Assurance Systems Engineering Symposium (HASE'07), pp. 367–368. IEEE (2007)
3. Chen, X., Jia, S., Xiang, Y.: A review: knowledge reasoning over knowledge graph. Expert Syst. Appl. **141**, 112948 (2019)
4. Cheng, Y., Leon-Garcia, A., Foster, I.: Toward an autonomic service management framework: a holistic vision of soa, aon, and autonomic computing. IEEE Commun. Mag. **46**(5), 138–146 (2008)
5. Ding, J., Zhang, D., Hu, X.H.: A framework for ensuring the quality of a big data service. In: International Conference on Services Computing (SCC), pp. 82–89. IEEE (2016)
6. E, X., Han, J., Wang, Y., Liu, L.: Big data-as-a-service: definition and architecture. In: 15th IEEE International Conference on Communication Technology, pp. 738–742 (2013)

7. Hamilton, W.L., Ying, R., Leskovec, J.: Inductive representation learning on large graphs (2017). arXiv preprint arXiv:1706.02216
8. Hinton, G.E., Salakhutdinov, R.R.: Reducing the dimensionality of data with neural networks. Science **313**(5786), 504–507 (2006)
9. Huang, L., Zhao, Q., Li, Y., Wang, S., Sun, L., Chou, W.: Reliable and efficient big service selection. Inf. Syst. Front. **19**(6), 1273–1282 (2017)
10. Jatoth, C., Gangadharan, G., Fiore, U., Buyya, R.: Qos-aware big service composition using mapreduce based evolutionary algorithm with guided mutation. Future Gener. Comput. Syst. **86**, 1008–1018 (2018)
11. Kephart, J.O., Chess, D.M.: The vision of autonomic computing. Computer **36**(1), 41–50 (2003)
12. Kim, B., Kim, J., Chae, H., Yoon, D., Choi, J.W.: Deep neural network-based automatic modulation classification technique. In: International Conference on Information and Communication Technology Convergence, pp. 579–582. IEEE (2016)
13. Landset, S., Khoshgoftaar, T.M., Richter, A.N., Hasanin, T.: A survey of open source tools for machine learning with big data in the hadoop ecosystem. J. Big Data **2**(1), 24 (2015)
14. Lin, Z., et al.: A structured self-attentive sentence embedding (2017). arXiv preprint arXiv:1703.03130
15. Liu, M., Tu, Z., Xu, X., Wang, Z.: A data-driven approach for constructing multilayer network-based service ecosystem models (2020). arXiv:2004.10383
16. Mezni, H., Sellami, M., Aridhi, S., Ben Charrada, F.: Towards big services: a synergy between service computing and parallel programming. In: Computing, pp. 1–36 (2021)
17. Mukherjee, T., Nath, A.: Big data analytics with service-oriented architecture. In: Exploring Enterprise Service Bus in the Service-Oriented Architecture Paradigm, pp. 216–234. IGI Global (2017)
18. Papazoglou, M.P., Traverso, P., Dustdar, S., Leymann, F.: Service-oriented computing: state of the art and research challenges. Computer **40**(11), 38–45 (2007)
19. Sellami, M., Mezni, H., Hacid, M.S.: On the use of big data frameworks for big service composition. J. Netw. Comput. Appl. **166**, 102732 (2020)
20. Siddiqa, A., et al.: A survey of big data management: taxonomy and state-of-the-art. J. Netw. Comput. Appl. **71**, 151–166 (2016)
21. Sun, S.: A survey of multi-view machine learning. Neural Comput. Appl. **23**(7–8), 2031–2038 (2013)
22. Taherkordi, A., Eliassen, F., Horn, G.: From iot big data to iot big services. In: International Symposium on Applied Computers, pp. 485–491. ACM (2017)
23. Wang, Q., Mao, Z., Wang, B., Guo, L.: Knowledge graph embedding: a survey of approaches and applications. IEEE Trans. Knowl. Data Eng. **29**(12), 2724–2743 (2017)
24. Wang, S., Su, W., Zhu, X., Zhang, H.: A hadoop-based approach for efficient web service management. J. Web Grid Serv. **9**(1), 18–34 (2013)
25. Wang, X., Yang, L.T., Feng, J., Chen, X., Deen, M.J.: A tensor-based big service framework for enhanced living environments. IEEE Cloud Comput. **3**(6), 36–43 (2016)
26. Xu, X., Sheng, Q.Z., Zhang, L.J., Fan, Y., Dustdar, S.: From big data to big service. Computer **48**(7), 80–83 (2015)
27. Yang, L.T., et al.: A multi-order distributed hosvd with its incremental computing for big services in cyber-physical-social systems. IEEE Trans. Big Data **6**, 666–678 (2018)

28. Yang, Y., Xu, J., Xu, Z., Zhou, P., Qiu, T.: Quantile context-aware social iot service big data recommendation with d2d communication. IEEE Internet Things **7**, 5533–5548 (2020)
29. Zhou, J., et al.: Graph neural networks: a review of methods and applications (2018). arXiv preprint arXiv:1812.08434

Towards a Deep Learning-Driven Service Discovery Framework for the Social Internet of Things: a Context-Aware Approach

Abdulwahab Aljubairy[1,2(✉)], Ahoud Alhazmi[1,2], Wei Emma Zhang[3], Quan Z. Sheng[1,2], and Dai Hoang Tran[1,2]

[1] Macquarie University, Sydney, NSW 2109, Australia
{abdulwahab.aljubairy,ahoud.alhazmi,dai-hoang.tran}@hdr.mq.edu.au,
michael.sheng@mq.edu.au
[2] Intelligent Computing Laboratory (ICL) Lab, Macquarie University, Sydney, Australia
[3] The University of Adelaide, Adelaide, South Australia 5005, Australia
wei.e.zhang@adelaide.edu.au
https://icl-mq.weebly.com/

Abstract. The Social Internet of Things (SIoT) is a new paradigm that enables IoT objects to establish their own social relationships without human intervention. A fundamental perspective of SIoT is to make socially capable objects, wherein objects can automatically share their services capability and exchange their experience with each other for the humans' benefit. Service discovery is a crucial task that requires fast, scalable, dynamic mechanisms. This paper aims to investigate the feasibility of adopting state-of-the-art deep learning techniques to build a social structure among IoT objects and design an effective service discovery process. To achieve this goal, we propose a framework that includes three phases: *i*) collecting information about IoT objects; *ii*) constructing a social structure among IoT objects using; and *iii*) developing an end-to-end service discovery model using the language representation model BERT. We conducted extensive experiments on real-world SIoT datasets to validate our approach, and the experimental results demonstrate the feasibility and effectiveness of our framework.

Keywords: Social Internet of Things · Service discovery · Graph neural networks · Natural language processing

1 Introduction

Service Discovery in the Internet of Things (IoT) is the process to detect and find the desired service that satisfies a given query to complete a particular task [3]. Diverse search discovery techniques have been constructed for this purpose, but these techniques are faced with numerous challenges which reduce their

© Springer Nature Switzerland AG 2021
W. Zhang et al. (Eds.): WISE 2021, LNCS 13081, pp. 480–488, 2021.
https://doi.org/10.1007/978-3-030-91560-5_35

performance quality [9]. In particular, the nature of IoT resources is dynamic and heterogeneous. Moreover, the exponential growth of IoT objects increases the burden on current approaches. Therefore, it has become essential to have suitable mechanisms to adapt and cope with these crucial challenges.

Recently, there has been a plethora of independent research studies to bring the next evolutionary step of the IoT paradigm by creating socially aware objects. This refers to creating a new generation of IoT objects that manifests themselves and can socialize with the surrounding peers for the sake of, but not limited to, discovering new services, exchanging experience, and benefiting from each other's capabilities. This new paradigm is referred to as the Social Internet of Things (SIoT) [1]. In this paradigm, the key idea is to build a network structure among IoT objects, wherein this structure can be shaped to ensure network navigability, increase the level of trustworthiness, and allow an efficient way of discovering services.

The service discovery process requires fast, scalable, dynamic mechanisms. In the IoT paradigm, resources are discovered using keywords, where sometimes the whole search space is used to satisfy the request. In the SIoT paradigm, the discovery process can occur when an object issues a text query, where the social structure among IoT objects can be utilized to respond to this request. Such a process can foster resource availability and make the discovery more efficient when exploiting objects relationships [1]. To boost resource discovery in SIoT, it is desired to have a mechanism that can proactively utilize advanced techniques. These techniques should consider contextual awareness and the similarities of IoT resources to offer a good discovery experience.

Some works investigated service discovery using relationships among the owners of IoT objects [13]. However, in the SIoT paradigm, objects can establish relationships with one another regardless of their owners' relationships. In this work, we propose a framework that constructs a social structure among IoT objects via their movement data wherein IoT objects can exploit this social structure to look up required services. Our framework includes three phases: i) collecting and crawling information about IoT objects; ii) constructing a social structure among IoT objects using a two-step approach; and iii) developing a service discovery model based on the language representation model BERT [4]. Overall, the main contributions of this paper are summarized as follows: designing and implementing a framework for discovering IoT resources, developing a greedy algorithm to extract social relationships among IoT objects for building the social structure among them, developing an end-to-end prediction model for predicting the future social relationships that might form among IoT objects, and developing the *parser* component based on the language representation model BERT for identifying the intent behind any query issued from an IoT object.

2   Preliminaries: Basic Definitions and the Problem Statement

Definition 1 (Stop Point). It is the time period that an object spends at a particular location, represented by <ObjectID, longitude, latitude, ts, te>.

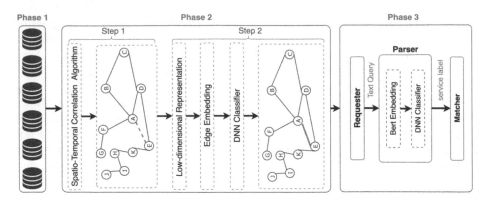

Fig. 1. An overview of our framework for discovering services in SIoT.

Longitude and latitude refer to the location. *ts* and *te* are time start and time end of the stop point at that location.

Definition 2 (Encounter). It is an overlap between any two IoT objects intervals at the same location approximately.

Definition 3 (Social Relationship). It is a relationship that can be established between any two IoT objects when they come into contact for some time on several occasions (i.e., several encounters occurred between two IoT objects according to Definition 2).

Definition 4 (Social Structure). It is an undirected graph denoted as $G = (V, E)$ where V represents the set of IoT objects in the graph and E represents the set of edges among IoT objects. Each edge represents an established social relationship (Definition 3) between any two IoT objects.

Given the stop points information of IoT objects, the goal is to build a social structure among IoT objects utilizing these stop points. Then, the resulted social structure can be exploited for resolving any search query issued by an IoT object in the community. Therefore, let O be the universal collection of IoT objects in the SIoT graph, for each search query q, there is a set of objects o_q that are relevant to the query q. The task of the discovery process is constructing the result set $\hat{o}_q$ that approximates the unknown o_q. This task is done by conducting the following procedure. First, we establish a social structure among IoT objects via the utilization of stop point information. Second, we evaluate the relevance of the discovered objects against the given query q with a relevance function $f(o, q)$.

3   Deep-Learning Driven Service Discovery Framework

This section explains our proposed framework for service discovery in SIoT. Figure 1 gives an overview of the major steps of our approach. After obtaining and collecting SIoT datasets, we firstly construct a social structure among

IoT objects using a two-step approach. Then, we elucidate the service discovery process in our framework. In the following, we will provide more details.

3.1 Phase 1: Data Collection

In the first phase, we collect information about IoT objects. This information includes traces of the locations of these objects. Each trace record reflects IoT objects' stop points (Definition 1), particularly those with motion capability (i.e., bus, taxi, etc.). Other objects are static, and their locations are known (i.e., POI, restaurant, theater, museum, etc.). In addition, this information includes services offered by these objects and the Application of interest to these objects.

3.2 Phase 2: Building a Social Structure Among IoT Objects

The essential perspective of the SIoT paradigm is to have a social structure (Definition 4), wherein this structure shows the internal relationships (Definition 3) among IoT objects. For that, we use a two-step approach as follows:

Step 1: *Constructing social structure using spatio-temporal encounters.* The goal is to construct a social structure among IoT objects using the obtained stop points from *Phase 1*. The core idea is to extract the spatio-temporal encounters that occurred among IoT objects using the Algorithm 1. The input of Algorithm 1 is the set of stop points P for all IoT objects. The output of this algorithm is a list of newly generated edges between SIoT objects is returned. These edges represent the social relationships among IoT objects. We develop this algorithm according to the greedy technique which follows the problem-solving heuristic of making the locally optimal choice at each stage. For example, in our problem, the local optimal strategy is to select the next stop point pj to compare if and only if the start time of the stop point pi falls before the end time of the previous stop point pi as line (4–6). If that succeeded, an overlap between the two objects is reported. To guarantee the performance of this algorithm, it is required to sort the set of stop points according to the time start of each p (line 2).

Step 2: *Boosting current social structure by predicting future social relationships.* The social structure of IoT objects is dynamic, and this structure expands when new relationships among IoT objects form. The goal of this step is to learn the SIoT graph obtained from *Step 1* in order to predict future social relationships that would form among IoT objects. We develop an end-to-end prediction model to anticipate future relationships. First, this model employs GraphSAGE [6] as embedding step to produce low-dimensional representation for IoT objects. Then, to anticipate if a relationship may form between any two IoT objects, we calculate the inner product of their embeddings and feed it into a classifier to determine the probability of the potential relationship.

3.3 Phase 3: Service Discovery Process

After creating a social structure among IoT objects, in this section, we explain the service discovery process exploiting the resulted social structure for finding

---

Algorithm 1: Spatio-Temporal Correlation Algorithm

    Input   : Stop Points for all IoT objects $P = \{p_1, p_2 \ldots p_n\}$
    Output: List of Edges (obj_id$_i$, obj_id$_j$, ts, te, Duration)
1 **begin**
2   **Sort** P according to ts (start time) of all stop points.
3     $i \leftarrow 0$
4   **while** $(i < length(P) - 1)$ **do**
5         $j \leftarrow i + 1$
6   **while** $(j < length(P)$ and $ts_{pj} <= te_{pi})$ **do**
7   **Find_Overlap** (pi, pj)
8             $j + +$
9   **end**
10         $i + +$
11   **end**
12   **return** Edges
13 **end**

---

the relevant service. This process involves three main components: requester, parser, and matcher.

The requester represents an IoT object that submits a text query. The parser is the component that captures this query, processes it, and unveils the intent behind it. Its main function is to recognize the required service from the query issued by this IoT object (requester). This component consists of two parts. The first part employs the BERT model to produce the corresponding contextualized representation for the query. The second part contains the DNN classifier fed with the obtained contextualized representations to label the query. This label refers to the required service.

When feeding the input query as token sequence $x = \{w_1, \ldots, w_n\}$ to BERT, it has two steps. Firstly, it packs the input features as $H_0 = \{e_1, \ldots, e_n\}$ using the combination of the token embedding, position embedding as well segment embedding (because the request query might have more than one sentence). As shown in Fig. 2, E represents the input embedding, H_i^L represents the contextual representation of token i, [CLS] is the special symbol added in front of every input example for classification output, and [SEP] is the special symbol to separate non-consecutive token sequences (i.e., two sentences). Secondly, the transformer layers are introduced to refine the token-level features layer by layer until it outputs the contextualized representations of the input tokens as $H^L = \{h_1^l, \ldots, h_n^l\}$. After obtaining the BERT representations H^L, we design a neural layer to recognize the label of the targeted service. We build this layer on the top of BERT embedding layer for solving the task as shown in Fig. 2. We investigate various designs for the DNN classifier layer namely linear layer, CNNLSTM [12], and Gated Recurrent Unit (GRU) [2]. This label is sent to the matcher component.

The matcher is a component that includes a graph search technique to resolve the request query submitted from Requester by using the label produced from

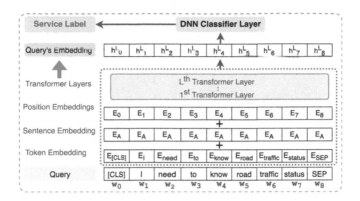

Fig. 2. An overview of the Parser component.

the parser. This component uses friends and friends of friends of the object requester from the social structure established in *Phase 2*. So, this component will be fed with two parameters: the requester object that sent the query and the service label obtained from the parser. According to this, Matcher will start searching the local neighborhood of the object requester. It will return the list of friend objects that offer the requested service. In case a friend object has not been found, friends of friend objects are searched.

4   Experimental Evaluation

Experiments were conducted to assess our framework's effectiveness and efficiency. In the following, we present the SIoT datasets and the query dataset used, show the baselines, and discuss the results.

Datasets. We used SIoT real-world datasets [8]. These datasets contain raw movement information about IoT objects, their services, and their Application of interest. We combined it with another dataset [11] that contains text queries about services similar to the services provided by IoT objects.

Datasets. To demonstrate the effectiveness of our framework, we compared the two parts of our framework with several baselines. For *Phase 2*, we used node2vec [5] and GCN [7]. For *Phase 3*, we used Glove [10].

Results and Discussion. Our framework can predict social interactions between IoT objects effectively. Table 1 demonstrates Accuracy, F1-score, AUCROC and AUCPR results of predicting social relationships. Our framework outperforms the two baseline approaches. In addition, we use AUCROC and AUCPR curves to show each embedding method's performance using different thresholds. Figure 3a and Fig. 3b show the performance curves of the embedding

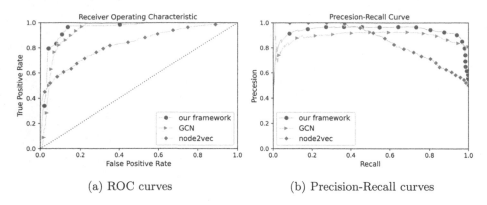

(a) ROC curves   (b) Precision-Recall curves

Fig. 3. The ROC and Precision-Recall curves for social relationships prediction model against baseline.

Table 1. The prediction model performance. Bold values indicate the best results.

Methods	Metrics			
	Accuracy	F1-Score	AUCROC	AUCPR
node2vec	0.736	0.755	0.825	0.840
GCN	0.878	0.885	0.955	0.939
Our framework	**0.903**	**0.902**	**0.966**	**0.968**

Table 2. The Parser performance. Bold values indicate the best results.

	Model	Accuracy	Precision	Recall	F1-Score
Glove	Linear	0.392	0.466	0.393	0.426
	CNNLSTM	0.791	0.783	0.791	0.787
	GRU	0.560	0.553	0.560	0.557
BERT	Linear	0.832	0.840	0.820	0.830
	CNNLSTM	0.854	0.87	0.85	0.85
	GRU	**0.890**	**0.870**	**0.870**	**0.870**

methods for baseline methods and our framework. From these figures, we can observe that our framework is highly stable and gives the best performance on average across all thresholds. Regarding the effectiveness of our Parser Component for Unveiling the Intent behind Request Queries, Table 2 shows the results of our *Parser* component against baseline techniques. We observed that using BERT with any DNN classifier layer outperforms their peers when using Glove. For example, we found that using only BERT with a simple classifier (BERT with Linear layer) achieves high results compared to advanced methods using Glove such as CNNLSTM. Comparing DNN classifier methods that used BERT as embedding layer, we found that the GRU classifier layer achieves the best results.

5   Conclusion

In this paper, we examined the impact of the latest advances in graph and language representation deep learning models on enhancing resources discovery in the Social Internet of Things (SIoT) paradigm. Specifically, we introduced our envisioned framework, which includes three phases: collecting the required data of IoT objects, building a social structure among IoT objects using a two-step approach, and developing an end-to-end model based on BERT to discover IoT resources using a query issued from an IoT object. The experimental results show the superiority and effectiveness of our framework. One of the future directions for this research is to extend the framework to incorporate historical preferences of IoT objects, which will enable the framework to find desired services for IoT objects proactively.

References

1. Atzori, L., Iera, A., Morabito, G., Nitti, M.: The Social Internet of Things (SIoT) - when social networks meet the internet of things: concept, architecture and network characterization. Comput. Netw. **56**(16), 3594–3608 (2012)
2. Cho, K., et al.: Learning phrase representations using RNN encoder-decoder for statistical machine translation. In: Proceedings of the Conference on Empirical Methods in Natural Language Processing (EMNLP), pp. 1724–1734 (2014)
3. Datta, S.K., Da Costa, R.P.F., Bonnet, C.: Resource discovery in internet of things: current trends and future standardization aspects. In: Proceedings of the World Forum on Internet of Things (WF-IoT), pp. 542–547 (2015)
4. Devlin, J., Chang, M., Lee, K., Toutanova, K.: BERT: pre-training of deep bidirectional transformers for language understanding. In: Proceedings of the Conference of the North American Chapter of the Association for Computational Linguistics: Human Language Technologies (NAACL-HLT), pp. 4171–4186 (2019)
5. Grover, A., Leskovec, J.: node2vec: scalable feature learning for networks. In: Proceedings of the 22nd International Conference on Knowledge Discovery and Data Mining (SIGKDD), pp. 855–864 (2016)
6. Hamilton, W., Ying, Z., Leskovec, J.: Inductive representation learning on large graphs. In: Proceedings of the Conference on Neural Information Processing Systems (NIPS), pp. 1024–1034 (2017)
7. Kipf, T.N., Welling, M.: Semi-supervised classification with graph convolutional networks. In: Proceedings of the 5th International Conference on Learning Representations (ICLR) (2017)
8. Marche, C., Atzori, L., Pilloni, V., Nitti, M.: How to exploit the social internet of things: query generation model and device profiles' dataset. Comput. Netw. **174**, 1–13 (2020)
9. Pattar, S., Buyya, R., Venugopal, K.R., Iyengar, S.S., Patnaik, L.M.: Searching for the IoT resources: fundamentals, requirements, comprehensive review, and future directions. IEEE Commun. Surv. Tutor. **20**(3), 2101–2132 (2018)
10. Pennington, J., Socher, R., Manning, C.D.: Glove: global vectors for word representation. In: Proceedings of the Conference on Empirical Methods in Natural Language Processing (EMNLP), pp. 1532–1543 (2014)

11. Rastogi, A., Zang, X., Sunkara, S., Gupta, R., Khaitan, P.: Towards scalable multi-domain conversational agents: the schema-guided dialogue dataset. In: Proceedings of the Conference on Artificial Intelligence (AAAI), pp. 8689–8696 (2020)
12. Wang, J., Yu, L.C., Lai, K.R., Zhang, X.: Dimensional sentiment analysis using a regional CNN-LSTM model. In: Proceedings of the Annual Meeting of the Association for Computational Linguistics (ACL), pp. 225–230 (2016)
13. Yao, L., Sheng, Q.Z., Ngu, A.H., Li, X.: Things of interest recommendation by leveraging heterogeneous relations in the internet of things. ACM Trans. Internet Technol. (TOIT) **16**(2), 1–25 (2016)

Tutorial and Demo

Graph Data Mining in Recommender Systems

Hongxu Chen$^{(\boxtimes)}$ (iD), Yicong Li, and Haoran Yang

University of Technology Sydney, Sydney, NSW, Australia
Hongxu.Chen@uts.edu.au, {Yicong.Li,Haoran.Yang-2}@student.uts.edu.au

Abstract. With the rapid development of e-commerce, massive data is generated from various e-commerce platforms. Most of the generated data can be represented in the forms of graph, which is capable to demonstrate the complicated relations among various entities, for example, graphs describe the interactions history between users and items. It is critical for the platforms to mine graph data to formulate recommendation strategy to gain more profits. For instance, in a user-item interaction graph, we can utilize graph data mining techniques to capture users' behavioral patterns to make personalized recommendation strategies. Graph data mining in recommendation is currently a research topic attracts more and more attentions from industry and academic fields. In this half-day tutorial, we will present some key graph data mining methods and its applications in recommendation. We hope to find out the directions for the future work and that more theoretical models can be applied under real-world scenarios.

Keywords: Graph data mining · Recommender systems · Graph neural networks · Explainable machine learning · Self-supervised learning

1 Target Audience, Prerequisites, and Benefits

- **Prerequisites.** This tutorial is aimed at algorithm designers and practitioners interested in graph data mining and recommendation and academic researchers in these domains. The audiences' acquaintance should cover basic knowledge of graph data, machine learning, and recommender systems.
- **Benefits.** After the tutorial, we expect that the attendees could 1) gain a high-level understanding of the overview of graph mining for recommendation, 2) know some key concepts in the above topics, 3) get familiar to some state-of-the-art graph-based recommendations.

1.1 Significance

Graph-based data mining approaches [5,11,18] have caught much attention since graphs can capture complicated relation among data (nodes and edges). These graph-based methods have shown promise for some real scenarios thanks to the

© Springer Nature Switzerland AG 2021
W. Zhang et al. (Eds.): WISE 2021, LNCS 13081, pp. 491–496, 2021.
https://doi.org/10.1007/978-3-030-91560-5_36

graph data structure, like natural language processing [11], healthcare [18], recommendation [5] and so on. For recommendation, traditional approaches mostly focus on the features of users and items, and the users and items' matrix transformation [7]. However, they ignore the potential graph-like relation among users and items. For example, adjacent nodes in recommendation graph may represent similar meaning, and diverse paths on graph tend to display specific relations among nodes. Therefore, in order to capture the above relations, some recent recommendations model the data to graph structures.

Although there were some related forums discussing graph-based recommendation in recent years, ours are different from them, and we add some state-of-the-art graph-based recommendations. Our tutorial mainly includes three parts, graph representation learning and its application for recommendations, graph-based explainable recommendations, and graph contrastive learning and future application in recommendation. Details are shown in Sect. 2.

1.2 Relevance

As can been seen in the reference list, the tutorial will summary and categorize the graph-based data mining for recommendation approaches in the last ten years. Therefore, this topic is perfectly suitable for The Conference on Information and Knowledge Management (CIKM), which is one of the top forums of research on data science and knowledge management in the world. Specifically, the main topic conforms the area of "Neural Information and Knowledge Processing". Since we will introduce knowledge-aware recommendation, our content also meets the area "Integration and Aggregation".

2   Outline

The tutorial contains three main parts, graph representation learning and its application in recommendation, reasoning for graph-based recommendation, and graph contrastive learning and promising future application in recommendation.

- **Graph Representation Learning in General and Its Application in Recommendation. (60 min)**
    1. Graph Representation Learning in General. (30 min)
    2. When Graph Meets Recommender Systems. (30 min)
- **Reasoning for Graph-based Recommendation. (60 min)**
    1. Reasoning is Important for Graph-based Recommendation. (10 min)
    2. Explainable Recommendations via Graph Modelling. (40 min)
    3. Conclusion and Future Directions. (10 min)
- **Graph Contrastive Learning for Recommendation. (60 min)**
    1. Introduction to Contrastive Learning. (10 min)
    2. Introduction to Graph Contrastive Learning. (20 min)
    3. Applications of Graph Contrastive Learning in Recommendation. (30 min)

3 Important References

3.1 Related Tutorials

Some closely related tutorials were presented in some forums. We will list some of them here and give short introductions for them:

- The 13th ACM International WSDM Conference, in Houston, Texas, February 3–7 2020. **Learning and Reasoning on Graph for Recommendation**[1]. This tutorial mainly focus on graph based recommendations, giving details about traditional graph data mining techniques more than sole GNNs based methods. Our proposed tutorial takes the scenario-specific taxonomy to introduce different graph mining techniques and their applications in recommendation. Moreover, we will introduce more up-to-date methods, including graph contrastive learning [14], to further enrich audiences' knowledge about this domain.
- International Joint Conference on Artificial Intelligence - Pacific Rim International Conference on Artificial Intelligence 2020, in Yokohama, Japan, January 8 2021. **Next-Generation Recommender Systems and Their Advanced Applications**[2]. This tutorial introduce the next-generation recommender systems from three aspects: session-based recommendation, graph based recommendation, interactive and conversation based recommendation, in a scenario-specific manner. However, our proposed tutorial will introduce related knowledge from different aspects: general graph learning methods, reasoning for graph based recommendation, and contrastive learning for recommendation.
- The Web Conference 2021, in Ljubljana, Slovenia, April 17 2021. **Deep Recommender System: Fundamentals and Advances**[3]. This tutorial systematically introduced deep recommendation systems from multiple aspects. One related aspect is the introduction to GNNs based recommendation systems. Comparing to our proposed tutorial, we both emphasize the roles of GNNs in recommendation. However, we pay more attentions to their application scenarios.

3.2 Related Literature

- Survey papers [1,9,19].
- Graph random walk and its applications in recommender systems [3,12,13].
- Graph embedding and its applications in recommender systems [2,4,15,21].
- Graph neural networks and its applications in recommender systems [6,8,16,22].
- Graph contrastive learning [14,17,23].
- Contrastive learning in recommendation [10,20]

---

[1] https://next-nus.github.io/.
[2] https://sites.google.com/view/shoujinwanghome/home/talks/ijcai-pricai-2020-tutorial.
[3] https://deeprs-tutorial.github.io/.

4   Bios of Presenters

- **Hongxu Chen** is a Data Scientist, now working as a Postdoctoral Research Fellow in School of Computer Science at University of Technology Sydney, Australia. He obtained his Ph.D. in Computer Science at The University of Queensland in 2020. His research interests mainly focus on data science in general and expend across multiple practical application scenarios, such as network science, data mining, recommendation systems and social network analytics. In particular, his research is focusing on learning representations for information networks and applying the learned network representations to solve real-world problems in complex networks such as biology, e-commerce and social networks, financial market and recommendations systems with heterogeneous information sources. He has published many peer-reviewed papers in top-tier high-quality international conferences and journals, such as SIGKDD, ICDE, ICDM, AAAI, IJCAI, TKDE. He also serves as program committee member and reviewers in multiple international conference, such as CIKM, ICDM, KDD, SIGIR, AAAI, PAKDD, WISE, and he also acts as invited reviewer for multiple journals in his research fields, including Transactions on Knowledge and Data Engineering (TKDE), WWW Journal, VLDB Journal, IEEE Transactions on Systems, Man and Cybernetics: Systems, Journal of Complexity, ACM Transactions on Data Science, Journal of Computer Science and Technology.
- **Yicong Li** is currently a PhD student of Data Science and Machine Intelligence (DSMI) Lab of Advanced Analytics Institute, University of Technology Sydney. She obtained her Master's degree from National University of Defense Technology in 2019. Her research interests mainly focus on data science, graph neural networks, recommender systems, natural language processing and so on. In particular, her current research is focusing on the explainable machine learning, especially the application in recommendation area. She has published papers in international conferences and journals, such as WSDM, KSEM and IEEE Access. She have also reviewed submitted papers in many top-tier conferences and journals, like AAAI, KDD, WWW, IJCAI, WSDM, ICONIP and so on. In addition, she has been invited to review manuscripts in IEEE Transactions on Neural Networks and Learning Systems (TNNLS), which is a top-tier journal in artificial intelligence.
- **Haoran Yang** is currently a Ph.D. student of Data Science and Machine Intelligence (DSMI) Lab under Advanced Analytics Institute, University of Technology Sydney. He obtained his B.Sc. from Nanjing University in 2020. His research interests include but not limited to data mining, graph neural networks, and recommender systems. Haoran's PhD research mainly focuses on formulating efficient graph neural networks and applying them in real-world problems. He had published a paper in top-tier data mining conference, ICDM. And he was invited to serve as a reviewer in CIKM.

References

1. Cai, H., Zheng, V.W., Chang, K.C.C.: A comprehensive survey of graph embedding: problems, techniques, and applications. IEEE Trans. Knowl. Data Eng. **30**(9), 1616–1637 (2018). https://doi.org/10.1109/TKDE.2018.2807452
2. Chen, C., Tsai, M., Lin, Y., Yang, Y.: Query-based music recommendations via preference embedding. In: Sen, S., Geyer, W., Freyne, J., Castells, P. (eds.) Proceedings of the 10th ACM Conference on Recommender Systems, Boston, MA, USA, 15–19 September 2016, pp. 79–82. ACM (2016). https://doi.org/10.1145/2959100.2959169
3. Cooper, C., Lee, S.H., Radzik, T., Siantos, Y.: Random walks in recommender systems: Exact computation and simulations. In: Proceedings of the 23rd International Conference on World Wide Web, WWW 2014 Companion, pp. 811–816. Association for Computing Machinery, New York (2014). https://doi.org/10.1145/2567948.2579244
4. Grover, A., Leskovec, J.: node2vec: scalable feature learning for networks. In: Krishnapuram, B., Shah, M., Smola, A.J., Aggarwal, C.C., Shen, D., Rastogi, R. (eds.) Proceedings of the 22nd ACM SIGKDD International Conference on Knowledge Discovery and Data Mining, San Francisco, CA, USA, 13–17 August 2016, pp. 855–864. ACM (2016). https://doi.org/10.1145/2939672.2939754
5. Guo, Q., Zhuang, F., Qin, C., Zhu, H., Xie, X., Xiong, H., He, Q.: A survey on knowledge graph-based recommender systems. IEEE Trans. Knowl. Data Eng. (2020)
6. Hamilton, W.L., Ying, Z., Leskovec, J.: Inductive representation learning on large graphs. In: Guyon, I., et al. (eds.) Advances in Neural Information Processing Systems 30: Annual Conference on Neural Information Processing Systems 2017, Long Beach, CA, USA, 4–9 December 2017, pp. 1024–1034 (2017). https://proceedings.neurips.cc/paper/2017/hash/5dd9db5e033da9c6fb5ba83c7a7ebea9-Abstract.html
7. He, X., Chua, T.S.: Neural factorization machines for sparse predictive analytics. In: Proceedings of the 40th International ACM SIGIR conference on Research and Development in Information Retrieval, pp. 355–364 (2017)
8. Kipf, T.N., Welling, M.: Semi-supervised classification with graph convolutional networks. In: 5th International Conference on Learning Representations, ICLR 2017, Toulon, France, Conference Track Proceedings, 24–26 April 2017. OpenReview.net (2017). https://openreview.net/forum?id=SJU4ayYgl
9. Liu, J., Duan, L.: A survey on knowledge graph-based recommender systems. In: 2021 IEEE 5th Advanced Information Technology, Electronic and Automation Control Conference (IAEAC), vol. 5, pp. 2450–2453 (2021). https://doi.org/10.1109/IAEAC50856.2021.9390863
10. Liu, Z., Ma, Y., Ouyang, Y., Xiong, Z.: Contrastive learning for recommender system. CoRR abs/2101.01317 (2021). https://arxiv.org/abs/2101.01317
11. Nastase, V., Mihalcea, R., Radev, D.R.: A survey of graphs in natural language processing. Nat. Lang. Eng. **21**(5), 665–698 (2015)
12. Nikolakopoulos, A.N., Karypis, G.: RecWalk: nearly uncoupled random walks for top-n recommendation. In: Culpepper, J.S., Moffat, A., Bennett, P.N., Lerman, K. (eds.) Proceedings of the 12th ACM International Conference on Web Search and Data Mining, WSDM 2019, Melbourne, VIC, Australia, 11–15 February 2019, pp. 150–158. ACM (2019). https://doi.org/10.1145/3289600.3291016

13. Perozzi, B., Al-Rfou, R., Skiena, S.: DeepWalk: online learning of social represen-
tations. In: Macskassy, S.A., Perlich, C., Leskovec, J., Wang, W., Ghani, R. (eds.)
The 20th ACM SIGKDD International Conference on Knowledge Discovery and
Data Mining, KDD 2014, New York, NY, USA, 24–27 August 2014, pp. 701–710.
ACM (2014). https://doi.org/10.1145/2623330.2623732

14. Qiu, J., et al.: GCC: graph contrastive coding for graph neural network pre-
training. In: Gupta, R., Liu, Y., Tang, J., Prakash, B.A. (eds.) The 26th ACM
SIGKDD Conference on Knowledge Discovery and Data Mining, Virtual Event,
KDD 2020, CA, USA, 23–27 August 2020, pp. 1150–1160. ACM (2020). https://
doi.org/10.1145/3394486.3403168

15. Tang, J., Qu, M., Wang, M., Zhang, M., Yan, J., Mei, Q.: LINE: large-scale information network embedding. In: Gangemi, A., Leonardi, S., Panconesi, A. (eds.)
Proceedings of the 24th International Conference on World Wide Web, WWW
2015, Florence, Italy, 18–22 May 2015, pp. 1067–1077. ACM (2015). https://doi.
org/10.1145/2736277.2741093

16. Velickovic, P., Cucurull, G., Casanova, A., Romero, A., Liò, P., Bengio, Y.:
Graph attention networks. In: 6th International Conference on Learning Representations, ICLR 2018, Conference Track Proceedings, Vancouver, BC, Canada,
30 April–3 May 2018. OpenReview.net (2018). https://openreview.net/forum?
id=rJXMpikCZ

17. Velickovic, P., Fedus, W., Hamilton, W.L., Liò, P., Bengio, Y., Hjelm, R.D.: Deep
graph infomax. CoRR abs/1809.10341 (2018). http://arxiv.org/abs/1809.10341

18. Wang, F., Cui, P., Pei, J., Song, Y., Zang, C.: Recent advances on graph analytics
and its applications in healthcare. In: Proceedings of the 26th ACM SIGKDD
International Conference on Knowledge Discovery & Data Mining, pp. 3545–3546
(2020)

19. Wu, S., Zhang, W., Sun, F., Cui, B.: Graph neural networks in recommender
systems: a survey. CoRR abs/2011.02260 (2020). https://arxiv.org/abs/2011.02260

20. Xie, X., Sun, F., Liu, Z., Gao, J., Ding, B., Cui, B.: Contrastive pre-training for
sequential recommendation. CoRR abs/2010.14395 (2020). https://arxiv.org/abs/
2010.14395

21. Yang, J., Chen, C., Wang, C., Tsai, M.: HOP-rec: high-order proximity for implicit
recommendation. In: Pera, S., Ekstrand, M.D., Amatriain, X., O'Donovan, J. (eds.)
Proceedings of the 12th ACM Conference on Recommender Systems, RecSys 2018,
Vancouver, BC, Canada, 2–7 October 2018, pp. 140–144. ACM (2018). https://
doi.org/10.1145/3240323.3240381

22. Ying, R., He, R., Chen, K., Eksombatchai, P., Hamilton, W.L., Leskovec, J.: Graph
convolutional neural networks for web-scale recommender systems. In: Guo, Y.,
Farooq, F. (eds.) Proceedings of the 24th ACM SIGKDD International Conference
on Knowledge Discovery & Data Mining, KDD 2018, London, UK, 19–23 August
2018, pp. 974–983. ACM (2018). https://doi.org/10.1145/3219819.3219890

23. You, Y., Chen, T., Sui, Y., Chen, T., Wang, Z., Shen, Y.: Graph contrastive learn-
ing with augmentations. In: Larochelle, H., Ranzato, M., Hadsell, R., Balcan, M.,
Lin, H. (eds.) Advances in Neural Information Processing Systems 33: Annual
Conference on Neural Information Processing Systems 2020, NeurIPS 2020, 6–12
December 2020, virtual (2020). https://proceedings.neurips.cc/paper/2020/hash/
3fe230348e9a12c13120749e3f9fa4cd-Abstract.html

Emerging Applications in Healthcare and Their Implications to Academia and Practice

Raj Gururajan[1], Xiaohui Tao[2(✉)], Yuefeng Li[3], Xujuan Zhou[1],
Soman Elangovan[4], Srinivas Kondalsamy Chennakesavan[5],
and Revathi Venkataraman[6]

[1] School of Business, University of Southern Queensland, Springfield, Australia
{raj.gururajan,xujuan.zhou}@usq.edu.au
[2] School of Sciences, University of Southern Queensland, Toowoomba, Australia
xiaohui.tao@usq.edu.au
[3] School of Computer Science, Queensland University of Technology,
Brisbane, Australia
y2.li@qut.edu.au
[4] Belmont Private Hospital, Brisbane, Australia
soman.elangovan@healthcare.com.au
[5] Rural Clinical School, University of Queensland, Brisbane, Australia
s.kondalsamychennakes@uq.edu.au
[6] School of Computing, SRM Institute of Science and Technology-Kattankulathur
Campus, Chennai, Tamil Nadu, India
revathin@srmist.edu.in

Abstract. Data is important for health applications. Clinicians and academics use structured and unstructured data to arrive at both diagnosis and prognosis. Artificial Intelligence (AI) plays a significant role in identifying various clinical diagnostic elements using data. These clinical diagnostic elements are not readily visible to humans. In this tutorial, we bring together academics, clinicians, health administrators from both public and private health and research scholars to discuss their experiences in developing and implementing AI-based cutting-edge healthcare and medical applications.

Keywords: Artificial intelligence · Machine learning/deep learning ·
Medical image analysis · Healthcare · Clinical practice

1 Introduction

Artificial Intelligence (AI) are "agents that receive percepts from the environment and perform actions." (Russell and Norvig 2020). Many artificial intelligence systems are powered by machine learning and deep learning. In the big data era, AI can be used to analyze huge, disparate data sets to identify patterns, trends, correlations, and other useful information that can harness in support

© Springer Nature Switzerland AG 2021
W. Zhang et al. (Eds.): WISE 2021, LNCS 13081, pp. 497–500, 2021.
https://doi.org/10.1007/978-3-030-91560-5_37

of better decision-making. AI plays a significant role to identify various clinical diagnostic elements that are not visible to humans. In domains such as population health, medicine, the nature of data sources, viz., Big Data, is conducive to pattern recognition and machine learning.

However, there appears to be some form of misalignment in clinicians understanding the full power of Artificial Intelligence and the academicians understanding the full gamut of medical applications and their nuances. Further, due to the differences in the respective domains of medicine and computing, the way clinicians understand how AI can work in health, and the way computing professionals see how the algorithms could be implemented in health does not match. In order to develop a shared understanding, these two groups of people should work together to understand the problem domain, associated operational issues, underlying technological infrastructure and various procedural frameworks. These interactions are challenging due to limited knowledge of the other domains, poor understanding of what others are doing and lack of understanding of the other's domain. Health and computing professionals can then develop solutions after having taken these constraints into consideration.

In this tutorial, we brought together academics, clinicians, health administrators from both public and private health, research scholars who work in these domains to discuss their experiences in developing and implementing cutting edge healthcare and medical applications.

2   AI Application for Health and Medicine

Machine learning enables computer systems to learn general concepts of inference from data samples. Deep learning (LeCun et al. 2015) is a machine learning technique based on representation learning where the system automatically learns and discovers the features needed for classification from processing multiple layers of input data. Many deep learning techniques have been explored for classification tasks and applied in health and medical research contexts. For example: the new development of advanced algorithms such as three-way decisions based classification (Li et al. 2017) and granule mining (Sewwandi et al. 2021), a nested ensemble method (Abdar et al. 2020), deep CNN model (Zhou et al. 2020) for breast cancer diagnosis and a multi-deep model (Zia et al. 2020) for lung cancer prediction, deep learning methods for pain intensity detection (Bargshady et al. 2020a) (Bargshady et al. 2020b), Decision tree model (Abdar et al. 2019) and hybrid particle swarm optimization (Zomorodi-moghadam et al. 2021) for diagnosis of coronary artery disease. Sentiment analysis and opinion mining for depression detection (Tao et al. 2016). Some AI applications for telehealth and for Dementia and Alzheimer prediction (Lafta et al. 2020), (Alkenani et al. 2020), (Alkenani et al. 2021). Recently, deep learning methods have been applied to predict sepsis (Kok et al. 2021) and Covid-19 (Barua et al. 2021).

3   Remarks

In this tutorial, the panel consists of international academics and health practitioners drawn from Australia, India, Singapore among other countries. The panel discussed the development of AI algorithms, providing technical solutions to health consumers in domains such as mental health, difficulties in providing cutting edge AI applications to rural areas, poor understanding from health customers as to what AI can offer, cost issues in developing AI solutions, infrastructure and resource requirements, developing proof of concepts into working solutions, regulatory environments and security issues.

Acknowledgement. We gratefully acknowledge the contributions from Dr Rashmi Gururajan (Royal Brisbane and Women's Hospital, Brisbane), Dr Kym Boon (Royal Brisbane and Women's Hospital, Brisbane), Professor Rajendra Udyavara Acharya (Ngee Ann, Singapore University of Social Science, Singapore), Dr Vaishnavi Moorthy (School of Computing, SRMIST-Kattankulathur Campus, Tamil Nadu-India), Dr. Dharini Krishnan (D. V. Living Sciences Enterprise Pvt. Ltd, India) and Dr. Aparna Kasinath (Syngene International Ltd, India). Without their kind support, this tutorial would not be possible.

References

Russell, S., Norvig, P.: Artificial Intelligence: A Modern Approach. Prentice Hall, Upper Saddle River (2020)

Deep learning. nature, 521(7553), pp. 436–444

Abdar, M., Nasarian, E., Zhou, X., Bargshady, G., Wijayaningrum, V.N., Hussain, S.: 2019 Performance improvement of decision trees for diagnosis of coronary artery disease using multi filtering approach. In: Proceeding of 2019 IEEE 4th International Conference on Computer and Communication Systems (ICCCS), pp. 26–30. IEEE

Abdar, M., Zomorodi-Moghadam, M., Zhou, X., Gururajan, R., Tao, X., Barua, P.D., Gururajan, R.: A new nested ensemble technique for automated diagnosis of breast cancer. Pattern Recogn. Lett. **132**, 123–131 (2020)

Zomorodi-moghadam, M., Abdar, M., Davarzani, Z., Zhou, X., Pławiak, P., Acharya, U.R.: Hybrid particle swarm optimization for rule discovery in the diagnosis of coronary artery disease. Expert Syst. **38**(1), e12485 (2021)

Lafta, R., Zhang, J., Tao, X., Zhu, X., Li, H., Chang, L., Deo, R.: A general extensible learning approach for multi-disease recommendations in a telehealth environment. Pattern Recogn. Lett. **132**, 106–114 (2020)

Alkenani, A.H., Li, Y., Xu, Y., Zhang, Q.: Predicting prodromal dementia using linguistic patterns and deficits. IEEE Access **8**, 193856–193873 (2020)

Alkenani, A.H., Li, Y., Xu, Y., Zhang, Q.: Predicting prodromal dementia using linguistic patterns and deficits. *IEEE Access*, **8**, 193856–193873 (2020)

Li, Y., Zhang, L., Xu, Y., Yao, Y., Lau, R.Y.K., Wu, Y.: Enhancing binary classification by modeling uncertain boundary in three-way decisions. IEEE Trans. Knowl. Data Eng. **29**(7), 1438–1451 (2017)

Sewwandi, M.A.N.D., Li, Y., Zhang, J.: Automated granule discovery in continuous data for feature selection. Inf. Sci. **578**, 323–343 (2021)

Zhou, X., Li, Y., Gururajan, R., Bargshady, G., Tao, X., Venkataraman, R., Barua, P.D. Kondalsamy-Chennakesavan, S.: A new deep convolutional neural network model for automated breast cancer detection. In proceeding of 2020 7th International Conference on Behavioural and Social Computing (BESC), pp. 1–4. IEEE (2020)

Zia, M.B., Zhou, J.J., Zhou, X., Xiao, N., Wang, J., Khan, A.: Classification of malignant and benign lung nodule and prediction of image label class using multi-deep model. Int. J. Adv. Comput. Sci. Appl. **11**(3), 35–41 (2020)

Bargshady, G., Zhou, X., Deo, R. C., Soar, J., Whittaker, F., Wang, H.: Enhanced deep learning algorithm development to detect pain intensity from facial expression images. Expert Syst. Appl. **149**, 113305 (2020)

Bargshady, G., Zhou, X., Deo, R. C., Soar, J., Whittaker, F., Wang, H.: The modeling of human facial pain intensity based on Temporal Convolutional Networks trained with video frames in HSV color space. Appl. Soft Comput. **97**, 106805 (2020)

Barua, P.D., et al.: Automatic COVID-19 detection using exemplar hybrid deep features with X-ray images. Int. J. Environ. Res. Public Health **18**(15), 8052 (2021)

Kok, C., et al.: Automated prediction of sepsis using temporal convolutional network. Comput. Biol. Med. **127**, 103957 (2020)

Tao, X., Zhou, X., Zhang, J., Yong, J.: Sentiment analysis for depression detection on social networks. In: Li, J., Li, X., Wang, S., Li, J., Sheng, Q.Z. (eds.) ADMA 2016. LNCS (LNAI), vol. 10086, pp. 807–810. Springer, Cham (2016). https://doi.org/10. 1007/978-3-319-49586-6_59

GPUGraphX: A GPU-Aided Distributed Graph Processing System

Qi Li, Kai Zou, Deyu Kong, Huhao Guan, and Xike Xie[✉]

University of Science and Technology of China, Hefei, China
{likamo,slnt,cavegf,hhguan}@mail.ustc.edu.cn, xkxie@ustc.edu.cn

Abstract. There are two major challenges for large-scale graph analytic processing, computational intensiveness caused by complex graph primitives and distributed data management caused by data of massive scales. Existing works on graph data management with CPU-based distributed systems or GPU-based single-node systems only partially solve the problem. Hence, it is desired to have a general graph processing system for both scaling out and scaling up. In this paper, we demonstrate GPUGraphX, a GPU-aided distributed graph processing system which utilizes computation capacities of GPUs for efficiency while taking the advantages of distributed systems for scalability. Results on representative graph algorithms on real datasets evaluate our proposals.

1 Introduction

Graphs have been widely used for representing complex relationship between entities, like social networks [12,16], knowledge graphs [20], road networks [21, 22], and Web analysis [10]. Graph analytics in big data era are often with diversified applications, large data volumes, and intensive computation overheads, posing challenges to devising general, scalable, and efficient graph data infrastructures.

There are two general ways of forging such infrastructures, scaling up and scaling out. There have been works for scaling up with multi-core CPU solutions, such as BGL [6] and Ligra [17], and with single-node GPU systems, such as Medusa [23] and Gunrock [18]. There have also been systems for scaling out to numbers of distributed nodes, such as Pregel [13], PowerGraph [5], GraphX [4]. Most scaling out solutions pay little attention to the computation intensiveness of large-scale graph processing. To achieve both scaling-up and scaling-out for large-scale graph operations, a natural idea is to combine HPC (high performance computing) hardwares, e.g., GPUs, with mature solutions of distributed graph systems to integrate the merits from both.

In this work, we demonstrate GPUGraphX[1], a GPU-aided distributed graph processing system based on GraphX. GraphX [4] provides a well-accepted API set including Pregel interfaces, and makes an efficient wrapper between graph-like operations and set operations in Spark, and retains the properties, such

---

[1] Code Repository: https://github.com/Kamosphere/spark-GPUGraphX.

© Springer Nature Switzerland AG 2021
W. Zhang et al. (Eds.): WISE 2021, LNCS 13081, pp. 501–509, 2021.
https://doi.org/10.1007/978-3-030-91560-5_38

as fault-tolerance, automatic failure recovery, and task re-execution of Spark. Based on GraphX, the system enables a seamless integration of different runtime environments with shared memory and data packaging techniques. GPU computation kernels are designed to be adaptive to existing GraphX execution flows to minimize the inter-framework costs, and also to make intermediate processing transparent to system users. More, the system utilizes a multi-layered structure to distribute workloads to heterogeneous hardware-based distributed systems, establishing platform portability of different runtime environments. It can be demonstrated that GPUGraphX achieves good efficiency and scalability.

To our best knowledge, Lux [8] is the only other work on multi-node GPU graph processing framework. Lux is based on Legion [2], which does not manage persistent resources like file systems [19], and does not automatically offer fault tolerance, letting alone the portability issues [3]. Our system is based on GraphX, which is mature for being deployed in practical and commercial environments, with dynamic and reliable resource management and load-balancing mechanism.

The rest of the paper is presented as follows. Section 2 makes a bird view for the system architecture. Section 3 investigates key system components. Section 4 introduces a set of system interfaces to implement graph algorithms in GPU-GraphX. Section 5 shows the details of the demonstration and reports experimental results for evaluating the system. Section 6 concludes the paper.

2   System Design

2.1   Overview

In general, the architecture of GPUGraphX consists of three layers, namely GraphX layer, bridge layer, and computation layer, as shown in Fig. 1.

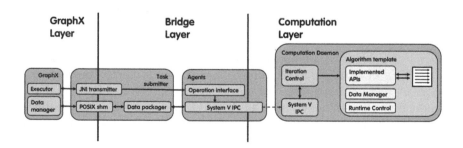

Fig. 1. System architecture

- The GraphX layer is the upper layer of the entire system. Within this layer, a big graph is partitioned into subgraphs which are stored in distributed nodes. During the graph computation process, the partitioned subgraphs are passed to GPU kernels via the bridge layer.

- The computation layer handles heavy-weighted computational tasks aided by GPU kernels. The computation layer makes GPU kernels hot-plug for better computational efficiency, by separating CUDA environment initialization from frequent JNI calls from the GraphX layer.
- The bridge layer connects the two environments of JVM (GraphX layer) and GPU CUDA C++ (computation layer), with the help of the daemon-agent framework (Sect. 2.2).

2.2 Daemon-Agent Framework

Daemon. Computation daemon is the part where graph algorithms execute. In our implementation, a daemon is combined with a computation resource, e.g., GPU. For a distributed node with multiple GPUs, we can deploy multiple daemons, in accordance to the number of GPUs, which enables transparency of hardware details to upper systems. Daemon hold the algorithm template and iteration logic controlling, as shown in Fig. 1. On running, the daemon creates an implemented instance of the algorithm template. For flexibly utilizing computation resources and workload balancing, system developers can simply control the amount of daemons created in a distributed node, while algorithm engineers only need to work on the implementation of programming interfaces.

Agent. Agents are designed as a bridge for upper distributed systems to communicate with daemons by exchanging data, invoking implemented algorithm interfaces, and controlling lifecycles of daemons. For a distributed node, agent is necessary to connect the upper system and daemons, while there can be multiple daemons to represent different computation resources deployed in the distributed node.

The interactions between daemons and agents are shown in Fig. 1.

3 Key Techniques

Synchronization Skipping. GraphX follows the BSP model, whereas graph algorithms are expressed as a set of consecutive iterations. The synchronization of each iteration is done by gathering and scattering intermediate messages for ensuring the data consistency. We implement a synchronization skipping mechanism to optimize the system runtime performance. At the end of computation of an iteration, agents check for *active vertices*, which are updated in this iteration and are resided in the same partition with all its outer edges. The property of active vertices allows for saving the costly synchronization overheads between partitions. This way, multiple computation iterations can be merged as a single iteration, so that the overall graph processing can be accelerated.

Shared-Memory Based Data Transferring. To accelerate the data transferring between different layers, we develop shared-memory based data transferring mechanism. Conventionally gathering graph data through JNI requires multiple rounds of data format conversion, which is costly on both time and space. To tackle this, we build a data transferring module based on Memory Mapped File, so that POSIX-based shared-memory can be accessed in an easy way, avoiding the conversion overhead in loading to JNI environment.

The data transferred from agents can not be directly accessed by daemons and their associated GPU kernels, since they belong to different system processes with no common memory space. To tackle this, we resort to kernel functions aided by UNIX SystemV. Data from upper systems or GPU devices is kept in the SystemV-based shared memory by holding a SystemV memory space. Data updates between agent and daemons can be immediately perceived, without extra transfers, leading to the minimum data transferring between the two ends.

GPU Initialization Detachment. During iterations, the invoking and releasing of agent by upper systems happen multiple times, so that the GPU environment is also initialized multiple times. Due to the long initialization process of GPU environments, it is not wise to use the mode of traditional function calls for the agent to control daemon execution. To overcome the dilemma, we detach the initialization from direct function calls by separating daemons from other parts of the system. The communication between agents and daemons is customized by a concise set of interfaces.

4   Programming Interfaces

The interfaces are of two categories, global in GraphX layer and local in computation layer. For ease of presentation, we showcase Bellman-Ford algorithm, as specified shown in Listings 1.1 and 1.2.

```
class pregel_SSSP {
  type SPMap = Map[VertexId, Double]
  def lambda_initMessage: SPMap
  def lambda_globalVertexMerge
  (v1: VertexId, v2: SPMap, v3: SPMap): SPMap
  def lambda_globalReduce
  (v1: SPMap, v2: SPMap): SPMap
}
```

Listing 1.1. SSSP-BF in GraphX Layer

For global interfaces, we first define SPMap, which is a map with the vertex ID of landmark as key and the distance from defined landmarks to the vertex value. Then, lambda_initMessage function is implemented to initialize the messages in Pregel framework in the form of SPMap. To define how to merges two SPMaps and updates the destination vertex's attribute with the smaller distance, lambda_globalVertexMerge function is required. lambda_globalReduce function in SSSP-BF merges the minimum distance with the same destination vertex ID among different parts of messages in a single iteration.

```
1  int BellmanFordGPU::MSGGenMerge_array(const VertexSet,
       const EdgeSet, const VertexValueSet, MessageValueSet
       )
2  cudaError_t MSGGenMerge_kernel_exec(VertexSet, EdgeSet,
       VertexValueSet, MessageValueSet)
3  int BellmanFordGPU::MSGApply_array(VertexSet,
       VertexValueSet, MessageValueSet)
4  cudaError_t MSGApply_kernel_exec(VertexSet,
       VertexValueSet, MessageValueSet);
```

Listing 1.2. SSSP-BF in Computing Layer

`MSGGenMerge_array` and `MSGApply_array` consist the local interfaces in computation layer. In particular, `MSGGenMerge_array` is required to copy data into GPU memory. `MSGGenMerge_kernel_exec` is implemented to handle the edge and generate locally merged messages in CUDA environment. `MSGApply_array` and `MSGApply_kernel_exec` need to be implemented to apply local messages on local vertex data, which are used in synchronization skipping.

5 Demonstration Details

5.1 Setting

We use five real datasets, Wiki-topcats [9], LiveJournal [9], Orkut [14], WRN [15], and Twitter [9], containing 1.79M, 4.84M, 3.07M, 23.9M, 41.65M vertices, and 28.51M, 68.99M, 117.18M, 28.9M, 1.468B edges, respectively.

Our demonstration is running on a cluster, which has 6 physical computing nodes, each of which has a CPU Xeon Scale 6248 (2.50 GHz, 20 cores), 2 NVIDIA V100 GPUs (32 GB GPU memory on each GPU), 384 GB physical memory. We build 12 virtual nodes for simulating the distributed environment using NVIDIA Docker, which isolates and configures GPU resources for every virtual node. We conduct experiments on a set of graph algorithms that are commonly accepted for the evaluation of graph systems [1,7,11], such as Bellman-Ford (SSSP-BF), Connected Components (CC), PageRank (PR), and Label Propagation (LP), with varied data scales.

5.2 Runtime Analyzer Demonstration

In our demonstration, we will show the working process of the system in running algorithm. In order to show more details, we also demonstrate the runtime analyzer of GPUGraphX in Fig. 2, which supports interactive analysis on logs for exploring insights of system. Figure 2(a) shows the breakdown of the total system runtime cost, which mainly consists of four parts, *computation*, *GraphX synchronization*, *GraphX maintenance*, and *result aggregation*[2]. From the pie

---

[2] The four parts correspond to the four stages of a GraphX iteration during the execution, aggregateMessageswithActiveSet@GraphImpl.scala, shipVertexAttributes@VertexRDDImpl.scala, shipVertexIds@VertexRDDImpl.scala, and count@VertexRDDImpl.scala.

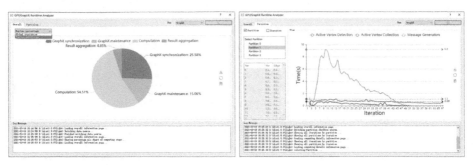

(a) System Runtime Analysis.   (b) Computation Time Analysis.

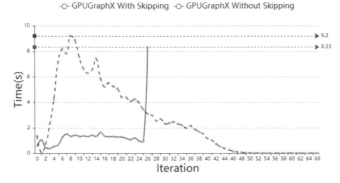

(c) GPUGraphX vs. GraphX (message generation).

Fig. 2. Runtime analyzer (an example on SSSP-BF, Wiki-topcats)

chart, it can be observed that the computation time is the major part of the system cost. Therefore, it necessitates the GPU acceleration for improving the overall system efficiency. Analysts can also switch to a finer-granular view to observe the runtime cost of every iteration, with the top-left drop-down menu.

The task is executed in distributed environments, i.e., a set of partitions. In Fig. 2(b), we show the GraphX computation time breakdown on a selected partition (e.g., Partition 1), where *Message Generation* is the dominant part. GPUGraphX improves the computation efficiency in two ways: 1) using GPU acceleration to reduce the cost of message generation; 2) using synchronization skipping to scale down the number of iterations. In Fig. 2(c), the message generation time of GPUGraphX is superimposed to that of GraphX. It can be observed that, the peak time is reduced from 9 to 8, and the number of iterations is reduced from 66 to 25. The last iteration of GPUGraphX incurs a sudden rise, because skipping accumulates the cost of condition checking to the last iteration, which is worthwhile considering the overall system overhead deduction.

5.3   Results

We show the improvement of GPUGraphX over GraphX not only in our demonstration, but also in Fig. 3. To show the acceleration of GPUs on the computation optimization, we use GraphX(comp) and GPUGraphX(comp) to indicate their corresponding time costs solely spent on the computation.

We test the scale-up performance in Fig. 3(a). It shows that GPUGraphX achieves up to 5x acceleration for the total running time, and up to 15x acceleration for the computation time, in all testing algorithms. We then test the scale-out performance in Fig. 3(b). The time costs of both GraphX and GPU-GraphX increase w.r.t. the data size, but the trend of GPUGraphX is more moderate. In particular, on dataset Orkut whose number of edges equals 117M, GPUGraphX achieves 3x acceleration on the computation time.

To understand the improvement, we make the breakdown of computation time, which consists of three parts, active vertex detection, active vertex collection, and message generation, in Fig. 3(c). It shows that GPUGraphX saves about 90% of the total computation time. Notice that it takes a bit additional overhead for GPUGraphX to execute active vertex collection. But it is worthwhile, considering the total improved efficiency. Figure 3(d) shows the results of with and without synchronization skipping mechanism on real datasets. We can find that the mechanism achieves 25%–50% decrease of the total running time.

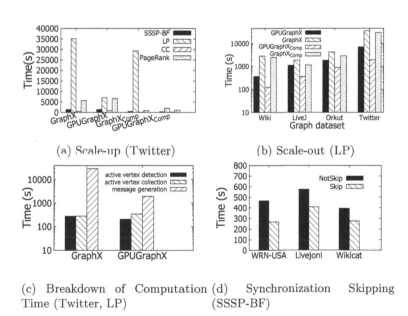

(a) Scale-up (Twitter)

(b) Scale-out (LP)

(c) Breakdown of Computation Time (Twitter, LP)

(d) Synchronization Skipping (SSSP-BF)

Fig. 3. Experiment results

6   Conclusion

In this demo, we showcase GPUGraphX, a GPU-aided distributed graph processing system built on top of GraphX. It achieves high-performance on large-scale graph data. Meanwhile, our system retains the merits of mature distributed systems for graph processing, in terms of task scheduling and fault recovery, etc. Experiment results show that our system achieves significant speed-ups in computation compared to the results on Spark.

Acknowledgments. This work is partially supported by NSFC (No. 61772492, 62072428) and the CAS Pioneer Hundred Talents Program. The numerical calculations in this paper have been done on the supercomputing system in the Supercomputing Center of University of Science and Technology of China.

References

1. Batarfi, O., El Shawi, R., et al.: Large scale graph processing systems: survey and an experimental evaluation. Clust. Comput. **18**(3), 1189–1213 (2015)
2. Bauer, M., Treichler, S., et al.: Legion: expressing locality and independence with logical regions. In: SC, p. 66 (2012)
3. Gill, G., Dathathri, R., et al.: Abelian: a compiler for graph analytics on distributed, heterogeneous platforms. In: Euro-Par, pp. 249–264 (2018)
4. Gonzalez, J.E., Xin, R.S., et al.: Graphx: graph processing in a distributed dataflow framework. In: OSDI, pp. 599–613 (2014)
5. Gonzalez, J.E., Low, Y., et al.: Powergraph: distributed graph-parallel computation on natural graphs. In: OSDI, pp. 17–30 (2012)
6. Gregor, D., Lumsdaine, A.: The parallel BGL: a generic library for distributed graph computations. In: POOSC (2005)
7. Han, M., Daudjee, K., et al.: An experimental comparison of pregel-like graph processing systems. PVLDB **7**(12), 1047–1058 (2014)
8. Jia, Z., Kwon, Y., et al.: A distributed multi-GPU system for fast graph processing. VLDB **11**(3), 297–310 (2017)
9. Leskovec, J., Krevl, A.: SNAP datasets: Stanford large network dataset collection (2014). http://snap.stanford.edu/data
10. Liu, Y., Xie, X.: Xy-sketch: on sketching data streams at web scale. In: WWW, pp. 1169–1180 (2021)
11. Lu, Y., Cheng, J., et al.: Large-scale distributed graph computing systems: an experimental evaluation. PVLDB **8**(3), 281–292 (2014)
12. Luo, J., Cao, X., Xie, X., Qu, Q., Xu, Z., Jensen, C.S.: Efficient attribute-constrained co-located community search. In: ICDE, pp. 1201–1212 (2020)
13. Malewicz, G., Austern, M.H., et al.: Pregel: a system for large-scale graph processing. In: SIGMOD, pp. 135–146 (2010)
14. Mislove, A., Marcon, M., et al.: Measurement and analysis of online social networks. In: IMC (2007)
15. Rossi, R.A., Ahmed, N.K.: The network data repository with interactive graph analytics and visualization. In: AAAI (2015). http://networkrepository.com
16. Saleem, M.A., Kumar, R., Calders, T., Xie, X., Pedersen, T.B.: Location influence in location-based social networks. In: WSDM, pp. 621–630 (2017)

17. Shun, J., Blelloch, G.E.: Ligra: a lightweight graph processing framework for shared memory. In: PPoPP, pp. 135–146 (2013)
18. Wang, Y., Davidson, A.A., et al.: Gunrock: a high-performance graph processing library on the GPU. In: PPoPP, pp. 11:1–11:12 (2016)
19. Watkins, N.: Programmable storage. Ph.D. thesis, University of California, Santa Cruz, USA (2018)
20. Wu, F., Xie, X., Shi, J.: Top-k closest pair queries over spatial knowledge graph. In: DASFAA, pp. 625–640 (2021)
21. Xie, X., Mei, B., Chen, J., Du, X., Jensen, C.S.: Elite: an elastic infrastructure for big spatiotemporal trajectories. VLDB J. **25**(4), 473–493 (2016)
22. Xie, X., Yiu, M.L., Cheng, R., Lu, H.: Scalable evaluation of trajectory queries over imprecise location data. TKDE **26**(8), 2029–2044 (2014)
23. Zhong, J., He, B.: Medusa: a parallel graph processing system on graphics processors. SIGMOD Rec. **43**(2), 35–40 (2014)

SQL2Cypher: Automated Data and Query Migration from RDBMS to GDBMS

Shunyang Li[✉], Zhengyi Yang[✉], Xianhang Zhang[✉], Wenjie Zhang[✉], and Xuemin Lin[✉]

UNSW, Sydney, Australia
{sli,zyang,xianhangz,zhangw,lxue}@cse.unsw.edu.au

Abstract. There are many real-world application domains where data can be naturally modelled as a graph, such as social networks and computer networks. Relational Database Management Systems (RDBMS) find it hard to capture the relationships and inherent graph structure of data and are inappropriate for storing highly connected data; thus, graph databases have emerged to address the challenges of high data connectivity. As the performance of querying highly connected data in relational query statements is usually worse than that in the graph database. Transforming data from a relational database to a graph database is imperative for improving the performance of graph queries. In this paper, we demonstrate *SQL2Cypher*, a system for migrating data from a relational database to a graph database automatically. This system also supports translating SQL queries into Cypher queries. *SQL2Cypher* is open-source (https://github.com/UNSW-database/SQL2Cypher) to allow researchers and programmers to migrate data efficiently. Our demonstration video can be found here: https://www.youtube.com/watch?v=eGaeBrVTJws.

Keywords: SQL · Cypher · RDBMS · GDBMS · Data migration

1 Introduction

Graphs have played an increasingly important role in data management with the prevalence of graph data in different application domains such as social networks, road networks and protein-protein interaction networks. Graph database management systems (GDBMS), are among the most fundamental infrastructure when managing graph data and have received a lot of attention from researchers and programmers globally [3]. GDBMS have the unique advantages of modelling and querying complex relationships, capturing and navigating complex data relationships and recursive path querying when handling graph data.

However, relational database management system (RDBMS) still comprise the majority share of the database market for legacy reasons, even when storing highly connected data [11]. Querying highly connected data in a RDBMS usually requires complex join operations and significant system overhead which can lead to a long execution times [12]. Hence, there naturally emerges the

© Springer Nature Switzerland AG 2021
W. Zhang et al. (Eds.): WISE 2021, LNCS 13081, pp. 510–517, 2021.
https://doi.org/10.1007/978-3-030-91560-5_39

demand for migrating from RBDMS to GDBMS. In this paper, we demonstrate *SQL2Cypher*, an automated tool for migrating data from RBDMS to GDBMS.

Migrating data from RDBMS to GDBMS involves redefining data schema, mapping relations and rewriting queries. The migration process is often time-consuming and labour-intensive. The high time and labour costs are one of the major reasons why companies choose to keep their legacy RDBMS. To address this problem, several automated tools [1,4,9] have been proposed to migrate data from RDBMS to GDBMS, however, we find that they are either outdated or incomplete. For example, [4] focuses on XQuery and [1] focuses on RDF data, while ignoring the nowadays more widely adopted property graph model [3]. The open-source tool Neo4j-ETL [9] allows the user to import data from relational databases to the popular graph database Neo4j. However, it does not provide automatic query translation, and users have to rewrite all previous SQL queries to Cypher queries manually [2] (the graph querying language used by Neo4j). More critically, Neo4j-ETL is not well maintained and has many issues at present[1] (e.g. error loading large dataset and error when mapping relations).

Motivated by the above reasons, in this paper, we develop and demonstrate *SQL2Cypher*. *SQL2Cypher* focuses on graph databases adopting the property graph model, in which each vertex/edge in the graph can have an arbitrary number of key-value pairs to represent its properties. *SQL2Cypher* allows users to migrate from relational databases to property graph databases automatically. It automatically derives the relationships among migrated tables, maps the relational model to the property graph model, imports the data from RDBMS to GDBMS, and finally translate the corresponding SQL queries to Cypher queries. It supports Open Database Connectivity (ODBC) [10] compliant relational databases (e.g., MySQL, PostgreSQL and Microsoft SQL Server) and Cypher-based graph databases (e.g., Neo4j, SAP HANA Graph [6] and PatMat [7]). In addition, several optimization strategies are implemented based on predictive interaction framework [8] and duplication detection [5] to further improve the speed and the quality of the migration.

Outline. The remainder of the paper is organised as follows. In Sect. 2, we introduce the overall system of *SQL2Cypher*, and we will present a brief performance evaluation. In Sect. 3, we demonstrate its basic workflow on a real-world scenario.

2   Overall System

SQL2Cypher consists of three-layer as shown in Fig. 1. The first is the user services layer, which is comprised of user interface (both graphical and command-line) where user can operate the system. Second, the application layer will receive commands from the user services and complete background processing. Finally, in the configuration layer, we efficiently store and manage different configurations of the system on the disk. We present the details of these three layers as follows.

---

[1] https://github.com/neo4j-contrib/neo4j-etl/issues.

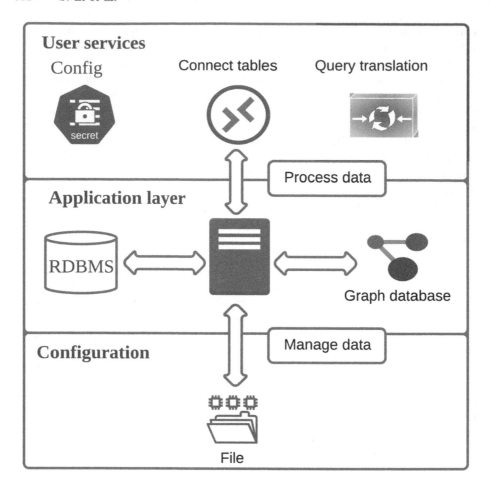

Fig. 1. SQL2Cypher architecture

2.1 User Services

Regarding the user services, as shown in the top of Fig. 1, we build several user-friendly graphical interfaces. User services passes the information entered by the user to the application layer and receives data from the application layer that needs to be displayed. User services are comprised of three sections which are configuration, table connections and query translations. In the configuration section, users can configure the essential information of the RDBMS and GDBMS (i.e., username, password and database URL). In the table connection section, the relationship between the tables is presented to the user in a graph. Users are able to modify the relationship between tables if any are incorrect or missing. Lastly, after the application layer processed the relationships among tables, users can utilize the system to translate SQL queries to Cypher queries. We use the $d3$[2]

---

[2] https://d3js.org/.

and *layui*[3] libraries to provide graph visualization and *CodeMirror*[4] to provide code highlighting. To improve usability, a command line interface (CLI) is also provided with has the same workflow.

2.2 Application Layer

The application layer connects the RDBMS and GDBMS. All background tasks are processed by the server in this layer. In order to adapt our system flexible and allow it to adapt to different database systems (e.g. MySQL, PostgreSQL and Microsoft SQL Server), the application layer connects RDBMS via Open Database Connectivity (ODBC) [10]. ODBC achieves database independence by using the ODBC driver as the translation layer between the application and RDBMS. Applications written using ODBC can be ported to other platforms on the client and server side without changing the data access code. In the following section we will explain in detail how the data migration is done in our system.

Algorithm 1: Generate appropriate edges

Input: *tuples* : table relationship tuples
Output: E
1 $PE \leftarrow \emptyset$
2 $GE \leftarrow \emptyset$
3 **while** *tuples* $\neq \emptyset$ **do**
4   **if** *isJoinTable(src, dst)* **then**
5          $e \leftarrow$ Make a property edge with *src, dst* and *label*
6          $PE.add(e)$
7   **else**
8   **if** *isConnect(src, dst)* **then**
9              $e \leftarrow$ Make a edge without property with *src* and *dst*
10             $GE.add(e)$
11   *tuples* $\leftarrow next(tuples)$
12 $E \leftarrow GE \cup PE$

Parsing Table Relationship. Table relationships can be seen as graph structures where tables are vertexes and edges are relationships. Our system can extract the relationships between tables based on the RDBMS schema; these relationships will be displayed as a graphical user interface or as a relationship path in the CLI. After data migration, the relationships between tables are stored in the configuration layer. For the naive approach, we store all the tables in the RDBMS as nodes in our GDBMS, then create edges for all the connected tables (connected by foreign keys). However, this approach can lead to high memory

---

[3] https://www.layui.com/.
[4] https://codemirror.net/.

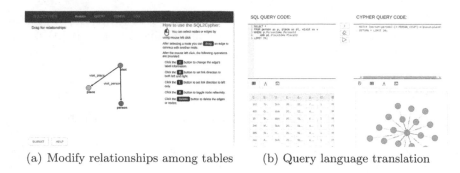

(a) Modify relationships among tables   (b) Query language translation

Fig. 2. Components of user interface.

usage and we will analyze memory usage and focus on optimization strategy in the following section (optimization analysis).

Algorithm 1 describes the transformation of the RDBMS to the GDBMS. The input value *tuples* is a set of tuples consisting of source, destination and label. Edges and vertexes are accessed via several functions: *IsJoinTable* and *IsConnect*. *IsJoinTable* is to check whether there exists a *join table* (a table that contains two foreign keys) between connected tables. *IsConnect* is to check whether the source table and destination table are connected. The processing of generating edges can encounter two cases.

Case 1. In this case, we implement an optimization strategy based on [8]. The strategy is to convert the *join table* to an edge, and the attributes in the *join table* as properties of the edge. This strategy will not only save memory usage but will also make the data more adaptable for the nature of the graph database. In the Algorithm 1 lines 4–6 can convert the *join table* to an edge with properties. We will explain this process with an example.

Optimization Analysis. Our optimization strategy will reduce not only the amount of vertex storage but also edge storage. We will show this result with a theoretical analysis. In the IMDB database, the *principal* is connected to the *name* and *title* tables respectively. Suppose *name* has n rows of data, *title* has m rows of data, and *principal* has e rows of data. For the naive approach, we need to connect *principal* with *name* and *principal* with *title*, so we need to store $n + m + e$ nodes and $2 * e$ edges. For our optimization strategies, we can connect *principal*, *name* and *title* together without saving the *principal* nodes. We only need to store $n + m$ nodes and e edges. Thus, we will save $2 * e$ of storage compared to the naive approach. In the real world, the edges of the graph are usually very numerous so our system can help users save more storage space after data migration.

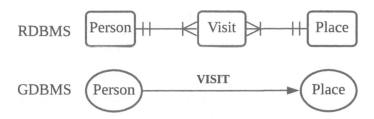

Fig. 3. Example of extracting table as edge

Example 1. In the top area of Fig. 3, there are three tables named *Person*, *Visit* and *Place*. The *Visit* table connects the *Person* table and the *Place* table and the *Visit* table also has two attributes which are *startTime* and *endTime* (the start time and end time of visiting a place) besides two foreign keys. In our system, the *Visit* table will be seen as a *join table*, so it will be converted to an edge and *startTime* and *endTime* attributes will be converted as the properties of that edge. In Cypher query pseudocode, this is like: (Person)-[:VISIT{startTime: value, endTime: value}]→(Place)

Case 2. In Algorithm 1 lines 8–10, if there is no *join table*, the system will detect the connection of two tables (at line 8). If *src* and *dst* are connected, the system will build an edge without properties (line 9).

Translating Query Language. We also provide conjunctive queries (e.g. selections, join, projection and insertion), which are expressed in SQL that translate to Cypher queries based on relationships path traversal operations. In our mechanism, we parse SQL into a list of tokens by building an abstract syntax tree (AST), and then traverse different operations to translate separately based on the relationships that are stored in the configuration layer.

2.3 Configuration

In this layer, all the configuration, databases information and the relationship between the tables is stored in a *pickle*[5] file. The relationships are then used to translate the query language. The system keeps the relationships and configuration updated based on the user input.

3 Demonstration

We will use two real-life example scenarios to demonstrate the overall experience of *SQL2Cypher*. The first scenario is COVID-19 spread which contains information about the COVID-19 test day of a person, the places visited by an infected

---

[5] https://docs.python.org/3/library/pickle.html.

person and the corresponding time. The purpose of the data migration is to effectively find people who have come into contact with an infected people to avoid COVID-19 spreading. The second scenario is IMDB, which contains basic movie information, title, crew, rating and person name, etc. The second scenario aims to demonstrate the optimization strategy of the system.

SQL2Cypher processes the schema in the RDBMS to form a graph structure, then the graph will be presented in the user interface as shown in Fig. 2a. Figure 2b presents queries translation and the execution result. In our demo, we will conduct the following sections to explain how to use our system to migrate data and translate queries.

3.1   Data Migration

Figure 2a displays the original relationships between person, visit and place tables from the first scenario. Users can modify (delete, add, and change) the connections between tables. After submission, our system will process the *join table* automatically.

Our second scenario demonstrates relationship modification during the data migration processing, as some tables need to be connected by reference key form. For example, the IMDB dataset contain a *principal* table, and the principal table connects with *title* and *name* tables by using *reference key*. Unfortunately, RDBMS can not detect *reference key*, therefore, users need to add edges among *principal*, *title* and *name* tables. To ensure that the migrated data is accurate, we do several queries on MySQL and Neo4j to compare the results. The queries contain the number of nodes/tables, values in nodes/tables and the value of relationships between tables.

3.2   Query Translation

This section will use the first scenario to demonstrate how to use our system to find potentially infected persons. A person is considered potentially infected if the person stays in the same place as an infected person at the same time. We can use the following SQL join operations to accomplish this task.

SELECT * **FROM** *person AS p, place AS pl, visit AS v*
WHERE *p.PersonID = v.PersonID*
AND *pl.PlaceID = V.PlaceID* **AND** *P.Healthstatus = "Sick"*;

After translating, *SQL2Cypher* will generate a graph pattern to find the potentially infected people using the following code:

MATCH $(p : person\{Healthstatus : "Sick"\}) - [r : VISIT]- > (pl : place)$
RETURN *;

In addition, users are able to execute both SQL and Cypher queries in the user interface and we provide several different forms of displaying the result. As shown in Fig. 2b, we demonstrate SQL queries result with tabular data and demonstrate graph structure for Cypher queries result. Graph structure makes it easier to find out relationships between data than tabular data.

4   Conclusion

In this proposal, we demonstrate how to use *SQL2Cypher* to migrate data from RDBMS to GDBMS and to translate SQL queries to Cypher queries automatically. In future works, we will refine the technique proposed to suit more different type of source and target database models.

Acknowledgements. Wenjie Zhang is supported by DP200101116. Xuemin Lin is supported by DP200101338.

References

1. Bizer, C.: D2R map-a database to RDF mapping language (2003)
2. Cypher (2021). https://neo4j.com/developer/cypher
3. Database trend (2020). https://db-engines.com/en/ranking_categories
4. De Virgilio, R., Maccioni, A., Torlone, R.: Converting relational to graph databases. In: 1st International Workshop on Graph Data Management Experiences and Systems, pp. 1–6 (2013)
5. Elmagarmid, A.K., Ipeirotis, P.G., Verykios, V.S.: Duplicate record detection: a survey. IEEE Trans. Knowl. Data Eng. **19**(1), 1–16 (2006)
6. Färber, F., Cha, S.K., Primsch, J., Bornhövd, C., Sigg, S., Lehner, W.: SAP HANA database: data management for modern business applications. ACM SIGMOD Rec. **40**(4), 45–51 (2012)
7. Hao, K., Yang, Z., Lai, L., Lai, Z., Jin, X., Lin, X.: PatMat: a distributed pattern matching engine with cypher. In: Proceedings of the 28th ACM International Conference on Information and Knowledge Management, pp. 2921–2924 (2019)
8. Heer, J., Hellerstein, J.M., Kandel, S.: Predictive interaction for data transformation. In: CIDR (2015)
9. Neo4j ETL. https://neo4j.com/developer/neo4j-etl/
10. ODBC (2021). https://docs.microsoft.com/en-us/sql/odbc
11. Sahu, S., Mhedhbi, A., Salihoglu, S., Lin, J., Özsu, M.T.: The ubiquity of large graphs and surprising challenges of graph processing. Proc. VLDB Endow. **11**(4), 420–431 (2017). https://doi.org/10.1145/3186728.3164139
12. Wang, R., Yang, Z., Zhang, W., Lin, X.: An empirical study on recent graph database systems. In: Li, G., Shen, H.T., Yuan, Y., Wang, X., Liu, H., Zhao, X. (eds.) KSEM 2020. LNCS (LNAI), vol. 12274, pp. 328–340. Springer, Cham (2020). https://doi.org/10.1007/978-3-030-55130-8_29

MOBA Game Analysis System Based on Neural Networks

Kangwei Li[1], Mengwei Li[1], Jia Tian[2], Xiaobo Cao[2], Tiezheng Nie[1], Yue Kou[1], and Derong Shen[1(✉)]

[1] Northeastern University, Shenyang 110819, China
shendr@mail.neu.edu.cn
[2] Beijing System Design Institute of the Electro-Mechanic Engineering, Beijing 100039, China

Abstract. The domain of knowledge contained in Multiplayer Online Battle Arena (MOBA) is quite complex, which is of great research value. With the rapid development of E-sports, the impact of data analysis on MOBA games is increasing. For example, data mining and deep learning methods can be used to guide players and develop appropriate strategies to win games. This paper proposes a novel MOBA game analysis system. The system includes three individual modules, namely lineup recommendation, real-time win rate prediction, and trend forecasting. The lineup module is implemented using NSGA-II algorithm to recommend hero combinations according to the enemy lineup. Win rate module is a neural network for predicting the quantitative advantage between teams. Trend module is a sequence-to-sequence model that forecasts the future team gold and exp. Finally, the system is applied to Dota 2, one of the most popular MOBA games. Experiments on a large number of professional match replays show that the system works well on arbitrary matches.

Keywords: MOBA game analytics · Deep learning in E-sports · Real-time win rate prediction · Time-series forecasting · Lineup recommendation

1 Introduction

MOBA game is currently one of the most popular genres of electronic games in the world. In this type of game, players are divided into two teams. Each team generally has five players, and each player controls a unique game character called hero. Heroes can gain exp and gold by killing minions, neutral creatures, enemy heroes, and structures. The goal of victory is to destroy the main structure of the enemy base while defending their own. Among championships of MOBA globally, Defense of the Ancients 2 (Dota 2) has the most generously awarded tournaments. Relying on the complex gameplay mechanics, the domain knowledge contained in MOBA games has great research value.

In recent years, artificial intelligence has achieved tremendous development and is widely applied in various fields. Indeed, sports analytics has been an emerging field in many professional sports games to help decision-making. It is foreseeable that E-sports analytics will also be useful for players in professional leagues. Great efforts

W. Zhang et al. (Eds.): WISE 2021, LNCS 13081, pp. 518–526, 2021.
https://doi.org/10.1007/978-3-030-91560-5_40

had previously been made on MOBA games, and some of the outstanding work has been applied. However, these applications usually need subscriptions and suffer from the limitation of scenarios. Moreover, there have been few efforts to learn and forecast the game-evolving trends during the game.

To address the main demands of players, we propose a novel MOBA game analysis system, which is able to recommend hero combinations according to the enemy lineups, predict the real-time win rate, and forecast the game-evolving trends. The main contribution of this paper is listed below.

- We introduce a novel MOBA game analysis system. The system integrates three individual modules through an interactive interface, namely lineup recommendation, real-time win rate prediction, and trend forecasting.
- We propose three models for the above modules. The lineup module is implemented using NSGA-II algorithm to recommend a combination of five heroes. Win rate module is a neural network for predicting the quantitative advantage between teams. Trend module is a sequence-to-sequence model that forecasts the future team gold and exp.
- We apply the system to Dota 2 and prove the effectiveness with experiments. We collect a large number of game data by parsing 30,697 professional match replays to train our model. Experiments show that the models perform effectively on arbitrary matches.

2 Related Works

Due to the complicated gameplay mechanism, MOBA games provide a brand-new platform for deep learning and other technologies. Research related to MOBA games mainly consists three aspects: strategy analysis, result prediction, and AI developing.

Strategy analysis aims at recommending appropriate heroes or items according to the enemy lineup and players' playstyle. Some researchers regard the order of picking heroes as sentences and predict the choice of a certain hero based on the context [1, 2]. These methods can only recommend one hero at a time and ignore the players' proficiency. Another type of methods uses Apriori or Monte Carlo Tree Search to generate hero combinations [3], while they can't model the synergy of specific heroes.

Predicting the results of MOBA games in real-time is one of the most important tasks in Esports analytical research. At present, this technology has been applied to the streaming of major professional leagues. In recent years, many researchers have used machine learning to solve this problem. Yang et al. [4] propose a sliding window to capture the dynamic changes during the game. Yu et al. [5] propose a quantitative evaluation method called discounted evaluation function and design a neural network model to fit the function. However, these models either suffer from insufficient features or ignores the dynamic changes of prior information. What's more, existing methods are mostly designed to predict the game result on specific time points, which however could deliver very little information about the game states. In real-world applications, it is arguably even more crucial to learn and forecast the game-evolving trends.

Artificial intelligence of MOBA games has attracted many researchers' attention. This complex game environment has promoted the development of game AI to a new level. Most AI studies are based on deep reinforcement learning [6–8], the key of which is the design of the reward function. Our system can work as the reward function in reinforcement learning models and the evaluation function in Monte Carlo planning models.

3   System Architecture

The system consists of three individual modules: lineup recommendation, real-time win rate prediction, and trend forecasting. Through the analysis of a large amount of game data, we respectively propose three models for these modules.

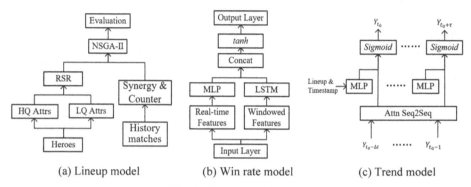

(a) Lineup model   (b) Win rate model   (c) Trend model

Fig. 1. System architecture

3.1   Lineup Recommendation

We intend to design a recommendation model which takes the enemy lineup as input and outputs the combination of five heroes. There are two factors that we need to take into consideration, which are the objective synergy and counter between individual heroes. We have designed a quantitative evaluation method through statistical analysis for the above factors [9]. For any two heroes i and j, we call $P_{counter}(i|j)$ as the counter coefficient of j against i, and $p_{synergy}(i,j)$ as the synergy coefficient between j and i:

$$P_{counter}(i|j) = p_j^i - p^i \tag{1}$$

$$p_{synergy}(i,j) = p_{i,j} - p^i \tag{2}$$

where p^i represents the win rate of hero i, p_j^i represents the win rate of i when hero i and hero j are picked as opponents, and $p_{i,j}$ represents the win rate of hero i and hero j are picked as teammates.

We assume that the appropriate lineup should meet the demands of both synergy and counter. So we summarizes this situation as a multi-objective optimization problem, and propose a lineup recommendation model based on NSGA-II [10]. This algorithm can coordinate various objective functions and find the optimal solution set that makes each objective function reach a relatively large (or relatively small) function value as much as possible. The recommendation process is shown in Fig. 1(a).

First, the hero attributes are divided into high-quality attributes (HQ Attrs) and low-quality attributes (LQ Attrs), and the rank-sum ratio (RSR) is used to evaluate the abilities of heroes. At the same time, the quantitative evaluation method is used to quantify the synergy and counter coefficient between individual heroes. Then we use NSGA-II to generate the lineup according to the enemy heroes. The main optimization goals of our model are as follows:

- The lineup has the largest sum of synergy and counter coefficient.
- The heroes are clustered by roles, only one hero can be selected from each role.
- The lineup consists of five heroes.
- The combination ability of the lineup is greater than the enemy.

Finally, the pre-trained classification models are used to evaluate the recommendation. We select Logistic Regression (LR), Gradient Boosting Decision Tree (GBDT) and eXtreme Gradient Boosting (XGB) as the binary classifiers.

3.2   Real-Time Win Rate Prediction

Win rates are considered an important quantitative indicator for evaluating real-time game situations. Most studies treat win rate prediction as a binary classification problem. However, the results of games should not be used as a real-time indicator for the entire game. Inspired by the discounted evaluation [5], we extend this definition to our model. For a specific time slice t, we call DE_t as the discounted evaluation:

$$DE_t(R, t, T) = \frac{tR}{T} \tag{3}$$

where $R = \pm 1$, represents the winner of the game, t is the current timestamp, and T is the duration of the entire game. The absolute value of DE_t is proportional to t, which can reflect the relative advantage between teams.

As a regression problem, we propose a supervised learning model to predict the value of DE_t. The structure of prediction model is shown in Fig. 1(b).

The model takes two types of features as input, namely real-time features and windowed features. Real-time features include the observable information of all characters at specific timestamps, such as lineup, health, gold, exp, and so on. These features are then fed in an MLP, which are feed-forward neural networks activated by *relu* and applied dropout. Windowed features include five-minute time-series data before the specific time point, which are the difference in gold, exp, kills, available buildings between two teams. In this part, we use LSTM to better adapt to the time-series data. Finally, the output is calculated by one neuron activated by *tanh* to get a result in the range $[-1,1]$. We can distinguish the final result by the sign of DE_t. Mean Absolute Error (MAE) is chosen as the loss function.

3.3 Trend Forecasting

Since there is no scoring mechanism in MOBA games, the situation in these games is usually evaluated by team gold and exp, which are typical time-series data. In order to improve the quality of these time-series data, we propose the following algorithm:

$$z = Diff * \frac{\min(R_r, R_d)}{\max(R_r, R_d)} \tag{4}$$

$$y = 10 \times \frac{(z - TD_{\min})}{(TD_{\max} - TD_{\min})} - 5 \tag{5}$$

$$X = \frac{1}{(1 + e^{-y})} \tag{6}$$

where $\min(R_r, R_d)$, $\max(R_r, R_d)$, $Diff$ respectively represent the minimum value, maximum value, difference between two teams, TD_{max} and TD_{min} are artificially set thresholds. Formula (4) can effectively reduce the number of samples where the difference is too large. Formula (5, 6) uses the $Sigmoid$ function to rescale the variable to [0,1].

We assume that the future trends are influenced by lineups, recent gold states, and recent exp states. Generally speaking, our model is based on sequence-to-sequence structure applied with the attention layer, as illustrated in Fig. 1(c). The goal of this model is to predict the gold and exp differences of the next τ time steps.

The lineup features are fed in MLPs activated by *relu* to learn the hidden representation at specific timestamps. We use an encoder-decoder framework based on LSTM to forecast the target sequence directly. And an attention layer is used to adjust the weights of different time-slices. Finally, the output layer takes the concatenation of the two parts and outputs the sequence of the next τ timesteps. We use MAE as the loss function.

4 Experiments

We choose Dota 2 as a typical MOBA game to apply our system. Dota 2 generates a replay file to record all the information in a match. An open-source parser called Clarity can parse the replay file and generate interval messages every second to record the state of each character. We use OpenDota's API to download 30,697 match replays of professional leagues from July 2018 to June 2020. Then we build the dataset by sampling time slices every 30 s. And the information contained in time slices includes: fort states, tower states, barrack states, hero name, strength, agility, intellect, health, mana, current_level, current_XP, damage, gold, last hits, denies, kills, deaths, assists, items. The main interface of the system is shown in Fig. 2. Since we cannot obtain the data of the ongoing game in real time, this system only supports the analysis of game replays. We provide an additional preprocessor to parse the replays into structured data. Then the users just need to import the preprocessed files and the system will output the prediction result.

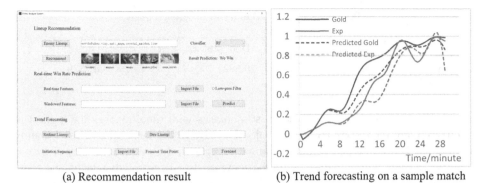

(a) Recommendation result   (b) Trend forecasting on a sample match

Fig. 2. Main interface of the system

4.1 Experiments on Lineup Recommendation Model

Since there is no direct evaluation method for lineup recommendation, we trained additional binary classifiers to verify the result of our recommendation. The classifiers take lineups of Dire and Radiant as input and predict the winner. The results of prediction accuracy are shown in Table 1. It is observed that our model performs even better than the real lineup. This may be because the real lineup is limited by factors such as Ban/Pick and player proficiency.

Table 1. Prediction accuracy of different lineups

	LR	GBDT	Xgboost
Real lineup	54.85%	52.95%	53.64%
Ours	**59%**	**62%**	**58%**

4.2 Experiments on Real-Time Win Rate Prediction Model

The binary prediction accuracy is selected as an evaluation metric. For the experiments, MLP has 5 hidden layers of 64 neurons and LSTM has 64 units. Although there is a lot of work on result prediction of MOBA games, the different datasets used have caused great differences. To evaluate the corresponding ability of our model, we compare it with two similar methods. MOBA-Slice is a real-time prediction model trained by Dota 2 professional leagues. Dota Plus is the official monthly subscription service of Dota 2. As one part of DotA Plus service, Plus Assistant provides a real-time win probability graph for every match. As the result in Table 2, the average accuracy of our model is much higher than the two other methods.

To further understand the performance of our model, we conduct separate tests to evaluate the effectiveness of different components of our model, as shown in Table 3. We divide the time slices into intervals to calculate the accuracy of the model at a specific

Table 2. Prediction accuracy of different models

	MOBA-slice	Dota plus assistant	Ours
Accuracy	0.7191	0.6821	**0.7830**

period. We can see that the accuracy first increases with time, reaches the highest in 20–40 min, and finally begins to gradually decrease in the late game. This is mainly because the beginning part of a game might be erratic, the initial attributes of heroes are more influential at this stage. With the accumulation of team advantages, the game usually ends or has clear gaps at about 30 min. Finally, the gaps will gradually decrease in the late game due to the upper limit of hero abilities and increase the difficulty of prediction.

Table 3. Prediction accuracy of different components at different time periods

Time/min	0–10	10–20	20–30	30–40	40–50	50–60	> 60	Avg
MLP	0.7302	0.7854	0.8189	0.8088	0.7667	0.7411	**0.6753**	0.7777
LSTM	0.6236	0.7247	0.7734	0.7757	0.7417	0.7124	0.6532	0.7123
Ours	**0.7423**	**0.787**	**0.8193**	**0.8125**	**0.7786**	**0.7512**	0.6727	**0.7830**

4.3 Experiments on Trend Forecasting Model

We evaluate the performance in terms of mean absolute error (MAE) of team gold and exp. Lower MAE indicates higher performance.

In our experiments, the input and output lengths are both 10 timesteps. Each MLP has 3 hidden layers of 64 neurons and LSTM has 64 units. Since there is no relative work on trend forecasting of MOBA games, we only conduct a series of tests on different components of our model, as shown in Table 4. We can learn that the prior sequence plays a crucial role to forecast the future series. With the help of the attention layer, the model can select relevant encoder hidden states across all time steps and achieve better performance. Although the lineup part performs poorly in the individual test, it still contributes to the whole model.

Table 4. Trend (gold and exp) forecasting results of different models

	Lineup part	Seq2Seq part	Seq2Seq (no Attn)	Ours
MAE_G	0.25365198	0.02089647	0.05620221	**0.01693125**
MAE_E	0.25567913	0.02920840	0.04728998	**0.02174589**

In real games, the earlier we know about the future evolving trends, the better for our team strategies. Since the output length of the existing methods is not greater than the input length, we test our model in a state space way. In order to simulate the observed information in the early game, we only feed the model lineups and the first five-minute features to forecast the target sequence of 5 to 10 min. Then we timestamp the predicted sequence and treat it as a new input for our model. Repeat this action to get a longer forecast sequence until we get enough information. Figure 2(b) shows the performance of our model on a sample match.

5 Conclusion

This paper introduces a novel MOBA game analysis system, which includes three individual modules: lineup recommendation, real-time win rate prediction, and trend forecasting. Before the game starts, the lineup recommendation model can recommend the combination of five heroes according to the enemy lineup. In the win rate part, we design an LSTM based neural network to fit the function of DE_t. In the trend part, we propose a sequence-to-sequence model to forecast the target series. Through applying the system to Dota 2, we prove its effectiveness with experiments on a large number of match replays.

Acknowledgment. .This work is supported by the National Natural Science Foundation of China (62072084, 62072086, 62172082), the National Defense Basic Scientific Research Program of China (JCKY2018205C012), and the Fundamental Research Funds for the Central Universities (N2116008).

References

1. Summerville, A., Cook, M., Steenhuisen, B.: Draft-analysis of the ancients: predicting draft picks in dota 2 using machine learning. In: Twelfth Artificial Intelligence and Interactive Digital Entertainment Conference (2016)
2. Zhang, L., et al.: Improved dota 2 lineup recommendation model based on a bidirectional LSTM. Tsinghua Sci. Technol. **25**(6), 712–720 (2020)
3. Chen, Z., et al.: The art of drafting: a team-oriented hero recommendation system for multiplayer online battle arena games. In: Proceedings of the 12th ACM Conference on Recommender Systems, pp. 200–208 (2018)
4. Yang, Y., Qin, T., Lei, Y.H.: Real-time esports match result prediction. arXiv preprint arXiv: 1701.03162 (2016)
5. Yu, L., et al.: Moba-slice: a time slice based evaluation framework of relative advantage between teams in moba games. In: Workshop on Computer Games, pp. 23–40. Springer, Cham (2018)
6. Berner, C., et al.: Dota 2 with large scale deep reinforcement learning. arXiv preprint arXiv: 1912.06680 (2019)
7. Wu, B.: Hierarchical macro strategy model for moba game AI. Proc. AAAI Conf. Artif. Intell. **33**(01), 1206–1213 (2019)
8. Ye, D., et al.: Mastering complex control in MOBA games with deep reinforcement learning. Proc. AAAI Conf. Artif. Intell. **34**(04), 6672–6679 (2020)

9. Wang, N., et al.: Outcome prediction of dota 2 using machine learning methods. In: Proceedings of 2018 International Conference on Mathematics and Artificial Intelligence, pp. 61–67 (2018)

10. Deb, K., et al.: A fast and elitist multiobjective genetic algorithm: NSGA-II. IEEE Trans. Evol. Comput. **6**(2), 182–197 (2002)

FedAggs: Optimizing Aggregate Queries Evaluation in Federated RDF Systems

Ningchao Ge, Peng Peng$^{(\boxtimes)}$, Zheng Qin, and Mingdao Li

Hunan University, Changsha, China
{ningchaoge,hnu16pp,zqin,limingdao}@hnu.edu.cn

Abstract. With the increasing scale of RDF datasets and the security requirements of data management, federated RDF systems are becoming a research hotspot. However, the existing federated RDF systems only support basic queries in SPARQL 1.0, and cannot be compatible with complex queries in SPARQL 1.1 well, such as aggregate queries. In this demo, we develop a demonstration system, named *FedAggs*, which can support efficient aggregate queries in federated RDF system. It implements five common aggregate operators in federated RDF system: MAX, MIN, AVG, SUM and COUNT. *FedAggs* improves the efficiency of aggregate query processing by adopting a query decomposition optimization method, employing a cost-based optimal query plan generation algorithm and an incremental execution strategy according to the characteristics of five aggregate operators. Experimental results show that *FedAggs* has good query performance.

1 Introduction

The Resource Description Framework (RDF) [9] is a family of World Wide Web Consortium (W3C) specifications originally designed as a metadata data model. Up to now, many data providers have published their own RDF datasets in their own sites, and they often also provide query interfaces to allow users to submit SPARQL [5] queries. A site with SPARQL query interfaces is called an *RDF source* in this demo. We can query a person's travel information through the SPARQL interfaces of traffic RDF source, and also query a person's financial information through the SPARQL interfaces of bank RDF source. However, we cannot query the travel information of people under certain financial conditions. Because there is no data communication between traffic RDF source and bank RDF source. In order to solve the queries of across RDF sources problem (federated query), federated RDF systems have been proposed [4,6,8] to integrate multiple RDF sources.

A federated RDF system consists of a control site and some RDF sources. These RDF sources have data relevance, but they can't communicate with each other directly, and only provide SPARQL query interface to the control site. At first the control site decomposes the federated query submitted by the user into multiple subqueries. Then, those subqueries are forwarded to the corresponding

W. Zhang et al. (Eds.): WISE 2021, LNCS 13081, pp. 527–535, 2021.
https://doi.org/10.1007/978-3-030-91560-5_41

RDF sources to execute. Finally, those subqueries results are joined to form the final query results of federated query.

As a widely used query type, aggregate queries can reflect the overall characteristics of RDF data and is an important update part of SPARQL 1.1. However, existing federated RDF systems only support and optimize basic queries in SPARQL 1.0, and cannot handle aggregate queries in SPARQL 1.1 efficiently. To remedy this defect, this demo implements a federated RDF system, named FedAggs, which optimizes the evaluation of aggregate queries over federated RDF systems.

In summary, FedAggs has the following characteristics:

– FedAggs implements five common aggregate operators in federated RDF systems: MAX, MIN, AVG, SUM and COUNT.
– FedAggs proposes a query decomposition optimization method to reduce the number of remote requests and improve the query efficiency.
– FedAggs builds a cost model, and adopts different optimal query plan generation and incremental execution strategies according to the characteristics of five aggregate operators to minimize the overall query cost.
– FedAggs shows excellent query performance over both real datasets and synthetic datasets.

2   System Overview

The system overview for aggregate query processing in FedAggs is shown in Fig. 1. For a federated SPARQL query, its query result can be obtained through four main steps: *query decomposition, query cost evaluation, optimal query plan generation* and *subqueries execution and post-processing.*

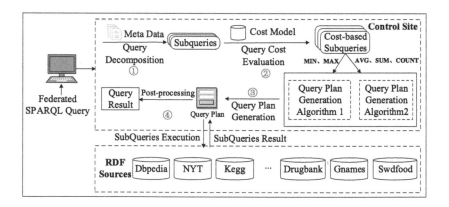

Fig. 1. System overview for aggregate query processing in FedAggs

2.1   Query Decomposition

Because SPARQL is the query language designed for centralized RDF system, it cannot be directly applied to federated RDF system. For a federated SPARQL query, it is necessary to be decomposed and source localization first. Previous methods only combine triple patterns with the same single RDF source into one subquery, but the triple patterns with the same multi-sources were not. However, we find some multi-sources with multi-source triples can be combined into one subquery without affecting the correctness of the query results. Therefore, FedAggs proposes a query decomposition optimization method, which allows combine triple patterns with the same multi-sources into one subquery. The schema can reduce the number of remote requests to improve the query efficiency by reducing the number of subqueries. The set of subqueries can be expressed as: $\mathcal{Q} = \{subQ_1@S_1, subQ_2@S_2, ..., subQ_n@S_n\}$. Among them, the $subQ_i@S_i$ represents that S_i is the RDF sources of subquery $subQ_i$. The subqueries of the example federated query Q of *Example* 1 after query decomposition is shown in Fig. 2.

Example 1. Example federated query Q: query the city with the largest population and its related attributes.

```
select (MAX(?population) as ?p) where {
        ?geonameplace dp:capital ?capital.
        ?geonameplace dp:foundingDate ?foundingDate.
        ?place g:countryCode ?countryCode.
        ?place g:population  ?population.
        ?place g:long   ?longitude.
        ?place g:lat    ?latitude.
        ?place owl:sameAs  ?geonameplace.
}
```

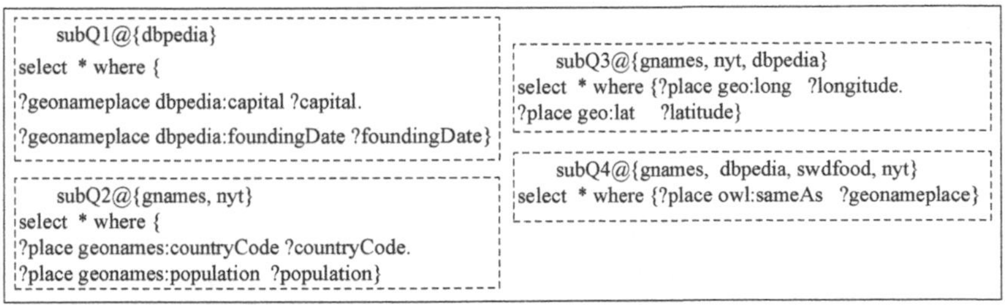

Fig. 2. Query decomposition result of *Example* 1

2.2 Query Cost Evaluation

There are huge differences of query overhead between different subqueries execution orders. A subqueries execution order is called a query plan. It is particularly important to obtain the optimal query plan for the subqueries, and the optimal query plan should have the lowest query cost. In order to quantitatively evaluate the query cost of a query plan, FedAggs designed a cost model, and the query costs and join cost of each subquery can be evaluated by it.

The cost model mainly depends on the triple patterns statistical information of each RDF source in the federated RDF system offline. We use $C(P_{ij})$ to represent the count of triple patterns whose predicate is P_{ij} in RDF source S_i, and $S(P_{ij})$ to represent the number of subject of those triple patterns. For a subquery q with two triple patterns P_1 and P_2, the query cost of q can be expressed as:

$$cost(q) = \begin{cases} C(P_1) \times C(P_2) & if \ P_1 \ full \ join \ P_2; \\ \dfrac{C(P_1)}{S(P_1)} \times \dfrac{C(P_2)}{S(P_2)} \times Min\{S(P_1), \ S(P_2)\} & otherwise. \end{cases}$$

For two subqueries q_1 and q_2, their join cost can be expressed as:

$$cost(q_1 \bowtie q_2) = \begin{cases} cost(q_1) \times cost(q_2) & if \ q_1 \ full \ join \ q_2; \\ Min\{cost(q_1) \times \dfrac{C(P_2)}{S(P_2)}, \ cost(q_2) \times \dfrac{C(P_1)}{S(P_1)}\} & otherwise. \end{cases}$$

where P_1 represents the triple pattern in q_1 that can be joined with q_2, and P_2 represents the triple pattern in q_2 that can be joined with q_1.

2.3 Query Plan Generation

Based on above cost model, FedAggs proposes an optimal query plan generation algorithm with dynamic programming. We use $C[dp[i]]$ to express the cost of optimal query plan with i subqueries. Then, its recurrence can be represented as:

$$C[dp[i]] = Min\{C[dp[i-2] \bowtie (q_l \bowtie q_i)],$$
$$C[(q_i \bowtie q_r) \bowtie dp[i-2]]\}$$

where q_l and q_r are two subqueries in $dp[i-1]$, which can left join or right join with q_i, respectively. Especially, $C[dp[1]] = cost(q_1)$, $C[dp[2]] = min\{cost(q_1 \bowtie q_2), \ cost(q_2 \bowtie q_1)\}$.

Generally, for SUM, COUNT and AVG of aggregate query types, the final result can be obtained with querying all the qualified triple patterns, but MIN and MAX can not. On this basis, FedAggs optimizes query plan generation algorithm according to the characteristics of these two kinds of aggregate query operators. Figure 3 shows the example query plan of the subqueries in Fig. 2.

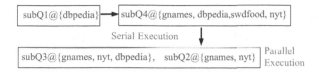

Fig. 3. Query plan of the subqueries in Fig. 2

2.4 Subqueries Execution and Post-Processing

The execution order and execution model of subqueries are determined by the query plan, including serial execution and parallel execution. As shown in Fig. 3, subquery $subQ_1$ is executed first. Then, subquerie $subQ_4$ is executed, and the matching range of subgraphs can be narrowed by the query results of $subQ_1$ to improve the query efficiency. Finally, the subqueries $subQ_2$ and $subQ_3$ can be executed in parallel. In particular, FedAggs proposes an incremental subqueries execution method to obtain the final result of MIN and MAX without querying all the triple patterns that meet the conditions. The incremental subqueries execution of query plan in Fig. 3 is shown in Fig. 4. In the end, the control site post-processes the results of subqueries to generate the final query result and return it to the user.

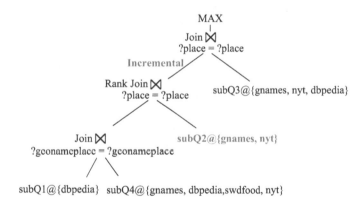

Fig. 4. Incremental subqueries execution of query plan in Fig. 3

3 Evaluation

In this section, we will introduce the composition, performance and main demonstration pages of federate RDF system FedAggs.

3.1   Setting of FedAggs

FedAggs consists of six Alibaba Cloud servers, of which five servers are used to manage RDF sources by Sesame [2] and the other one is used as the control site. RDF sources of FedAggs contain two famous comprehensive RDF benchmark suites FedBench [7] and WatDiv [1].

FedBench. FedBench [7] is a comprehensive benchmark suite for testing and analyzing the performance of federated query processing strategies on semantic data. There are more than 200 million triple patterns in FedBench, which contains 11 datasets, involving *CrossDomain*, *LifeSciences* and *SP²Bench*.

WatDiv. WatDiv [1] is a benchmark that enables diversified stress testing of RDF data management systems. In WatDiv, instances of the same type can have the different sets of attributes. We generate three datasets varying sizes from 100 million to 300 million triples and divide the schema graph of the collection into several connected subgraphs with METIS [3].

3.2   Performance of FedAggs

To assess the performance of our approach, we removed the optimization scheme from FedAggs to form the version Basic. We design five aggregate queries both for MAX, MIN, AVG, SUM, COUNT (Q_1–Q_5) on FedBench. As shown in Fig. 5, the number of remote access and the total run time of FedAggs is obviously better than Basic for federated aggregate queries.

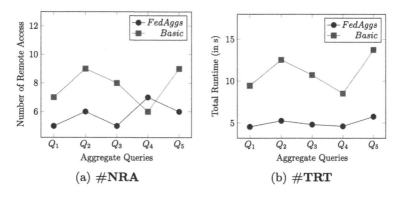

Fig. 5. Effectiveness comparison between FedAggs and Basic.

Among that, we can find that the number of remote access of FedAggs is more than Basic for Q_4. It is due to the cyclic execution of incremental subquery execution method. In addition, we tested the robustness of the proposed method on different sizes of WatDiv datasets, as shown in Fig. 6. Experimental results show that the average query time of FedAggs increases linearly with the increase of dataset size.

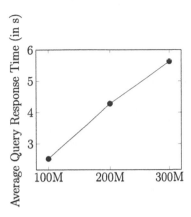

Fig. 6. Scalability test of FedAggs.

3.3 Demonstration of FedAggs

FedAggs is an online federated RDF system, which provides an interactive Web interface. Figure 7 and Fig. 8 are the main pages of FedAggs.

Figure 7 demonstrates the home page of FedAggs. Users can select an example aggregate query from the drop-down box or enter an query. Then, the query execution process, including query decomposition result, optimal query plan and the times of each part are presented in the top of Fig. 8. Finally, the query result is shown in the bottom of Fig. 8. More detailed demonstrations can be referred with http://47.92.169.107:8080/FedAggs/Demo/index.html.

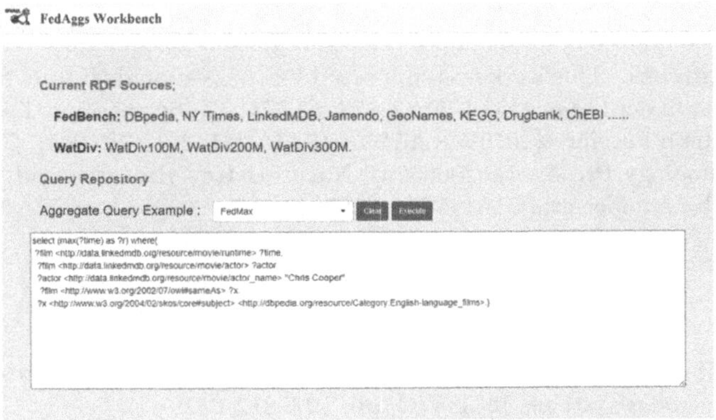

Fig. 7. Aggregate query home page of FedAggs

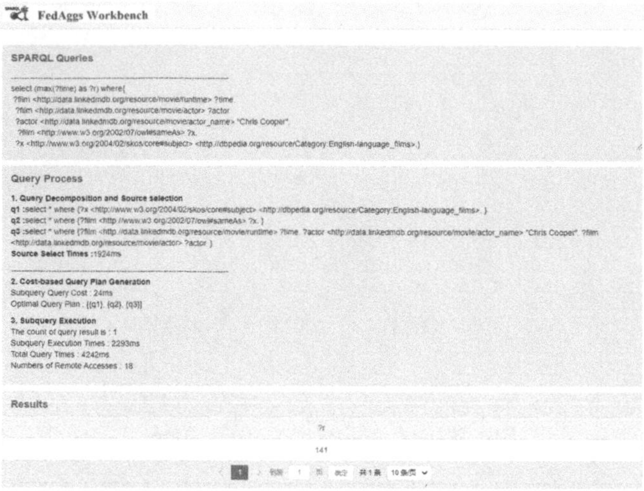

Fig. 8. Aggregate query result page of FedAggs

4 Conclusion

In this demo, a federated RDF system is built, named FedAggs, which realizes and optimizes five common federated aggregate operators in SPARQL 1.1: MAX, MIN, SUM, COUNT and AVG. FedAggs improves the aggregate query efficiency by adopting a query decomposition optimization method, a cost-based optimal query plan generation algorithm and an incremental subquery execution strategy. Finally, it shows a well performance on both realdatasets and synthetic datasets.

Acknowledgments. This work was supported by the National Natural Science Foundation of China under Grant (No. U20A20174, 61772191), Science and Technology Key Projects of Hunan Province (2019WK2072, 2018TP3001, 2015TP1004), ChangSha Science and Technology Project (kq2006029), National Key Research and Development Program of China under grant 2019YFB1406401.

References

1. Aluç, G., Hartig, O., Özsu, M.T., Daudjee, K.: Diversified stress testing of RDF data management systems. In: ISWC, pp. 197–212 (2014)
2. Broekstra, J., Kampman, A., Harmelen, F.V.: Sesame: a generic architecture for storing and querying RDF and RDF schema. In: Towards the Semantic Web: Ontology-Driven Knowledge Management (2003)
3. Karypis, G., Kumar, V.: Multilevel graph partitioning schemes. In: ICPP, pp. 113–122 (1995)
4. Peng, P., Ge, Q., Zou, L., Özsu, M.T., Xu, Z., Zhao, D.: Optimizing multi-query evaluation in federated RDF systems. In: TKDE (2019)

5. Prudhommeaux, E.: Sparql query language for RDF (2008). http://www.w3.org/TR/rdf-sparql-query/
6. Saleem, M., Ngomo, A.N.: HiBISCuS: hypergraph-based source selection for SPARQL endpoint federation. In: ESWC, pp. 176–191 (2014)
7. Schmidt, M., Görlitz, O., Haase, P., Ladwig, G., Schwarte, A., Tran, T.: FedBench: a benchmark suite for federated semantic data query processing. In: ISWC, pp. 585–600 (2011)
8. Schwarte, A., Haase, P., Hose, K., Schenkel, R., Schmidt, M.: FedX: optimization techniques for federated query processing on linked data. In: ISWC, pp. 601–616 (2011)
9. Swick, R.R., World Wide Web Consortium: Resource Description Framework (RDF) model and syntax specification. W3C Recommendation (1998)

JUST-Studio: A Platform for Spatio-Temporal Data Map Designing and Application Building

Yuan Sui[1], Ruiyuan Li[1,2(✉)], Xu Wang[1], Jun Liu[1], and Juncheng Tang[1]

[1] JD Intelligent Cities Research, Beijing, China
{suiyuan,wangxu649,liujun513,tangjuncheng3}@jd.com,
liruiyuan@whu.edu.cn
[2] Chongqing University, Chongqing, China

Abstract. With the proliferation of GPS (Global Position System) and other space sensors, there emerged many spatial-temporal (a.k.a. ST) urban applications, such as trajectory visualization and geofence analysis. It is time-consuming and tedious to build these applications from scratch by writing codes only. This paper introduces JUST-Studio, a holistic platform that analyzes spatio-temporal data and builds urban applications with minor efforts. JUST-Studio consists of two main components: service manager and ST-App designer. By this platform, users could: 1) upload their own ST data to the built-in data store, or connect to an existing ST data store and register it as a data source; 2) create ST models using ST data and publish ST services; 3) make ST base map and build ST applications based on ST services. JUST-Studio is not only capable of making commonly used static map scenes, but also good at rendering dynamic data. In this paper, we will use vehicle trajectories with geofence analysis to build an application for monitoring restricted areas, which is very common in urban traffic applications.

Keywords: Data virtualization · Web mapping · Location analysis · Application builder

1 Introduction

With the development of spatial mapping and sensors, spatio-temporal data (a.k.a. ST data) has attracted extensive attention and been frequently applied in a variety of industries. For instance, by using status information and trajectories of taxies, a series of drop-off points can be excavated [1, 2]. We can recommend these drop-off points to taxi drivers to increase vehicle utilization rate and make more money [3]. However, with the popularization of ST data, people started to realize the limitation of desk map applications. Demands of map sharing and collaboration are becoming stronger. As a result, large-scale web-based cloud GIS (Geographic Information System) system starts to come into the market.

From the web 1.0 era in 1990, web mapping has appeared. With the iteration of Internet technology, from the initial static mapping to the cloud mapping to the recent intelligent mapping, cartography has gone through nine stages of development [4]. Nowadays,

© Springer Nature Switzerland AG 2021
W. Zhang et al. (Eds.): WISE 2021, LNCS 13081, pp. 536–546, 2021.
https://doi.org/10.1007/978-3-030-91560-5_42

many outstanding online GIS platforms give us excellent online map experience. With vector tile technology, people could not only browse the map, but also dynamically set map styles as they like. Mapbox [5] is undoubtedly a good application case, who has a ton of customization features for personalized maps. Mapbox is tile-focused, so its map display is smooth and pleasing to the eyes. However, Mapbox has two shortcomings for terminal users. First, its services only solve the base map presentation problem of GIS. It still needs a lot of coding if users want to interact with the map or have data analysis requirements. Second, the studio provided by Mapbox only supports static data. It does not provide visualization layers for dynamically changing data, e.g., dynamic trajectories. Compared to Mapbox, CARTO [6], equipped with spatial analysis services such as path planning and geocoding, provides a better experience in interactive analysis and map designing. Through CARTO builder, the results of spatial analysis can be fluently displayed on the map. Furthermore, CARTO allows users to upload data and store it in its cloud PostGIS database [6]. However, CARTO suffers from the following problems: 1) it can only define basic interactive map operations, such as `mouse click` and `drag`. 2)Like Mapbox, it does not support dynamic data analysis.

To this end, as an important part of our JUST (**J**D **U**rban **S**patio-**T**emporal data engine) project [7–9], we build a holistic platform, i.e., JUST-Studio, which can not only provide basic collaborative map operations, but also:

(1) Support the incorporation and analysis of dynamic spatio-temporal data. Dynamic spatio-temporal data could be displayed on the front-end map in real-time. At the same time, further complex spatio-temporal analysis operations can be performed on dynamic data.
(2) Provide analysis capabilities. We provide many out-of-the-box spatio-temporal analysis services, by which users can set their own analysis styles, drag and drop different components to build various fancy map applications without any codes.

To the best of our knowledge, JUST-Studio is the first system to build spatio-temporal urban applications for both static and dynamic scenarios without writing any codes. We also provide an online demo: http://just-studio.urban-computing.com/.

The outline of this paper is as follows. In Sect. 2, we give the system overview of JUST-Studio. The main components of JUST-Studio are described in detail in Sect. 3 and Sect. 5. Finally, we demonstrate the practical effects of JUST-Studio by building a map application in two scenarios in Sect. 6.

2 System Overview

Figure 1 presents the system overview of JUST-Studio, which consists of two main components: **Service Manager** and **ST-App Designer**. Service Manger includes three layers: 1) data store layer, which stores and manages both static and dynamic ST data, such as geographic data, remote sensing data, trajectory data and mobile phone data; 2) ST-model layer, which offers a variety of ST application models on ST data, such as basic layer model, road network model, GOI (Geometry of Interests) model and trajectory model; 3) service layer, which provides two types of services, i.e., layer services and control services. Based on the services provided by Service Manager, users can construct various fancy urban applications with ST-App Designer in a pulling-dragging manner.

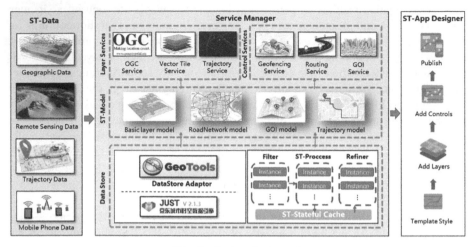

Fig. 1. System framework

3   Service Manager

Service Manager provides a wealth of services for ST-App Designer to call. It acts as a solid foundation for map application construction. For the convenience of users, Service Manager simplifies the whole service creation process into three steps: 1) Data Source Register, 2) Model Creating, and 3) Model Publishing. They are used to correlate data, define algorithmic models, and publish algorithmic capabilities, respectively.

3.1   Data Source Register

Data source configuration is the first step to use ST data. Classified by the means of data access, there are two modes of configuration. First, users upload local files in ShpFile or GeoPackage formats; second, users register existing data sources, which can be both static and dynamic. Static data source refers to the data that is not accessed and updated in real time. It includes traditional relational databases, such as PostgreSQL, and distributed NoSQL databases, such as HBase [10]. Dynamic data sources refer to data sources that are accessed in a real-time fashion, such as Kafka [11], a message-oriented middleware.

Data Source Register has a strict inspection process. It includes: 1) *data connectivity testing*, which examines whether the target data source is connectable or whether the uploaded data is complete; 2) *spatial property checking*, which inspects whether the target data source contains correct spatial fields, such as longitude, latitude, or WKT (Well-Know Text) [12] fields; 3) *special data spatial check*, which is related to specific models. For example, the road network model needs to check the connectivity of road networks, because an unconnected road network will lead to wrong path planning results. Users will be ready to use these data sources after the inspection process. Next, we will briefly explain the static and dynamic data supported by JUST-Studio.

Static Data. In order to access multiple heterogeneous data sources, we choose JUST [7] to store static data. JUST provides a unified JustQL language for multiple data sources, which makes the management and analysis of massive ST data easier. Users can switch different execution engines for both spatio-temporal query scenarios and spatio-temporal mining cases. When users only conduct applications and analysis in small or medium-sized data scenarios, they can use local engine mode, which can not only meet the requirements of computing efficiency, but also does not occupy too many resources. When it comes to city-level massive data scenarios, they can switch to the distributed Spark [13] computing engine. This scalable mode can cover almost all ST analysis and computing scenarios. JUST-Studio extended its OGC (Open Geospatial Consortium) [14] adapter to achieve a unified layer management mode based on GeoTools framework [15].

Dynamic Data. In order to better access and process real-time data, we design a real-time data management framework, as shown in the right part of Data Store in Fig. 1. The framework consists of two components. One is the upper ST processing engine, and the other is the underlying ST state storage. The processing engine is responsible for accessing real-time ST data. Currently, JUST-Studio supports access to distributed Kafka messages. After access, the engine filters out unwanted data in real time according to parameters set by users. Remained data will flow into the processor, where ST data will be transformed and assembled according to predefined logic. The results will be stored into stateful storage. Stateful storage is a cache component of ST data that receives information from the processor in real time and updates the old information. Cache data will be purged periodically at a preset time. When a user accesses real-time data, the request would be sent to Refiner. Refiner extracts the latest data from the stateful storage, refines the data based on the user's request parameters, and returns the final result to the caller.

Real-time data management framework is a predetermined flow framework. Different data and different processing logic will get different results. We will show you in Sect. 3.2 and Sect. 3.3 how to implement specific real-time data services through dynamic model definitions.

3.2 Data Model Creating

Data models are the bridges between data sources and spatial-temporal services. A data model must be associated with at least one data source. Parameters are then set according to different model types, which together with the data determine the final service result. We can think of the model as a kind of ST algorithm. For example, if we want to build a path planning application, we first need to create a road network model. When we associate a road network data source, we need to set the parameters of the model, such as geometry field, direction field, speed limit field. After that, we have a path planning algorithm model based on the specific road network data.

ST-Static Model. There are three static temporal and spatial models in JUST-Studio: layer model, road network model and GOI model. Layer model mainly orients to map presentation scenarios. Relatively, the configurations of road network models and GOI models are more abundant. The configuration of road network models has been briefly introduced before. Users need to map some data source fields to the parameters required by our path planning algorithm. However, field mapping is not mandatory. Users can create models without any configuration. In that way, JUST-Studio would use default parameters to provide path planning services. The GOI model is used to serve keyword retrieval for general geometry objects. Users must specify at least one keyword field.

ST-Dynamic Model. Dynamic ST data refers to the data whose properties change over time. For instance, trajectory data, whose GPS point positions are time-varying. Dynamic ST model is an algorithm package of ST dynamic data. Before creating the trajectory model, the user needs to associate a Kafka data source that receives GPS points. JUST-Studio would partition Kafka's incoming data based on the user-defined trajectory key, and then allocates a thread to each partition processor. With each thread, filtering and processing operations would be conducted. As shown in Fig. 2, taking the vehicle trajectories as an example, using license plate number as the key, GPS points are assigned to different Kafka cluster partitions. Consuming thread pool in downstream system would subscribe GPS points of all partitions and perform filtering and processing operations in the form of Pipeline. This strategy of partitioned processing can greatly improve the throughput of real-time processing. It is important for time-sensitive scenarios, such as real-time visualizations. Users can configure the way of trajectory processing, such as setting the map matching [16] processor. If the accessed Kafka data is the original GPS points, notable errors may occur. Setting map-matching can rectify GPS points to the correct position in real-time. Since map-matching requires a benchmark road, a road network model must be associated with it. In addition to the processing methods, the system also provides indexed configurations. Users can choose the latest point of trajectory data or the trajectory line as the index, as shown in Fig. 3. When the point index is used, JUST-Studio would create a `gps_ point` column for the trajectory table, which stores the last GPS point for each vehicle. When querying the trajectories, JUST-Studio

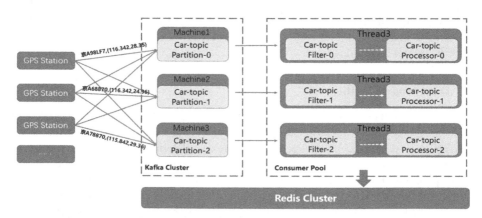

Fig. 2. ST-dynamic data process architecture

would filter GPS points by the nearest point of the vehicle. When users choose the line index, it will establish a `traj_ line` column to store the latest trajectory line object, as shown in Fig. 4. In this case, the trajectory will be filtered by the whole vehicle line. These two storage methods both have their own application scenarios. In the next section we will give a more detailed description.

Key	Value	
gps_point	traj_series	properties
Latest GPSPoint$_1$	LinkedList<GPS Point>	number$_1$ speed$_1$ ...
Latest GPSPoint$_2$	LinkedList<GPS Point>	number$_2$ speed$_2$ ...
...	...	...
Latest GPSPoint$_n$	LinkedList<GPS Point>	number$_n$ speed$_n$ ...

Fig. 3. Point indexed trajectory

Key	Value	
traj_line	traj_series	properties
Latest LineString$_1$	LinkedList<GPS Point>	number$_1$ speed$_1$ ...
Latest LineString$_2$	LinkedList<GPS Point>	number$_2$ speed$_2$ ...
...	...	...
Latest LineString$_3$	LinkedList<GPS Point>	number$_n$ speed$_n$ ...

Fig. 4. Line indexed trajectory

3.3 Service Publishing

Services are outlets for model capabilities and bridges between data and visual interfaces when building ST applications. Classified by ST application scenarios, JUST-Studio provides two categories of services. One is the ST map configuration service, called Layer Service. The other is the ST application configuration service, also known as Control Service. This kind of service capabilities are displayed in JUST-Studio through ST analysis components.

Layer Service. Static ST data supports two types of layer services. One is the standard OGC service, including WMS (Web Map Service) [17], WFS (Web Feature Service) [18] and WMTS (Web Map Tile Service) [19]. The other is a vector tile service. We extend the algorithm based on the vector tile standard structure [20] proposed by Mapbox. We add a pixel-based spatial thinning algorithm to balance the integrity of data expression and display efficiency. In addition, dynamic ST data can also publish layer services. For example, the trajectory service can provide real-time trajectory visualization. When visiting from the front end, to obtain the trajectories of all vehicles within a spatial range at a specific moment, users only need to dynamically input the current map space range. Note that the trajectory visualization displays the latest location information of the vehicle, so the trajectory model depends on the needs to be built with the point index.

Control Service. Unlike Layer Service, Control Service concentrates on ST analysis scenarios. They are the service basis for ST-App Designer (simplify Designer later). In Designer, one or more control services are packaged into an analysis component, allowing users to build applications easily. Here we focus on the control service of dynamic ST data, i.e., geofencing service. A geofence is a virtual perimeter for a real-world geographic area [21]. The use of geofence is called geofencing. There are many applications involved in geofencing. When used to children's location service, geofence can notify parents if a child leaves a designated area [22]. Another case is the vehicle tracking. When a car drives into a user- predefined perimeter, it would send some warning messages immediately. With JUST-Studio, users can add or remove geofence geometry objects for the geofencing service, and then associate a trajectory model with it. Once geofencing service is online, the system would monitor the positions of vehicles in real-time. If a vehicle enters the fence, one record will be added to the monitoring table. The system will continuously monitor the entered vehicles until they leave the fence. Eventually the information such as entering time, leaving time, driving routes would be recorded in the monitoring table. Given that fence monitoring has a short time difference, the geofencing needs to capture the trajectory routes of all vehicles. Therefore, the trajectory model associated with the geofencing service must be stored as a line index.

3.4 Massive Data Optimization

With the support of JUST, JUST-Studio could access huge amounts of ST data. However, on the other hand, it poses a big challenge to the Studio's processing power. Taking map service as an example, the visualization of 1,000 geometries and 100,000 geometries has completely different requirements for the processing efficiency of the server side. Users usually do not care about the amount of data in the database, and they just want to see the results on the map in seconds. JUST-Studio has made many optimizations for this. Say the user wants to display 100,000 geographical elements on a map. We first partition the geographic spatial area of the user request based on the slicing strategy of map tiles, and then retrieve the data in the database in parallel using the divided sub-areas (data thinning parameters can be set here). After obtaining the data of each region, we simplify the data using Douglas-Peucker algorithm [23], of which the simplification coefficient is calculated according to the pixel value of each region. After that, we transform the geographic coordinate system into a local coordinate system. Finally, we compress the data using zigzag compression method [24]. The procedure is shown in Fig. 5. By this algorithm, JUST-Studio can easily render millions of geographical elements in the front end in seconds. If we switch our execution engine to Spark mode, the amount of data we can handle will be even larger.

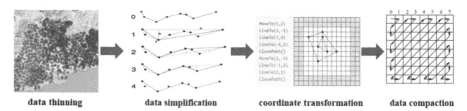

data thinning   data simplification   coordinate transformation   data compaction

Fig. 5. Massive ST data map service optimization

4   ST-App Designer

ST-App Designer is a configuration module of JUST-Studio in the front end. There are mainly two functions. One is to configure map styles, and the other is to construct ST applications based on the configured map styles. Users can publish the applications. Other users can directly access the designed ST applications, or use the front-end SDK provided by Designer to parse the map and process a secondary development.

4.1 Map Designing

ST-App Designer is built based on Mapbox GL JS framework. We choose Mapbox because of its high rendering efficiency for vector tile services. As mentioned earlier, Service Manager provides real-time update vector tile services. ST-APP Designer can synchronously read all vector tile services published by the user. Based on the layer services released by Service Manager, the system supports various types of rendering effects. 1) *Point style rendering*. For the point data, ST-APP Designer can set the point icon style and label information. 2) *Line style rendering*. It provides line type, line width, color and other style settings; 3) *Surface style rendering*. It supports the texture, color, boundary style and other configurations. 4) 2.5D *surface rendering*. It provides a three-dimensional stretching effect for surface data. 5) *Trajectory rendering*. It supports real-time display of trajectory data and can change the styles of trajectory line and trajectory icon.

4.2 Application Designing

In addition to the configuration of the map style, Designer also provides users with an environment for constructing spatial-temporal application scenarios. We define an ST application as the combination of a map and some spatial-temporal controls. The user selects one of several designed map styles as the base map of an ST application, and then adds the required controls on top of this. JUST-Studio provides three categories of controls. The first is heat map control, which allows user to build a map with high and low render effects. The second is analysis control, which is the most important element of scenario construction. At present, the system provides POI query, path planning, geographical fence monitoring and other capabilities. All of these controls allow users to drag and drop to change positions and sizes. The analysis capabilities of the controls come from the control services published by Service Manager. Users only need to associate interface controls with service models and configure parameters to add corresponding functions to the applications. The third is general control, such as map scale, north finger, layer control and other common elements in the map.

4.3 Sharing

Both map styles and scenario styles can be shared with other users. There are two modes of sharing: 1) Webpage sharing, which is the fastest sharing method. Other users can access ST applications directly when they get the sharing link. A common application is to nest iframes into their own web pages. 2) Style file sharing, which is a more flexible sharing method. After users get the link, they can use JUST-Map-SDK (simplify SDK later) to restore the application quickly. At the same time, they can use the interface provided by SDK for a secondary development.

5 Demonstration

As shown in Fig. 6, JUST-Studio initial interface is a dashboard structure with the configured map templates provided by the system at the top. By clicking the cards in the area of **System Default Templates**, users can quickly create a map style based on a template and enter the map design interface. Users can also create maps in the **Map Style Dashboard** area, which also supports the import of existing native styles. As shown in the figure, the designer allows users to create a new scenario application. Users click **New Scene Style** or select a basic map style to enter the scene editing area. Below, we will show how a designer works through a geofence monitoring scenario.

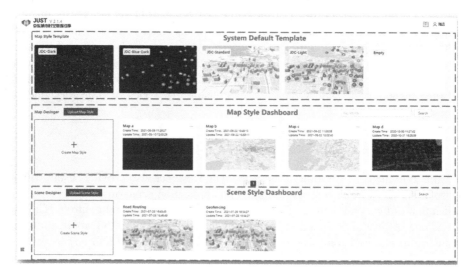

Fig. 6. Main page of ST-AP designer

Geofence Monition Scene. Geofencing monitors vehicle information, so we need to receive vehicle trajectory data first. We configure a Kafka data source to receive real-time vehicle data by Service Manager. The Kafka data source needs to configure the topic of the message, as well as the longitude, latitude and direction fields. We then create a dynamic trajectory model associated with the previous Kafka data source, as shown in Fig. 7. As mentioned in Sect. 3.3, in order to monitor the vehicle's path within the fence, we use the line index to store trajectory data. In this demo, the vehicle GPS position Kafka got is the original position, which may have large errors. Therefore, we need to open the map matching option and associate it with a road network model. Then, we need to create a geofencing model, which is configured by associating a trajectory model simply. Finally, by launching the geofencing model, we have all services associated with geofencing applications.

After completing the geofencing service configuration, we need to build the application in Designer. First, we select a map template and enter a workbench. Next, we

Fig. 7. Scene designer workbench for geofencing

add another geofencing control, resize and style it, associate a geofencing model in the parameters box, and then publish the application. By doing so, we complete the construction of a geofencing application. Now we have a trajectory, but where is the fence? Note that JUST-Studio creates a geofencing application, so the fence must be created by the users using the application. Various areas of interest can be created through our fence tool. Once saved, these areas begin to monitor the vehicles. The warnings of entering the fence will be displayed in real time in the results bar.

Acknowledgments. This work is supported by National Key R&D Program of China (2019YFB2103201) and National Natural Science Foundation of China (61976168).

References

1. Hu, Y., et al.: SALON: a universal stay point-based location analysis platform. In: ACM SIGSPATIAL (2021)
2. Ruan, S., et al.: Cloudtp: a cloud-based flexible trajectory preprocessing framework. In: ICDE, pp. 1601–1604. IEEE (2018)
3. Yuan, J., et al.: Where to find my next passenger. In: Proceedings of the 13th International Conference on Ubiquitous Computing, pp. 109–118 (2011)
4. Veenendaal, B., Brovelli, M.A., Li, S.: Review of web mapping: Eras, trends and directions. ISPRS Int. J. Geo Inf. **6**(10), 317 (2017)
5. Mapbox Studio (2021). https://studio.mapbox.com/
6. Carto (2021). https://carto.com/
7. Li, R., et al.: Just: JD urban spatio-temporal data engine. In: ICDE, pp. 1558–1569. IEEE (2020)
8. Li, R., et al.: Trajmesa: a distributed NoSQL storage engine for big trajectory data. In: ICDE, pp. 2002–2005. IEEE (2020)
9. Li, R., et al.: TrajMesa: a distributed NoSQL-based trajectory data management system. TKDE (2021)

10. Vora, M.N.: Hadoop-HBase for large-scale data. In: ICCSNT, vol. 1, pp. 601-605. IEEE (2011)

11. Thein, K.M.M.: Apache Kafka: Next generation distributed messaging system. Int. J. Sci. Eng. Technol. Res. **3**(47), 9478–9483 (2014)

12. Well-known   text   (2021).   https://en.wikipedia.org/wiki/Well-known_text_representa tion_of_geometry

13. Zaharia, M., et al.: Resilient distributed datasets: a fault-tolerant abstraction for in-memory cluster computing. In: NSDI, vol. 12, pp. 15–28 (2012)

14. OGC (2021). https://en.wikipedia.org/wiki/Open_Geospatial_Consortium

15. GeoTools Project (2021). https://www.geotools.org/

16. Newson, P., Krumm, J.: Hidden Markov map matching through noise and sparseness. In: Proceedings of the 17th ACM SIGSPATIAL International Conference on Advances in Geographic Information Systems, pp. 336–343 (2009)

17. WMS (2021). https://en.wikipedia.org/wiki/Web_Map_Tile_Service

18. WFS (2021). https://en.wikipedia.org/wiki/Web_Feature_Service

19. WMTS (2021). https://en.wikipedia.org/wiki/Web_Map_Tile_Service

20. Mapbox Vector Tiles. https://www.mapbox.com/developers/vector-tiles/. Accessed 12 Mar 2015

21. Abbas, A.H., et al.: GPS based location monitoring system with geo-fencing capabilities. AIP Conf. Proc. **2173**(1), 020014 (2019)

22. LaMarca, A., De Lara, E.: Location systems: an introduction to the technology behind location awareness. Synth. Lect. Mob. Pervasive Comput. **3**(1), 1–122 (2008)

23. Hershberger, J.E., Snoeyink, J.: Speeding Up the Douglas-Peucker Line-Simplification Algorithm. University of British Columbia, Department of Computer Science, Vancouver (1992)

24. Protocol Buffers Encoding (2021). https://developers.google.com/protocol-buffers/docs/enc oding

Author Index

Printed in the United States
by Baker & Taylor Publisher Services